VERIFICATION, VALIDATION, AND TESTING OF ENGINEERED SYSTEMS

WILEY SERIES IN SYSTEMS ENGINEERING AND MANAGEMENT

Andrew P. Sage, Editor

A complete list of the titles in this series appears at the end of this volume.

VERIFICATION, VALIDATION, AND TESTING OF ENGINEERED SYSTEMS

AVNER ENGEL

WILEY
A JOHN WILEY & SONS, INC., PUBLICATION

Published by John Wiley & Sons, Inc., Hoboken, New Jersey
Published simultaneously in Canada

Editorial contribution—Dr. Peter Hahn

Library of Congress Cataloging-in-Publication Data:

Engel, Avner.
Verification, validation, and testing of engineered systems/Avner Engel.
p. cm.—(Wiley series in systems engineering and management)
Includes bibliographical references and index.
ISBN 978-0-470-52751-1 (cloth)
1. Quality assurance. 2. Quality control. 3. Systems engineering. 4. System failures (Engineering)–Prevention. 5. Testing. I. Title.
TS156.6.E53 2010
658.5′62—dc22
2009045885

Printed in Singapore

M WEP354446 030425

To my parents:
Josef Engel, Lea Engel and Tova Engel

and my revered teachers:
Dr. Itzhak Frank, Professor Jerry Weinberg and Professor Miryam Barad

Contents

Preface

Systems testing is carried out one way or another in all development and manufacturing projects, but seldom is this done in a truly organized manner and no book currently available describes the process in a comprehensive and implementable form. Along the same line of thinking, virtually no systems Verification, Validation, and Testing (VVT) research is conducted throughout the academic world. This is especially odd, since some 50–60 percent of a systems development cost is expended on either performing VVT activities or correcting system defects during the development process or during the life of the developed product.

This book attempts to put together a comprehensive compendium of VVT activities and corresponding VVT methods for implementation throughout the entire lifecycle of systems (i.e. Definition, Design, Implementation, Integration, Qualification, Production, Use/Maintenance and Disposal). In addition, the book strives to alleviate the fundamental testing conundrum, namely: What should be tested? How should one test? When should one test? And, when should one stop testing? In other words, how should one select a VVT strategy and how should it be optimized? Although early quality pioneers (e.g., Juran in the 1950s) proposed a conceptual quality cost model, no one proposed a quantitative and credible model which can be used to answer the above questions. This book provides such a model, together with data from a real-life project, which show significant potential savings in either cost, time or both. The book is organized in three parts:

The first part (Chapter 1) provides introductory material about systems and VVT concepts. This part presents a comprehensive explanation of the role of VVT in the process of engineered systems throughout their lifetime and explains the essence of systems VVT and the linkage between VVT and systems development, manufacturing, use/maintenance and retirement.

The second part (Chapters 2–5) is essentially a reference guide, describing typical systems VVT activities which may be conducted during an engineered systems lifetime. A reciprocal and comprehensive set of methods for carrying out these VVT activities is also provided. More specifically, the second part describes 40 systems development VVT activities (Chapter 2) and 27 systems post-development activities (Chapter 3). Corresponding to these activities, this part also describes 17 non-testing systems VVT methods (Chapter 4) and 33 testing systems methods (Chapter-5). In-text citations are provided wherever needed, usually within theoretical sections of the book. In addition, subchapters contain a set of citations for further reading. Readers will undoubtedly be able to absorb and implement some or all of this information in their daily work-life as systems or test engineers.

The third part of the book (Chapters 6–8) describes ways to model systems quality cost, time and risk (Chapter 6), as well as ways to acquire quality data and optimize the VVT strategy in the face of funding, time and other resource limitations and in accordance with different business objectives (Chapter 7). Finally, this part describes the methodology used to validate the quality model along with examples describing a system's quality improvements (Chapter 8). Readers will be able to learn how to collect and aggregate quality data within their organizations. In addition to becoming familiar with this significant information, readers will be introduced to four Cost, Time and Risk Models. Systems engineers are encouraged to use these models in order to optimize their VVT strategies, thereby realizing as much as ten percent reduction in engineering manpower or schedule in the development of engineered systems. The software is freely available to readers via an Internet site (**http://www.adisw.com**) in binary form under Beta release conditions.

Fundamentally, this book is written with two categories of audience in mind. The first category is composed of VVT practitioners, including Systems, Test, Production and Maintenance engineers as well as first and second line managers. These people may be employed by development and manufacturing industries (e.g., Aerospace, Automobile, Communication, Healthcare equipment, etc.), by various civilian agencies (e.g. NASA, ESA, etc.) or with the military (e.g., Air force, Navy, Army, etc.). This book may also be used as a supplemental graduate level textbook in courses related to systems VVT. Typical academic readers may be graduate school students or members of Systems, Electrical, Aerospace, Mechanical, and Industrial Engineering faculties. This book may be fully covered in two to three semesters (although parts of the book may be covered in one semester). University instructors will most likely use the book to provide engineering students with knowledge about VVT, as well as to give students an introduction to formal modeling and optimization of VVT strategy.

ACKNOWLEDGMENTS

Many friends and colleagues have contributed generously to the writing of this book. To all of them, I would like to express my sincere gratitude and appreciation. In particular, I wish to thank Dr. Peter Hahn, who has been a tireless and devoted companion in the book-writing project from its inception. He edited the original manuscript and contributed numerous and valuable suggestions to improve the book.

The SysTest project, partially funded by the European Commission (see Appendix A), focused my attention onto systems verification, validation and testing. My appreciation goes to all the consortium members and in particular to professor Eduard Igenbergs of the Technical University of Munich, who provided both a philosophical foundation and ample encouragement, and to Professor Tyson Browning of the Texas Christian University, part of whose scientific writings and words of wisdom are embedded in this book. The Advanced System and Software Engineering Technology (ASSET) group at Israel Aerospace Industries (IAI) was a significant milieu for learning and expanding. My special gratitude goes to ASSET group leader, Dr. Michael Winokur. I am also grateful to Shalom Shachar of the IAI/Lahav Division, who conducted the SysTest pilot project at IAI, helped in collecting field data and became a sounding board and advisor regarding many aspects of the VVT quantitative model. In addition, I am beholden to Michael Garber of Adi Mainly Software (AMS), who developed the VVT-Tool software package which embodies the VVT model.

Several close friends were involved in creating this book. In particular, I would like to mention Avi Egozi and Arie Rokach, who suggested the book project in the first place and provided advice throughout the writing process. Also my sincere appreciation goes to Menachem Cahani (Pampam), who volunteered to illustrate several caricatures in the book. I also am genuinely indebted to Professor Miryam Barad of the Tel-Aviv University, an esteemed teacher who taught me how to conduct scientific research and write about it.

Most of all, my deepest thanks go to my wife, Rachel, and my children, Ofer, Amir, Jonathan and Michael, who encouraged my book efforts with advice, patience and love,

Avner Engel
Tel-Aviv, Israel

Part I

Introduction

Chapter 1

Introduction

1.1 OPENING

This chapter serves as motivation for learning about systems Verification, Validation and Testing (VVT) as well as a map for using the book as a reference source on this complex and multifaceted process. We emphasize here the multitude of reasons for applying VVT. It sets the tone for the subject matter we hope to cover. It gives the reader insight into the attitudes of the author and the care with which the book was prepared. A clear statement is made of the purpose for which the book has been written.

The book is a compendium of facts about systems VVT. In fact, we think little has yet been published that is as comprehensive on this subject. By listing the potential audience for the book, we hope to encourage its wide distribution and to increase among engineers, managers, academicians and students an appreciation of the benefits of rigorously applying VVT to almost every endeavor involving a product or service, be it for purposes commercial, private or public. This chapter contains the following elements:

Opening. This part provides a background, purpose and the intended audience of the book. In addition, it describes its structure and contents as well as the scope of application and some terminology descriptions.

VVT systems and process. This part introduces VVT systems and processes as components of engineered systems. In addition, it describes basic VVT definitions and elaborates on the fundamental VVT dilemmas. Also, this part describes modeling of systems and VVT lifecycle as well as modeling of VVT processes and risks as cost and time drivers.

Verification, Validation, and Testing of Engineered Systems, Avner Engel

Canonical systems VVT paradigm. This part introduces the concept of canonical systems VVT paradigm which includes phases of systems' lifecycle, views of systems and VVT aspects of systems.

Methodology application. This part introduces methodology application including VVT methodology overview, VVT tailoring and typical VVT documentation.

1.1.1 Background

The manufacturing industry used to be concerned with the design, development, production and maintenance of stand-alone products, whether simple or complex. Today, however, manufacturing has broadened its scope to include products, services or solutions that include a variety of components, integrate a large mix of technologies and involve both people and machines. It is this broad range of complex entities that we address in this book. The basic term we use for these complex entities is *engineered systems*. However, throughout this book, when appropriate, we will freely use terms such as *products* or *services*. The term engineered systems is distinguished from *systems* in the sense that the former is created by engineers who apply science and mathematics to find suitable solutions to problems.

Traditional and high-technology manufacturing industries are responding to the challenge to satisfy consumer needs and ensure competitive and sustainable growth by reducing time to market and customizing products (or expanding product ranges) while producing the required goods in the quantities demanded with the appropriate quality at reduced costs. For instance, in the automobile sector, the lead time for manufacturing a car at the beginning of the 1990s was five to six years, whereas today it is about two to three years and is estimated to be only 18 months in the near future. Therefore, controlling schedules, costs and quality in product development, manufacturing and maintenance remains a major challenge for today's industries. Increases in complexity, decreases in development budgets and shortened time to market for new products, services and solutions are leading developers to search for new ways of improving the quality of what they deliver by improving their technologies, processes, methodologies and tools.

The overall development process is only as strong as its weakest link. A critical and largely ignored link in this process is system VVT, which comprise vital activities and involve processes. A tool of systems engineering, VVT focuses on ensuring that engineered systems are delivered as error free as possible, are functionally sound and meet or exceed the user's needs. Often VVT is carried out as merely a vehicle for finding and eliminating errors. It can do much more than that. Today, many system developers perform VVT only in the test phase of the project, a late and highly constrained period in the product development cycle. As a result, increases in overall development time and costs associated with product rework often exceed 20% of expanded engineering efforts (Capers, 1996). Admittedly, balancing testing cost and schedule with quality is difficult. However, quality problems discovered later by the user can necessitate expensive repairs and are likely to damage the reputation of the system or, worse, damage the reputation of the system's developer.

Given the fundamental role of VVT in achieving product quality and reducing waste, this book aims at rectifying two critical current VVT problems, namely, lack of comprehensive system VVT methodology and lack of a practical, quantitative VVT process model for selecting a VVT strategy to optimize testing cost, schedule and economic risk. This book, which to a large measure is based on the European Commission–supported SysTest project, was written in order to rectify these problems.

1.1.2 Purpose

One of the central objectives of this book is the creation of generic VVT methodology. This *VVT methodology* consists of a selection of VVT activities and methods which can be applied throughout the system lifecycle in different industrial application fields and can be tailored according to the individual project needs.

The VVT methodology delivers generic means for comprehensive cost-effective VVT in the industry. In addition, the objectives of this methodology are as follows:

- To cover the entire product lifecycles from the definition to the disposal of the system
- To supply tailoring rules for different industry domains (e. g. electronics/avionics, control systems, automobile, food packaging systems, steel production), development cycles and project types
- To specify activities and methods for VVT on the system level together with their interrelationship
- To define VVT strategies that can be used in a broad variety of industrial applications

1.1.3 Intended Audience

The VVT methodology described in this book is applicable to all regional and industrial sectors. Although system VVT is performed throughout industry, it has not become a topic for research within the international community either in industry or in academia. Therefore, the definition of a generic VVT methodology will provide comprehensive knowledge for many students and practitioners. This book was written for the reader who has a background knowledge of project management, systems engineering and quality assurance. Those who participate in system development will benefit from the material covered in this book. These include:

1. *Project Managers and VVT Managers.* This book can guide project and VVT managers in the methods they select, adapt and tailor for planning, control and tracking of projects.

2. *Quality Assurance (QA)/Quality Control (QC) Staff.* For QA and, QC staff, this book offers an overview of the system QA activities and methods available and their principal advantages and disadvantages. Quality assurance staff can apply the VVT methodology guidelines for the selection of VVT procedures and the estimation of process and product risks.
3. *Members of a VVT Team.* This book serves as an aid for test teams by providing them with an overview of useful procedures for conducting a VVT process within the context of system development projects and beyond. Thus, the VVT methodology guidelines of this book become a useful tool for categorizing VVT activities within the system lifecycle overall context and by referencing further information.
4. *System Developers and Maintainers.* This book is relevant for system developers in that they deliver insight into the measures of error avoidance and error detection. Developers can draw important conclusions about the functional domains of the system developed that are critical where VVT are concerned.
5. *Mechanical, Electronics and Software Designers.* Other specialists need this book in order to take VVT aspects into account when they determine structures and select the technologies for system development, production and maintenance. This book can be an important basis for this, as it shows not only the possibilities but also the limitations of VVT procedures.
6. *Component and Subsystem Suppliers.* A clear definition and a specification with respect to VVT measures are essential, especially for system development projects that involve supplier companies. This book forms a convenient basis for those projects since it provides a mutual definition, nomenclature and techniques as well as a body of VVT methods.
7. *Auditors.* To evaluate the maturity of a development project, auditors and auditing agencies can also apply the VVT methodology. Adherence to standards, deployment of established procedures, as well as the maturity of the processes' implementation can be evaluated in this way.
8. *Regulatory and Standardization Agencies.* Material presented in this book may be helpful in forming and updating national or international standards and regulations of standardization committees in which certain procedures for defined system classes are classified as binding or just recommended. Of course, it is not the aim of this book to define or force standardization. However, it could provide important suggestions with regard to such an endeavor.

1.1.4 Book Structure and Contents

This book is divided into three parts and a set of appendices as described below.

Part I: Introduction Part I of this book contains basic introductory material organized in one chapter. It starts by describing the purpose, the intended audience, the structure and the content of the book, the scope of the applications and the terminology and notation used throughout this book. It continues by providing basic introduction to systems theory, relevant background on systems and software VVT as well as risk and uncertainty theory. In addition, this chapter introduces VVT concepts and discusses the modeling of systems and the VVT lifecycles. It then defines generic phases, views and aspects of the system lifecycle that are used in this book. Finally, the chapter provides a VVT methodology overview, typical VVT documents and a methodology for VVT tailoring.

Part II: VVT Activities and Methods Part II of this book describes the VVT activities typically associated with each phase of the system lifecycle. For each VVT activity, the book describes one or more methods for carrying out those activities:

- *Chapter 2, System VVT Activities: Development,* describes typical VVT activities which may be conducted during system development, that is, during the Definition, Design, Implementation, Integration and Qualification phases of the system's lifecycle.
- *Chapter 3, System VVT Activities: Postdevelopment,* describes typical VVT activities which may be conducted during system postdevelopment, that is, during Production, Use/Maintenance and Disposal phases of the system's lifecycle.
- *Chapter 4, System VVT Methods: Nontesting,* describes a set of VVT nontesting methods, complementing the VVT activities described in the VVT activities chapters. In particular this chapter describes the following nontesting system VVT methods: preparing VVT products, performing VVT activities and participating in reviews.
- *Chapter 5, System VVT Methods: Testing,* describes a set of VVT testing methods, complementing the VVT activities described in the VVT activities chapters. Specifically, this chapter describes a collection of system testing methods grouped into the following categories: white-box testing and black-box testing; the latter is further divided into basic testing, high-volume testing, special testing, environment testing and phase testing.

Part III: Modeling and Optimizing VVT Process Part III of this book describes ways to model system quality cost, time and risk as well as ways to acquire quality data and optimize the VVT strategy in accordance with different business objectives. In addition, Part III describes the methodology used to validate the quality models along with examples describing a system's quality improvements.

- *Chapter 6, Modeling Quality Cost, Time and Risk,* describes system quality modeling—in particular, VVT cost and risk modeling, VVT time and risk modeling and fuzzy VVT cost modeling.
- *Chapter 7, Obtaining Quality Data and Optimizing VVT Strategy,* presents typical quality data of engineered systems from various industries as well as practical ways and means to elicit and aggregate quality data (i.e., cost, time and risks of VVT activities). The chapter continues by describing various techniques to optimize VVT strategies in order to reduce cost, time and system risks.
- *Chapter 8, Methodology Validation and Examples,* describes a validation process which compares actual measurements of system quality cost and time with model prediction. Finally, this chapter provides several examples of the entire system quality improvement process.

Appendices This portion of this book contains a collection of appendices as follows:

- *Appendix A—SysTest Project*
- *Appendix B—VVT Master Plan (VVT-MP)*
- *Appendix C—Acronyms*
- *Index*

Figure 1.1 will help the reader to navigate this book.

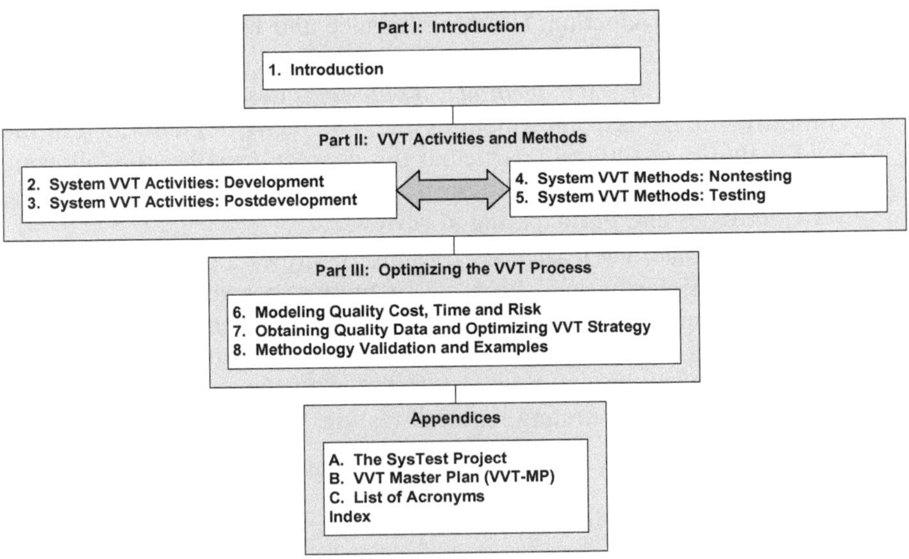

Figure 1.1 Book structure and navigation.

1.1.5 Scope of Application

This book covers system VVT, hopefully, without bias toward a specific application. The VVT methods described are applicable to a broad spectrum of system requirements: whether safety critical or non–safety critical, whether mission critical or non–mission critical or whether the requirements are hard real time or nontemporal. The VVT methodology described herein supports the quality assurance phases all the way from system requirements definition to system disposal. Furthermore, it supports different system hierarchy levels of quality measures, from component testing to system testing. The book's VVT methodology guidelines can be applied to mass-produced systems as well as to small production quantities or few-of-a-kind paradigms.

The present book is applicable to system developments in various industrial sectors. They may be regarded as recommendations only. Or, they can be considered binding for an individual project if the stakeholders for that project agree upon this course of action.

1.1.6 Terminology and Notation

In this book, when we use the terms *has to/must*, *shall* and *should* we mean the following:

- *Has To/Must.* This is the highest level of recommendation and describes cases where the described process, procedure or approach *works only in this way*.
- *Shall.* At this level, the user is *strongly recommended* to use the described process, procedure or approach in this way.
- *Should.* This level of recommendation describes cases where this author has experienced that this process, procedure or approach is the *best*.

Each VVT activity or method described in this book is presented, as much as possible, in a common format, thus facilitating the orientation and presentation of more detailed information on each activity.

1.2 VVT SYSTEMS AND PROCESS

1.2.1 Introduction—VVT Systems and Process

This section serves as an introduction to the VVT process. It starts with the definition of an engineered system, that is, a man-made artifact that depends upon scientifically based and experiential processes that are logically applied. VVT attempts to help these systems achieve their full potential in terms of performance, efficiency and economy of precious resources. What follows is a detailed discussion of what is meant by VVT in all its manifestations. This includes a variety of definitions, as given by various experts, industries, engineering organizations and government agencies.

As a discipline VVT is an outgrowth and expansion of the earlier disciplines quality assurance and quality control. It is an evolving concept and thus will continue to be redefined with time and with the development of new techniques for design and evaluation of engineered systems. Thus, it is not surprising that there would be disagreement in the engineering and business community on just what comprises a VVT program.

Here, we attempt to give an overview of the many perceptions about VVT from the various stakeholders in the VVT process, that is, customers, manufacturers, regulators, professional organizations and government. Thus, we break down the differences between VVT definitions as seen by various technical disciplines: electrical and electronics engineering, telecommunications, artificial intelligence and the modeling and simulation community. The definitions and perceptions of VVT, as seen by the systems engineering community and more specifically by the International Council on Systems Engineering (INCOSE), are also covered, as are the VVT definitions used by the author in this book.

We attempt to give an appreciation of the difficulties of applying VVT to large and complex systems. Since VVT efforts should begin early in the lifecycles of a system and are not completed until the system is decommissioned and its components recycled, the issues are complex and manifold. Thus, we bring a section describing the stages of the system lifecycle and relate it to complementary VVT lifecycle phases.

Measuring VVT performance is key to good VVT planning. There is a delicate balance between the risks avoided by good system VVT and the risks to a system's development and deployment by too much VVT.

1.2.2 Engineered Systems

General Systems The term system (from Latin *systema*) has emerged in the twentieth century as a key building block of systems theory, an area of study that predominantly refers to the *science of systems* that resulted from Bertalanffy's general system theory (Bertalanffy, 1976).

An intuitive description of a "system" is that it is composed of separate elements organized in some fashion with certain interfaces among the elements and between the system and its environment. In addition, a system tends to affect its environment and be affected by it. This involves some type of input and output (e.g., materials, energy, information). Most importantly, a system produces results not obtainable from the collection of its individual elements.

Based on this notion, we can adopt either an elementary definition, "*A system is an interdependent group of items forming a unified whole*" (*Webster's* dictionary), or a more sophisticated definition, "*A system is a combination of components that act together to perform a function not possible with any of the individual parts*" [Institute of Electrical and Electronics Engineers (IEEE) Electronic Terms].

Engineered Systems The goal of engineering processes is to develop and produce efficient and reliable systems (products, services or solutions) that meet a specific need under a defined set of constraints. To achieve this, the system will follow a typical creation lifecycle, whose phases could be defined as Definition, Design, Implementation, Integration, Qualification and Production. During its useful lifetime, a system will go through a Use/Maintenance phase, culminating in the disposal of the system.

According to Braha et al. (2006), the classical engineering process has several notable characteristics: (1) a search for a single solution, namely, engineers tend to seek a single solution, which often revolves around a unique design concept, for the specified problem, (2) the desire for a well-behaved system, that is, engineers prefer systems whose behavior can be predicted and encapsulated by precise description and (3) the application of a top-down problem-solving approach, which fundamentally depends on the assumption that any system can be described wholly by describing the behavior of its parts and their interactions. Therefore, according to Braha et al. (2006), classically engineered systems have the following attributes: (1) predictability, that is, the system works in predictable ways; (2) reliability, that is, the system is able to perform a required function under stated conditions for a stated period of time; (3) transparency, that is, the structure of the system and its processes can be described explicitly; and (4) controllability, that is, the system can be directly governed according to stated instructions under stated conditions.

We can now accept either the definition of the Council on Systems Engineering (INCOSE) organization: "A system is an integrated set of elements to accomplish a defined objective" adopted in 1995, or a rather sophisticated definition, attributed to Dr. Eberhardt Rechtin (1990):

> A system is a construct or collection of different elements that together produce results not obtainable by the elements alone. The elements, or parts, can include people, hardware, software, facilities, policies, and documents; that is, all things required to produce systems-level results. The results include system level qualities, properties, characteristics, functions, behavior and performance. The value added by the system as a whole, beyond that contributed independently by the parts, is primarily created by the relationship among the parts; that is, how they are interconnected.

We further accept the distinction that an engineered system is often composed of "enabling products" required to provide lifecycle support in addition to the "end products", which performs the required operational functions (see Figure 1.2). The end product may be a single manifestation of the system or may be produced in small or large quantity.

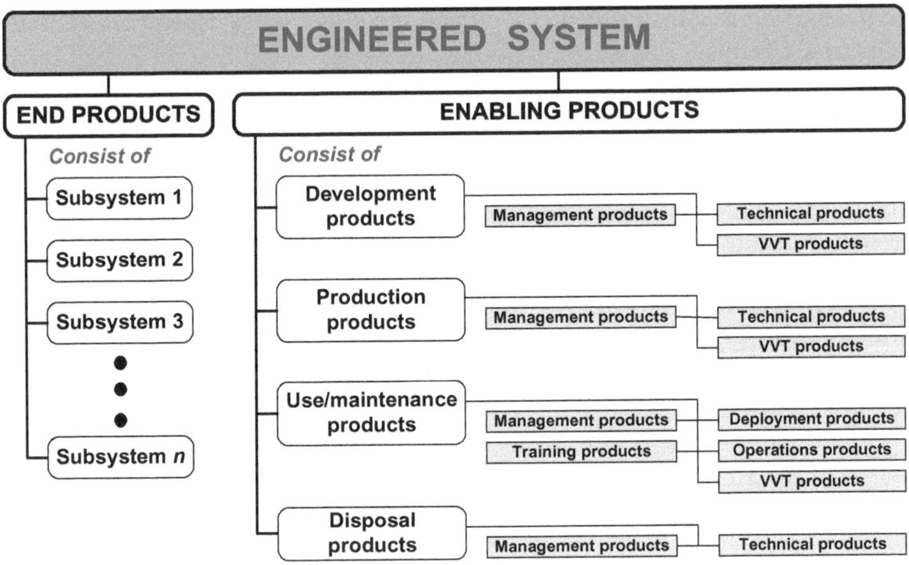

Figure 1.2 Typical structure of engineered system.

1.2.3 VVT Concepts and Definition

The acronym VVT stands for Verification, Validation and Testing. These terms have some common significance. The purpose of this discussion is to explain and encapsulate the unique meaning of each term. This section contains the following topics:

- The on-going VVT terminology debate and the general purpose of the VVT process
- The various definitions of the terms verification, validation and testing as reflected in the scientific and engineering literature
- The VVT principle and definition trends and the specific VVT definition adopted for this book

VVT Terminology and Objectives This section discusses the on-going VVT terminology debate and the general purpose of the VVT process as reflected in the scientific and engineering literature.

VVT Terminology Debate It seems that no published article on the evaluation of systems is written without first defining VVT. Many authors choose to define this term by citing some of the more popular definitions. Others, realizing the lack of clarity in those definitions, come up with their own definitions. As a result, there is confusion about exactly what VVT is and how it can be implemented in different systems.

The mere existence of confusion and the debate over definitions indicates that the VVT discipline is still in its infancy and the intent of this discussion is to dispel some of this confusion.

Purpose of the VVT Process Another question that confronts us is what should be the final purpose of the VVT process? Should it serve to eliminate errors or serve as a means to certify that a system is free of errors? Following are the arguments.

Elimination of errors is akin to debugging a computer program. The program is exercised to discover an incorrect behavior, and then the bug causing the incorrect behavior could be identified and removed. This is necessary, not only for computer programs, but also in many other fields where systems are expected to be dependable. This book reflects the author's opinion that VVT must first strive to eliminate errors if it is to be useful. On the other hand, there is a significant commercial value in being able to say that a system is free of errors and works as intended. Unfortunately, this is merely wishful thinking. To guarantee that a system is free of errors is logically impossible unless a truly exhaustive way of evaluating its functionality can be implemented. This would not be feasible for all but the most trivial systems. We conclude that the purpose of VVT should be to eliminate as many defects as possible within existing constraints of available time, money and other resources.

What is to be achieved by VVT? Fairley (1985) indicates that the goal is to assess and improve the quality of the system. He also provides quality attributes to evaluate the VVT process. These attributes, which have been altered to suit the systems arena, are presented in Table 1.1.

TABLE 1.1 VVT Quality Attributes

Function	Responding to the Following Queries
Correctness	Given valid inputs, does the system perform its tasks as expected?
Completeness	Does the system meet all of the requirements that have been placed on it?
Consistency	Are similar things handled in a similar manner? Is the system consistent with another system that is part of the same family?
Reliability	Does the system perform reasonably well in all cases, even, for instance, in the presence of pathological conditions?
Usefulness	Does the system provide a useful service?
Usability	Is the system convenient to use when carrying out its designated task?
Efficiency	Is the system efficient in its use of resources, such as time, memory, network bandwidth, and peripherals?
Standards conformance	Does the system conform to standards, both notational and external standards of interface to the outside world?
Overall cost-effectiveness	Is the system a cost-effective solution to the problem?

VVT Definitions in Various Fields The following discussion presents different definitions for the terms verification, validation and testing as reflected in the scientific and engineering literature.

1. *Nontechnical Community.* The nontechnical *Merriam-Webster's* dictionary defines the term *verify* as (1) "to confirm or substantiate in law by oath" and (2) "to establish the truth, accuracy, or reality of." It defines the term *validate* as (1) "to make legally valid," (2) "to grant official sanction to by marking," (3) "to confirm the validity of (an election)" and (4) "to support or corroborate on a sound or authoritative basis." It provides 55 different definitions for the term *test*. The most relevant nontechnical ones are (1) "a critical examination, observation, or evaluation," (2) "the procedure of submitting a statement to such conditions or operations as will lead to its proof or disproof or to its acceptance or rejection" and (3) "a basis for evaluation." The intuitive understanding of the above terms corresponds well with the nontechnical dictionary definition. The technical definition of VVT is another matter.
2. *IEEE Community.* The IEEE defines validation and verification for engineered hardware and software systems as follows (*IEEE-610*):
   - *Verification* is the process of evaluating a system or component, to determine whether the products of a given development phase satisfy the conditions imposed at the start of that phase.
   - *Validation* is the process of evaluating a system or component during or at the end of the development process, to determine whether it satisfies specified requirements.
3. *Telecommunication Community.* In its Telecom Glossary 2000, the American National Standard for Telecommunications defines the terms as follows:
   - *Verification.* (1) Comparing an activity, a process, or a product with the corresponding requirements or specifications. (2) [The] process of comparing two levels of an information system specification for proper correspondence (e.g., security policy model with top-level specification, top-level specification with source code or source code with object code).
   - *Validation.* (1) Tests to determine whether an implemented system fulfills its requirements. (2) The checking of data for correctness or for compliance with applicable standards, rules, and conventions.
   - *Testing.* Physical measurements taken (1) to verify conclusions obtained from mathematical modeling and analysis or (2) for the purpose of developing mathematical models.
4. *Artificial Intelligence Community.* Gonzalez and Barr (2000) suggest the following definitions for these terms in the Artificial Intelligence (AI) community:

- Verification is the process of ensuring that the intelligence system (1) conforms to specifications and (2) its knowledge base is consistent and complete within itself. The intent of this definition is that the process of verification represents an internal benchmark, rather than an external one. Making it internal is highly significant, as errors can be found without the need to exercise the system with test cases.
- Validation is the process of ensuring that the output of the intelligence system is equivalent to that of human experts when given the same input.

5. *Modeling and Simulation Community.* The Department of Defense (DoD) Defense Modeling and Simulation Office (DoDD-5000.59) gives a formal definition. It defines Verification and Validation (V&V) as follows:

   - Verification is the process of determining that a model implementation accurately represents the developer's conceptual description and specification.
   - Validation is the process of determining the degree to which a model is an accurate representation of the real world from the perspective of intended uses of the model.

   Balci (1998), a noted researcher in the Modeling and Simulation (M&S) field, and later Balci et al., (2000) extend the DoD definition for VVT as follows:

   - *Model verification* is substantiating that the model is transformed from one form into another, as intended, with sufficient accuracy. Model verification deals with building the model correctly. The accuracy of transforming a problem formulation into a model specification or the accuracy of converting a model representation from a micro flowchart form into an executable computer program is evaluated in model verification.
   - *Model validation* substantiates that the model, within its domain of applicability, behaves with satisfactory accuracy, consistent with the M&S objectives. Model validation deals with building an *accurate* model. An activity of accuracy assessment can be labeled as verification or validation based on an answer to the following question: In assessing the accuracy, "Does the model's behavior compare well to the corresponding system behavior?" Even if the answer to the question of accuracy is "yes," that does not answer the question of whether the model is the right one.
   - *Model testing* is determining whether inaccuracies or errors exist in the model. In model testing, the model is subjected to test data or test cases to determine if it functions properly. Test failure implies the failure of the model, not the test. A test is devised, and testing is conducted to perform either validation or verification or both. Some tests

are designed to evaluate the behavioral accuracy or validity of the model, and some other tests are intended to determine the accuracy of model transformation from one domain into another (verification). Sometimes, the whole process is called model VV&T or, for short, VVT.

VVT Concepts in System Engineering Lake (1999) explains the formal definition and intuitive meaning of V&V in system engineering (see Figure 1.3):

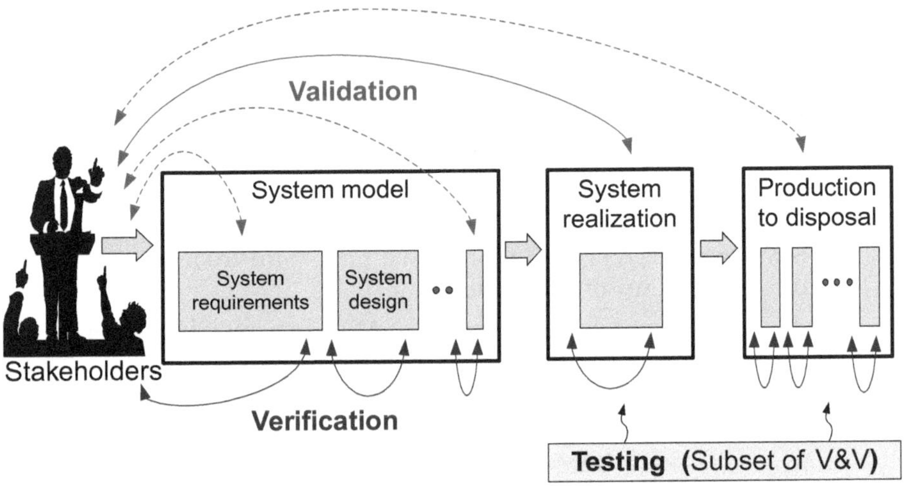

Figure 1.3 Verification and validation in system engineering perception.

- Verification is the process of evaluating a system to determine whether the products of a given development phase satisfy the conditions imposed at the start of that phase.
- Validation is the process of evaluating a system to determine whether it satisfies the stakeholders of that system.

These terms will now be further elaborated:

1. *System Verification.* The meaning of the term verification is to evaluate a realized product against specified requirements. The intent is to determine whether the finished product satisfies the specific requirements for which it was built. In addition, the verification responds to the question: "Was the product built (written, built, coded, assembled and integrated) correctly"? There are two formal definitions of verification:
   - Confirmation by examination and provision of objective evidence that the specified requirements to which a product was built, coded or

assembled has been fulfilled (American National Standards Institute/Electronics Industries Association ANSI/EIA-632)

- The process of evaluating a system or component to determine whether the products of a given development phase satisfy the conditions imposed at the start of that phase (IEEE-610)

According to Lake (1999), verification failure (i.e., lack of confirmation) typically reveals the following types of design or implementation errors:

- Specified requirements (specifications, drawings, parts lists) have not been documented adequately.
- Developers/builders have not followed the specified requirements for the product.
- Procedures, workers, tools and equipment are improper or have been improperly used for building the product.
- Procedures and means have been improperly planned for verification.
- Verification procedures have been improperly implemented.

2. *System Validation.* The meaning of validation is evaluating a realized product against specified (or unspecified) requirements in order to determine whether the product satisfies its stakeholders. In other words, validating a product is determining whether the product does what it is supposed to do in the intended operational environments. In addition, the validation responds to the question: "Was the right product built?" There are two formal definitions of the term validation:
   - Confirmation by examination and provision of objective evidence that the specific intended use of a product (developed or purchased), or aggregation of products, is accomplished in an intended usage environment (ANSI/EIA-632)
   - "The process of evaluating a system or component during or at the end of the development process to determine whether it satisfies specified requirements" (IEEE-610)

According to Lake (1999) typical validation errors stem from:

- Input requirements not adequately identified
- Design process incorrectly executed
- Input requirement changes not communicated
- Procedures and means improperly planned for validation
- Validation procedures improperly implemented

3. *System Testing.* The meaning of the term *testing* is operating or activating a realized product or system under specified conditions and observing or recording the exhibited behavior. Here are two formal definitions of this term:

- "An activity in which a system or component is executed under specified conditions, the results are observed or recorded, and an evaluation is made of some aspect of the system or component" (IEEE-610)
- "The process of operating a system or component under specified conditions, observing or recording the results, and making an evaluation of some aspect of the system or component" (IEEE-610).

VVT Definition in This Book This section concludes this VVT presentation. It provides the author's view as to the trends in VVT definitions. These trends form the basis for the VVT definition which has been adopted for this book.

1. *Trends in VVT Definitions.* It should by now be obvious that we really do not have a single concept regarding the meaning of the VVT of systems, at least from the standpoint of the technical community. Some say that validation and verification are one and the same thing, others say verification deals with specifications, others say it is validation that deals with specifications while still others say that they both do. Furthermore, some authors relate consistency and completeness to verification while others do so with validation. Nevertheless, some trends have emerged (see Table 1.2). These trends are not universally accepted but simply were observed.

TABLE 1.2 Trends in VVT Definition

Trend Number	Description
1	Verification deals with satisfying the written specifications of systems.
2	Verification involves the internal structural correctness of systems.
3	Verification relates to the evolving lifecycle processes of systems.
4	Validation compares the system to the needs of stakeholders. These needs may vary in time.
5	In order to validate a system, the requirements of the stakeholders, whether formally specified or not, must be known.
6	Testing involves some type of exercising the system. This is a static and dynamic process that evaluates functional correctness.
7	Testing can be accomplished as a subset of either verification or validation.

2. *Principles of VVT.* Balci (1998) suggests a set of principles for carrying out verification and validation properly. This information, in a condensed form, is provided in Table 1.3 with some adjustments to account for the systems environment.

TABLE 1.3 Principles of VVT

Principle Number	Description
1	VVT has to be conducted throughout the entire system lifetime and faults should be detected as early as possible in the system life.
2	VVT has to be planned, documented and conducted by unbiased parties.
3	Performing complete system VVT is not possible and a successful VVT of each subsystem does not imply overall system credibility.

3. *VVT Definition in This Book.* This book has adopted the systems engineering VVT definition based on the 15 VVT principles suggested by Balci (1998). Specifically, this is the collection of VVT definitions set forth in IEEE-610 and elaborated upon by Lake (1999) (see Table 1.4). The general acceptance of these definitions by the system engineering community was a factor in this decision.

TABLE 1.4 VVT Definition in This Book

Term	Definition
Verification	**The process of evaluating a system to determine whether the products of a given lifecycle phase satisfy the conditions imposed at the start of that phase.**
Validation	**The process of evaluating a system to determine whether it satisfies the stakeholders of that system.**
Testing	**An activity in which a system is activated under specified conditions, the results are observed or recorded, and an evaluation is made of some aspect of the system.**

1.2.4 The Fundamental VVT Dilemma

It is well understood that it is impossible to prove that a system actually meets all it functional capabilities as well as all standards, statuary directives, and ethical values and at the same time adheres to business objectives. The main limiting factors other than plain physics are the cost and time to market, which is required in order to bring products into common use. Therefore it is the domain of the system VVT engineer and management to strive for an optimal solution of the VVT process. As this issue is a central theme in system VVT, the book addresses the issues of cost, risk and time of the VVT process in great detail. Figure 1.4 depicts the fundamental balancing and optimizing of the VVT process. Highlighted are the business objectives emphasized in this book.

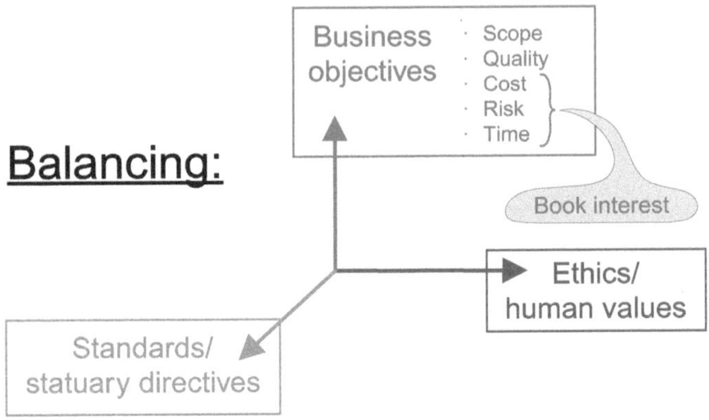

Figure 1.4 Balancing and optimizing the VVT process.

1.2.5 Modeling Systems and VVT Lifecycle

This section describes major system lifecycle models and in particular systems' lifecycle definitions used by U.S. government and commercial organizations. A generic system lifecycle adopted for this book is also presented.

Major System Lifecycle Models An overall system lifecycle model describes a cradle-to-grave paradigm of engineered systems. Different organizations [e.g., the National Aeronautics and Space Administration (NASA), DoD] and industries (e.g., automobile, electronics, telecommunication, aerospace) define various system lifecycle models. For example, the DoD acquisition lifecycle process has 4 major phases and 22 minor phases, as defined in Table 1.5.

TABLE 1.5 Major System Lifecycle Phases as Defined by U.S. DoD

Major Systems Lifecycle Phase			
0	**I**	**II**	**III**
Concept Exploration **(CE)**	Program Definition & Risk Reduction **(PD&RR)**	Engineering & Manufacturing Developmen **(EMD)**	Production, Fielding/Deployment & Operational Support **(PFD&OS)**
1. System analysis	6. Concept design update	11. Detail design	17. Production rate verification
2. Requirements definition	7. Subsystem trade-off	12. Development	18. Operational test & evaluation
3. Conceptual design	8. Preliminary design	13. Risk management	19. Deployment
4. Technology & risk assessment	9. Prototyping, test, & evaluation	14. Development test and evaluation	20. Operational support & upgrade

TABLE 1.5 ***Continued***

Major Systems Lifecycle Phase			
0	I	II	III
5. Preliminary cost, schedule & concept	10. Integration of manufacturing & supportability considerations	15. System Integration, test & evaluation	21. Retirement
		16. Manufacturing process & verification	22. Replacement planning

0. *Concept Exploration.* The CE phase begins with a definition of project or product objectives, mission definition, definition of functional requirements, definition of candidate architectures, allocation of requirements to one or more selected architectures and concepts, trade-offs and conceptual design synthesis and selection of a preferred design concept. An important part of this phase is the assessment of concept performance and technology demands and the initiation of a preliminary risk management process.

I. *Program Definition and Risk Reduction.* The PD&RR phase is oriented to a risk management strategy in order to prove that the system will work prior to committing large amounts of resources to its full-scale engineering and manufacturing development. This is the first phase in the development cycle where significant effort is allocated to developing tangible products such as top-level specifications, decomposing and allocating system requirements and design constraints to lower levels, supporting preliminary design, monitoring integration of subsystem trade-offs and designs and detailed project plans.

II. *Engineering and Manufacturing Development.* During the EMD phase, detailed design and test of all components and the integrated system are accomplished. This may involve fabrication and testing of engineering models and prototypes in order to check that the design is correct. The hardware and software design for the EMD usually differ from those of the PD&RR phase. This is usually justified to minimize the PD&RR phase costs and to take advantage of lessons learned during PD&RR in order to improve the EMD design. Thus, most of the analysis, modeling, simulation, trade-off and synthesis tasks performed during CE and PD&RR are repeated at a higher fidelity. A requirement validation process should be conducted before the EMD hardware and software is produced. This will ensure that the entire system will function as envisioned.

III. *Production, Fielding/Deployment and Operations and Support.* During production, deployment and operational use, the focus is on solving problems that arise during manufacturing, assembly, integration and verification as well as the transition into its deployed configuration. Additionally, attention is given to customer orientation, validation and acceptance testing. During the phase of operations and support, systems are usually under the control of the purchasers/operators. This involves a turnover of the system from experienced developers into less experienced operators. This leads to a strong operations and support presence by the developers in order to train and initially help operate the system. During this period, there may be upgrades to the system to achieve higher performance levels.

Government and Commercial Program Phases INCOSE (2007) further illustrates and compares several typical lifecycle phases of government and commercial organizations (see Figure 1.5). This figure emphasizes that system lifecycles in different domains are fundamentally similar in that they move from requirements, definition, and design through manufacturing, deployment, operations and support (and sometimes to deactivation), but they differ in the vocabulary used and nuances within the sequential process.

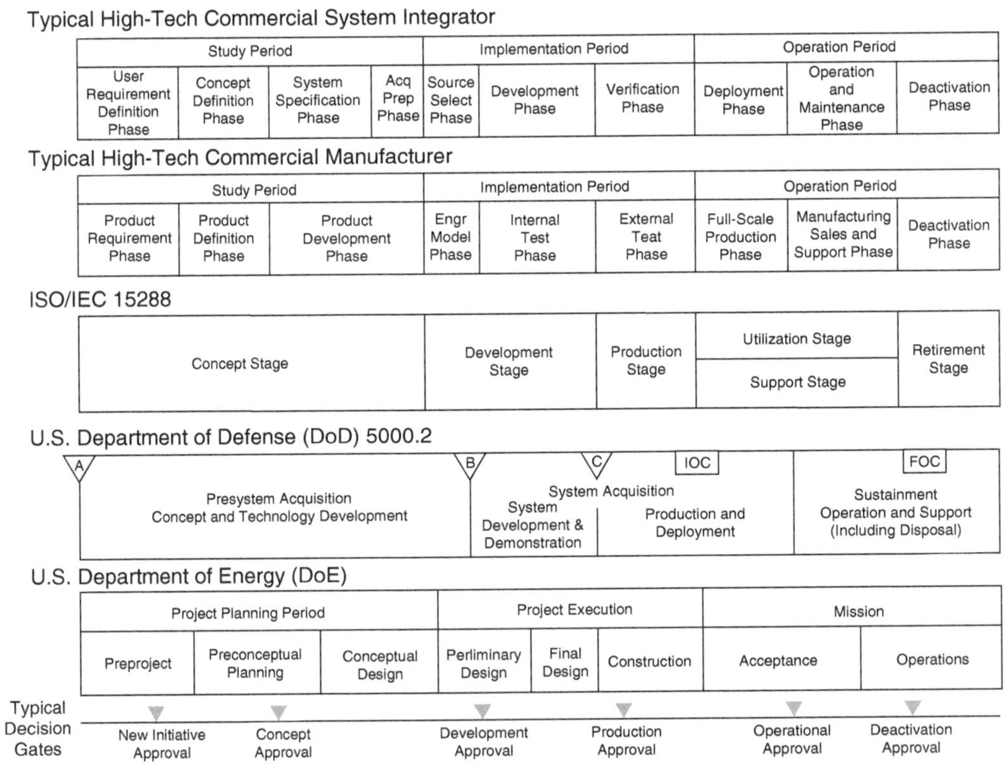

Figure 1.5 System lifecycle phases as illustrated in INCOSE, 2007.

Generic System Lifecycle Adopted for This Book This book has adopted the generic system lifecycle model (see Table 1.6) that is used in the SysTest project due to its generality and practicality. It is a generic extension of the model of system lifecycle phases and VVT activities suggested by Addy (1999) and Boehm (2001). This system lifecycle model extends the well-established V-Model (Martin and Bahill, 1996), which portrays project evolution during the development portion of the system lifecycle.

TABLE 1.6 Generic System Lifecycle Definition Model

Phase	Purpose
	Development
Definition	Formulate the system operational concepts and develop the system requirements.
Design	Create a technical concept and architecture for the system.
Implementation	Create the elements of the system. Each element is built or purchased, then tested to ensure its stand-alone compliance with its allocated requirements.
Integration	Connect the implemented elements into a complete system.
Qualification	Perform formal and operational tests on the completed system to assure the quality of the system as a whole.
	Postdevelopment
Production	Produce the completed system in appropriate quantities.
Use/Maintenance	Operate the system in its intended environment in order to accomplish intended functionality, maintain the system and correct any defects.
Disposal	Properly dispose of the system and its elements upon completion of its life.

Figure 1.6 depicts the V-Model as a part of the overall generic system lifecycle model developed during the SysTest project and adopted for this book (Engel et al., 2001).

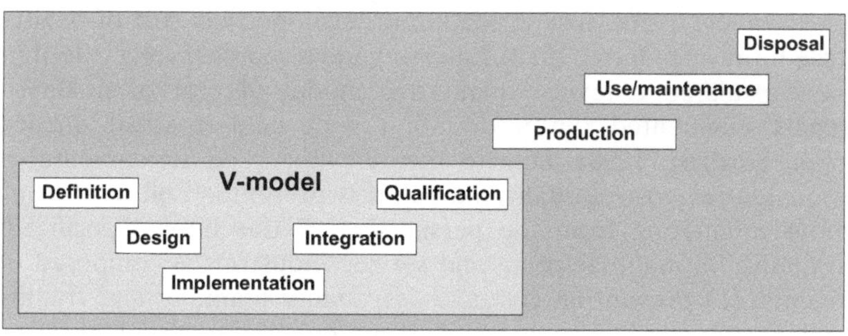

Figure 1.6 V-Model as part of overall generic system lifecycle model.

The left-hand side of the V-Model corresponds to satisfying stakeholders' requirements and the design of the desired system and its components. The right-hand side of the V-Model consists of building the individual components, integrating them and then verifying and validating the whole system. Figure 1.6 depicts the V-Model as a part of the overall generic system lifecycle model developed during the SysTest project and adopted for this book (Engel et al., 2001). Figure 1.7 depicts a generic system lifecycle model together with the corresponding generic VVT lifecycle, with which it is associated.

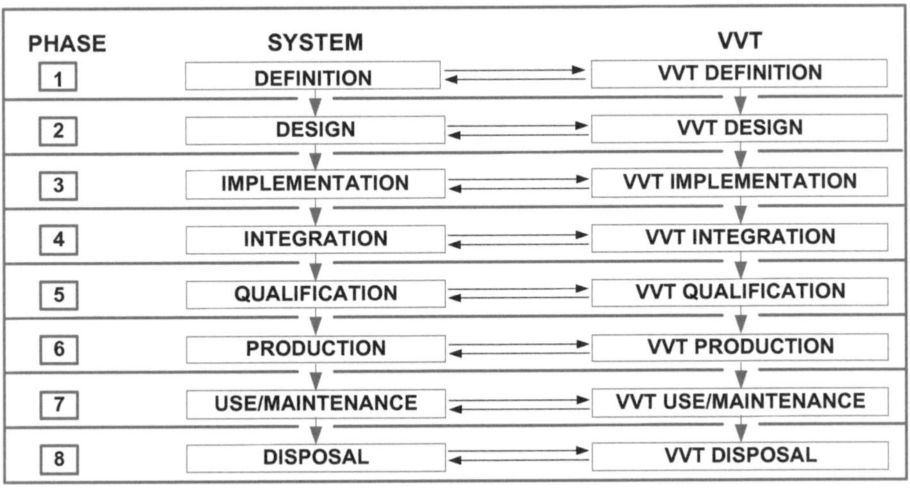

Figure 1.7 Modeling generic systems and VVT lifecycles.

1.2.6 Modeling VVT and Risks as Cost and Time Drivers

Traditional Modeling Quality Cost The cost of quality is the overall cost associated with ensuring the quality of products or services delivered to customers. In the 1950s, Joseph M. Juran developed his cost-of-quality concepts (see Juran and Gryna, 1980). Later, several researchers (e.g., Montgomery, 2001) encapsulated a lexical qualitative model of cost of quality. Some researchers augmented the information with field-obtained quality cost data (e.g., Sörqvist, 1998). Due to the relevancy and fundamental nature of this qualitative cost-of-quality model, it is presented below with relevant alterations emanating from the perspective of this book. Specifically, the cost of quality in manufacturing and service industries is composed of four components: (1) prevention cost such as quality planning and training, (2) assessment cost such as product inspection and testing, (3) internal failures cost such as scrap, rework and retest and (4) external failure costs such as warranty charges, liability cost and indirect cost. We will now map system quality costs to this model.

1. *Prevention Costs.* Prevention costs are costs expanded on the prevention of nonconformance to specifications during system development, manufacturing and maintenance. Important subcategories of prevention costs are shown in Table 1.7.

TABLE 1.7 Subcategories of Prevention Cost

Subcategories	Type
Quality Planning. Costs associated with the creation of various quality plans (e.g., inspection plan, reliability plan).	VVT cost
Product/Process Design. Costs incurred during the quality evaluation of system development and production processes which are intended to improve the overall quality of products as well as costs incurred during the evaluation of the development and manufacturing effectiveness (e.g., input versus output, return on investment)	VVT cost
Process control. The cost of process control activities, such as collecting samples and generating control charts which monitor the development or the manufacturing process in an effort to reduce variation and create quality within system.	VVT cost
Burn-in. The cost of preshipment exercising and evaluation of system in order to minimize early-life defects in the field.	VVT cost
Training. The cost of developing, implementing, operating, and maintaining training programs in order to achieve system quality.	VVT cost
Quality Data Acquisition and Analysis. The cost associated with creating, purchasing, and operating quality of data collection and distribution system as well as the cost of running the quality data system to obtain information about systems and process quality performance and analyzing and publishing it for management, customers and other stakeholders.	VVT cost

2. *Assessment Costs.* Assessment costs are those costs associated with measuring and evaluating purchased materials, components and subsystems as well as verifying, validating and testing systems (i.e., end products and enabling products) to ensure conformance to specified requirements and standards. The major subcategories of assessment costs are described in Table 1.8.

TABLE 1.8 Subcategories of Assessment Cost

Subcategories	Type
Inspection and Test of Incoming Material. Costs associated with the inspection and testing of appropriate vendor's supplied raw material, components and subcategory either at the vendor's facility or at the receiving station of the firm. In addition, this subcategory includes verification of all vendor-supplied documentation as well as periodic audit of the vendor's quality assurance system.	VVT cost
Systems Verification, Validation and Test. The cost of checking the conformance of the systems throughout the various stages of development and manufacturing, including final acceptance testing, packing and shipping checks and any test done at the customer's facilities prior to turning systems over to the customer. In general, assessment cost also covers tests and evaluation associated with system maintenance activities as well as verification and validation of appropriate disposal process.	VVT cost
Consumed Materials and Products. The cost of material and products consumed in destructive quality tests or devalued by reliability tests.	VVT cost
Maintaining Accuracy of Test Equipment. The cost of ensuring that the measuring instruments and equipment are calibrated on an ongoing basis.	VVT cost

3. *Internal Failure Costs.* Internal failure costs are incurred when materials, components, subsystems or systems do not meet quality requirements and these failure are discovered prior to delivery of the systems to customers. The major subcategories of internal failure costs are described in Table 1.9.

TABLE 1.9 Subcategories of Internal Failure Cost

Subcategories	Type
Scrap. The net loss of labor, material and overhead resulting from defective product or systems that cannot economically be repaired or used.	Risk cost
Rework. The cost of correcting system chronic or sporadic defects so that they meet specifications. This process may transpire once or several times.	Risk cost
Retest. The cost of repeated verification, validation and testing of systems that have undergone rework or other modifications.	Risk cost
Failure Analysis. The cost incurred to determine the global causes of recurring system failures. Note that this subcategory is not referring to a regular testing process but to a wider phenomenon of persistent system failures.	Risk cost

TABLE 1.9 *Continued*

Subcategories	Type
Downtime. The cost associated with idle development or production facilities and manpower that result from nonconformance to requirements. The development may be halted until certain information is obtained. A production line may be down while a defective system or product is evaluated or repaired.	Risk cost
Yield Losses. The cost of process yield that is lower than might be attainable by improved quality controls.	Risk cost
Downgrading. The cost associated with inferior products and systems that do not meet the entire customer's requirements. Downgrading implies that such products yield less profit relative to products that conform to specifications. In addition, inferior products adversely affect the reputation of the firm, causing loss of revenues.	Risk cost

4. *External Failure Costs.* External failure costs occur when systems do not perform satisfactorily and the problems are identified after these systems have been supplied to customers. The subcategories of external failure costs are described in Table 1.10.

TABLE 1.10 Subcategories of External Failure Cost

Subcategories	Type
Complaint Adjustment. All costs associated with the investigation and adjustment of either justified or not justified complaints attributable to the nonconforming product.	Risk cost
Handling Defective Products and Systems. All costs associated with either fixing systems at customers' premises or replacing nonconforming products and systems that are returned from the field.	Risk cost
Warranty Charges. All costs involved in service to customers of faulty systems under warranty contracts.	Risk cost
Liability Costs. All costs associated with defective products and systems incurred as a result of system liability litigations.	Risk cost
Indirect Costs. Costs incurred because of customer dissatisfaction with the level of quality of the delivered system. They include the costs of business reputation loss, future business loss and market share loss that may result from delivering defective systems that do not meet the customer's expectations.	Risk cost

Waste in Product Development The Lean Aerospace Initiative (LAI) was born out of declining defense budgets and military industrial overcapacity, prompting a new defense acquisition paradigm, that is, affordability rather than performance. The U.S. Air Force (USAF) and the Massachusetts Institute of Technology (MIT) launched this initiative in 1993.

Researchers dedicated to the philosophy called "lean" are interested in eliminating waste that occurs during systems' development phase of projects.

Womack and Jones (2003) classified all product-making activities into Value Adding (VA), to be continually perfected; Non–Value Adding (NVA), to be eliminated; and Required Non–Value Adding (RNVA), such as those required by contract or law, to be faithfully executed. No formal study is available on the relative amounts of NVA and RNVA waste in the aerospace programs (Oppenheim, 2004). Table 1.11 shows two sets of product development waste categories as classified by two studies.

TABLE 1.11 Two Sets of Product Development Waste Classifications

Classification by Millard (2001)	Classification by Morgan (2002)
1. Overproduction (creating unnecessary information) 2. Inventory (keeping more information than needed) 3. Transportation (inefficient transmittal of information) 4. Unnecessary movement (people having to move to gain or access information) 5. Waiting (for information, data, inputs, approvals, releases, etc.) 6. Defects (insufficient quality of information, requiring rework) 7. Overprocessing (working more than necessary to produce the outcome)	1. Hand off (transfer of process between parties) 2. External quality enforcement (including performance requirements) 3. Waiting 4. Transaction waste 5. Reinvention waste 6. Lack of system discipline 7. High process an arrival variation 8. System overutilization and expediting 9. Ineffective communication 10. Large batch sizes 11. Unsynchronized concurrent processes

In an ideal world, systems are created perfectly and VVT procedures would not be necessary. Therefore, performing VVT and incurring VVT appraisal and impact risks are clearly NVA activities. Obviously, optimizing the VVT strategy leads to less costly NVA results. Our world is not ideal and the VVT process is a necessary expenditure that is required to ensure the quality of systems. Therefore, one can say that just about all VVT activities lie on the border between VA and NVA activity regions.

Modeling Cost and Risk VVT cost can be considered a cost associated with classical prevention and assessment, while risk impact cost is usually associated with sustaining internal and external failures. Developing risk-based cost models involves three activities:

- Identifying VVT risks
- Estimating risk probability
- Estimating risk effects

In the literature, we find several methodologies dealing with these topics. The main ones are discussed below.

Methodology Based on Perception of Engineering Process A detailed approximation of the underlying cost and risk of a project can be obtained by viewing the engineering process as a tree structure and each node in the tree is an engineering activity. The standard engineering tool of Work Breakdown Structure (WBS) is an available vehicle to promote and support this methodology. Engineering process parameters such as cost/duration, including the VVT tasks, are first identified. Experts then assign valuations to them based on the experts' technical knowledge. To take into account uncertainties, rather than assigning only a best estimate of task cost and duration, these experts can assign a minimum, a most likely and a maximum estimate for each of these two quantities.

VVT activity costs and durations are fairly easy to predict, whereas the costs and durations of engineering processes are somewhat less predictable due to their physical nature. Fortunately, engineering experts are able to do a fairly good job at estimating risks, risk impact probabilities, and risk impact costs. Because expert opinions often differ, the cost estimates for normal engineering activities and the risk cost estimates are recognized to be probability functions across the different categories and expert opinions. The data are presented to participants and stakeholders as a range of values rather than a single value in terms of a cost–risk curve (e.g., a histogram of risk–cost density distribution). It should be noted that more sophisticated approaches for transforming the three estimate levels into probabilistic data are available, for example, with the aid of a beta distribution (Fente et al., 1999).

Methodology Based on Balancing Cost/Availability and Benefits Browning (1998, 1999) describes a method for identifying acceptable risks. The method balances product pricing and availability timing with the value of the product to the customer. The designers of systems must fit the design process to optimize this process. Browning's thesis first addresses the sources of risk of not meeting this optimization and classifies it into six categories: (1) cost, (2) schedule, (3) performance, (4) technology, (5) business and (6) market risks. Then he builds a framework and a model to represent the relationships between these risks. A stochastic simulation is then used to generate probability distributions of possible costs, schedules and performance outcomes. These distributions model uncertainty and are analyzed in relation to impact functions. The model provides the means to explore several management options for optimizing the above parameters.

Methodology Based on Holistic Philosophy of Risk Scenarios Haimes (1998) coined the term Hierarchical Holographic Modeling (HHM) to depict complex systems using multiple models created along different perspectives. Extending this concept, Haimes et al. (2002) proposed an analytic framework called Risk Filtering, Ranking, and Management (RFRM), which can identify, prioritize, assess, and manage risk scenarios of large-scale systems. In a nutshell, the risk assessment portion of RFRM follows these steps: First, the HHM must be developed to describe a multifaceted model of the system's "as-planned" scenario. Then, the set of risk scenarios is qualitatively filtered and ranked according to the system stakeholders' views. Finally, a quantitative filtering and ranking of possible risks must be carried out based on the likelihood of system failures and the consequences of such events. Lamm and Haimes (2002) use the HHM and RFRM methodologies to analyze the security of the U.S. national information infrastructures.

Methodology Based on System Safety Program Requirements Muessig et al. (1997) describe another methodology in the context of a risk–benefit analysis approach to the selection of an optimal set of Verification, Validation, and Accreditation (VV&A) activities. This risk modeling is based on an adaptation of the U.S. military standard MIL-STD-882C, System Safety Program Requirements. In the model, VVT risks are quantified in terms of probability of occurrence and impact or severity levels within the context of specific applications. Two variables are involved in modeling risks as cost drivers: (1) the uncertainty of risk occurrence and (2) the severity of risk impact.

1. *Uncertainty of Risk Occurrence.* The first element affecting risk is the uncertainty with which undesirable events occur. The risk model defines the probability of occurrence of a given risk factor in different ways, depending on the category of the risk factor that is being considered. The effect of undesirable events impacting the system can be measured by (1) the number of items affected in a population, (2) the number of events per unit of time or (3) the total number of events over the life of the system or product.

   The model of Muessig et al. (1997) divides the probability continuum into five bands and gives guidelines for selecting the appropriate band. Table 1.12, extracted from MIL-STD-882C, provides these guidelines in terms of the number of undesirable events over a lifetime and per number of items in a population.

TABLE 1.12 Probability of Risk Occurrence

Probability Description	Likelihood of Occurrence over Lifetime of Item	Likelihood of Occurrence by Number of Items
Frequent	Likely to occur frequently	Widely experienced
Probable	Will occur several times in life of item	Will occur frequently
Occasional	Likely to occur sometime in life of item	Will occur in several items
Remote	Unlikely but possible to occur in life of item	Unlikely but can reasonably be expected to occur
Improbable	So unlikely it can be assumed occurrence may not be experienced	Unlikely to occur but possible

The reader may substitute "system" or "product" for the word "item," as appropriate.

2. *Severity of Risk Impact.* The second element affecting risk is the severity of the impact of an undesirable event, should the event be experienced. The risk model developed by Muessig et al. (1997) expands the MIL-STD-882C while grouping the impact severity into four bands: catastrophic, critical, marginal and negligible. The criterion for assigning one of these impact bands to a particular risk depends on the category of that risk. The impact categories that are discussed in the model are personnel and equipment safety, environmental damage and occupational illness. Depending on the particular use of the system being considered, some of these impact categories might not apply, and additional categories might be added—for example, impact on end-user capability or effectiveness, cost, performance, schedule and political or public reaction. A set of criteria for determining the level of impact for each of the different impact categories is provided in Table 1.13 as an illustrative guideline.

TABLE 1.13 Severity of Risk Effects

	Risk by Impact Levels			
Categories	Catastrophic	Critical	Marginal	Negligible
Human safety	Death	Severe injury	Minor injury	Less than minor injury
Systems safety	Major equipment loss; broad-scale major damage	Small-scale major damage	Broad-scale minor damage	Small-scale minor damage
Environmental damage	Severe	Major	Minor	Some trivial
Occupational illness	Severe and broad scale	Severe or broad scale	Minor or small scale	Minor and small scale
Financial losses of program	Loss of program funds; 100% cost growth	Fund reductions; 50–100% cost growth	20–50% cost growth	<20% cost growth
Functional performance of product	Design does not meet critical thresholds	Severe design deficiencies but thresholds met	Minor design flaws but fixable	Some trivial "out of spec" design elements
Schedule slippage of product	Slip reduces overall capabilities	Slip has major cost impacts	Slip causes internal turmoil	Republish schedules

TABLE 1.13 *Continued*

	Risk by Impact Levels			
Categories	Catastrophic	Critical	Marginal	Negligible
Political or public impact of event	Impact widespread (Watergate)	Significant (Tailhook '91)	Embarrassment ($200 hammer)	Local
Negative impact due to unidentified stakeholders	Major stakeholder blocks program (Israeli AWACS sale to China)	Stakeholder requires product modifications (FAA disqualifies new aircraft)	Stakeholder requires minor system modifications	Upgrading sales campaign to cover newly recognized stakeholders
Future losses of potential revenues	Customers determined to abandon product	Major market share loss	Customers dissatisfied with product	Competitor plan to develop similar product

1.3 CANONICAL SYSTEMS VVT PARADIGM

1.3.1 Introduction—Canonical Systems VVT Paradigm

An engineered system does not appear suddenly in just an instant. Like any other entity, it needs to be brought into being, cared for and nourished, challenged and utilized and finally put to rest. Thus, the concept of a system life is appropriate. This section discusses that life and describes the role of VVT in its phases. This is presented in terms of the canonical system VVT paradigm composed of (1) phases of the systems lifecycle, (2) views of the systems and (3) aspects of the systems.

A system, in this context, is a set of interacting or interdependent entities, man made or otherwise, existing and forming an integrated whole that fulfills a certain purpose or set of objectives. For an engineered system to adequately meet its objectives, the goal should be to invent, develop, adapt or optimize system behavior within a set of required properties. The man-made parts of an engineered system can undergo development from different disciplines, such as mechanics, hydromechanics, electronics, computation and programming. Other parts, such as human operators or technicians, can also undergo development from other disciplines, such as education, training and work experience.

Figure 1.8 helps the reader to envisage the many interactions involved in the VVT process. It depicts the canonical system VVT paradigm as a three-dimensional object:

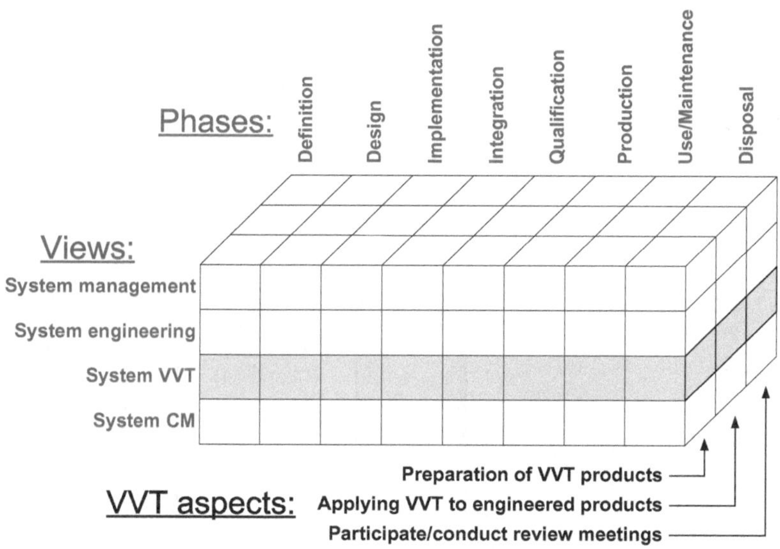

Figure 1.8 Canonical system VVT paradigm.

- *First Dimension.* Lifecycle phases include all the system lifecycle phases (i.e., Definition to Disposal).
- *Second Dimension.* System views include, among others, the following components: System management, Systems engineering, System VVT and System Configuration Management (CM).
- *Third Dimension.* Aspects of systems include the following components: Preparation of VVT products, Applying VVT to engineered products and Participating or conducting reviews.

Knowing the phases of the system lifecycle is essential for understanding how VVT is implemented throughout the life of a system. Thus, each phase is discussed separately and the appropriate VVT activities for that phase are described. During the entire lifecycle, from system definition to system disposal, there are at least four views of the system. Naturally, the most important view for this book is VVT. For completeness, short descriptions of the remaining views are also provided.

Here each activity of a system lifecycle can be categorized by placing each of them in one of the cubes depicted in the three-dimensional stack of cubes shown. These activities describe what has to be done in order to achieve the desired degree of quality in a system.

The VVT activities, however, indicate only what may be done to assure the quality of a system. Thus, for each VVT activity, this book provides one or more VVT implementation methods. These VVT methods describe how to perform an activity by defining a sequence of steps that should be performed.

From this perspective, a step within a method may indeed be a VVT activity unto itself. While some VVT activities are straightforward and may be implemented by only one method, others may be carried out using one of several methods. An example of a hierarchy depicting activities and methods is shown in Figure 1.9. Each element of the canonical system VVT paradigm (i.e., phases of the system lifecycle, views of the system and aspects of the system) will now be discussed in more details.

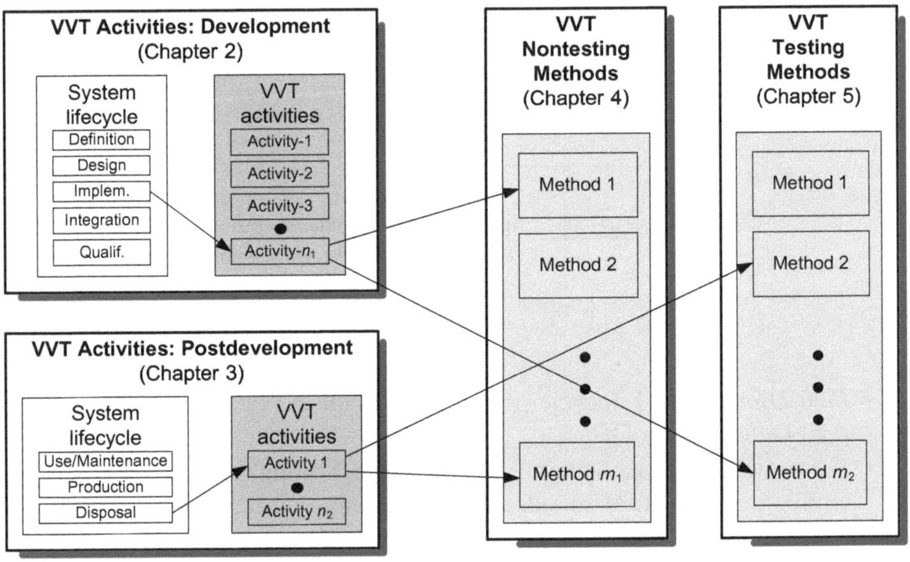

Figure 1.9 Hierarchy of VVT activities and methods.

1.3.2 Phases of the System Lifecycle

Each individual activity of a system lifecycle is allocated to one of the phases and works smoothly together with other activities to achieve the overall goals of that phase. There are several mostly overlapping phases, each describing a particular period of the overall system lifecycle. Depending on the system (hardware versus software development, safety-critical versus noncritical application, etc.), some of these phases are considered more relevant than others. As mentioned above, the canonical phases of a system's lifecycle are Definition, Design, Implementation, Integration, Qualification, Production, Use/Maintenance and Disposal.

In our system lifecycle framework, eight phases encompass the system lifecycle. Depending on the system under consideration, some of these phases may be more or less important. These eight phases pretty much cover the same areas as the five phases called out in the ISO/IEC 15288: Concept (Define/Design), Development (Implement/Integrate/Qualify), Production (Produce),

Utilization and Support (Use and Maintain) and Retirement (Disposal). The eight phases of a system lifecycle are described in the following.

System Definition During the system Definition phase, the requirements of the system are elaborated as completely and precisely as possible in terms of system, hardware and software requirements. Specifications that could constitute the actual system definition could take many forms. For instance, textual requirements, formal requirements, system models or prototypes can be artifacts of system requirements activity.

From the perspective of VVT, during this phase, a project should produce a set of system requirements that are complete, clear and consistent. VVT planning consists of defining forward-looking VVT-related concepts and goals. Specific details of VVT are few, but the planner should be looking at defining the overall VVT framework in general terms that support the emerging system architecture. For example, if the system requirements mandate built-in test capabilities, the VVT philosophy could emphasize intrinsic self-instrumentation capabilities within components in order to reduce the need for developing intrusive and expensive instrumentation.

In the Definition phase, allocation of requirements to hardware and software is usually incomplete; so many specifics of VVT cannot be fully developed. Once systems engineering begins to define the Technical Performance Measures (TPMs) that will assist in meeting system performance requirements, some of the details of VVT requirements can be established. The VVT philosophy during this phase must be forward looking and flexible, as this is the time that system definition is most fluid.

The primary objective in VVT planning in this phase is to define the framework for VVT throughout the program to the level of detail possible. Just as the system receives its architectural concepts during this phase, VVT develops its own architecture that supports the program needs. As system requirements are being analyzed and lower level specifications are being written, VVT planning focuses on the analysis of test requirements and influence of specifications from a test and instrumentation perspective. If self-test requirements are articulated at a top level, or if requirements analysis and derivation imply the need for self-instrument requirements, then the VVT planning can both influence and build upon these expected capabilities as they become defined.

System Design The technical concept of the system, the principles and the underlying system architecture for the implementation of the system are determined during the system Design phase. The total complex system is divided into manageable subsystems and components and the functions of the individual elements as well as their interrelations are described.

As requirements get refined and assigned into subsystems and components, VVT will now have a more concrete structure against which to direct specific test strategies. General TPMs will become allocated and apportioned to subsystems and components. The resulting greater specificity allows VVT planning efforts to be directed toward the implementation phase and integration phase needs.

System Implementation The design concept is realized during the system implementation phase. If the system is a hardware-based system, this implementation is only a prototype (i.e., the first instance of the system built) that must be reproduced during the system Production phase. At the completion of the system Implementation phase, all individual components of the overall system should be available and functioning.

During system implementation, VVT efforts are directed toward those emerging subsystems, their verification against system requirements and their refinement. As requirements are verified with respect to implemented components, they should also be validated against stakeholder needs. This validation should be a continuous process. Whenever subsystem or component definition and specificity permit, the associated requirements should be validated.

System Integration The focal point of this phase is the integration of the implemented subsystems with the aim of setting up the complete system.

VVT activities during system integration are directed at verifying that the interfaces between subsystems or components as well as between the system as a whole and external elements meet requirements and that the whole meets system requirements as well. VVT activity should also be focused toward validation of each requirement within the relevant integrated subsystem. VVT planning during this phase is directed toward preparing for qualification of the system.

System Qualification The system Qualification phase is a formal phase during which the system runs through a number of tests often prescribed by external agencies, customers or standards. The goal is to assure the quality of the system as a whole. Ideally, during this phase, no constructive developments on the system should be carried out. In practice, however, often certain parts of the system are being tested while other parts are still under various stages of development.[1]

At this point, the formal validation of the verified requirements ensures that the system meets the stakeholder true needs and that those needs are accurately reflected in the captured requirements. VVT activities include testing the system and ensuring that all requirements are verified using the proper method (i.e., analysis, inspection, demonstration, testing or certification). VVT planning consists of selecting appropriate qualification testing for inclusion in the Production phase as a subset of acceptance testing. VVT planning starts the preparations to support testing of purchased parts and conduct-

[1] Concurrent engineering is a methodology of developing different parts of a system in an unsynchronized manner so each part may, in parallel, be at a different stage of development (e.g., definition design, implementation, integration, qualification) at any given time. This approach, which attracted unsavory reputation, is under intensive scientific research and gaining due respect as a legitimate way to reduce elapsed time required to bring systems into the market.

ing component qualification before inclusion into the produced systems. VVT planning also includes developing an efficient production VVT strategy to assure good system components are delivered with a test subset that is viable and economical.

System Production Once the system is deemed ready, the next phase is to produce final products for sale or use. VVT activities include testing of purchased parts and the conduct of component qualification tests. VVT planning includes preparing to receive and process field failure data when the system is fielded.

System Use and Maintenance When regarding the overall system lifecycle one must also consider the VVT activities during the Use and Maintenance phase. The system is now fielded and under customer control. It operates in its intended environment and manned by operators who have been trained in its proper use. Maintenance should be performed in accordance with the policies and guidelines established during its development. Failures may occur due to component wear, operator error or unanticipated harsh environmental factors as well as defective design or poor manufacturing process. If these occur during the warranty period, the program/project team should have responsibility for correction and possibly additional rework if the failure has revealed a fundamental system deficiency. Also, during this phase, eventual improvements to the system functions are introduced, errors are eliminated, and systems are maintained.

System Disposal After the use of the system, its disposal becomes an important aspect which should have been planned from the earliest days of the system development. During this phase systems must be dismantled, recycled, if necessary, and/or finally disposed of.

In general, VVT activities are performed within this phase only for systems with public safety issues associated with the system disposal or for systems that had specific disposal-related requirements imposed during their development. In these cases, there are likely to be enabling technologies required (such as nuclear waste disposal) which will have VVT activities. If the program is of sufficiently long duration, the disposal-enabling technologies may require certification or validation that should be planned for in advance and executed when needed.

1.3.3 Views of the System

During the entire lifecycle, from system definition to system disposal, there are different views one could have on the system. Naturally, the most important view for this book is the "VVT" view, which focuses on all activities that are implemented to assure the required quality by means of verification, validation, and testing of the system or system components. Such activities should be performed during every lifecycle phase to assure the quality of intermediate or final lifecycle products. Beside this view, there are of course other views, such as system management, systems engineering and configuration management, which are related but of secondary importance for this book.

System Management View System management includes activities concerned with organizational issues associated with a system or a product. These include:

- The subdivision of the development and production process into phases and activities
- The division and definition of the work to be done
- The regulation of communication
- The organization and control of the work flow

The activities set out in system management comprise planning and controlling of various activities, the allocation of internal roles and the setting up of an interface to units outside the project (i.e., subcontractors, management, etc.). Typically, system management contains the following main tasks: project initialization, detailed planning, project control, reporting, cost–benefit analysis, phase reviews, risk management, resource management, contractor management and training.

System Engineering View System engineering is that set of activities which directly leads to the development, production, use and maintenance and finally disposal of a system, as opposed to other activities related to system management, quality assurance and configuration management, which (crucial though they are) play a supporting role from the perspective of system construction. The system development lifecycle covers the following main activities:

- System requirement analysis
- Software/hardware requirement analysis
- System and subsystem design
- Component and subsystem implementation (hardware/software units)
- System integration
- System qualification

In system development, all activities directly relevant to the system development lifecycle process and the respective documents are grouped together. A system development lifecycle encompasses the complete set of activities that generate and implement engineering decisions about a system:

- What it should do (and not do)
- Which technologies should be used and where
- How it should be structured into parts
- How parts should be obtained (design-and-build, reuse-and-adapt, acquire, etc.)

- How VVT should be done
- How integration should be performed
- How to produce systems (for mass market or a small number of products)
- How to maintain systems and dispose of obsolete ones

Verification, Validation, and Testing View Conventional wisdom says that to produce competitive products one must identify the requirements and proceed to meet these in an efficient and effective way. This is a quality assurance process, which can be separated into three different levels: the organizational level, the process level and the product level. The activities relevant to the VVT view serve as the basis for the detailed explanation of activities and methods in the following chapters of this book.

Configuration Management View Configuration Management (CM) comprises those activities that must be performed in order to manage all the parts and their relationships and to support systems engineers in maintaining the integrity of the system. It is a service function that allows the various participants involved in the system engineering process to perform their perspective role confidently.

1.3.4 VVT Aspects of the System

Each individual activity describes one block of work of the project's complex network of tasks. Each VVT activity may be assigned to one of the following VVT aspects:

- *Prepare VVT Products.* This VVT aspect encompasses VVT activities related to preparation of VVT products, such as developing a certain VVT plan and designing and fabricating certain VVT tools or simulations.
- *Perform VVT Activities.* This VVT aspect encompasses VVT activities related to actual VVT of various system engineering products, for example, verifying a system design document and testing a package of software.
- *Participate in Reviews.* This VVT aspect encompasses VVT activities related to either participating in or conducting a system review, for example, participating in a system Preliminary Design Review (PDR) and conducting a Test Readiness Review (TRR).

1.4 METHODOLOGY APPLICATION

1.4.1 Introduction

In this section we begin to get to the heart of the subject matter in this book. VVT has developed over the years into a set of tools that are tried and proven to save time and money and ensure success in the design and building of complex systems. Having covered the preliminaries in the previous sections, we concentrate here on the tools and techniques available for system VVT. We begin with an overview of the VVT methodology. The basis of this methodology is a process model that assists VVT planning by providing calculation of the cost and risk associated with the various VVT strategies. This process is a guide to modern VVT planning as performed by VVT practitioners, in coordination with the other stakeholders of the engineered system. As mentioned, a good VVT process does not "just happen." It is the product of thorough planning and strategy.

Since there is no such thing as a "typical" engineered system, what is good for one system in the way of VVT may not be good for another. So, we go on to show how VVT can be tailored to different kinds of systems, different organizations and different project parameters. Heuristics are described for tailoring VVT concepts to specific engineered systems based on project size/complexity and type (i.e., system or industry). Specific attention is paid to the electronics/avionics, aerospace, automotive, food packaging and steel production industries as representative of many other industries. Hints are given for ameliorating project risks by tailoring VVT.

An important issue is the means by which VVT can be monitored and stakeholders can be assured that VVT is properly applied. Remember the old adage, "The job is not complete until the paperwork is done." Of course, today paperwork does not necessarily imply the generation of paper documents. But, records do have to be kept and a trace of VVT steps and functions must be made. This is the only way to assure that the process works and that monies allocated for VVT have been properly spent. Among the necessary documents are the Project Management Plan (PMP), the Systems Engineering Management Plan (SEMP), the VVT Master Plan (VVT-MP), the Testability Program Plan (TPP), the Maintainability Program Plan (MPP), the Reliability Program Plan (RPP), the System Test Plan (SysTP), the Software Test Plan (STP, if appropriate), the First Article Inspection Plan (FAIP), the Production Plan (PP), the Maintenance Plan (MP), the Integrated Logistic Support Plan (ILSP) and the Disposal Plan (DP). While, for any specific system not all of these plans may be required, we provide fair details of what these documents consist. In summary, reading this section sets the stage for the following chapters, which cover the "how to" for implementing VVT.

1.4.2 VVT Methodology Overview

The basis of the VVT methodology is to apply an informed strategy and planning process to the selection and sizing of VVT activities. Through such a process, VVT activities, methods, tools and products are optimized to reduce project risk while improving cost, quality and development time. This book describes a process model that assists VVT planning by providing calculation of cost and risk associated with various VVT strategies. The effort required for performing the VVT strategy, planning, and modeling should be commensurate with the size of the project, so that the effort expended will be repaid in improved quality and reduced project cost, risk and development time.

Methodology for VVT Strategy and Planning The generic VVT process is depicted in Figure 1.10 (Lévárdy et al., 2004). It is an iterative process that can be applied to the entire system lifecycle, to a subset of the system lifecycle (e.g., system development) or to any of the individual lifecycle phases. The VVT process has four main segments: (1) VVT tailoring at the organization and project level, (2) Rough VVT planning at the system level, (3) Detailed VVT planning and (4) VVT execution.

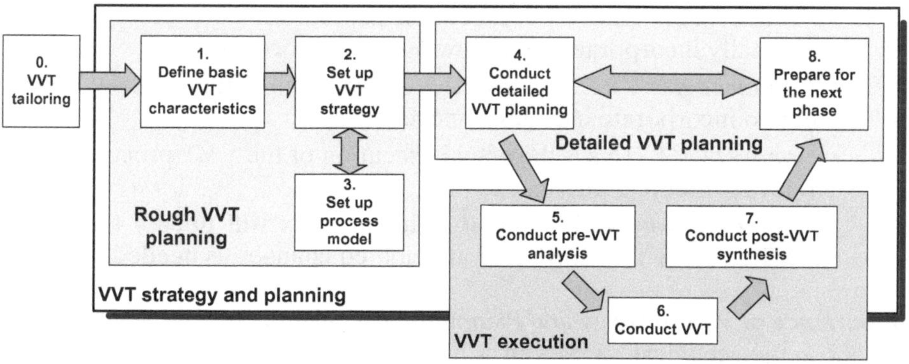

Figure 1.10 VVT methodology for strategy and planning (Lévárdy et al., 2004).

The VVT for strategy and planning encompass the following steps:

1. *VVT Tailoring.* Before starting a project, those managing the project should determine the factors that characterize the project and enterprise. Based on these factors, the project managers should tailor the VVT methodology to suit the project. Tailoring consists of high-level decisions about the use of this methodology and its parts based on knowledge of the organization and insights gained in earlier project.
2. *Rough VVT Planning.* At the outset of each project, it is necessary to plan the VVT process, at least in a rough manner, and establish a VVT strategy. The VVT strategy considers business objectives and their relationship to the project as well as issues related to programmatic and strategy risks. Strategy consists of creating a set of requirements and constraints that guide the VVT planning along with primary decisions about the VVT activities to follow. VVT rough planning uses the following three process groups:
   - *Define basic VVT characteristics.* This determines the basic characteristics that guide and bound the VVT strategy.
   - *Set up VVT strategy.* This codifies the strategy into a selection of activities and methods while also defining the requirement verification methods to be used.

- *Set up a VVT process model.* This uses the VVT process model to support the strategy definition by using calculation of cost, time and risk to explore alternative strategies.

3. *Detailed VVT Planning.* Throughout the system's lifecycle and especially at the beginning of each lifecycle phase, VVT engineers should reexamine or/and establish a detailed VVT plan. This plan should identify specific activities, methods, tools and products that will implement the actual VVT process. The VVT plan also identifies the types, formality and amount of effort to be applied to each VVT activity.
4. *VVT Execution.* The VVT execution process for each lifecycle phase will usually incorporate the following three process groups:
   - *Conduct a pre-VVT analysis.* This analysis will update the VVT strategy to incorporate changes as needed.
   - *Conduct VVT.* This is the actual execution of the VVT process for the relevant lifecycle phase.
   - *Conduct a post-VVT synthesis.* This analysis will update the future VVT strategy to incorporate anticipated changes as needed.

Importance of VVT Strategy and Planning A vital and effective VVT process enhances the technical success of a development program. A well-planned VVT strategy reduces program risk, whereas lack of adequate VVT planning can contribute to programmatic risks. Program costs are minimized when redundant testing is reduced or eliminated. Good VVT planning helps to eliminate redundant testing. Lowest risk is ensured when program strategy includes VVT at an early point in the program and provides continuous attention to VVT-related details. Figure 1.11 illustrates the areas where the implementation of the VVT methodology tends to improve the traditional company VVT processes.

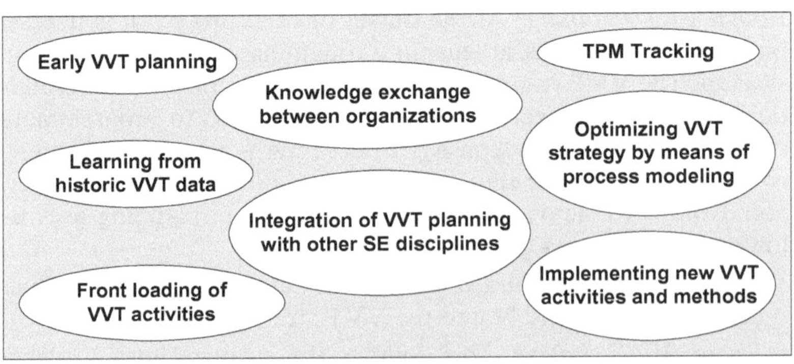

Figure 1.11 Key areas improved by using the VVT methodology (Lévárdy et al., 2004).

Philosophy for VVT Strategy and Planning A good VVT process does not just happen. It is the product of thorough planning and strategy. The philosophy driving VVT should be "Verify early, validate continuously." VVT must combine programmatic thinking with technical thinking. Ultimately, project success is determined in large measure by the effectiveness of its VVT. Technical success depends upon meeting or exceeding performance requirements. Good VVT supports both. A well-planned VVT will:

- Save money through reduced or eliminated test redundancy
- Protect the schedule by being efficient in demands for resources and time
- Assure technical success by identifying areas of performance risk
- Facilitate the Integration phase by ensuring robust component and subsystem interfaces
- Guarantee stakeholder delight by validating requirements against true needs early enough to effect timely change if needed

1.4.3 VVT Tailoring

The VVT methodology is intended to apply to a broad range of projects and enterprises. This section provides guidance and heuristic suggestions on how the unique factors of each project and enterprise may modify the strategy and planning process. Tailoring should be performed at two different levels:

- *VVT Tailoring for Each Organization/Industry.* This tailoring is usually performed once for the enterprise, with occasional updates. In addition, it can be performed on an organizational level for different product lines, thus establishing tailored VVT methodology for each product line. In the event a business undergoes major organizational changes, there might be a need to perform the tailoring again.
- *VVT Tailoring for Specific Projects.* This tailoring is usually performed at the beginning of each project or major replan as part of the VVT planning process.

Tailoring Parameters Three groups of tailoring parameters have been identified for tailoring the VVT methodology: (1) organization/project parameters, (2) programmatic risks and (3) product characteristics.

1. *Organization/Project Parameters.* Table 1.14 identifies three typical major organization and project parameters. These parameters are key discriminators between diverse organizations and product lines as well as projects and are used for both organizational and project VVT tailoring.

TABLE 1.14 Typical Organization/Project Parameters

Parameter	Characteristics
Project size	• Large—Multiteam projects usually more than several million dollars and more than one year duration • Small—Few staff members, limited budget (less than $1 million), few month schedule (less than one year)
Project complexity	• High—Involves many diverse entities or high projects requirements (e.g., performance requirements, aggressive schedule) • Low—Typically simple products manufactured in large quantities
Project type	• Concept exploration—Typically research projects • Technology demonstration—New concept/technology realization in a prototype (possibly limited) for customers' demonstration • Full-scale development/manufacturing—New product development and manufacturing • Maintenance—Improving existing products by fixing deficiencies or adding limited capabilities • Upgrade—Substantially improving existing products by introducing new capabilities

2. *Programmatic Risk Parameters.* Table 1.15 presents three typical programmatic risks that significantly affect VVT project tailoring and planning.

TABLE 1.15 Typical Programmatic Risk Parameters

Parameter	Characteristics
Unachievable schedule	Allocated time to completion is too short to deliver all required capabilities with required quality and maturity.
Insufficient budget	Allocated budget is too small to deliver all required capabilities with required quality and maturity.
Insufficient quality	Allocated resources (e.g., people, schedule, budget, facilities) are not sufficient to meet product quality requirements.

3. *Product Characteristic Parameters.* Table 1.16 presents six product characteristics affecting VVT activities, methods and tool selection.

TABLE 1.16 Typical Product Characteristic Parameters

Parameter	Characteristics
Critical	Mission-critical or safety/health-critical systems parts—Failure in these parts can cause significant human/financial/environmental damage.
Complex	Contains complex system requirements, architecture, real time, deployment, use, production or disposal. Complex systems can be defined as disproportionably large, intricate or convoluted.
Innovative	New technology/feature/capability that has not been previously proved and validated.
Changed	Existing system capability that must undergo limited upgrade/improvement.
Precise	Systems require meeting high-performance or precision requirements.
Need certification	System which requires formal approval/certification by regulatory agencies [e.g., Food and Drug Administration (FDA) and Federal Aviation Administration (FAA)]

Tailoring Heuristics: General Tailoring should always be done within a context and with the benefit of experience. While creating the VVT methodology, certain heuristics were identified. This section contains tailoring heuristics for each relevant parameter.

1. *Organization/Project Parameters.* Table 1.17 presents tailoring heuristics for project size/complexity.

TABLE 1.17 Heuristics for Tailoring Based on Project Size/Complexity

Parameter	VVT Heuristics
Large	• Use incremental or evolutionary VVT lifecycle. • Define detailed VVT process and schedule. • Use frequent informal and formal technical reviews. • Plan for concurrent and early integration activities. • Use formal detailed technical and management VVT documentation. • Use formal requirements and change control. • Adopt the following VVT methods: classification tree method, evolutionary testing, requirements tracing, hierarchical testing, defect tracing, regression testing, etc. • Automate VVT as much as practical. • Use high-end VVT tools and facilities.
Small	• Use less formal VVT process. • Consider merging VVT phases. • Use less formal reviews. • Focus on less formal and less detailed technical documentation. • Adopt VVT methods such as walkthrough.

2. *Project Type.* Table 1.18 presents tailoring heuristics for project type.

TABLE 1.18 Heuristics for Tailoring Based on Project Type

Parameter	VVT Heuristics
Concept exploration	• Use evolutionary VVT lifecycles. • Use less formal VVT process. • Use informal reviews. • Adopt the following VVT methods: simulation, model checking, benchmarking, etc.
Technology demonstration	• Use less formal VVT process. • Use less formal reviews. • Adopt the following VVT methods: prototyping, simulation, model checking, benchmarking.
Full-scale development/ manufacturing	• Use incremental or evolutionary VVT lifecycles. • Define detailed VVT process and schedule. • Use frequent informal and formal technical reviews. • Plan for concurrent and early integration activities. • Use formal detailed technical and management VVT documentation. • Use formal requirements and change control. • Adopt the following VVT methods: classification tree method, evolutionary testing, requirements tracing, hierarchical testing, defect tracing, regression testing, etc. • Automate VVT as much as practical. • Use high-end VVT tools and facilities.
Maintenance	Use regression testing, impact analysis, inspection and walkthrough.
Upgrade	Use regression testing, impact analysis, inspection and walkthrough.

3. *Industry Type.* Tables 1.19–1.22 present additional VVT tailoring characteristics and heuristics unique for each of the industry types examined in the SysTest project.

TABLE 1.19 Heuristics for Tailoring in Aerospace/Avionics Industry

- Mostly large projects evolving from previous or existing systems.
- Often projects involve large and critical systems of systems that require different tailoring for different subsystems.
- Mostly few-of-a-kind projects. Production is often in a few or tens of units (emphasizing development rather than production)
- Due to each customer's unique requirements, tailoring is required for essentially every project.
- Certification authorities are major VVT stakeholders.
- Real-life tests are generally mandatory.
- Many projects have aggressive schedule objectives leading to concurrent VVT and incremental lifecycles.
- Some customers require the transfer of technology and future support knowhow to their organizations. This implies delivering many enabling products to the customer and therefore requires their higher quality and increased VVT effort.
- Technology development projects require evolutionary lifecycles, prototyping, simulation, and Design Of Experiments (DOE) methods.
- Very long lifecycle (more than 30 years life span is not uncommon)

TABLE 1.20 Heuristics for Tailoring in Automotive Industry

- Production volumes vary between a few hundred cars in the top luxury segment to several hundred thousand in the economy class.
- Typical development cost for a new model lies between $100 million and $1 billion.
- New developments are usually introduced in the luxury car sector (because of cost as well as lower production volumes).
- Most automotive embedded systems are large distributed systems running on many central processing units (CPUs) and communicating via buses.
- Most projects impose hard time-to-market constraints resulting in aggressive schedules leading to concurrent VVT.
- High competition with other automobile manufacturers.
- Most projects involve a large number of subcontractors for the implementation of different components, e.g., software modules. This often implies close interaction with external processes and organizations.
- Worldwide distribution of products results in different components and subcontractors for different regions and in a widespread distribution of enabling products.
- Generally high-quality requirements.
- End-user/consumer products resulting in high usability requirements and corresponding VVT activities such as early simulations

TABLE 1.21 Heuristics for Tailoring in Food Packaging Industry

- Standard small–medium size product developments are based on previous knowledge, historical database and best practices.
- Standardized projects require tailoring only for the specific issued product properties. The other requirements must be comparable with the historical data.
- Large, complex and innovative equipment developments require particular attention to concept development and screening based on objective measurements.
- All products are human health critical. A set of procedural VVT activities must be applied in order to fulfill food production regulations.
- Large-scale tailoring is required only for innovative products.
- New products start with a technology demonstrations phase. This phase must be objectively assessed using appropriate metrics.
- Continuous VVT monitoring approach is essential for the final customer and the human health safety.
- Physical testing, particularly in the intended environment, is important but entails great expenditures. VVT tailoring may be appropriate in certain cases.

TABLE 1.22 Heuristics for Tailoring in Steel Production Industry

Steel production is a process of making steel slabs from iron ore. This industry presents several VVT tailoring characteristics:

- Massive production (e.g., 250.000 tons/year) with a few product critical parameters to be verified (e.g., weight and size of steel slabs as well as physical and chemical composition).
- Intensive production and speed rates that require production line monitoring and optimization.
- In general, faulty steel products can be corrected.
- Steel production lines are similar systems; therefore, VVT tailoring requirements are basically the same for most projects.

Tailoring Heuristics: Programmatic Risks This section contains some tailoring heuristics for ameliorating project risks (Table 1.23).

TABLE 1.23 Heuristics for Tailoring Based on Anticipated Project Risks

Risk	VVT Heuristics
Unrealistic schedule	• Negotiate the scope of VVT effort to reduce it to a realistic level. • Negotiate with the customer for a realistic schedule. • Use formal requirements/change control to avoid unauthorized scope increase. • Move some of the desired functionality into future versions. • Deliver the product in stages so VVT activities could be stretched over a longer period. • Use incremental VVT lifecycles. • Adapt less formal VVT process (less documentation, reviews, etc.). • Negotiate the quality of some parts—implement them to "just enough" quality degree, and not more. • Use testing facility in two or three shifts. • Get another testing facility and team for parallel testing in two facilities. • Start testing earlier with less mature subsystems.
Insufficient budget	• Use strict requirements/change control to avoid unbudgeted scope increase. • Negotiate the scope of VVT effort in order to reduce it. • Convince the customer to extend the schedule. • Transfer budget from less critical projects to a more critical project. • Negotiate acceptable quality. Identify ways to reduce VVT efforts spent on less critical requirements. • Adapt less formal VVT process (e.g., less documentation, reviews). • Start VVT with mature work products. • Conduct upstream requirements and design reviews (when it is least expensive to introduce change).
Insufficient quality	• Plan for increased VVT effort, schedule and budget. • Define Detailed VVT process. • Use domain experts for VVT of complex, risky or critical parts of the system. • Use frequent informal and formal technical reviews. • Build consensus about acceptable quality. • Adopt the following VVT methods: inspection, walkthrough, boundary value analysis, robustness testing, behavior testing, back-to-back testing, prototyping, etc. • Use high-end VVT tools and facilities.

Tailoring Heuristics: Product Characteristics This section contains some tailoring heuristics to accommodate product characteristics (Table 1.24).

TABLE 1.24 Heuristics for Tailoring Based on Product Characteristics

Characteristic	VVT Heuristics
Critical	• Perform criticality analysis and allocate more VVT effort for critical parts. • Conduct upstream requirements and design reviews, inspections and walkthroughs. • Use Independent Verification and Validation (IV&V) team. • Use hierarchical testing with caution not to leave out important tests. • Test enabling products more rigorously. • Adopt the following VVT methods: robustness testing, safety testing, model checking, boundary value analysis, Failure Modes and Effects Analysis (FMEA), etc. • Use high-fidelity models and simulations. • Use VVT automated tools to assure engineering data consistency.
Complex	• Use domain experts for VVT of complex parts. • Use formal inspections for requirements and design. • Use Model Checking, Simulations, and Back-to-back testing. • Emphasize interface VVT. • Use VVT automated tools to assure engineering data consistency.
Innovative	• Use evolutionary VVT lifecycle. • Emphasize validation activities with stakeholders. • Adopt the following VVT methods: prototyping, simulation, model checking and exploratory testing.
Changed	• Use waterfall VVT lifecycle strategy • Adopt the following VVT methods: regression testing and impact analysis.
Precise	• Test enabling products more rigorously • Adopt the following VVT methods: benchmarking, simulation and model checking.
Need certification	• Often certification requirements are not identified explicitly. The VVT cost and time required are very high and must be taken into account. • Employ regulatory domain experts.

1.4.4 VVT Documents

This section provides an overview of various strategy and planning documents that can be used in conjunction with the VVT methodology. In other words, these documents either are produced by VVT engineers or contain sections related to the VVT process. Documents that control the definition of the project from inception to conclusion should contain clear statements about the VVT strategy. The documents discussed below play specific roles in the project. Project management usually decides which documents are required for a specific project.

Project Management Plan (PMP)

1. *Review.* The PMP, which sometimes is identified as an Engineering Program Plan (EPP), identifies the activities, critical milestones and events in relationship to systems engineering management and schedule control and typically includes the following events as a minimum:

   - Formal technical review for the system(s), subsystem(s), and their corresponding configuration items
   - Trials and test releases (if applicable)
   - Engineering releases
   - Production release
   - Acceptance tests
   - Logistic support events
   - Formal audits
   - Formal progress reviews

   These data identify the major activities and events required by the Statement of Work (SOW) or similar contract document defining the scope of the work. Any planned program strategies and build planning are identified in detail appropriate to the information available. The project management plan contains the project schedule(s) and identifies the appropriate activities, showing when each activity is initiated, the availability of draft and final deliverables and other milestones, and the due date for the completion of each activity. In addition, entry and exit criteria should be defined for each activity, that is, the conditions that should exist for the activity to start and for the activity to stop.
2. *Plan Source Pointer.* IEEE 1058.1 provides guidance for software PMP preparation. While its utility for hardware-oriented or hybrid developments is not proven, it is nevertheless an excellent resource. It can be purchased from the IEEE.

   The European Cooperation for Space Standardization document ECSS-M-30A, Project Phasing and Planning, provides planning principles and guidance but no template for the plan itself. It is an initiative established to develop a set of user-friendly standards to be utilized in all European space activities. Another source of PMP templates is the DI-MGMT-80004 management plan and the older DI-A-5239B management plan, which was superseded by DI-MGMT-80004.

Systems Engineering Management Plan (SEMP)

1. *Overview.* The SEMP establishes the overall plan for the technical development of a specific project. The SEMP defines the system performance parameters and preferred system configuration to satisfy the technical requirements and provides the planning and control of technical program tasks. It includes integration of engineering specialties and management of the entire system development effort. This includes design engineering, computer software engineering, specialty engineering, test engineering, logistics engineering, quality evaluation, and production engineering. The ultimate objective of the SEMP is to provide a disciplined framework to meet cost, technical performance, and quality and schedule objectives for the project or program. It is important that the SEMP establish the VVT philosophy for the program.
2. *Plan Source Pointer.* There are several good sources for a model SEMP. The first is Appendix C of the INCOSE *Systems Engineering Handbook*. The second is from The European Cooperation for Space Standardization document ECSS-E-10, Part 1B, systems engineering (November 2004), Appendix A. Some online sources are available but are not always free to the public. For example, the military standard DI-MGMT-81024 System Engineering Management Plan (SEMP). Two older standards that provide useful templates are the Data Item Description DI-S-3618, System Engineering Management Plan (SEMP), and DI-E-7144, Simulator System Engineering Management Plan (SEMP), both of which were superseded by DI-MGMT-81024.

Test and Evaluation Management Plan (TEMP)

1. *Overview.* The TEMP defines the approach to test and evaluate the project from both a technical and a management perspective. The TEMP defines the system test program and preferred test infrastructure necessary to satisfy the VVT philosophy set forth in the SEMP and meets the verification requirements. The TEMP provides for the planning and control of test program tasks.
2. *Plan Source Pointer.* The TEMP is similar in concept to the SEMP in that it provides an overall plan for the development of the testing program for the project. It can follow the organization of the SEMP. Another source of document structure is the U.S. military specification Data Item Descriptions (DID). One, which could fulfill the needs of the TEMP, is DI-NDTI-81284, Test and Evaluation Program Plan (TEPP).

Verification Validation and Testing Master Plan (VVT-MP)

1. The Test and Evaluation (TEMP) issued by the U.S. DoD was designed to manage and plan system testing (in the narrow sense of the term) during the system qualification phase. It does not deal with the multitude of VVT activities which are nontesting by nature or occurring at other

system lifecycle phases. A proposed VVT-MP which deals with the strategic planning of the entire VVT process in a broader manner is provided in Appendix B.

Testability Program Plan (TPP)

1. *Overview.* The TPP identifies the performing activity approach for implementing a testability program. It is mostly used to provide the acquirer with a basis for review and evaluation of the testability program. It usually is applicable for all systems and equipment development programs.
2. *Plan Source Pointer.* The TPP should be prepared in accordance with MIL-HDBK-2165, *Testability Handbook for Systems and Equipment.* Data item description and documentation guidance can be found in DI-MNTY-81604, Maintainability/Testability Demonstration Test Plan.

System Test Plan (SysTP)

1. *Overview.* The SysTP elucidates how to implement a system testing program. The purpose of the SysTP is to assure attainment of the requirements of the acquisition as stated in the system/subsystem specification.

   Requirement compliance may be proven through one of five methods, that is, analysis, inspection, demonstration, testing or certification. The SysTP describes the approach to using all five methods throughout the program life in a coordinated and efficient fashion. The SysTP considers resource allocation, facilities planning and overall scheduling of test activities as they support the overall project schedule.
2. *Plan Source Pointer.* See Section 2.6.1 on how to generate a qualification/acceptance SysTP.

Software Test Plan (STP)

1. *Overview.* The STP identifies the performing activity approach for implementing an organized software verification program. The purpose of the STP is to assure attainment of the requirements of the software system as stated in the System/Subsystem Specification. Requirement compliance may be proven at different levels during the software development process. Requirements proven through an instrumented "test" at a module or unit level may be verified using a demonstration of performance at higher levels. The STP describes the approach to use the appropriate verification methods (analysis, inspection, demonstration, testing or certification) throughout the software development in a coordinated and efficient fashion. The STP considers resource allocation, facilities planning and overall scheduling of test activities as they support the overall software development and integration schedule.

2. *Plan Source Pointer.* The STP structure should follow the software development approach. Object-oriented software is tested and integrated differently than modular or functional software implementations. Military standards templates appropriate for STP documentation are DI-IPSC-81438A, Software Development and Documentation, and the family of documents it superseded—DI-NDTI-80808, Test Plans/Procedures; DI-MCCR-80307, Software General Unit Test Plan; DI-MCCR-80308, Software System Integration and Test Plan; and DI-MCCR-80309, Software System Development Test and Evaluation Plan—all of which provide templates for STP. The legacy DIDs may be found to be useful with software projects using modular, functional code architectures. The now-superseded MIL-STD-498, Software Development and Documentation, had a well-organized software approach, which can be found in IEEE/EIA 12207, Standard for Software Lifecycle Processes.

Reliability Program Plan (RPP)

1. *Overview.* The RPP identifies the performing activity approach for implementing a reliability program. The purpose of the RPP is to assure attainment of the reliability requirements of the system as stated in the system/subsystem specification.

   Reliability should be stated initially in development specifications as a goal with a lower minimum acceptable requirement. In this case, realistic requirements are determined and incorporated later in the development specification together with the requirements for system demonstration. In general, both reliability and performance should be considered of similar importance, although this view may vary from one project to another.
2. *Plan Source Pointer.* The RPP should be prepared in accordance with MIL-STD-785. Additional details can be obtained using MIL-HDBK-781A, *Handbook for Reliability Test Methods, Plans, and Environments for Engineering, Development Qualification, and Production.*

Maintainability Program Plan (MPP)

1. *Overview.* The MPP identifies the performing activity approach for implementing a maintainability program to support the fielded system. The purpose of the MPP is to improve operational readiness, reduce maintenance manpower needs, reduce system lifecycle cost and provide data essential for management. In addition, the MPP should assure attainment of the maintenance requirements of the system as stated in the system/subsystem specifications. These usually include:
   - Time (e.g., turnaround time, time to repair, time between maintenance actions)

- Rate (e.g., maintenance hours per operating hours, frequency of preventative maintenance)
- Complexity (e.g., number of people and skill levels, variety of support equipment)

The expectation of carrying out repairs by substitution of components is also defined in the MPP.

2. *Plan Source Pointer.* An MPP should be prepared in accordance with the MIL-STD-470B. Additional guidance can be obtained from MIL-HDBK-2084, *Handbook for Maintainability of Avionic and Electronic Systems and Equipment.* Another resource for producing the maintenance plan is MIL-T-81821 (3), General Specification for Trainers, Maintenance, Equipment and Services.

First Article Inspection Plan (FAIP)

1. *Overview.* The FAIP identifies the performing activity approach for implementing first article inspection. The purpose of the FAIP is to fulfill Physical Configuration Audit (PCA) requirements of the acquisition as articulated in the SOW or other overarching program requirement documentation. The requirements are usually fulfilled by the drawings and supporting lists.
2. *Plan Source Pointer.* The FAIP can draw guidance from DI-QCIC-81110, Inspection and Test Plan, and either DI-NDTI-81307A, First Article Qualification Test Plan, or the older DI-T 5315, First Article Qualification Test Plan. Another source is the Society of Automotive Engineers, Standard SAE-AS9102A, Aerospace First Article Inspection Requirement.

Production Plan (PP)

1. *Overview.* The PP identifies the performing activity approach for implementing production of the system that is being developed and is being taken into a production phase. The PP defines the planning and control of production tasks. It includes integration between the production organization and engineering specialties and the management of an integrated effort. This includes design engineering, computer software engineering, specialty engineering, test engineering, logistics engineering, quality evaluation, and production engineering with the goal of improving production. The ultimate objective of the PP is to provide a disciplined framework to meet production cost and quality and schedule objectives for the system in a production environment. The PP should establish the VVT philosophy for production.
2. *Plan Source Pointer.* This plan should be written in accordance with the specific requirement of the project.

Integrated Logistic Support Plan (ILSP)

1. *Overview.* The ILSP identifies the approach the performing activity should take for implementing a logistic program to support the fielded system. The purpose of ILSP is to assure attainment of the logistic requirements of the system as stated in the system/subsystem specification in a manner that is integrated into all aspects of the program. This addresses the inclusion of design features, which facilitates logistic support, including maintenance, transportation and repair.
2. *Plan Source Pointer.* The European Cooperation for Space Standardization document ECSS-M-70A 19 (April 1996), Integrated L ogistic Support, provides general information and guidance of integrated logistic support and planning principles but no template for the plan itself. ECSS-M-70A 19 is available at the ECSS website (http://www.ecss.nl). Other online resources of this nature are available but are not free to the public. Military standards provide a broad spectrum of ILSP material to considerable depth if the investment is warranted. The U.S. Department of the Army standard DA PAM 700-50, Integrated Logistic Support: Developmental Supportability Test and Evaluation Guide, currently provides top-level guidance on ILSP.

Disposal Plan (DP)

1. *Overview.* The DP identifies the performing activity approach for disposing of the system. The purpose of the DP is to fulfill requirements of the acquisition with respect to an orderly and safe disposal of a system whose components or subsystems impose a public safety hazard or serious environmental threat. A DP is not ordinarily required in non-dangerous procurements.
2. *Plan Source Pointer.* This plan should be written in accordance with the specific requirement of the project. The DP could be based on the DoD 4160.21-M, *Defense Materiel Disposition Manual*, dated August 18, 1997 (see http://www.dtic.mil/whs/directives/corres/html/416021m.htm).

1.5 REFERENCES

Addy, A. E., Verification and Validation in Software Product Line Engineering, Dissertation, Department of Computer Science and Electrical Engineering, College of Engineering and Mineral Resources, West Virginia University, 1999.

ANSI/ITAA EIA-632, Processes for Engineering a System, American National Standards Institute/Information Technology Association of America, Sept. 1, 2003.

Balci, O., Verification, Validation, and Accreditation, in *Proceedings of the 1998 Winter Simulation Conference*, Washington, DC, Dec. 13–16, Piscataway, NJ, 1998, pp. 41–48.

Balci, O., Ormsby, F. W., Carr, T. J., and Saadi, D. S., Planning for Verification, Validation, and Accreditation of Modeling and Simulation Applications, in *Proceeding of the 2000 Winter Simulation Conference*, Orlando, FL, Dec. 2000.

Bertalanffy, V. L., *General System Theory: Foundations, Development, Applications*, George Braziller. 1976.

Boehm, B., Software Defects Reduction Top 10 List, *IEEE Computer*, 34(1), Jan. 2001.

Braha, D., Minai, A. A., and Bar-Yam, Y. (Eds.), *Complex Engineered Systems: Science Meets Technology*, Springer, 2006.

Browning, R. T., Modeling and Analyzing Cost, Schedule, and Performance in Complex Systems Product Development, Ph.D. Thesis, Massachusetts Institute of Technology, Cambridge, MA, 1998.

Browning, R. T., Sources of Performance Risk in Complex Systems Development, paper presented at INCOSE1999, Brighten England, June 1999.

Capers, J., *Applied Software Measurement: Assuring Productivity and Quality*, Mcgraw-Hill, New York, 1996.

DA PAM 70050;DA PAM 700-50, Integrated Logistic Support: Developmental Supportability Test and Evaluation, Department of the Army, Washington, DC.

DI-E-7144, Data Item Description, System Engineering Management Plan (SEMP), superseded by DI-MGMT-81024, June 1984.

DI-IPSC-81438A, Data Item Description, Software Test Plan (STP), Dec. 1999.

DI-MCCR-80307, Data Item Description, Software General Unit Test Plan (STP).

DI-MCCR-80308, Data Item Description, Integration and Test Plan.

DI-MCCR-80309, Data Item Description, Development Test and Evaluation Plan.

DI-MGMT-81024, Data Item Description, System Engineering Management Plan (SEMP), Aug. 1990.

DI-MNTY-81604, Data Item Description, Maintainability/Testability Demonstration Test Plan, Feb. 2001.

DI-NDTI-80808, Data Item Description, Test Plans/Procedures, May 1989.

DI-NDTI-81284, Data Item Description, Test and Evaluation Program Plan (TEPP), Sept. 1992.

DI-NDTI-81307A, Data Item Description, First Article Qualification Test Plan and Procedures, Nov. 2006.

DI-QCIC-81110, Data Item Description, Inspection and Test Plan, Dec. 1990.

DI-S-3618, Data Item Description, Systems Engineering Management Plan (SEMP), U.S. Department of Defense, Feb. 1970.

DI-T-5315, Data Item Description, First Article Qualification Test Plan, U.S. Department of Defense.

DoD 4160.21-M, Defense Materiel Disposition Manual, U.S. Department of Defense, Washington, DC, Aug. 1997.

DoD 5000.59, Modeling and Simulation (M&S) Management, Department of Defense Directive, August 2007.

ECSS-E-10, Part 1B, *European Cooperation for Space Standardization*, System Engineering branch, Nov. 2004.

ECSS-M-70A, Integrated Logistic Support, *European Cooperation for Space Standardization*, Apr. 1996.

Engel, A., et al., Developing Methodology for Advanced Systems Testing—SYSTEST, research grant proposal for the European Commission, Research Proposal Office, GRD1-2001-40487, May 2001.

Fairley, E. R., *Software Engineering Concepts*, McGraw Hill, New York, 1985.

Fente, J., Knutson, K., and Schexnayder, C., Defining a Beta Distribution Function for Construction Simulation, in *Proceedings of the 1999 Winter Simulation Conference*, Vol. 2, Squaw Peak Resort, Phoenix, AZ, Dec. 1999, pp. 1010–1015.

Gonzalez, A., and Barr, V., Validation and Verification of Intelligent Systems—What Are They and How Are They Different? *J. Exper. Theor. Artif. Intell.*, 12(4), Oct. 2000.

Haimes, Y. Y., *Risk Modeling, Assessment, and Management*, Wiley-Interscience, New York, 1998.

Haimes, Y. Y., Kaplan, S., and Lambert, J. H., Risk Filtering, Ranking, and Management Framework Using Hierarchical Holographic Modeling, *Risk Anal.*, 22(2), 383–398, 2002.

IEEE 6101991IEEE 610-1991, IEEE Computer Dictionary—Compilation of IEEE Standard Computer Glossaries, Institute of Electrical and Electronics Engineers, New York, 1991.

IEEE/EIA 12207IEEE/EIA 12207, Standard for Software Lifecycles Processes, Institute of Electrical and Electronics Engineers/Electronic Industries Association, 1996.

INCOSE-TP-2003-002-03.1, C. Haskins (Ed.), *Systems Engineering Handbook—A Guide for System Lifecycles Processes and Activities*, Version 3.1, INCOSE, Aug. 2007.

ISO/IEC 15288ISO/IEC 15288, Systems and Software Engineering—System Lifecycles Processes, International Organization for Standardization/International Electrotechnical Commission, 2008.

ISO/IEC 15288ISO/IEC 15288, Systems and Software Engineering—System Lifecycles Processes, International Organization for Standardization/International Electrotechnical Commission, 2008.

Juran, J. M., and Gryna, F. M., *Quality Planning and Analysis: From Product Development Through Use*, 2nd ed., McGraw-Hill, New York, 1980.

Lake, J., *V & V in Plain English*, INCOSE, Brighton, UK, June 1999.

Lamm, A. G., and Haimes, Y. Y., Assessing and Managing Risks to Information Assurance: A Methodological Approach, *Syst. Eng. J.*, 5(4), 286–314, Nov. 2002.

Lévárdy, V., Hoppe, M., and Honour, E., Verification, Validation & Testing Strategy and Planning Procedure, in *Proceedings of the 14th Annual International Symposium of INCOSE*, Toulouse, France, June 20–24, 2004.

Martin, N. J., and Bahill, A. T., *Systems Engineering Guidebook: A Process for Developing Systems and Products*, CRC Press, Boca Raton, FL, 1996.

Millard, R. L., Value stream analysis and mapping for product development, Master's thesis in Aeronautics and Astronautics, Massachusetts Institute of Technology, Cambridge, MA, June 2001.

MIL-HDBK-781A, *Handbook for Reliability Test Methods, Plans, and Environments for Engineering, Development, Qualification, and Production*, Revision A.

MIL-HDBK-2084, *Handbook for Maintainability of Avionic and Electronic Systems and Equipment*, July 1995.

MIL-HDBK-2165, *Testability Handbook for Systems and Equipment, Naval Sea Systems Command*, July 1995.

MIL-STD-470B, Maintainability Program for Systems and Equipment, May 1989.

MIL-STD-498, Software Development and Documentation, Dec. 1994.

MIL-STD-785-Rev B, Reliability Program for Systems and Equipment, Sept. 1980.

MIL-STD-882c, System Safety Program Requirements, U.S. Department of Defense, Jan. 19, 1993.

MIL-T-81821 (3), Trainers, Maintenance, Equipment and Services General Specification, Mar. 1983.

Montgomery, C. D., *Introduction to Statistical Quality Control*, 4th ed., Wiley, New York, 2001.

Morgan, J. M., High performance product development: a systems approach to a lean product development process, Ph.D. thesis, University of Michigan, 2002.

Muessig, R. P., Laack, R. D., and Wrobleski, W. J., Optimizing the Selection of VV&A Activities—A Risk/Benefit Approach, paper presented at Winter Simulation Conference, Atlanta GA, Dec. 7–10, 1997, pp. 60–66.

Oppenheim, W. B., Lean Product Development Flow, *Syst. Eng.*, 7(4), 352–376, 2004.

Rechtin, E., *Systems Architecting*, Prentice-Hall, Englewood Cliffs, NJ, 1990.

SAE-AS9102A, Aerospace First Article Inspection Requirement, Society of Automotive Engineers, January, 2004.

Sörqvist, L., On Poor Quality Costing, Ph.D. Thesis, Department of Production Engineering, Royal Institute of Technology, Stockholm, Sweden, Mar. 1998.

Womack, P. J., and Jones, T. D., *Lean Thinking: Banish Waste and Create Wealth in Your Corporation*, Free Press; 2nd edition, 2003.

Part **II**

VVT Activities and Methods

Chapter 2

System VVT Activities: Development

2.1 STRUCTURE OF CHAPTER

This chapter describes a set of VVT activities that typically occur within the system development lifecycle phases. We provide detailed information for each VVT activity in a standard format designed to aid the reader in determining the activity's applicability to a specific system. As mentioned before, one should (1) tailor the VVT methodology by using the tailoring guidelines and (2) consider using the VVT process model for optimizing the VVT strategy. Also, at the beginning of each system lifecycle phase, one should consider updating the VVT planning document.

2.1.1 Systems Development Lifecycle Phases and VVT Activities

Typically, each VVT activity may be carried out within one of the system development lifecycle phases, reviewed here:

1. *Definition.* This formulates the system operational concepts and develops the system requirements. The overall VVT strategy is determined and the engineering products of this phase are assessed.
2. *Design.* This creates a technical concept and architecture for the system. The engineering products of this phase are assessed.
3. *Implementation.* This creates the elements of the system. Each element is built or purchased, then evaluated or tested to ensure its stand-alone compliance with its allocated requirements.

Verification, Validation, and Testing of Engineered Systems, Avner Engel

4. *Integration.* This combines the implemented elements into a complete system. Throughout the integration process the emerging system is assessed on a step-by-step basis against requirements and stakeholders' desire.
5. *Qualification.* This performs formal and operational tests on the completed system to assure the quality of the system as a whole. The entire system is assessed against requirements and stakeholder needs.

2.1.2 VVT Activity Aspects

In general, each VVT activity is related to one of three aspects:

1. *Preparation of VVT Products*[2]. This aspect of VVT activities involves:
   - Identifying the VVT stakeholders and managing issues related to them
   - Planning the VVT process
   - Tailoring the VVT process to specific projects and systems
   - Preparing various VVT strategic documents [e.g., Verification, Validation and Testing Management Plan (VVT-MP)] and tactical documents [e.g., System Test Plan/Description/Report (SysTP, SysTD, SysTR)]
   - Defining, designing, building or purchasing the infrastructure and supporting equipment required for the VVT process
2. *Applying VVT to Engineered Systems.* This aspect of VVT activities involves assessing the various system engineering plans [e.g., System Engineering Management Plan (SEMP)] and other system engineering documents [e.g., System Requirements/Design Specifications (SysRS, SysDS)]. In addition and most important, this involves performing actual assessment of components, subsystems, and enabling products as well as of systems.
3. *Participating in or Conducting Technical Reviews.* This aspect of VVT activities involves participating in and sometimes leading informal and formal system reviews [e.g., System Requirement Review (SysRR), System Design Review (SysDR), Test Readiness Review (TRR)]. Technical reviews are performed to provide visibility into the systems' functional and technical characteristics as well as to establish management controls for assessing project cost, schedule, and quality.

[2] From a book organization standpoint, we opted to insert "preparation of VVT products" activities at the same phase they are going to be utilized. The reader should be aware that by and large such activities take a long time to completion and therefore must be started at earlier stages.

2.1.3 VVT Activity Format

In general, each VVT activity in this book is described using the following elements:

1. *Objective.* This describes the objective of the pertinent VVT activity.
2. *Description.* This describes, in some detail, the purpose, implementation and essence of the pertinent VVT activity.
3. *Methods and Further Literature.* This points to one or more relevant VVT methods which explain how to carry out the pertinent VVT activity. The reader can find a detailed description of each VVT method in either Chapter 4 or Chapter 5. In addition, this section provides reference material for gaining a better understanding of the pertinent VVT activity.

2.2 VVT ACTIVITIES DURING DEFINITION

The purpose of the system Definition phase is to formulate the system operational concepts and create the system requirements, usually documented in the form of specifications or models. One purpose of VVT activities during the system Definition phase is to ensure that the system requirements and system concepts accurately reflect the real-world operational needs. VVT activities also lay the foundation for further VVT planning based on fully understanding the system requirements and concepts. The VVT tailoring process and the VVT strategy determination typically occur at the beginning of the system Definition phase. The VVT process model should be initialized with known or estimated parameters. The following sections define specific VVT activities that are appropriate for the system Definition phase.

2.2.1 Generate Requirements Verification Matrix (RVM)

Objective The objective of this VVT activity is to determine (1) the method of verifying each system requirement, (2) when it will be done within the lifecycle of the system and (3) the specific procedure according to which the verification will be accomplished.

Description Creating or updating the Requirements Verification Matrix (RVM) is an ongoing activity that may start as early as the creation of a response to a Request For Proposal (RFP) or with the first release of the test and verification plan. The RVM is a table listing the following elements (see example in Figure 2.1):

Requirement ID	Requirement traceability	Verification method						Verification stage					Procedure ID
		None	Analysis	Inspection	Demonstration	Test	Certification	Definition	Design	Implementation	Integration	Qualification	
SL-1	A.1.1	X											
SL-2	A.1.2		X					X					DD-45
SL-3	A.1.3			X	X							X	XZS-0
SL-4	B.5						X	X					VT-00
SL-5	B.6	X											-
SL-6	K.22				X	X				X	X		VT-02
SL-7	K.23		X			X		X	X				VT-03
SL-8	Z.1.2					X						X	RN-33

Figure 2.1 Example of RVM.

Requirement ID. Identifies a name or an identification number for each requirement.

Requirement Traceability. Provides traceability to an appropriate document (i.e., usually customer document) and specific requirement.

Verification Method. Typically, there are five types of verification methods: analysis, inspection, demonstration, testing or certification. In addition "no verification" is also an option. The following is a short description of each verification method:

- *Analysis.* Verification that specification requirements have been met by technical evaluation of system descriptions, charts, reduced performance data and so on. Typical analysis utilizes mathematical models, simulations, test algorithms, calculations, charts, graphs and so on.
- *Inspection.* Verification by physical and visual examination of an item and comparing appropriate characteristics of the item with referenced standards in order to determine compliance with requirements. Typical inspection techniques are visual, auditory, olfactory, touch, physical manipulation, mechanical or electrical gauging or measurement and so on.
- *Demonstration.* Functional confirmation that a specification requirement is met by observing the qualitative results of an operation or through an exercise performed under a specific condition.
- *Testing.* Verification of the specification or requirements through the application of established test procedures within specified environmental conditions as well as subsequent compliance confirmation through analysis of the generated test data.
- *Certification.* Verification based on a signed certificate of compliance (from the producer) stating that a delivered item is a standard product that meets all procurement specifications, standards, and other requirements.

Verification Stage. Indicates when the verification is to be conducted. Basically there are two orientations to specifying this information: (1) by event, for example, First Article Verification (FAV), or (2) by lifecycle phase (i.e., Definition, Design, Implementation, Integration, Qualification, Production, Use/Maintenance or Disposal).

Verification Procedure. The specific procedure required to accomplish verification [e.g., System Test Description (SysTD), First Article Acceptance Plan (FAAP), Production Acceptance Plan (PAP)]. Note that this element of the RVM is normally dealt with at a later phase.

Normally a skeleton RVM is created at the beginning of a project identifying each requirement along a single and sometimes multiple assigned verification methods. At a later time, the verification stage (or stages) is added and finally, the specific verification procedure is identified.

Methods and Further Literature

Section 4.2.1, Requirements verification matrix (RVM)
Section 4.2.3, Hierarchical VVT optimization
Section 4.3.2, Compare images and documents
Section 4.3.3, Requirements testability and quality
Section 4.3.4, System test simulation
Section 5.7.6, Qualification testing

- Engel (2008)
- INCOSE-TP-2003-002-03.1 (2007)
- Mooz et al. (2003)

2.2.2 Generate VVT Management Plan (VVT-MP)

Objective The objective of this activity is to thoroughly plan the VVT strategic process for a given project[3]. The management plan should deal with all relevant resources and risks concerning technical and management issues and covering both end products and enabling products.

Description VVT planning constitutes the definition of all VVT activities, determining budgets and other needed resource and scheduling the entire VVT process. The planner must identify which development products should be assessed and to what degree. The VVT process should be scheduled so that the VVT effort is balanced and the VVT documentation and test articles become available when they are needed. The optimized VVT plan should offer VVT termination criteria and timing. For this purpose, it must be decided in which lifecycle phase a given system property should be assessed. Creating

[3] Readers are directed to Appendix B for more information.

or updating the VVT-MP is an ongoing activity that should start at the beginning of a project.

The VVT-MP (described in Appendix B) is an expansion of the *Test and Evaluation Management Plan* (TEMP), U.S. Department of Detense (DoD) directive 5000.2-R[4]. As a tool for planning the overall VVT process, the TEMP is unsatisfactory as it concentrates almost exclusively on testing in the narrow sense of the term and only during the Qualification phase (*test and evaluation* in DoD lingo) and is rife with military acronyms.

The VVT-MP provides users with guidance concerning the preparation of a management plan for performing VVT throughout the development stage of systems. It contains the following key elements:

- *System Introduction.* Describes the following: (1) project applicable documents, (2) mission description, (3) system description and (4) critical technical parameters.
- *System VVT Processes.* Describes the following: (1) integrated VVT program schedule, (2) VVT program management, (3) VVT strategy, (4) planning the VVT activities and (5) VVT limitations.
- *VVT Resources.* Describes the following: (1) test articles, (2) Test sites and instrumentation, (3) test support equipment, (4) test expendables, (5) operational force test support, (6) simulations, models and test beds, (7) manpower/personnel training and (8) budget summary.

The VVT-MP generation process is presented in Figure 2.2.

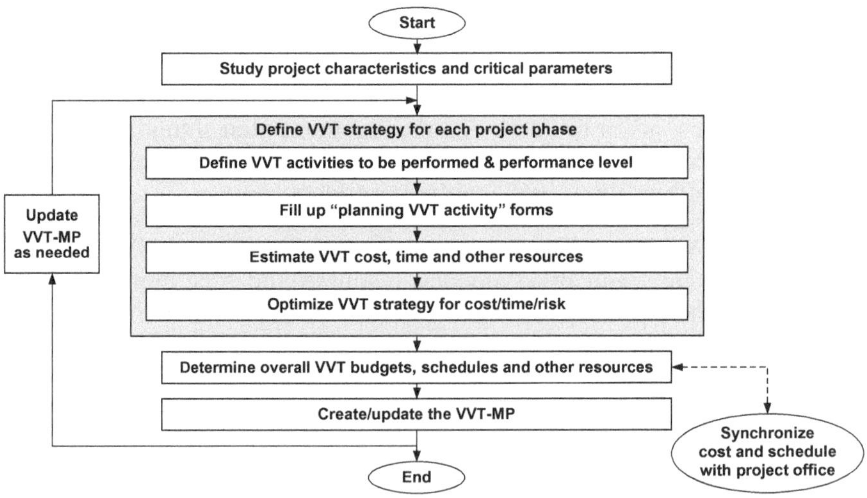

Figure 2.2 VVT-MP generation flow chart.

[4]Mandatory Procedures for Major Defense Acquisition Programs (MDAPS) and Major Automated Information System (MAIS) Acquisition Programs, DoD, 2001.

After understanding the project characteristics and the critical parameters that must be verified, the planner defines the VVT strategy, that is, the set of activities to be performed and the performance level[5] of each VVT activity within each development phase.

A specific "VVT planning form" shall be filled out for each VVT activity which is to be performed. This form contains description of the VVT activities, required budgets, schedule estimates and other resource needed (e.g., infrastructure and supporting equipment). The specific VVT strategy shall take into account the project characteristics and translate them into specific VVT tasks that must be performed by the VVT organic team and other engineers performing VVT activities as part of their regular activities. Finally, the VVT planner creates the VVT-MP and updates it as needed.

All these VVT resource requirements must be negotiated and coordinated with the project manager or the project office. However, very often, the budget or schedule allocated to the VVT planner is less than originally required and he or she must optimize the VVT strategy for the project at hand. This usually takes time and often can be achieved only during the Design phase.

Methods and Further Literature

Section 4.3.1, VVT process planning   Appendix B: VVT-MP

- Beizer (1990)
- DeMillo et al. (1987)
- Koomen and Pol, (1999)
- Spillner et al. (2007)

2.2.3 Assess the Request For Proposal (RFP) Document

Objective The objective of assessing the completeness and consistency of the RFP or a comparable customer document is to verify that the organization is able to meet all RFP requirements. Additionally, one must verify internal consistency within the RFP as well as consistency between the RFP requirements and existing regulations, laws, societal values and standards, avoidance of negative environmental impact and full adherence to the organization's regulations and ethics.

Description A RFP is an invitation for system or subsystem producers, often through a bidding process, to submit a proposal on a specific system or service. Similarly, a less formal request for system development may be initiated within the organization itself. Assessment of such documents brings structure to the procurement decision and allows the risks and benefits of the potential project to be identified.

[5] Generally, the VVT process is abbreviated in order to reduce costs, meet tight schedules or eliminate the need for expensive or scarce resources. Obviously, a certain level of risk is involved in eliminating any VVT step and the planner of the VVT process and the stakeholders in the project must be aware of these risks.

The following describes a practical assessment of an RFP, which typically has the following structure: (1) background and objectives, (2) services requested, (3) required documentation, (4) time estimates and fees, (5) bidder qualifications and (6) submission information.

- *Background and Objectives.* Assess whether the RFP provides sufficient information and background about the customer or entity issuing the RFP. In addition, assess whether the RFP lists the objectives of the specific contract work being solicited. Generally, the RFP should include sufficient information for bidders to appropriately assess customers' needs and write a proposal detailed enough that can evaluate the suitability of the proposed system.
- *Services Requested.* Assess this most important part of an RFP, the outline of services requested. Specifically check for internal inconsistencies or if some requirements are vague in describing what is expected of the contractor. Obviously, the more specific the RFP is, the more likely responses will be relevant and thorough. An RFP calling for the development and production of a system must be very specific about the exact system performance requirements, the expected level of VVT, the desired schedule and the required scale of production.
- *Required Documentation.* Assess the specific documentation required by the RFP as a part of executing the project. Also verify that management is aware, ready and able to provide the needed level of documentation. Also, verify that the organization's Intellectual Property (IP) will be protected if the project is undertaken. For example, make sure everyone involved in the proposal process has signed a confidentiality agreement covering proprietary information that needs to be protected.
- *Time Estimates and Fees.* Assess the RFP for expected timelines and payment schedule. The RFP should give bidders sufficient information to decide if they can realistically fulfill the needs outlined in the RFP. Inclusion of a fee schedule in an RFP makes it possible to determine whether the project can be completed for reasonable cost or if the cost of the project will outweigh the benefits.
- *Bidder Qualifications.* Usually, an RFP asks for documentation to demonstrate the qualifications of bidders to perform the required tasks. In general, company qualifications should demonstrate the ability to meet the managerial and technical requirements outlined in the RFP. Assess these requirements to ensure that your organization is not expected to divulge confidential or privileged information whose release would hurt the company, legally, financially or competitively.
- *Submission Information.* Virtually all RFP documents include a deadline for proposal submission. Assess the company's ability to generate a complete RFP response package within the allotted time. Submission of an incomplete proposal or failure to meet the proposal deadline could indicate that the company might be unable to deliver the system on time.

Methods and Further Literature

Section 4.3.2, Compare images and documents
Section 4.4.1, Expert team reviews
Section 4.4.3, Group evaluation and decision

- Porter-Roth (2001)

2.2.4 Assess System Requirements Specification (SysRS)

Objective The objective of this activity is to verify the SysRS or comparable customer document. Specifically, each requirement in that document should be assessed with regard to consistency and traceability to the RFP, verifiability, clarity, attainability, integrity and future-ability (see definitions below). In addition, each requirement should have the following supporting information: necessity, assumptions and accountability (definitions below).

Description The SysRS generated by the engineering staff is evaluated against the RFP or a similar customer document. It is important to note that the term "system" includes both enabling products and end products. Ideally, each requirement should be discussed with the customer of the system as well as other stakeholders in order to ensure the following:

- *Consistency with RFP.* Verify that each system requirement stated in the SysRS appears, in one form or another, within the RFP or is directly derived from it. Also ensure that the intent and meaning of the original requirement are maintained.
- *Traceability to RFP.* Verify that each system requirement in the SysRS is traced to one or more paragraphs or sections in the RFP or similar customer document.
- *Verifiability.* Ensure that each system requirement is verifiable or testable. This means that requirements must be stated in rigorous terms without ambiguities. For example, requirements containing phrases such as "maximize", "minimize", "support", "adequate", "but not limited to", "user friendly", "easy" and "sufficient" are often not verifiable[6]. Thus, it will be necessary to clarify with the customer what is really meant by such requirements.
- *Clarity.* Verify that each requirement is stated in an understandable language, preferably employing short sentences that contain no ambiguities.
- *Attainability.* Verify that each system requirement can be implemented, with full awareness of the limitations of the organizations that will be doing the work. Requirement attainability should be verified from

[6]Nevertheless, engineers should not automatically snub at nonverifiable requirements. For example, industrial designers often generate crucial, "difficult-to-verify" requirements which deal with aesthetic and alluring qualities of products and systems.

multiple points of view, including technical, financial, legal, environmental, ethical and programmatic.

- *Integrity.* Verify the overall integrity of the entire system requirement set. This entails ensuring that all requirements are complete and no requirement duplicates or contradicts another requirement.
- *Future-ability.* Assess the SysRS relative to future lifecycle phases. Specifically, verify that, in addition to meeting design and test requirements, the system meets (1) production, (2) use and maintenance and (3) disposal requirements.
- *Necessity.* Verify that for each system requirement there exists an associated statement justifying the need for the requirement (e.g., by customer requirement or other reason).
- *Accountability.* Verify that for each system requirement there is a name of the author (owner) associated with that specific requirement. This person should be willing and able to defend the requirement and should be available to assess how a design change may impact a given requirement.
- *Assumptions.* Verify that for each system requirement there exists a statement of assumptions made by the author (owner) of the requirement.

Methods and Further Literature

Section 4.3.2, Compare images and documents
Section 4.3.3, Requirements testability and quality
Section 4.4.1, Expert team reviews
Section 4.4.3, Group evaluation and decision

- INCOSE-TP-2003-002-03.1 (2007)
- Mooz et al. (2003)

2.2.5 Assess Project Risk Management Plan (RMP)

Objective The objective of this activity is to assess the Risk Management Plan (RMP) of a project. In general, this assessment covers four elements: (1) risk identification, (2) risk quantification, (3) risk responses and (4) risk control.

Description A risk is described in terms of an undesirable event that, were it to happen, would have an adverse impact on a project or the system. The phrase "were it to happen" implies a probability P, $0\% > P > 100\%$, and the phrase "would have an adverse impact" implies some cost C. The expected cost E of that risk is commonly calculated as $E = PC$. Assessing the project risk management plan entails checking the following elements:

- *Assess Risk Identification Element.* Evaluate the risk management plan to verify that all reasonable risks have been identified by name and described to a sufficient level of detail. In addition, check that each risk has been assigned an appropriate category. For instance, a new technology that must be verified under field conditions would be assigned to Technical

Risk, a delay in delivery of a key component would be assigned to Schedule Risk, project cost overruns would be assigned to Financial Risk, lack of qualified system testers would be assigned to Management Risk, and so on.

Also verify that all identified risks include two qualitative components. The first is the cause of the risk (e.g., shortage of programmers within the organization) and the second is a description of a potential impact (e.g., milestones may not be achieved).

- *Assess Risk Quantification Element.* Risks need to be categorized into bends of criticality (e.g., high-, mid- and low-level risks). Verify therefore that the risk management plan contains a general risk level mapping similar to the example provided in Figure 2.3. In the figure, risks need to be quantified in two dimensions, namely, (1) the probability of undesirable event occurrence and (2) the cost impact if the undesirable event will in fact materialize. It is important to note that all impacts, regardless of risk category, should be evaluated from a strict monetary point of view (e.g., a delay in delivery of a system leads, usually, to some added cost). Verify therefore that the RMP identifies probability (P) and cost (C) for each identified risk.

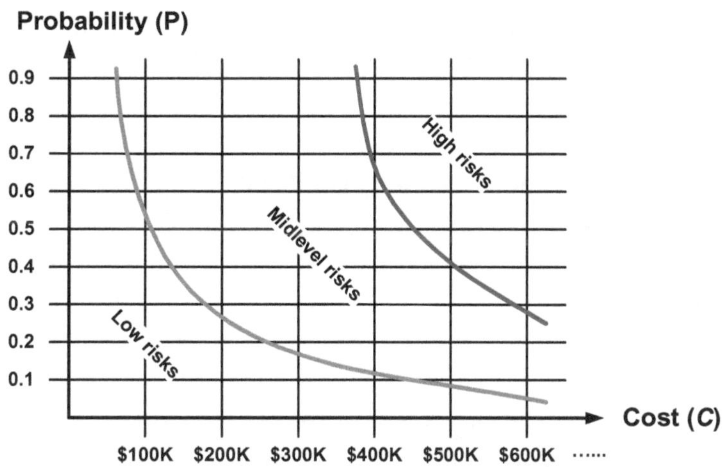

Figure 2.3 Example of a risk categorization graph.

- *Assess Risk Response Element.* Verify that each identified risk points to a description of a specific risk response strategy. This strategy should be evaluated to verify that it identifies (1) what needs to be done, (2) who is responsible for this action and (3) what should be the scheduled time for this action.

  In general, response strategies map into one of three categories. Verify that each identified risk has been assigned to one of these categories:

  (a) *Transfer the risk.* The responsibility for a risk may be transferred to someone else. For example, a dedicated and expert subcontractor can be assigned to handle or mitigate a particularly risky part of a project.

(b) *Mitigate the risk.* An action to lessen either the impact or the probability of the risk may be identified. For example, a risk that relates to lack of available engineers within the organization may be mitigated by rescheduling lower priority projects or modifying the system design to eliminate a not-so-necessary high-technology development.

(c) *Ignore the risk.* A risk may be small enough due to either a very small probability or a small potential impact. Therefore, mitigation activity may not be warranted[7].

- *Assess Risk Control Element.* Verify that the risk management plan identifies how the ensemble of risks will be monitored and controlled. The assessor of the plan should check for specific mechanisms (e.g., regular risk reviews with all cognizant individuals) to identify actions outstanding, risk probability and impact, removal of obsolete risks and identification of newly determined or suspected risks.

Methods and Further Literature

Section 4.3.1, VVT process planning
Section 4.4.1, Expert team reviews
Section 4.4.3, Group evaluation and decision

- Cooper et al. (2004)

2.2.6 Assess System Safety Program Plan (SSPP)

Objective The objective of this activity is to assess the System Safety Program Plan (SSPP) of a project. This assessment is carried out to ensure that all systems, subsystems and their interfaces operate effectively, without sustaining failures or jeopardizing the safety and health of operators, maintenance personnel or others in the vicinity.

Description Professor Nancy Leveson of the Massachusetts Institute of Technology (MIT) suggests in her yet-unpublished book, "Engineering a Safer World" (to be published by MIT Press), that safety accident models and techniques of modern engineered systems need to change but in reality are not. This need stems from the following (partial quote from Leveson):

- *Fast Pace of Technological Change.* Technology is changing faster than the engineering techniques developed to cope with an undesirable event

[7] Nevertheless, catastrophic risks must be carefully assessed even if the probability of the undesirable event is so small that the expected risk cost (E) seems negligible. For instance, while the probability (P) of a well-designed and carefully constructed dam collapsing may be extremely low, the potential harm (C) of such an event is enormous. Thus, on the surface, the risk cost ($E = PC$) may seem insignificant. However, one should never permit a catastrophic risk to be placed in the "ignore" category.

or accident. Lessons learned about designing to prevent accidents may become ineffective for new technologies.

- *Changing Nature of Accidents.* Digital technology has created a revolution in many fields of engineering, but system safety engineering techniques have not kept pace.
- *New Types of Hazards.* The increasing dependence on information systems is creating a potential for loss or incorrect information that may lead to physical, scientific or financial loss.
- *Increasing System Complexity and Coupling.* Complexity is increasing in today's systems, particularly the interactions between subsystems and between the system and its environment. We are designing systems with potential interactions that cannot be thoroughly understood, anticipated or guarded against, leading to many new failure modes.
- *More Complex Relationships between Humans and Automation.* Humans are increasingly sharing control of systems with various levels of automation. These changes are leading to new types of human errors and accidents.
- *Increasing Potential Loss from Accidents.* Our new scientific and technological discoveries have created new and increased environmental hazards. Such systems can harm increasing numbers of people and impact future generations through pollution, genetic damage and the like.
- *Changing Regulatory and Public Views of Safety.* In today's complex and interrelated societal structure, responsibility for safety is shifting from the individual to governments. Individuals are demanding that government assume greater responsibility for controlling system behavior through laws and various forms of oversight and regulation.

A SSPP is a widespread means for identifying potential hazards during the development process and preventing hazards by addressing their root causes. As a rule, hazards must be eliminated or reduced to a tolerable level, provided that the penalties, in terms of cost, time and effort, are not disproportionate to the improvements gained. This principle, called ALARP (As Low As Reasonably Practicable), forms the basis for safety management (see Figure 2.4).

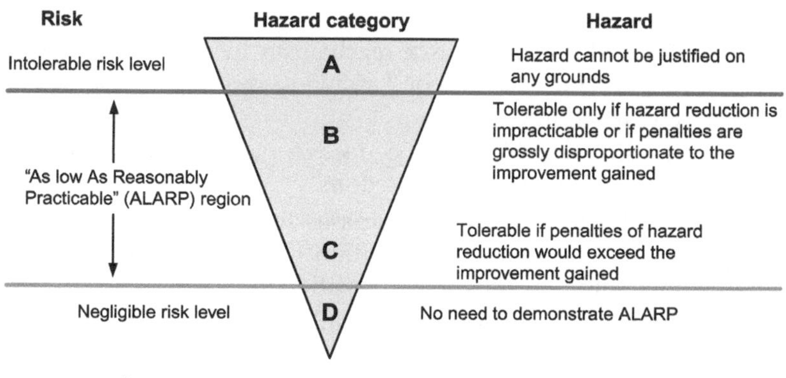

Figure 2.4 The ALARP Triangle: Example of hazard concern category model.

The risk associated with a hazard is a product of the severity and probability (or frequency) of the hazard and is often split into four concern categories, A, B, C and D. Table 2.1 shows how a hazard concern category is assigned based on frequency and severity of a given hazard. Note that the hazard concern category D is never given to a disastrous or catastrophic risk event, no matter what its probability.

TABLE 2.1 Example definition of hazard categories: A, B, C and D

	Hazard Severity Category				
Frequency	Disastrous	Catastrophic	Critical	Severe	Minor
Frequent	**A**	**A**	**A**	**A**	**B**
Probable	**A**	**A**	**A**	**B**	**C**
Occasional	**A**	**A**	**B**	**C**	**C**
Remote	**A**	**B**	**C**	**C**	**D**
Improbable	**B**	**C**	**C**	**D**	**D**
Non-credible	**C**	**C**	**D**	**D**	**D**

Assessment of the SSPP should include checking:

- Whether the SSPP improves the level of safety by identifying hazards, introducing hazard control measures and making sure that potential hazards are continually reviewed and dealt with using ALARP throughout the life of the system.
- Whether the SSPP establishes and maintains a safety culture among all persons involved with the project, thus ensuring that safety becomes a routine part of everybody's work.
- Whether the SSPP establishes safety reviews throughout the life of a project and that every effort is made to achieve as high a level of safety as possible.
- Whether the SSPP establishes a mechanism to allow undesirable incidents, accidents, near misses or "accidents waiting to happen" to be reported and acted upon.
- Whether the SSPP establishes procedures for identification and recording of hazards and taking mitigating actions.
- Whether the SSPP establishes processes for "top-down" and "bottom-up" hazard analyses with the intention of determining how accidents could happen and how they may be avoided.
- Whether the SSPP provides an audit trail for all safety-related decisions.

Methods and Further Literature

Section 4.3.1, VVT process planning
Section 4.3.7, Model-based testing
Section 4.4.1, Expert team reviews
Section 4.4.3, Group evaluation and decision

- Brauer (2005)
- Hollnagel et al. (2006)

2.2.7 Participate in System Requirements Review (SysRR)

Objective The objective of the SysRR is to assess the status of the system requirements and check that the producer, purchaser and other stakeholders of the system agree on the intent of the specification and program requirements of the proposed system.

Description The SysRR is normally conducted during the system concept exploration stage. This is generally the first review, during which the producer presents his or her preliminary views of the system and the development process. Such review may take place after agreement on the functional analysis and preliminary requirement allocation to work clusters such as operations, maintenance and training as well as concord on the initial direction and progress of the producer's system engineering management effort and his or her concurrence with a balanced and complete system configuration.

Often, there will be a need for an Internal Software Requirement Review (I-SRR) and an Internal System Requirement Review (I-SysRR) followed by a formal Software Requirement Review (SRR) and a formal SysRR. Reviews, in all cases, should be assessed against the RFP as well as the Software Requirement Specification (SRS) and the SysRS or equivalent documents.

Methods and Further Literature

Section 4.3.3, Requirements testability and quality
Section 4.4.2, Formal technical reviews
Section 4.4.3, Group evaluation and decision

- INCOSE-TP-2003-002-03.1 (2007)
- MIL-STD-1521B (1995)
- Roetzheim (1990)

2.2.8 Participate in System Engineering Management Plan (SEMP) Review

Objective The objective of this review is to assess the SEMP. The SEMP describes the contractor's or the developer's proposed efforts for planning, controlling and conducting a fully integrated engineering effort.

Description The SEMP is used to encapsulate (1) the technical program planning and control and (2) the planned system engineering process. It should be assessed along the following lines:

Format and General Components. The SEMP document should be structured in a manner and format appropriate to the organization[8] and should include some general components. The SEMP assessment should include:

- Verification that the SEMP is constructed according to a defined manner acceptable to the organization and other relevant stakeholders.
- Verification that the SEMP identifies the specific program or project and its purpose. In addition the SEMP should contain an introduction and a summary of the SEMP document itself.
- Verification that the SEMP identifies all the applicable and referenced documents which are required for the specific program or project.

Engineering Management. The SEMP should define appropriate project management requirements for the definition, design, implementation, integration, qualification, production, use/maintenance and disposal of the engineered system. The SEMP assessment should include:

- Verification that the SEMP identifies organizational responsibilities and authority for system engineering management, including control of subcontractors
- Verification that the SEMP explains the integration and coordination of the program efforts for engineering specialty areas in order to achieve a best mix of the technical/performance values
- Verification that the SEMP identifies levels of control established for performance and design requirements as well as the method used
- Verification that the SEMP identifies plans and schedules for all technical program reviews
- Verification that the SEMP identifies technical program assurance and configuration control methods for the engineering products and documentation as well as appropriate mechanisms for approval and certification

Engineering Processes. The SEMP should provide detailed description of the engineering process to be used, including the specific tailoring of the process to the characteristics of the system or project. The SEMP assessment should include:

- Verification that the SEMP identifies all the procedures to be used in implementing the engineering processes
- Verification that the SEMP identifies all relevant mathematical or simulation models to be used during the development of the system

[8]For example, in accordance with U.S. DoD, Data Item Description DI-MGMT-81024, Draft MIL-STD-499C, Engineering Management, revised March 24, 2005.

Methods and Further Literature

Section 4.3.1, VVT process planning
Section 4.4.2, Formal technical reviews
Section 4.4.3, Group evaluation and decision

- DI-MGMT-81024 (2005)
- Sage and Rouse (1999)

2.2.9 Conduct Engineering Peer Review of the VVT-MP Document

Objective The objective of this activity is to assess the VVT-MP document by means of a disciplined engineering practice for detecting and correcting defects.

Description Engineering Peer Review (EPR) refers to a type of review in which documents and similar work products are examined by the author and several of his or her "peers"[9] in order to evaluate its technical content and quality. EPRs are focused, in-depth technical reviews used to provide confirmation and offer options by bringing in experts early and at appropriate points throughout the system's lifecycle. These reviews are most effective when accomplished with a small group of reviewers working intimately with the developers. As much as possible, reviewers should be experts independent of the executing team. They are responsible for the actual execution as well as all subsequent closure of issues resulting from the review.

Verifying system work products by means of peer reviews increases the likelihood that weaknesses will be identified. In fact, this approach is considered to be the most effective method for document assessments. Peer reviews are distinct from management reviews, which are conducted by management representatives, as well as from formal project reviews, which are often conducted in the presence of customers. They are also distinct from audit reviews, which are conducted by personnel external to the project, usually in an adversarial position.

The assessment of the VVT-MP document in a peer review setting is typically conducted along the following stages: (1) planning the peer review, (2) preparing for the peer review on an individual basis, (3) conducting the peer review and (4) performing peer review follow-up activity.

Methods and Further Literature

Section 4.4.1, Expert team reviews
Section 4.4.3, Group evaluation and decision

- Wiegers (2001)

[9] Peers are persons or colleagues who have equal standing within an organization. Management and especially line managers are typically not involved in the conduct of a peer review.

2.3 VVT ACTIVITIES DURING DESIGN

The purpose of the system Design phase is to develop a technical concept and architecture for the target system. The architecture identifies the system elements and their interactions as they will be implemented, with sufficient detail to minimize the risk on the development or purchase of those elements. Creating this detail requires allocating requirements to each element and performing enough analysis and preliminary design effort to ensure the feasibility of meeting the requirements.

The remainder of this section covers VVT activities that are appropriate for the system Design phase.

2.3.1 Optimize the VVT Strategy

Objective The objective of this activity is to optimize the VVT strategy, thereby reducing the quality cost or quality time with minimal detrimental effect on the actual quality of the engineered system. Quality cost consists of VVT costs plus failure costs, whereas quality time is the duration, on the critical path of the system lifecycle, required to develop, manufacture, maintain and dispose of the engineered system as well as perform VVT activities and remove defects from engineered systems.

Description Generally, there is a correlation between VVT investment and system quality. Early in the 1950s, Joseph Juran (1998) proposed a qualitative model defining "quality cost" as the sum of VVT costs plus failure costs. He suggested that there is an optimal VVT strategy that will yield minimum overall quality cost (see Figure 2.5).

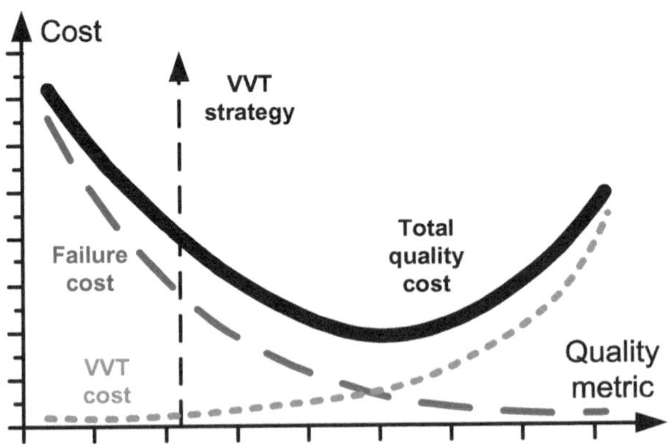

Figure 2.5 Juran's quality cost model.

Juran's quality cost model makes a lot of sense. There is a cost to product failures, but there is also a cost to avoiding product failure. The idea for most systems is to minimize total expected quality cost. The main weakness of Juran's model is that it is qualitative and therefore does not help in designing practical VVT strategies. Furthermore, even if an optimal VVT strategy cost were to be ascertained, large numbers of VVT strategies of equal optimal cost are possible. This problem was addressed by designing a set of quantitative models to compute the quality cost as well as quality time as a function of the VVT strategy and other relevant parameters (for more information, see Chapters 6, 7 and 8).

Using a quantitative modeling approach can yield quality cost/time savings of 10–20% of development cost. Since quality cost/time may consist of 50–60% of engineering system development cost/time, the return on investment, especially in medium to large projects, could be substantial (Engel and Shachar, 2006). The process of optimizing the VVT strategy is depicted in Figure 2.6 and explained below.

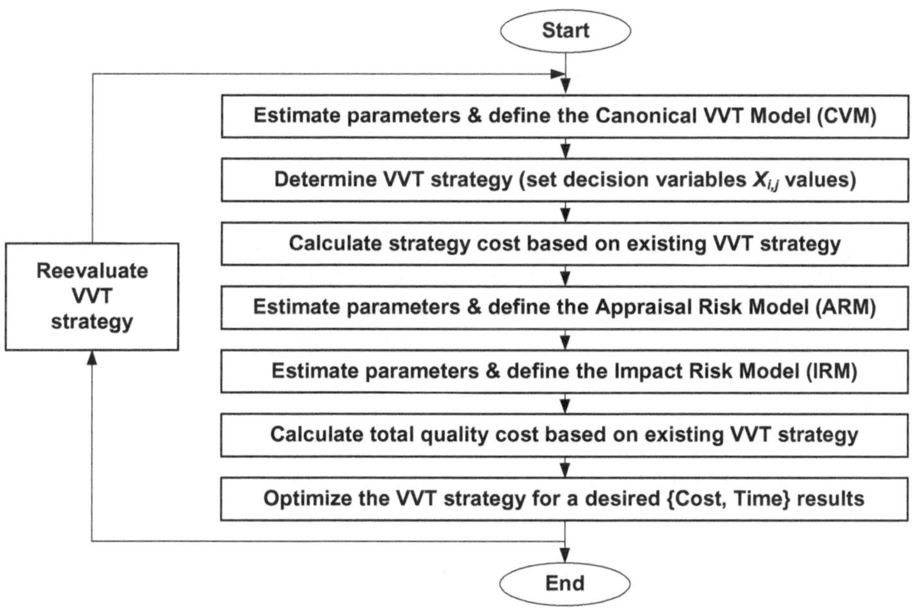

Figure 2.6 Optimizing the VVT strategy to desired cost or time targets.

- *Step 1: Estimate Parameters and Define Canonical VVT Model (CVM).* An exhaustive and comprehensive set of possible VVT activities must be created. Then, an estimated cost and time associated with each activity should be generated. This CVM is a hypothetical framework encapsulating the performance of a "complete and ideal" set of VVT activities designed to verify, validate and test a system throughout its lifecycle (see Chapter 6).

- *Step 2: Determine VVT Strategy (Set decision variables $X_{i,j}$ values).* A set of *decision variables* must be determined in order to enable realistic qualitative and quantitative modeling of costs, times and risks associated with carrying out an incomplete set of VVT activities. A decision variable $X_{i,j}$, $0 \leq X_{i,j} \leq 1$, defines the VVT activity performance level such that the entire set defines the VVT strategy (see Chapter 6).
- *Step 3: Calculate Strategy Cost Based on Existing VVT Strategy.* Multiplying the cost of each VVT activity in the CVM by its corresponding performance level and summing the results yield a practical and realizable VVT strategy cost. For a given VVT strategy, this cost can be estimated by summing the individual VVT activity costs. For this purpose, it is permissible to make the simplifying assumption that each VVT activity is independent of any other VVT activity (see Chapter 6).
- *Step 4: Estimate Parameters and Define Appraisal Risk Model (ARM).* A set of parameters must be estimated in order to calculate the *Expected Appraisal Risk* cost. This is the cost of rework and retesting associated with the discovery of failures during the performance of the VVT activities. This cost is stochastic and is highly dependent on the competency of people and the quality of processes within the organization (see Chapter 6).
- *Step 5: Estimate Parameters and Define Impact Risk Model (IRM).* Another set of parameters must be estimated in order to calculate the *Expected Impact Risk* cost. This cost is associated with failures emanating from partial (or not) performing VVT activities (undertaking a risk). These risks have a stochastic effect on the system and are discernible only subsequent to the partial performance or nonperformance of the VVT activity. Impact costs are generated based on "failure scenarios" suggested by risk and domain experts (see Chapter 6).
- *Step 6: Calculate Total Quality Cost Based on Existing VVT Strategy.* Calculate the total quality cost based on the existing VVT strategy by summing (1) VVT strategy cost, (2) appraisal risk cost and (3) impact risk cost (see Chapter 6).
- *Step 7: Optimize VVT Strategy for Desired {Cost, Time} Results.* As mentioned in Chapter 1, it is not possible to perform a complete VVT process (e.g., execute every procedure in the CVM) due to resource constraints: chiefly time and money. Therefore, optimization (i.e., cost or time minimization) of the VVT strategy is desired. The optimization decisions must consider, on the one hand, the controllable variables associated with investments in VVT activities and, on the other hand, the outcome of these decisions, which are associated with risk impacts and system failures. In addition, certain real-life constraints must be placed on the optimized solution, for example, contractual obligations, company policies and environmental concerns. As an initial approximation, one can assume independence of risk impacts and decompose the decision

process into separate decisions for each VVT activity. It is possible to use a variety of optimization techniques with the objective of getting optimal VVT performance levels $X^*_{i,j}$ which minimize the total expected VVT cost or time (see Chapter 7).

- *Step 8: Reevaluate VVT Strategy.* Whenever possible, reevaluate the assumptions leading to the various parameter estimates and consider modifying the optimal VVT strategy.

Methods and Further Literature

Section 4.2.3, Hierarchical VVT optimization
Section 4.2.5, Classification tree method
Section 4.2.6, Design of experiments (DOE)
Section 4.3.5, Failure mode effect analysis
Section 4.3.6, Anticipatory failure determination
Section 4.3.8, Robust design analysis
Chapters 6, 7, 8, Obtaining quality data and optimizing VVT strategy

- Barad and Engel (2006)

2.3.2 Assess System/Subsystem Design Description (SSDD)

Objective The objective of this activity is to assess the System/Subsystem Design Description (SSDD). The SSDD should be evaluated at both system and subsystem levels, checking for (1) harmony with system concepts embodied, for example, in the RFP and the SysRS, and (2) content and structure sufficient to implement the desired system.

Description The SSDD, as the primary instrument of system design, should fulfill its role as a bridge between the conceptual system as envisioned by its sponsors and the actual one. Therefore the assessment of the SSDD should verify the following:

- *Consistency.* The consistency of the system design versus the system functional requirements and system interface requirements.
- *Feasibility.* The feasibility of system design within the framework bounds of the contract (e.g., funding, schedule and other resources).
- *Policy and Ethics.* That the system design meets company policies and ethics as well as existing standards, laws and environmental statutes. Finally that the system design fulfills any licensing and certification requirements.

The purpose of the SSDD is to describe the system-wide or subsystem-wide design. The assessment of the SSDD should verify that it fulfill its role as an instrument of design containing the elements required to embody a sound system. This verification process includes the following:

- *Scope.* Verify that the SSDD contains a full identification of the system to which it applies and its purpose as well as identification of all relevant stakeholders (e.g., project sponsors, acquirers, users, developers and relevant support agencies).
- *Referenced Documents.* Verify that the SSDD identifies all the documents referenced within the SSDD.
- *Systemwide Design Decisions.* Verify that the SSDD presents system design decisions, including definition of (1) inputs the system must accept and outputs it should produce, (2) system behavior in response to each input or condition and handling of improper inputs, (3) handling and meeting requirements for controlled degradation, safety, security and privacy and (4) construction choices for the hardware or software.
- *System Components.* Verify that the SSDD contains the system architectural design. More specifically, verify that it (1) identifies the components of the system and their relationships with other components, (2) states the purpose of each component and identifies the system requirements and systemwide design decisions allocated to it and (3) provides computer resource data for each computer subsystem or other aggregate of computer hardware.
- *Concept of Execution.* Verify that the SSDD describes the concept of execution among all system components.
- *Interface Design.* Verify that the SSDD describes the interface characteristics of each system element. More specifically, it should identify each internal and external system interface, the elements it is connected to and its unique characteristics.
- *Requirements Traceability.* Verify that the SSDD contains a set of two-way traceability between each system element identified in this SSDD and the system requirements allocated to it.

Methods and Further Literature

Section 4.3.2, Compare images and documents
Section 4.3.5, Failure mode effect analysis
Section 4.3.6, Anticipatory failure determination
Section 4.3.7, Model-based testing
Section 4.4.1, Expert team reviews
Section 4.4.3, Group evaluation and decision

- Sage and Rouse (1999)

2.3.3 Validate System Design by Means of Virtual Prototype

Objective The objective of this activity is to validate, by means of a virtual prototype, whether a given design meets the system requirements. A further objective of using a simulated system is to evaluate the selected design for robustness under a variety of input values as well as assessing the sensitivity of system behavior to modifications in critical design parameters.

Description This activity is based on simulating the system in order to validate the system design against the system requirements, capture its weaknesses and strengths and detect system design failures. Technological advances make it possible today to virtually define system designs in completely integrated and associative parametric representations that are directly suitable for functional verification and accurate sensitivity design studies. Accurate system modeling permits identification of how external parametric changes affect not only a single component of the system but also the integration of the various components into the final assembly. This new ability to define design objectives in terms of quantifiable system outputs (when the system is subject to expected functional constraints) can support true design optimization.

This activity should continue into later system lifecycle phases, including Integration and Qualification. The validation of intermediate and final products may be obtained by comparing the system behavior with the virtual prototype results. Using the virtual prototype instead of the final product may even eliminate some physical tests and their corresponding cost. In some cases, it is appropriate to extend this activity throughout the useful life of the system. Planned improvements to the real system can first be tried on the virtual system without the devastating cost of failure should something go wrong.

Today, a number of commercially available, software-based, simulation tools support such virtual validation. Such tools also include sensitivity and optimization capabilities, which may be used to assess system robustness as well. They are built to discover some constraints on the system or to obtain the system behavior under external conditions.

System design verification by simulation must be handled with care. In fact, many pitfalls are concealed behind apparently realistic graphical images. A complex system's behavior is difficult to simulate correctly, especially if features belonging to different disciplines have to be considered. Quite often parameters relevant to very important system characteristics, such as material behavior, are not well known, and the level of uncertainty may significantly affect the quality of results. For these reasons, it is recommended that simulation models are kept as simple as possible in order to have control over their response and to allow an easier interpretation of the results. In addition, such design tools should always be validated prior to being used in an industrial or research setting.

Methods and Further Literature

Section 4.3.4, System test simulation
Section 4.3.5, Failure mode effect analysis
Section 4.3.6, Anticipatory failure determination
Section 4.3.7, Model-based testing

- Karnopp et al. (1990)
- Matko et al. (1992)
- Ogata (2003)
- Zienkiewicz and Morgan (2006)

2.3.4 Validate System Design Tools

Objective The objective of this activity is to validate that system design tools will produce correct results. A typical design tool may be a software simulation package, a database management system, a hardware test bench and the like. Tool results may be deduced from different perspectives (e.g., simulation, visualization, output data).

Description Systems engineers use a variety of support tools to accomplish the system design process. Such tools encompass a wide range of functionalities. Simple database management tools are used for capturing, for example, the structure, relationships and functionality of systems and produce a set of documents or printed lists. However, higher echelon design tools use simulation and other techniques to help designers in analyzing complex engineering problems, visualizing the result, answering typical "what if" questions and so forth.

Design tools, especially the more sophisticated ones, using simulating and virtual prototyping of the target system, should be validated prior to widespread usage. We use the term "tool validation" to mean that (1) a given tool works properly and (2) operators of the tool have sufficient training to ensure both proper operation of the tool and correct interpretation of its outputs. Using invalidated or improperly validated tools could result in a design that does not fulfill requirements or discovery of failures in later lifecycle phases, either of which is costly.

The basic strategy for validating a design tool is to evaluate it using a set of "reference cases". A reference case is a set of input data as well as the needed tool operation steps and corresponding expected results that have been computed manually or are known from existing system experience. The design tool is operated with these reference cases and the real results are then compared with the expected ones in order to check if the tool is performing correctly (see Figure 2.7).

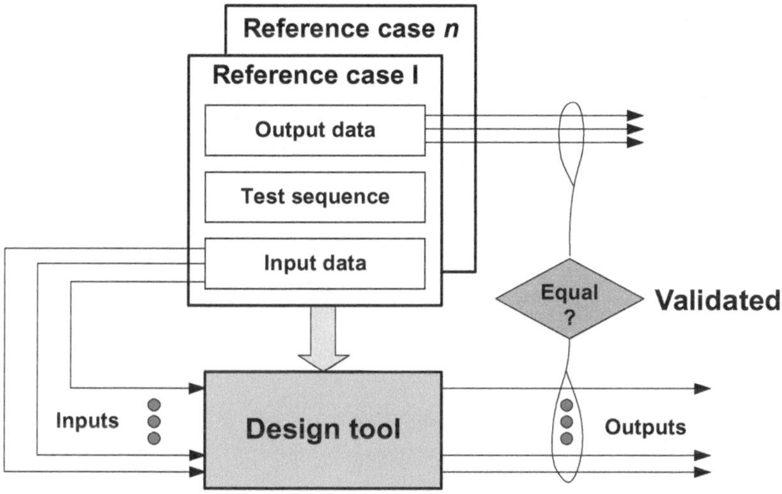

Figure 2.7 Strategy of validating system design tools.

Initial validation should be made using well-known cases. For a simulation tool, textbook cases should be used. For example, consider that we wish to validate a tool for designing airplane structures, such as wing or tail parts. Having it design a Timoshenko beam could validate certain aspects of such a tool. One can check the resulting design by performing a finite-element analysis of the designed beam to prove that it is structurally sound, thus (partially) validating the design tool.

Methods and Further Literature

Section 4.3.5, Failure mode effect analysis

- Pichler et al. (1996)

Section 4.3.6, Anticipatory failure determination

- Schertz and Whitney (2001)

2.3.5 Assess System Design for Meeting Future Lifecycle Needs

Objective The objective of this activity is to assess the existing design and verify that it considers not only the current system's requirements but all future system lifecycle phases, in particular the Production, Use/Maintenance and Disposal lifecycle phases.

Description Some systems engineers, especially the ones employed in the "few-of-a-kind" (e.g., aerospace) industries, where few identical products are manufactured, tend to design systems considering only the development segment of the entire lifetime of the system. That is, their design responsibility ends once the system passes its qualification process. Other systems engineers, often employed in the "many-of-a-kind" (e.g., automobile, consumer electronics) industries, which manufacture thousands and sometimes hundreds of thousands of nearly identical products (though often different variants of products to different customers), seem to be well aware that their design responsibility extends to the entire system lifecycle (see Figure 2.8).

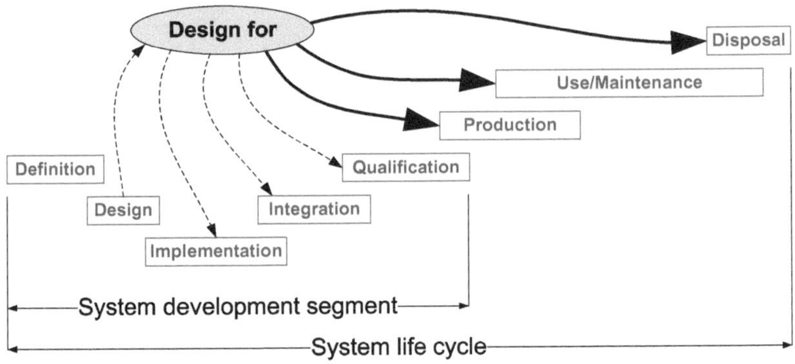

Figure 2.8 Designers should consider all future system lifecycle phases.

The verification of the system design should consider not only whether or not the system qualifies in its design review but also all other system lifecycle phases with particular emphasis on the Production, Use/Maintenance and Disposal lifecycle phases. The verification concerns should therefore include:

Production Verification Needs

- Verification that the system design considers the complexity and cost of components, subsystems and system fabrication and integration as well as production facilities construction. Optimal design, from a production standpoint, entails inexpensive system elements which are simple and cheap to manufacture and assemble in the appropriate quantities.
- Verification that the system design utilizes, to the extent possible, components and subsystems that have been already designed, manufactured and used in other past and present systems. Optimal design, from a production standpoint, entails modular component strategy striving to minimize the overall repertoire of manufactured components and subsystems as much as possible.
- Verification that the system design considers the need to obtain raw materials as well as other resources such as production tools, floor space and warehouses. The design should rely, as much as reasonably possible, on easily obtained raw materials and manufacturing facilities.
- Verification that the system design considers the need to validate system elements after fabrication and integration. The design should support easy means for manufacturing validation.

Use/Maintenance Verification Needs

- Verification that the system design considers the need to use the system on a continuous basis with high degree of reliability and dependency. The design should consider long-term durability, sometimes under adverse environmental conditions, with suitable resilience to recurring users' mistakes and abuse.
- Verification that the system design considers the need to maintain the system on a regular basis. The design should support easy access to all parts of the system for examination and parts replacement. In addition, the design should seek to maximize the use of common elements and minimize the need for spare parts.
- Verification that the system design considers the need to use the system on a continuous basis without incurring negative environmental impact or health or injury risks for users, operators, maintenance crews and others affected by the presence of the system. The design should consider long term-consumer safety and refrain, as much as reasonably possible, from utilizing dangerous materials, exposure to hazardous levels of radiation and the like.
- Verification that the system design considers possible unplanned future system upgrades and modifications. The design should strive to support flexible and adaptable system architecture permitting optimal clustering of components into modules while minimizing the transaction costs associated with internal interfaces.

Disposal Verification Needs

- Verification that the system design considers the need to dispose of the system in accordance with existing regulations with minimal adversity to the environment. The design should ensure, as much as reasonably possible, that the system contains minimal amount of hazardous materials.
- Verification that the system design considers the final disassembly at the end of the system's lifetime such that it should be achieved in a cost-effective manner, recovering as much raw material for recycling as possible.

Methods and Further Literature

Section 4.3.5, Failure mode effect analysis
Section 4.3.6, Anticipatory failure determination
Section 4.3.7, Model-based testing
Section 4.4.1, Expert team reviews
Section 4.4.3, Group evaluation and decision

- Engel and Browning (2008)
- Mumford (2000)
- Suh (1995)

2.3.6 Participate in the System Design Review (SysDR)

Objective The objective of this activity is to participate in the SysDR and, in general, ensure that (1) SSDD is adequate and cost effective in satisfying all system requirements, (2) allocated requirements to the subsystems represent a complete and optimal synthesis of the system requirements and (3) technical program risks are identified, ranked and avoided or reduced to a manageable level.

Description The SysDR is conducted in order to evaluate the overall system design against the total system requirements. Many organizations conduct the SysDR in two stages: Preliminary Design Review (PDR) and Critical Design Review (CDR). The PDR is usually a formal technical review of the basic design approach for the system. It is often conducted prior to a detailed design and summarized in a preliminary SSDD. The overall program risks associated with each part of the system should also be reviewed on a technical, cost and schedule basis. The CDR is normally also a formal technical review of the final design of the system. Ideally it should be conducted prior to the Implementation phase to ensure that the detailed design solutions, as reflected in the SSDD, have been stabilized. In reality CDR often occurs after the Implementation phase was initiated. The VVT engineer should verify that, at a minimum, implementation deals with well-established and familiar elements of the system.

The SysDR encompasses the total system requirements (i.e., hardware, computer software, VVT, operations, training, maintenance facilities, logistical support, etc.). Also included in the review are system engineering management activities (e.g., requirement allocation, manufacturing methods and processes, program risk analysis, system cost-effectiveness analysis, logistics support analysis, trade studies, internal and external interface studies, VVT planning, specialty engineering and configuration management).

Participation in the SysDR involves the following VVT activities:

- Verification that the SSDD is adequate and cost effective in satisfying validated mission requirements
- Verification that the allocated set of requirements to the subsystems and components represent a complete and balanced synthesis of the system requirements
- Verification that the technical program risks are identified, ranked and either avoided or reduced through (1) trade-off studies, (2) subsystem/component hardware proofing, (3) a responsive test program and (4) implementation of comprehensive engineering disciplines (e.g., worst case analysis, failure mode and effects analysis, maintainability analysis, produce-ability analysis standardization)
- Verification that the combination of operations, manufacturing, maintenance and logistics harmonizes with the overall program concepts (e.g.,

quantities and equipment, unit product cost, computer software, personnel, facilities)

- Verification that a technical understanding of the requirements and the design of the system has been reached by all responsible parties

Methods and Further Literature

Section 4.4.2, Formal technical reviews

Section 4.4.3, Group evaluation and decision

- INCOSE-TP-2003-002-03.1 (2007)
- MIL-STD-1521B (1995)

2.4 VVT ACTIVITIES DURING IMPLEMENTATION

The purpose of the system Implementation phase is to create the elements of the system. Some elements may be purchased from other producers and therefore may require purchase specifications. Other elements may require detailed engineering design. Each element, whether purchased or built by the system producer, should be verified against its design and then tested to ensure its stand-alone compliance with its allocated requirements.

VVT activities during the system Implementation phase include detailed planning of the testing process as well as performing simulation, analysis or actual testing, mostly at the subsystem level, in order to verify detailed designs/specifications against requirements.

2.4.1 Preparing the Test Cycle for Subsystems and Components

Objective The objective of this activity is to prepare the testing process for subsystems and components. This includes (1) planning the test process with the objective of specifying the elements necessary to perform and manage these tests, (2) preparing the infrastructure for executing the various tests, (3) designing the test cases for all relevant subsystems and components and (4) creating a test documentation infrastructure which will provide information to interested parties as test data accumulate throughout the test cycle.

Description Testing subsystems and components during the Implementation phase is an integral part of the system-building process. It is usually not a stand-alone activity but rather is performed in parallel with the development. For instance, when building an embedded component, the development teams build the hardware, write the software code and integrate the two into a working entity. Meanwhile, the test team plans the test process, designs and builds test cases and develops the infrastructure necessary to conduct the tests. Eventually, the test team performs the actual tests on the components submitted for formal testing. It then assesses and reports on the overall quality and feature completeness of the test article.

Preparing the test cycle for subsystems and components lays the foundation for the actual performance of testing activities. These activities are tightly interconnected and often iterative in execution. They include planning the test process, building test infrastructure, designing the test cases and creating test documentation infrastructure (see Figure 2.9). Preparation of the test cycle must take into account the management of the test articles for the different development products and related test cases. This includes the collection and storage of test cases, test data, expected values, actual values, other test and technical parameters as well as the rules regarding database access rights and resource distribution.

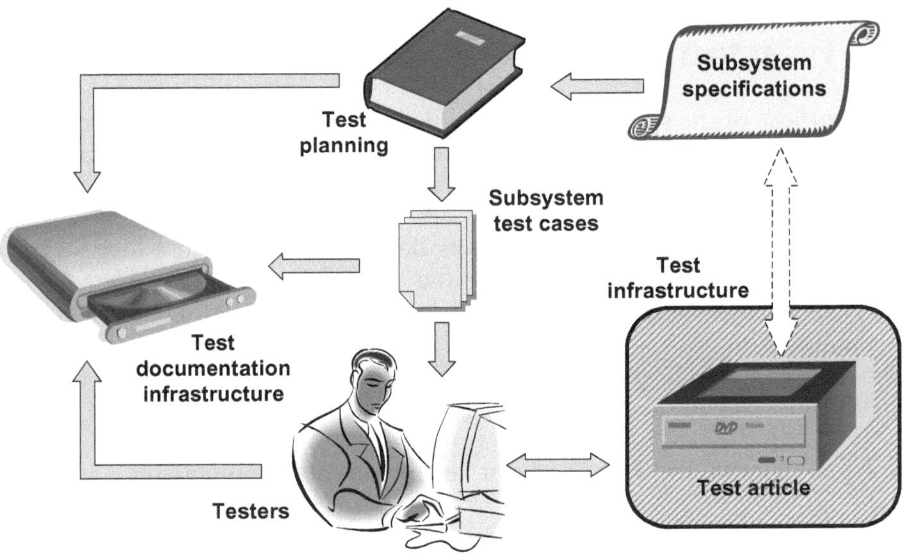

Figure 2.9 Preparing the test cycle for subsystems and components.

The test-planning document should define a specific policy regarding the level of testing required of products developed by subcontractors as well as Commercial Off The Shelf (COTS) products. A rather soft policy will mandate only a review of the testing documents produced by subcontractors and probably accepting COTS products without any functional testing.

1. *Planning Test Process.* Planning the test process is an important administrative and technical activity. Once it is completed, the test cases can be designed, built and then managed. Before testing can begin, the test environment must be established for each test article of the developed system and the enabling products. To test the subsystems, the simulation environments or test frame must be implemented. If the system component test is carried out bottom up, it is usually sufficient to create a test driver which provides the test article with the established test data. In other cases, it may be necessary to

imitate the behavior of system components which have not yet been implemented by means of so-called stubs. The implementation of a suitable test frame is the precondition for an extensive automation of the test.

Due to the close interaction of embedded systems with their application environment and their development in host–target environments, the provision of a test environment is more difficult than for conventional software systems. If the target system is, for example, created in parallel with the software development, or if the necessary hardware is exclusively on the customer's premises because it is permanently installed as part of a more extensive system, then early tests on the target system are impossible. The same is true when system testing may pose a possible danger to people, property or the environment; extensive tests on target systems are only conceivable with the aid of costly safety measures. In all cases where testing is prohibitively costly or profoundly dangerous, methods are necessary which allow for a test on a host system that is as close to reality as possible. For this purpose, comprehensive simulation environments should be substituted for direct testing of the system.

The fact that often the target system is inadequately equipped makes the test more difficult. The target system often lacks storage media, making it only possible to store actual values or monitoring results by means of the implementation of special communication mechanisms between the host and the target system.

In addition to the management of the data stocks accumulated during the test and the provision of the test environment, the test organization should also ensure that the tests are as reproducible as possible, so that regression tests can be carried out easily after changes have been made to the system. The repetition of the identical temporal sequences of input situations involves considerable organizational effort.

2. *Building Test Infrastructure.* Test infrastructure is the environment where test articles are activated during the physical testing process. Sometimes the test infrastructure is simply a common office environment: desk, power outlet, computer and so on. However, often the test infrastructure must provide multiple types of support to the test article, which may include specialized harnesses supporting environmental, mechanical, electrical, chemical, computing and other interfaces. Test infrastructure planning and building involves a multitude of concerns. Here are some of them:

- *Hardware and Software Infrastructure.* A decision must be made as to the specific hardware and software elements as well as tools that are needed for the infrastructure. This issue is naturally related to the fundamental nature of the planned testing, which may be either manual or automated in some way. Generally, infrastructure for manual testing is more appropriate for few-of-a-kind systems. Conversely, infrastructure needed to test large quantities of similar test articles, including embedded components, should support automatic testing.

- *Commercial Considerations.* Commercial considerations are paramount in designing test infrastructure. The initial purchase or development cost of the hardware and software test elements or tools could exceed available budgets and thus compromise the system procurement. In those cases, one should consider using COTS equipment, reusing available test equipment from previous test infrastructures or other innovative but sound testing alternatives. Maintenance of the testing infrastructure is also an important consideration. First, various elements of the infrastructure fail every now and then. Second, test article characteristics may change and therefore the infrastructure must be modified accordingly.
- *Standardization and Modularity.* A key design decision relates to the issue of infrastructure standardization and modularity. Long-term considerations dictate virtually always an optimal infrastructure design based on modular components using standard interfaces. This makes the maintenance more affordable and the resulting test infrastructure more suitable for reuse by future programs.
- *Safety Considerations.* Sometimes, safety issues are neglected in test infrastructure planning. In fact multifaceted test infrastructures may present hazardous conditions that risk the safety of testers and others in the test area. The test designers should consult safety experts as an integral part of test infrastructure planning and design.
- *Security and Confidentiality.* Infrastructure security and confidentiality, especially related to embedded systems, is also a sometimes neglected area. Test engineers should be cognizant of security threats such as hackers, scheming competitors, disgruntled employees and others who might be able to attack a system via the testing infrastructure. In the same vein, a system test report should be released only on a need-to-know basis. For example, competitors, customers and even some engineers of the provider should not be privy to such information. For systems containing private information about real people, the information must, by law, be kept from public view (including persons within the organization). Therefore the test infrastructure must be designed and built to support privacy requirements.

Different testing objectives dictate different test infrastructure, for example, some "special-purpose" infrastructures:

- *Infrastructure for Load/Capacity/Volume Testing.* This type of infrastructure supports the nonfunctional requirement validation of system performance. For example, it supports the validation of systems' ability to process expected load, capacity and volumes under defined production environment conditions as well as in peak business conditions. In addition, the temporal behavior of the system is also measured to evaluate whether the system is functioning within the specified acceptable parameters. Normally, the test infrastructure will present multiple-load scenarios to the system and will monitor the system's ability to process the various test loads.

- *Infrastructure for RF/EMI/EMC Testing.* This type of infrastructure is created to verify the Electromagnetic Compatibility (EMC) of a test article with a noisy, Radio Frequency (RF) environment, in other words, how an external Electromagnetic Interference (EMI) affects the proper functioning of test articles and how test articles affect other system elements or the environment through emitted radiation.
- *Infrastructure for Environmental Testing.* This is a test infrastructure for validating the behavior of the test article under extreme environmental conditions such as heat, cold, shock, vibration, humidity, rain and so on. Since infrastructure for environmental testing is expensive and is needed only on special occasions, most organizations use outside facilities or laboratories for environmental testing. These facilities or laboratories deliver a broad range of specialized experimental and analytical services. An added advantage in using outside organizations is that formal accredited testing enhances the validity to the test results.

Test engineers should remember that the test infrastructure is "a means to an end" and that end is to improve the probability of detecting potential faults. The idea is to find a failure before the customer does. In addition, test engineers must remember the costs of maintaining the test infrastructure. Every piece of software or hardware added to the infrastructure must also be maintained. Since the tested products will inevitably change over time, the infrastructure should be designed with the ability to be modified and expanded.

3. *Designing Test Cases.* A test case consists of a set of test data for the input parameters of the test article, additional conditions which are necessary for the execution of the test case, for example, triggering events (i.e., specifying the times for the occurrence of an input situations), as well as the expected values for the output parameters. Test cases should be created for each test article. They in turn direct the testing of the subsystems or the enabling products. Therefore, the test designer should take the test-planning specifications regarding the stipulated test strategy and test goals into account. If a certain internal system state is specified for a test case, then additional data should be provided in order to set the subsystem into the desired mode of operation before the actual test is carried out. A test case definition should explicitly state the goal of the test, for instance, the execution of a certain system function, the coverage of internal structures or the achievement of a certain state or mode. In addition, acceptance criteria must be defined for each test case so clear pass/fail determination may be achieved.

Test case design determines the quality of the test, because selecting the test data which are to be applied to a test article determines the type, scope and therefore performance of the test. If test cases which are relevant to a particular facet of a system are omitted or forgotten, the likelihood of detecting existing errors in the system decreases.

System and subsystem testing methods are described at length in Chapter 5. Nevertheless it is worth mentioning that test cases may be grouped into white-box and black-box tests. Test case design using white-box techniques tend to focus on the internal structure of the test article. However, by and large, white-box tests do not consider the functionality of the tested article and therefore the test article cannot be considered to be fully verified. In contrast, black-box testing methods often disregard the internal structure of the test article, seeking to discover errors in its functional behavior. Consequently, both white-box and black-box testing should normatively be used in industrial practice.

4. *Creating Test Documentation Infrastructure.* The test plan encompasses an in-depth explanation of the test strategy, goals and the detailed description of all further settings for test planning and organization. Test results also include a list of tested test articles (e.g., development releases and enabling products), the respective test environment and the corresponding test methods. Furthermore, the test cases should be documented with test data, expected values or acceptance criteria as well as by actual values. The test results are processed in such a way that discrepancies between expected and actual values, as well as functional and nonfunctional requirements, are clearly shown. As a result the fulfillment of test goals can be evaluated easily, and errors detected can statistically be summarized.

All of the above information and more should be collected, organized and made available for review. It is also important to archive such information as it can become valuable as a starting point for system upgrades or new similar projects.

Methods and Further Literature

Section 4.2.3, Hierarchical VVT optimization
Section 4.2.5, Classification tree method
Section 4.2.6, Design of experiments (DOE)
Section 4.3.8, Robust design analysis
Section 5.7.4, Component and subsystem testing

- Beizer (1990)
- Beizer (1995)

2.4.2 Assess Suppliers' Subsystems Test Documents

Objective The objective of this VVT activity is to assess the subsystem producers' test documents. This is a key step in verifying that the delivered subsystem has been adequately tested assuring that the subsystem performance complies with its specified requirements.

Description A complex system generally comprises components and subsystems. These components and subsystems take on a variety of forms, for example, mechanical devices, electronic hardware, firmware, software, chemical or physical processes and various combinations of these. Thus, the kind of testing involved and the resulting test documentation may differ greatly from subsystem to subsystem. Another consideration is the maturity of the specific subsystem. If, for instance, the subsystem being purchased has been widely distributed, utilized, stressed and tested under a variety of environmental conditions, the documentation for its performance may take on a very different character than the performance test data required for a newly designed subsystem or a subsystem with very little historical use.

Test data shall be reviewed to verify that the subsystem performs as required by its specification. For software, a technical understanding shall be reached on the validity and the degree of completeness of the software test reports and, as appropriate, of the enabling products, such as training simulators, various manuals (e.g., operator's manual, software user's manual, system diagnostic manual), subsystem packaging and so on.

For some subsystem products, especially those with a history of poor performance, test document assessment shall be a prerequisite to acceptance of the subsystem. For newer or more complex subsystems, this assessment may be conducted on a progressive basis throughout the subsystem's development and would culminate with the completion of the qualification testing of the subsystem. The qualification testing shall be conducted on a configuration of the subsystem that is representative (prototype or preproduction) of the configuration to be released for production. When a prototype or preproduction article is not produced, the review shall be conducted on a first production article. For cases where subsystem qualification can only be determined through integrated system testing, reviews for such subsystems will not be considered complete until completion of the integrated system testing.

Methods and Further Literature

Section 4.4.1, Expert team reviews
Section 4.4.3, Group evaluation and decision
Section 5.7.4, Component and subsystem testing

- Craig and Jaskiel (2002)
- Monczka et al. (2008)
- Pennella (2006)

2.4.3 Perform Acceptance Test Procedure—Subsystems/Enabling Products

Objective The objective of this activity is to perform an Acceptance Test Procedure (ATP) on subsystems and enabling products—more specifically, to (1) perform the specified dynamic test suite on the test article, (2) collect, save and analyze the parameters and behavior of the test article and (3) evaluate these values against the expected behavior of the test article in order to determine whether the test has passed or a failure has been detected (black-box objective). A secondary objective of this activity is mostly applicable to either hardware or software components. Hardware components are tested in terms of the quality of assembly and manufacturing. Software in embedded test articles is evaluated in terms of cyclomatic complexity[10], program coverage and meeting stated programming conventions (white-box objective).

Description Throughout the testing of subsystems, components and enabling products, tests are performed using the test information established during test case design. As a result, actual values are generated and the dynamic behavior of the test articles can be determined, monitored, recorded and compared to expected values and behavior. In addition, hardware components are tested in terms of the quality of assembly and manufacturing. Similarly, software code embedded in components and subsystems is analyzed to find errors and assess coverage and readability. If no errors are found and the coverage and readability criteria are met, the software can be tested further. The following describes the testing process in more detail:

1. *Execute Testing Process.* Following test case design and preparation, the test article is exercised with the selected test data. This activity is referred to as "test execution". The actual values found for the output parameters are saved for later evaluation.

As previously explained in the description of test planning, tests on the target system carried out in the real application environment should be as extensive as reasonable in order to be able to take all the qualities of the test object into account. Only on the target system is it possible to test functional and non-functional program behavior in the real application environment realistically and to recognize errors in the interplay of system hardware and software.

Due to the high level of specialization of the developed system and its enabling products and as they are closely intertwined with the real application environment, commercial testing tools will have limited role in the process. In-house development is time and cost intensive and is only possible for large projects. Often the target system lacks storage media for the storage of test information. Furthermore, regulating or controlling intervention on the part of the tester during test execution is costly and time consuming. The provision of test articles with test data capacity can itself become costly and time consuming. Therefore, if the real application environment is not available during the subsystem and component testing, as is often the case, it is necessary to implement an extensive environment simulation.

2. *Monitor System Testing Process.* Monitoring serves to supervise the test execution and collect appropriate test data. The behavior of the test article

[10] Cyclomatic complexity is a software metric developed by Thomas McCabe in 1976 (see McCabe, 1982). It measures the complexity of software code. We evaluate this set of parameters in order to verify that software is constructed in a simple and straightforwarded manner to support easy future modifications.

must be observed and recorded in order to create the prerequisites for a comparison between expected and actual values during test evaluation and in order to recognize deviations from specified behavior. For this purpose, infrastructure functions realized in hardware or software must be provided which allow the process to be recorded. For this, the system is usually created with an embedded monitoring technique which registers and records internal system signals. Such embedded functions can also serve diagnostic support roles during the system use and maintenance phase. For larger projects, external hardware monitors and logic analyzers are also employed.

If the system is instrumented to carry out some testing functions (e.g., physical characteristics, temporal behavior), then potential problems may arise. Such problems, which are termed "probe effects," always change the behavior of the system to some degree. For this reason, some tests should be repeated with a test article version that does not have instrumentation. Alternatively, it is possible to avoid probe effects by integrating capabilities for process monitoring in the test article from the outset. This is practical only when the target system has sufficient capacity to handle this additional permanent instrumentation. For system level evaluation, such permanent instrumentation has the added benefit that it can be used for a further recording of the process.

3. *Evaluate System Testing Results Against Expected Values.* During test evaluation, actual and expected values as well as actual and expected behavior are compared, taking the defined acceptance criteria into account and thus ascertaining the test results. A pass/fail decision must be made and recorded regarding the behavior of the test article during the testing process. An error is present if the demonstrated behavior does not correspond with the expected targets. Errors can be caused by three sources and the test engineer must be cognizant of this reality: (1) the test article is indeed malfunctioning, (2) the test case defines an incorrect prediction of expected values or expected behavior and (3) the test process did not occur exactly as it was meant to be, due to either an error in the test design or an error in the test execution. It is also an error if the test fails to meet the selected test goals and the test criteria to the desired extent. If the test goals defined during test planning have not yet been met by the test, the test may need to be supplemented with additional test cases.

4. *Perform Static Tests and Analysis.* This activity is generally applicable to hardware and software components. It is recommended that static hardware evaluation be performed as soon as a component is available for testing. This may be done either manually by simple inspection or automatically using commercially available tools (e.g., wire harness testing tools, printed circuit board testers). It is recommended that static software analysis should be performed as soon as the source code is available. This way, problems can be detected before functional verification, which naturally is more expensive. When the code is mostly hand written, performing this activity is recommended; however, programs created automatically by certified code generation tools should not be assessed in this manner.

Methods and Further Literature

Section 4.2.5, Classification tree method
Section 4.2.6, Design of experiments (DOE)
Section 4.3.4, System test simulation
Section 4.3.5, Failure mode effect analysis
Section 4.3.6, Anticipatory failure determination
Section 4.3.7, Model-based testing
Section 4.3.8, Robust design analysis
Section 5.2.1, Component and code coverage testing
Section 5.2.2, Interface testing
Section 5.3.1, Boundary value testing
Section 5.3.2, Decision table testing
Section 5.3.3, Finite State Machine testing
Section 5.3.4, Human-system interface testing
Section 5.4.1, Automatic random testing
Section 5.4.2, Performance testing
Section 5.4.3, Recovery testing
Section 5.4.4, Stress testing
Section 5.5.1, Usability testing
Section 5.5.2, Security vulnerability testing
Section 5.5.3, Reliability testing
Section 5.5.4, Search-based testing
Section 5.5.5, Mutation testing
Section 5.6.1, Environmental Stress Screening (ESS) testing
Section 5.6.2, EMI/EMC testing
Section 5.6.3, Destructive testing
Section 5.6.4, Reactive testing
Section 5.6.5, Temporal testing
Section 5.7.1, Sanity testing
Section 5.7.2, Exploratory testing
Section 5.7.3, Regression testing
Section 5.7.4, Component and subsystem testing

- Beizer (1990)
- Kaner (1996)

2.4.4 Assess System Performance by Way of Simulation

Objective The objective of this activity is to (1) test a virtual realization of subsystems or components in an environment that simulates how they would be exercised in the final complete system, thus determining if they meet design specifications, (2) provide an early determination of complete system performance in response to a variety of possible input and environmental conditions and (3) confirm that component and subsystem specifications were complete and without errors.

Description Simulation models permit virtual testing of system implementation under different conditions, from system concept through the various stages of implementation and often through deployment and maintenance phases. The general idea of virtual prototyping is to support development of complex systems. The main goal in simulations is to study operation and control of the developed system using computerized models. Furthermore, it is possible to use a collection of hierarchical models in order to simulate alternative sequences of the steps involved in the implementation phase, allowing an easier identification of possible sources of problems.

Early in the implementation phase, one can expect the simulation models to be almost entirely virtual models. That is, little actual system hardware and software would have been available. The exception is where prior versions of the system have been developed and possibly even deployed and decommissioned. Simulations with virtual models can give only certain approximate results. That is why virtual prototypes are not any substitute for the real physical or developed prototypes. Simulations can however support the concurrent development and design process of a system, be it purely hardware, software or a combination thereof. As system development progresses, virtual models are gradually replaced by early physical and real components and subsystem prototypes. At this point the simulations become more meaningful and the measurements made can be counted upon to be more realistic. Thus, design modification decisions would have a more factual basis and risks can be assessed more accurately.

At a later stage in the development, it is possible to explore the response of the system to different loading conditions and operating environments. This allows a deeper understanding of system behavior and a quicker selection of possible corrective action to unexpected or unwanted responses. At this stage of system development, the level of knowledge should be enough to allow the creation of fairly detailed models of the system, taking into account the experience already gained with simplified/partial models used in the previous phases. If a hierarchical modeling approach was used from the very beginning, the cost of modeling in terms of human effort and time should be kept at a low level; otherwise, due to the mature technical stage reached, the complexity of the virtual system may result in a very expensive modeling effort. High modeling costs can be mitigated if the design environment allows integration and information sharing among different tools.

Methods and Further Literature

Section 4.3.4, System test simulation
Section 4.3.7, Model-based testing

- Banks et al. (2004)
- Law and Kelton (2006)

Section 5.7.4, Component and subsystem testing

- Lehtonen (2001)

2.4.5 Verify Design Versus Implementation Consistency

Objective The objective of this activity is to verify the consistency between the design of the test article and its implementation. In addition, if contradictions are found, the objective of this activity is to ascertain whether the design or the implementation is the correct response to the requirements.

Description This activity calls for a comparison analysis of design versus implementation. The analysis will indicate whether the implemented test article has been built according to the current design and, if not, whether the design or the implementation needs correction.

In some domains, especially in software, the terms *design* and *implementation* appear to connote varying degrees of abstraction in the continuum between some details (design) and complete details (implementation). However, the amount of detail alone is insufficient to characterize the differences, because design documents often contain information that is not explicit in the implementation (e.g., design constraints, standards, performance goals) and therefore they cannot result from omission of details. Thus, we would expect a distinction to be qualitative as well as quantitative.

The comparison analysis between the design and the implementation of the test article should seek to find discrepancies between the two and, if detected, attempt to identify the correct and the erroneous ones. The analysis should cover the following areas:

- *Design Decisions.* Evaluate the design and the implementation of the test article regarding (1) inputs it accepts and outputs it produces, (2) behavior in response to each input or condition and handling of illegal inputs, (3) handling and meeting controlled degradation, safety, security and privacy requirements and (4) construction choices for hardware–software components.
- *Elements.* Evaluate the design and the implementation of the test article regarding (1) elements of the test article and their relationships with other elements, (2) the purpose of each element in relation to requirements allocated to it and (3) computer resource data for any aggregation of computer hardware.
- *Execution.* Evaluate the design and the implementation of the test article regarding the concept of execution among its elements.
- *Interfaces.* Evaluate the design and the implementation of the test article regarding the interface characteristics of each element, more specifically, each internal and external interface, the elements to which it is connected and its unique characteristics.

Methods and Further Literature

Section 4.4.1, Expert team reviews

- Cleland and Ireland (2006)

2.4.6 Participate in Acceptance Test Review—Subsystems/ Enabling Products

Objective The objective of this activity is to participate in Acceptance Test Reviews (ATRs) of subsystems and enabling products in order to ensure that the testing of specified components, subsystems and enabling products has been completed satisfactorily. Another objective is to reach a technical understanding of the test results and the validity and degree of completeness of the test documents.

Description This is sometimes an informal review that is normally conducted after the testing of components, subsystems and enabling products has been completed. It normally takes place toward the end of the Implementation phase. The subsystem and enabling product testing review should determine whether the testing process has been conducted in accordance with the test-planning document as well as with the appropriate test case designs. Several such reviews are sometimes required, in order to properly assess the entire set of components, subsystems and enabling products within a project. On the one hand, conducting multiple reviews has the advantage that each component, subsystem or enabling product is reviewed independently and as soon as it passed its individual functional tests. On the other hand, when there are multiple reviews, a final acceptance test review should be conducted in order to assess the overall interoperability of the entire ensemble of components, subsystems and enabling products.

As mentioned before, the test-planning document should define a specific policy regarding the level of testing required of products developed by subcontractors as well as COTS products. A rather soft policy will mandate only a review of the testing documents produced by subcontractors and probably accepting COTS product without any functional testing. VVT team participation in the review(s) is needed in order to ensure that the following activities have been accomplished during the review:

- Verification that the test planning document has been reviewed
- Verification that the relevant test case design documents used in conducting the component, subsystem and enabling product testing have been reviewed
- Verification that the results acquired during the relevant tests as depicted in the test result documents have been carefully reviewed
- Verification that the traceability between requirements and their associated component, subsystem and enabling product tests have been reviewed
- Verification that all test limitations (e.g., tests that have not been conducted, tests that failed) and their corresponding unverified capabilities have been identified and reviewed and an explicit action plan has been devised to deal with all such open issues

- Verification that all known component, subsystem and enabling product problems as well as test hardware and software infrastructure and tool problems have been identified and reviewed

Methods and Further Literature

Section 4.4.2, Formal technical reviews

Section 4.4.3, Group evaluation and decision

- Cleland and Ireland (2006)

2.5 VVT ACTIVITIES DURING INTEGRATION

The purpose of the system Integration phase is to combine the system components or subsystems into a complete system. Integration encompasses a series of planning tasks and activities that bring system elements together in an orderly manner while verifying that their relationships are in accordance with the architecture. Integration requires nearly continuous testing.

2.5.1 Develop System Integration Laboratory (SIL)

Objective The objective of this VVT activity is to design and build a System Integration Laboratory (SIL), otherwise known as a hardware-in-the-loop integration test facility. The purpose of the SIL is to validate the system during and after integration within a mixture of virtual and real subsystem environments. This is done by testing an evolving system using a combination of virtual models of subsystems and real subsystems.

Description The integration and testing of complex systems is normally achieved by an iterative succession of integration and testing steps. Initially a virtual prototype of the system is formed by creating a simulated system environment using a collection of virtual subsystems (software and hardware simulators) in lieu of the planned real subsystems.

The virtual prototype of the complete system is exercised to record inputs for the later more realistic assembly model of the complete system and to specify the desired subsystem outputs. The assembly model is then exercised using these inputs and tested against the desired outputs. If the design and implementation are correct, results of these tests should be identical to the ones obtained with the virtual prototype model. If the results are the same, there is a good chance that the actual system when first assembled will work correctly. All this implies that intermediate models of the subsystems should be designed with the same level of accuracy and compatibility of inputs and outputs as they would be in the final configuration. Clearly, this is an engineering challenge.

Finally, each virtual subsystem is replaced with a real subsystem and the prototype real system must be tested and verified to meet the relevant system requirements. At the end of the integration and testing process, the entire prototype system is composed of real subsystems (depicted as the final configuration in Figure 2.10).

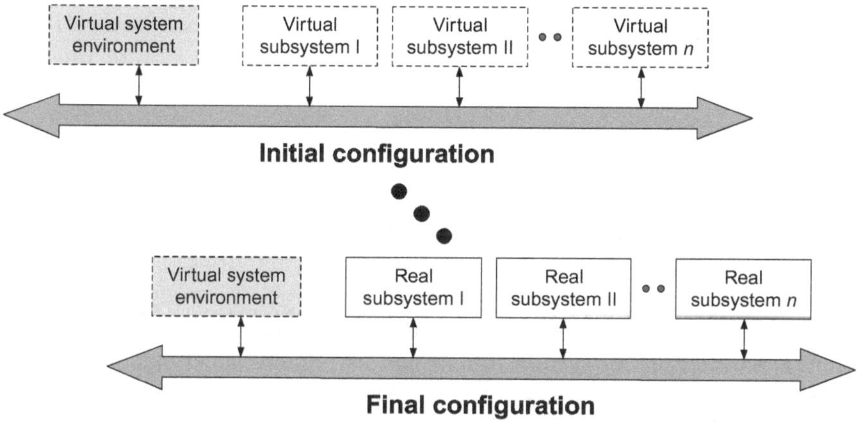

Figure 2.10 System integration using virtual and real subsystems.

This evolving setup is the SIL. As this activity uses models coming from the system Design phase and the system Implementation phase as well as from subcontractors, the SIL must be planned and created early in the development process.

A typical SIL consists of multiple simulators, emulators and test beds and a control center manned by VVT engineers who provide a range of test scenarios. A SIL can be used to dry run integration tests including Multielement Integration Testing (MEIT) and Flight Element Integration Testing (FEIT) as well as to conduct integrated software load testing and verify the system architecture. In addition, the SIL is also available to conduct early hardware/software integration testing as well as to facilitate system operator and user crew training.

Finally, the SIL will most probably carry risk reduction, since it can provide an integrated testing facility available throughout the life of the system. Specifically, it constitutes a platform to test interface compliance and interoperability capabilities and reduces the risk of failure during larger scale testing later in the system lifecycle (e.g., during destruction tests, flight tests, systems-of-systems tests).

Methods and Further Literature

Section 4.2.2, System integration laboratory (SIL)

- Booher (2003)

Section 5.7.5, Integration testing

- Grady (Ed.) (1994)

2.5.2 Generate System Integration Test Plan (SysITP)

Objective The objective of this activity is to develop a System Integration Test Plan (SysITP) that guides the verification process such that each component, subsystem and enabling product is integrated within a given system and works as intended. The objective of this plan is therefore to ensure that no major interface issues remain unresolved by the time for system functional testing.

Description The SysITP documents the level of testing necessary to validate the step-by-step integration of components, subsystems and enabling products into an overall functioning system. This plan helps the VVT team in comprehending the logical sequence of the test integration activities and assists project management in tracking the progress of the integration process. The outcome of this plan is that all relevant parties will agree on how to proceed before the system is handed off for system functional testing and acceptance testing. The following is a proposed structure for a SysITP (adopted and tailored from MIL-STD-498):

Proposed Structure: System Integration Test Plan

Section 1: Scope

1.1: *Identification.* A full identification of the system undergoing integration testing.

1.2: *System Overview.* A brief statement of the purpose of the system undergoing system integration testing. It shall also describe the general nature of the system, hardware and software; summarize its operation and maintenance as well as identify the project key stakeholders (e.g., system's sponsor, acquirer, user, developer, support agencies).

1.3: *Document Overview.* A summary of the purpose and contents of this document.

1.4: *Relationship to Other Plans.* A description of the relationship of this document to related project management plans and in particular to the System Integration Test Description (SysITD).

Section 2: Referenced Documents. This section shall list all documents referenced in this plan.

Section 3: Integration Test Strategy. This section shall describe the overall integration strategy. Integration tests required to verify that subsystem integration perform as expected must be described together with their expected results. At the lower levels, these tests may focus on testing of interfaces among components within given subsystems. As more of the system is put together, tests will focus on interfaces among subsystems and between the system and the environment.

3.1: *Integration Entry Criteria.* The criteria that must be met before integration of specific elements may begin.

3.2: *Integration Strategy.* The integration approach (e.g., top down, bottom up, functional groupings) and the rationale for choosing that approach.

3.3: *Subsystem Integration Sequence.* The order in which subsystems will be integrated.

3.4: *Integration Test Exit Criteria.* The criteria for determining that integration tests have been completed. In addition, this section shall describe the final set of functional tests to be run at the end of integration in order to verify overall functionality of the system. These functional tests are intended to confirm that the system has been successfully integrated and that the system is ready to undergo functional and acceptance testing.

Section 4: Integration Test Infrastructure and Logistics

4.1: *Tools and Test Equipment Required.* A list of all tools and test equipment needed to accomplish the system integration testing. Examples are computer workstations, measurement equipment and host operating systems.

4.2: *Participating Organizations and Personnel.* The organizations that will participate in the system integration testing and the roles and responsibilities of each. In addition, this subsection shall identify the number, type and skill level of personnel needed during the test period, the dates and times they will be needed and any special needs to ensure continuity and consistency in performing the test program.

Section 5: Planned Integration Tests

5.x (x = *1, 2, … ,* N*): Subsystems to be Integrated.* The subsystems to be integrated and tested. In addition, this subsection shall include the following elements to describe the scope of the planned testing:

- *Test Levels.* The levels at which testing will be performed, for example, subsystem level within a system or system level within external environment.

- *Test Classes.* The types or classes of tests that will be performed (e.g., functional tests, interface tests, timing tests, erroneous input tests, loading tests).
- *General Test Conditions.* The conditions that apply to all of the tests or to a group of tests.
- *Data Recording, Reduction and Analysis.* The identification and description of the data recording, reduction and analysis means to be used during and after the testing process.

Section 6: Test Schedules. This section shall contain or reference the schedules for conducting the tests identified in this plan.

Section 7: Requirements Traceability. This section shall contain traceability from each test identified in this plan to the subsystem requirements and vice versa.

Methods and Further Literature

Section 4.2.3, Hierarchical VVT optimization
Section 4.2.5, Classification tree method
Section 4.2.6, Design of experiments (DOE)
Section 4.3.1, VVT process planning
Section 4.3.8, Robust design analysis
Section 5.7.5, Integration testing

- MIL-STD-498 (1994)

2.5.3 Generate System Integration Test Description (SysITD)

Objective The objective of this activity is to develop a SysITD containing a set of test case procedures and associated information necessary to integrate components, subsystems and enabling products and to produce a whole system that will satisfy the system architectural design and the customers' expectations expressed in the system requirements.

Description System level integration testing focuses mainly on verifying both internal system interfaces and data flow among components, subsystems and enabling products as well as verifying external system interfaces (from/to external systems). In addition integration testing will verify the emerging system level functionalities.

The SysITD defines the procedure and environment for integrating and testing the elements (i.e., components, subsystems and enabling products) within the combined and evolving system. Integration of subsystems is an evolutionary process performed in several iterations. Within each iteration, an additional mature element is integrated and tested. The order in which elements are added depends upon their availability and the results of previous integration efforts. This process continues until all elements have been integrated and proven to be working properly within a real or a simulated environment.

Figure 2.11 depicts the logic of creating test descriptions. The system operational scenarios and especially the critical operational issues are analyzed together with the system key performance parameters in order to determine potential system failure modes. Using the findings of this analysis, a collection of test scenarios is planned leading to the creation of an appropriate number of test descriptions.

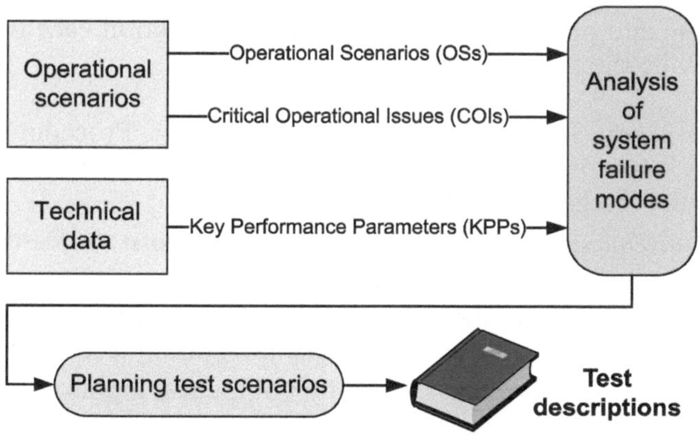

Figure 2.11 Logic of creating test descriptions.

During the development of the SysITD an integration strategy must be devised that specifies the integration approach (top down, bottom up, functional groupings, etc.), the integration rationale and the order in which the subsystems are integrated and tested. A proposed SysITD structure is provided below (adopted and tailored from MIL-STD-498):

Proposed Structure: System Integration Test Description

Section 1: Scope. This section shall be divided into the following paragraphs:

1.1: *Identification.* A full identification of the system and the software to which this document applies.

1.2: *System Overview.* The purpose of the system to which this document applies. In addition it shall describe the general nature of the system, operation and maintenance and identify the project stakeholders (e.g., system's sponsor, acquirer, user, developer and support agencies).

1.3: *Document Overview.* A summary of the purpose and contents of this document.

Section 2: Referenced Documents. This section shall list all the documents referenced in this document.

Section 3: Interface Test Descriptions. This section shall be divided into paragraphs, each describing a unique integration test case.

3.x *(*x = *1, 2, …,* N*):* Integration Test Identifier. These subsections shall identify a system integration test case by a unique identifier, state its purpose and provide a brief description. In addition each paragraph shall provide the following relevant information:

a. *Hardware, Software and Other Preparations.* Procedures necessary to prepare the hardware, software and other elements for the system integration test.
b. *Requirements Addressed.* System requirements addressed by the integration test case.
c. *Prerequisite Conditions.* Any prerequisite conditions that must be established prior to performing the integration test case.
d. *Integration Test Inputs.* Description of the test inputs necessary for the test case.
e. *Expected Integration Test Results.* All the expected test results for the test case. Both intermediate results as well as final test results should be provided, as applicable.
f. *Criteria for Evaluating Results.* The criteria to be used for evaluating the intermediate and final results of the test case.
g. *Integration Test Procedure.* Definition of the test procedure for the test case. The test procedure should be composed of a series of individual steps listed sequentially in the order of the planned actual execution.
h. *Assumptions and Constraints.* Any assumptions made and constraints or limitations imposed in the description of the test case due to system or test conditions, such as limitations on timing, interfaces, equipment, personnel and database/data.

Section 4: Requirements Traceability. This section shall contain traceability from each test case in this SysITD to the system requirements and vice versa.

Methods and Further Literature

Section 4.2.5, Classification tree method
Section 4.2.6, Design of experiments (DOE)
Section 4.3.8, Robust design analysis
Section 5.7.5, Integration testing

- MIL-STD-498 (1994)

2.5.4 Validate Supplied Subsystems in Stand-Alone Configuration

Objective The objective of this activity is to validate each subsystem in a stand-alone configuration prior to integration with other subsystems. It can be thought of as an acceptance test for the subsystem. Such validation determines whether or not the subsystem requirements have been met and often go further and fully stress the subsystem in order to determine under what conditions it would be likely to fail and how this failure is manifested.

Description It is recommended that, before physical integration, each supplied subsystem should be validated in a stand-alone configuration. This last-minute qualification activity should be performed by the integrator with appropriate support provided by the producers of the subsystem. Stand-alone validation is most appropriate when the overall system is based on a modular structure comprising a variety of subsystems and enabling products.

Stand-alone validation of a subsystem is important because in such a configuration many inputs are available for perturbing the subsystem and many more outputs are available to expose the true behavior of the subsystem. Therefore, this activity will improve considerably the reliability and effectiveness of the integrated system.

One should not be tempted to avoid this step by prematurely integrating the subsystem into a final system configuration, performing the testing on that configuration and assuming that if the final system works well then automatically the subsystem is perfect. Testing a subsystem in an integrated configuration could easily mask the existence of internal subsystem defects.

Methods and Further Literature

Section 4.3.4, System test simulation
Section 4.3.5, Failure mode effect analysis
Section 4.3.6, Anticipatory failure determination
Section 5.3.1, Boundary value testing
Section 5.3.2, Decision table testing
Section 5.3.3, Finite-state machine testing
Section 5.3.4, Human-system interface testing
Section 5.4.1, Automatic random testing
Section 5.4.2, Performance testing
Section 5.4.3, Recovery testing
Section 5.4.4, Stress testing
Section 5.5.1, Usability testing

Section 5.5.2, Security vulnerability testing
Section 5.5.3, Reliability testing
Section 5.5.4, Search-based testing
Section 5.5.5, Mutation testing
Section 5.6.1, Environmental Stress Screening (ESS) testing
Section 5.6.2, EMI/EMC testing
Section 5.6.3, Destructive testing
Section 5.6.4, Reactive testing
Section 5.6.5, Temporal testing
Section 5.7.1, Sanity testing
Section 5.7.2, Exploratory testing
Section 5.7.3, Regression testing

- Ogata (2003)
- Zienkiewicz and Morgan (2006)

2.5.5 Perform Components, Subsystem, Enabling Products Integration Tests

Objective The objective of this activity is to validate that the system, created from the aggregate of components, subsystems and enabling products, is functioning in accordance with its requirements and will fulfill its acquirer's expectations.

Description System integration testing is performed to demonstrate that the system requirements, as defined in the System/Subsystem Specifications (SSSs), have been met. The capabilities of the system and its enabling products are evaluated to assess the overall integrity, functionality, operability and conformance to the defined requirements. During this process, the system shall be evaluated using the SysITP and the SysITD.

Sometimes, portions of the tests may be postponed to a later date with prior approval of the project manager. The rationale for skipping portions of the test plan and updated test plan should be documented. Integration and test team members shall be drawn from the development team when possible, as their expertise and experience with the system are valuable. Exact team composition will be specified in the test plan. The infrastructure configuration relies on test environments which duplicate field hardware and system conditions. Any exceptions, such as simulated interfaces, shall be noted prior to test execution. Any requirements that cannot be tested prior to release shall be documented in the System Integration Test Report (SysITR).

It is recommended that an Integration Readiness Review (IRR) shall be conducted for critical systems (e.g., flight safety, financial transactions) to ensure that the system itself as well as the SysITP, SysITD and other documentation are all in order.

If a system is developed in multiple builds (i.e., building stages), integration testing of the last version of the system will not occur until the final build. System integration testing in each build should be interpreted to mean planning and performing tests of the current build of the system to ensure that the system requirements to be implemented in that build have been met. The following is a generic procedure for integration and testing the system and its enabling products (adopted and tailored from MIL-STD-498):

Proposed System Integration Testing Procedure

Section 1: Testing on Target System. The developer's system integration testing shall include testing on the target system or an alternative system.

Section 2: Preparing for System Integration Testing. The developer shall prepare the test data and procedures needed to carry out the integration test cases. In particular this refers to the SysITP and the SysITD.

Section 3: Performing System Integration Testing. The developer shall conduct system integration testing. This process shall be conducted in accordance with the SysITP and the SysITD.

Section 4: Analyzing and Recording System Integration Test Results. The developer shall analyze and record the results of the system integration testing. The result will be summarized in the SysITR.

Methods and Further Literature

Section 4.2.5, Classification tree method
Section 4.2.6, Design of experiments (DOE)
Section 4.3.5, Failure mode effect analysis
Section 4.3.6, Anticipatory failure determination
Section 4.3.8, Robust design analysis
Section 5.3.1, Boundary value testing
Section 5.3.2, Decision table testing
Section 5.3.3, Finite -state machine testing
Section 5.3.4, Human-system interface testing
Section 5.4.1, Automatic random testing
Section 5.4.2, Performance testing
Section 5.4.3, Recovery testing
Section 5.4.4, Stress testing
Section 5.5.1, Usability testing

Section 5.5.2, Security vulnerability testing
Section 5.5.3, Reliability testing
Section 5.5.4, Search-based testing
Section 5.5.5, Mutation testing
Section 5.6.1, Environmental Stress Screening (ESS) testing
Section 5.6.3, Destructive testing
Section 5.6.4, Reactive testing
Section 5.6.5, Temporal testing
Section 5.7.1, Sanity testing
Section 5.7.2, Exploratory testing
Section 5.7.3, Regression testing
Section 5.7.5, Integration testing

- MIL-STD-498 (1994)

2.5.6 Generate System Integration Test Report (SysITR)

Objective The objective of this activity is to document and publish the results of the system integration testing process. System integration testing verifies that the integration of the components, subsystems and enabling products was successful and that applications function correctly in an end-to-end testing. It is an opportunity to identify and solve both procedural and functional problems prior to formal qualification and acceptance tests of the system in the next phase.

Description The SysITR records the results of verifying the operation of each component when integrated into the system. It should include a purpose, introduction, test objectives, a description of how the tests were conducted and a summary of the test results. In addition the report should describe any follow-up testing that may be required as a result of problems encountered during the integration testing.

As a rule, all relevant requirements[11] identified in the SSS and/or the RVM should be tested during integration testing. Rigorous traceability between specifications and testing will increase the likelihood that the system satisfies all of the requirements and does not contain undesirable functionalities. Readers should note that a SysITR often reflects an expanded RVM developed during the Definition phase of the project.

At the completion of each cycle of integration testing, the integration test report should be updated. Thus documenting test results and listing any discrepancies that must be resolved before the emerging integrated system is used as the foundation for another integration cycle. A final test report is generated at the completion of integration testing, indicating any unresolved difficulties that require management attention. A proposed SysITR structure is provided below (adopted and tailored from MIL-STD-498):

[11] Often, some of the requirements will not be tested during integration, for example, certain physical automobile road tests under specific environmental conditions.

Proposed Structure: System Integration Test Report

Section 1: Scope. This section shall be divided into the following paragraphs:

1.1: *Identification.* A full identification of the system to which this document applies.

1.2: *System Overview.* A statement of the purpose of the system to which this document applies. It shall describe the general nature of the system; summarize the operations and maintenance and identify the project stakeholders (e.g., sponsor, acquirer, user, developer, support agencies).

1.3: *Document Overview.* A summary of the purpose and contents of this document.

Section 2: Referenced Documents. This section shall list all the documents referenced in this report.

Section 3: Overview of Test Results. This section shall be divided into the following paragraphs to provide an overview of the test results:

3.1: *Overall Assessment of System Tested.* This subsection shall:

a. Provide an overall assessment of the system based on the test results indicated in this report
b. Identify all the remaining deficiencies, constraints or limitations which were detected by the testing process
c. For each remaining deficiency describe:
   - Its impact on the system and system performance, including identification of requirements not met
   - Its impact on system and system design
   - A recommended solution/approach for correcting the deficiency.

3.2: *Impact of Test Environment.* An assessment of the manner in which the test environment may be different from the operational environment and the effect this difference would have on interpreting the test results.

3.3: *Recommended Improvements.* Any recommended improvements in the design, operation or testing of the system tested.

Section 4: Detailed Test Results. This section shall be divided into the following paragraphs to describe the detailed results for each test, often composed of a collection of test cases:

4.x *(*x = *1, 2, …,* N*):* Project-Unique Identifier of Test. These subsections shall describe each individual test. Each subsection shall identify a test by project-unique identifier and shall summarize the results of the test. The summary shall include the completion status of each test. When the completion status indicates a failure, its subsection shall be expanded to include the following information related to the problem(s) that occurred:

a. A description of the problem(s) that occurred
b. The deviation(s) if any, from the original test case/procedure (e.g., substitution of required equipment, procedural steps not followed, different input parameters) and the rationale for the deviation(s)
c. An assessment of the impact stemming from each deviation from the original test

Methods and Further Literature

Section 5.7.5, Integration testing

- MIL-STD-498 (1994)

2.5.7 Assess Effectiveness of the System Built In Test (BIT)

Objective The objective of this activity is to assess the effectiveness of the Built-In-Test (BIT) functionality within embedded systems. In particular, the objective of this activity is to evaluate whether the BIT meets its testability requirements in terms of level of fault detection, level of fault isolation as well as level of erroneous fault detection and erroneous fault isolation within embedded systems.

Description The BIT function is responsible for the automatic or manual monitoring, detection and isolation of internal system failures and the propagation of such information to a system component having responsibility for operator notification or for predefined automated error handling or recovery. BIT detection implies the ability of the BIT function to discover failures as they occur in real time. BIT isolation implies the ability of the BIT function to identify the failing element (hardware or software or both) when the failure does occur. Obviously, requirements for isolation resolution depend on the system at hand. When we deal with an entire vetronics (vehicle electronics) system, we seek to isolate the failure to a specific subsystem, whereas when we deal with a failed electronic board, we seek to isolate the failure to a specific electronic component.

Modern design includes BIT functionality in virtually all embedded systems from household equipment such as television sets to car and trucks and airplanes. For example, Figures 2.12 and 2.13 depict a Scania truck (Scania is a Swedish company) together with a block diagram of its vetronics system. Typical operational requirements for BIT performance in such systems are that 99% of all vetronics faults and 100% of faults relating to safety-critical elements must be detected.

Figure 2.12 Scania truck system (Scania AB).

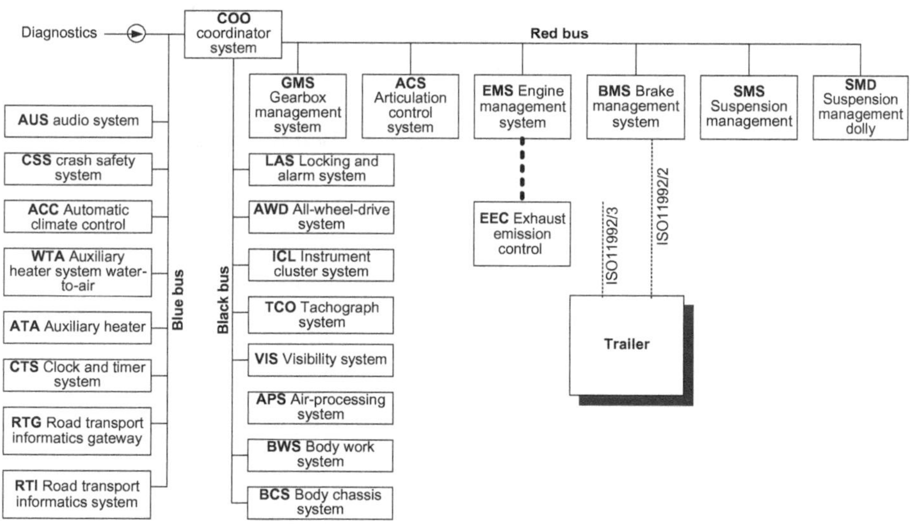

Figure 2.13 Scania truck embedded Vetronics system (Scania AB).

In addition, 100% of the failures must be isolated to the failing vetronics subsystem. As can be seen, BIT implications for testability, reliability, maintainability and product quality are significant.

Basic BIT Principles Figure 2.14 depicts basic BIT principles. The bit controller is the entity which receives external commands and transmits internal BIT results. It activates a test pattern generator that exercises the System Under Test (SUT). Data received from the SUT is evaluated, and if incorrect, then a fault is declared and isolated to a specific failed component.

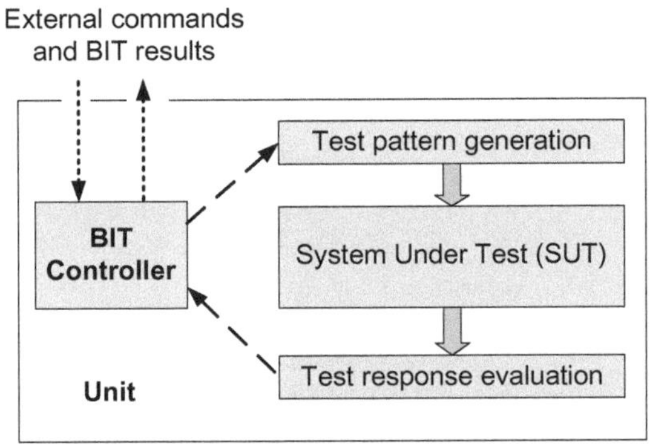

Figure 2.14 Basic BIT principles.

The BIT controller issues a set of test requests either upon a specific external command (i.e., initiated automatically on a power-up sequence or manually by the operator of the system) or continually on a time interval basis. A typical test case specifies the initial state of the SUT and its environment, the test inputs, the expected results and the criteria for declaring SUT failure. The overall BIT output consists of the returned test values, nature of the detected failures and a message identifying the failed component.

Categories of BITs Fundamentally, there are two main categories of BITs (see Figure 2.15).

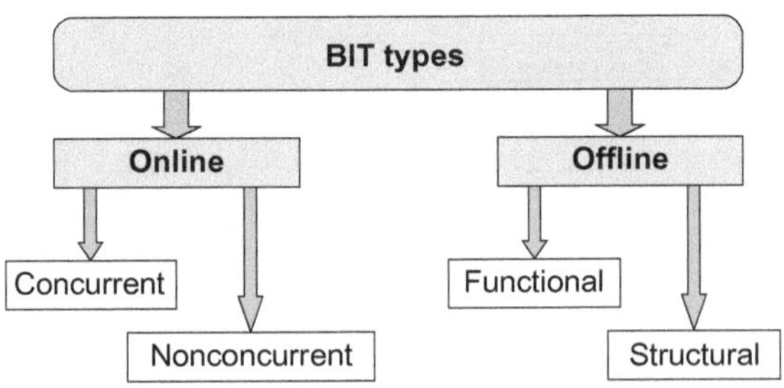

Figure 2.15 Categories of BITs.

1. Under *online BIT operation*, the BIT operation occurs concurrently with normal SUT operation. Here we distinguish between (1) concurrent online BIT in which testing occurs simultaneously with normal functioning of the SUT and (2) nonconcurrent online BIT where testing is carried out while the SUT is placed, for a very short time (measured in milliseconds), into a nonfunctioning state.
2. Under *offline BIT operation*, the BIT operation occurs when the SUT is in an idle operation. Here again we distinguish between (1) functional offline BIT, which is based on the functional behavior of the SUT (black-box testing), and (2) structural offline BIT, which is based on the structure of the SUT (white-box testing).

Levels of BITs We distinguish among several levels of BIT operations, that is, the specific environment in which the BIT operation takes place:

1. *Operational BIT.* This BIT is intended to diagnose a system during normal operation. The purpose of this BIT is to detect and isolate faults down to field-replaceable units.
2. *Production BIT.* This BIT is intended to diagnose the SUT during the manufacturing stage. Different BITs for newly manufactured microchips, electronic boards, components, subsystems and systems are used with the ability to detect and isolate faults down to the appropriately replaceable elements.
3. *Depot BIT.* This BIT is intended to diagnose a system during on-going storage in depot or storage. The purpose of this BIT is to detect and isolate faults down to the depot-replaceable boards and components.

Problems with BIT BIT contributes significantly to product quality during the Manufacturing as well as the Use and Maintenance phases of a system's lifecycle. Nevertheless, it also embodies some distinct liabilities. First, it invariably necessitates additional BIT hardware and software. This increases the development and manufacturing cost and time and often is accompanied by some operational overhead, degraded performance and timing problems within the SUT. A second liability is related to situations where the BIT detects an error when, in fact, none exists (type I, or alpha, error) and, conversely, sometimes the BIT does not detect an error when one does exist (type II, or beta, error). Yet another type of BIT liability stems from isolation of a fault to an incorrect component.

Methods and Further Literature

Section 4.2.6, Design of experiments (DOE)

Section 4.3.4, System test simulation

Section 4.3.5, Failure mode effect analysis

Section 4.3.6, Anticipatory failure determination

Section 4.3.8, Robust design analysis
Section 5.2.1, Component and code coverage testing
Section 5.2.2, Interface testing
Section 5.7.3, Regression testing
Section 5.7.4, Component and subsystem testing

- Archbald (1990)
- Bardell et al. (1987)

2.5.8 Conduct Engineering Peer Review of the SysITR

Objective The objective of this activity is to assess the SysITR document by means of a disciplined engineering practice for detecting and correcting defects.

Description Engineering Peer Review (EPR) refers to a type of review in which the author of the engineering product and a few of his or her peers examine documents and similar work products in order to evaluate their technical content and quality. Verifying system work products by means of peer reviews increases the probability that weaknesses will be identified. In fact, this approach is considered to be the most effective method for document assessments. Peer reviews are distinct from formal project reviews, which are often conducted by and in the presence of technical managers and sometimes customers.

The assessment of the SysITR document in a peer review setting is typically conducted along the following stages: (1) planning the peer review, (2) preparing for the peer review on an individual basis, (3) conducting the peer review and finally (4) performing peer review follow-up activity.

Methods and Further Literature

Section 4.3.2, Compare images and documents
Section 4.4.1, Expert team reviews
Section 4.4.3, Group evaluation and decision

- MIL-STD-498 (1994)

2.6 VVT ACTIVITIES DURING QUALIFICATION

The purpose of the system Qualification phase is to perform formal and operational tests on the integrated prototype system to assure the quality of the system as a whole. Ideally, during the system Qualification phase, no further construction activities are allowed. Generally, system qualification tests are made on a physical target system in a real (rather than simulated) environment. Nevertheless, it is possible to perform some verification on a virtual prototype when actual physical tests are too expensive or pose risk to humans, property or the environment. In such circumstances, system simulations help realize substantial cost savings; however, qualification tests should be considered not fully conclusive.

2.6.1 Generate a Qualification/Acceptance System Test Plan (SysTP)

Objective The objective of this activity is to develop a qualification/acceptance SysTP that guides the verification process such that the system and its enabling product work as intended. There are slight differences between a qualification system test plan and an acceptance system test plan. The objective of the first one is related to an internal developer's evaluation of the system, whereas the objective of the second one is related to demonstrating the system for the customer's evaluation.

Description The qualification/acceptance SysTP documents the level of testing necessary to validate the successful completion of the system development. As mentioned above, a qualification SysTP is usually an internal document, reflecting the producer's view of the system, whereas an acceptance SysTP is focused more on the customer's view of the system.

This plan helps the VVT team in comprehending the logical sequence of the qualification or acceptance test activities. The outcome of this plan is that all relevant parties will agree on how to proceed before the system is delivered to the customer. The following is a proposed structure for a qualification/acceptance SysTP. It was adopted and tailored from MIL-STD-498.

Proposed Structure: Qualification/Acceptance System Test Plan

Section 1: Scope. This section shall be divided into the following subsections:

1.1: *Identification.* A full identification of the system undergoing qualification/acceptance testing.

1.2: *System Overview.* A brief statement of the purpose of the system undergoing system qualification/acceptance testing. It shall also describe the general nature of the system, hardware and software; summarize its operation and maintenance as well as identify the project key stakeholders (e.g., system's sponsor, acquirer, user, developer, support agencies).

1.3: *Document Overview.* A summary of the purpose and contents of this document.

1.4: *Relationship to Other Plans.* The relationship of this document to related project management plans and in particular to the qualification/acceptance SysTD.

Section 2: Referenced Documents. This section shall list all documents referenced in this plan.

Section 3: Qualification/Acceptance Strategy. This section shall describe the overall system's qualification/acceptance test strategy. These tests are required to verify that the system performs as expected. Tests will therefore focus on the functional behavior of the system as well as interfaces between the system and its environment.

3.1: *Qualification/Acceptance Entry Criteria.* The criteria that must be met before qualification/acceptance of the specific system element may begin.

3.2: *Testing Strategy.* The testing approach and the rationale for choosing that approach (e.g., the environment in which the testing occurs: system integration laboratory, ground/flight tests, live fire tests, etc.).

3.3: *Testing Sequence.* The order in which qualification/acceptance tests shall be executed.

3.4: *Testing Exit Criteria.* The criteria for determining that tests have been completed.

Section 4: Test Infrastructure and Logistics

4.1: *Tools and Test Equipment Required.* This subsection identifies all the tools and test equipment needed to accomplish the system testing. Examples are computer workstations, measurement equipment, software and hardware tools and host operating systems.

4.2: *Participating Organizations and Personnel.* This subsection identifies the organizations that will participate in the system testing and the roles and responsibilities of each organization. In addition, This subsection shall identify the number, type and skill level of personnel needed during the test period, the dates and times they will be needed and any special needs to ensure continuity and consistency in performing the test program.

Section 5: Planned Qualification/Acceptance Tests

5.x *(*x = *1, 2, … ,* N*):* System Element to be Tested. These subsections shall identify each system element to be tested. Each subsection shall include the following aspects of the planned testing:

- *Test Levels.* The levels at which testing will be performed, for example, component level and system level.
- *Test Classes.* The types or classes of tests that will be performed (e.g., functional tests, interface tests, timing tests, illegal input tests, maximum capacity tests).

- *General Test Conditions.* The conditions that apply to all of the tests or to a group of tests.
- *Data Recording, Reduction, and Analysis.* The identification and description of the data to be recorded, reduced and analyzed during and after the testing process.

Section 6: Test Schedules. This section shall contain or reference the schedules for conducting the tests identified in this plan.

Section 7: Requirements Traceability. This section shall contain traceability from each test identified in this plan to the system requirements and vice versa.

Methods and Further Literature

Section 4.2.3, Hierarchical VVT optimization
Section 4.3.1, VVT process planning
Section 5.7.6, Qualification testing
Section 5.7.7, Acceptance testing
Section 5.7.8, Certification and accreditation testing
Section 5.7.10, Production testing
Section 5.7.11, Installation testing

- MIL-STD-498 (1994)

2.6.2 Create Qualification/Acceptance System Test Description (SysTD)

Objective The objective of this activity is to develop a qualification/acceptance SysTD. It contains a set of test case procedures and associated information necessary to verify that the system satisfies the architectural design and the customers' expectations expressed in the system requirements. There are slight differences between a qualification system test description and an acceptance system test description. The objective of the first one is related to an internal developer's evaluation of the system, whereas the objective of the second one is related to demonstrating the system for customer approval.

Description System level qualification/acceptance testing focuses mainly on verifying the functionality of the system together with its enabling products as well as verifying external system interfaces (from/to external systems).

The qualification/acceptance SysTD defines the procedure and environment for testing the systems and enabling products. This process continues until the system is proven to be working properly within a real or simulated environment.

During the development of the qualification/acceptance SysTD a testing strategy must be devised that specifies the testing approach (e.g., the setting in which the testing occurs: system integration laboratory, ground/flight tests, live fire tests, etc.), the testing rationale and the order in which the system elements should be tested. A proposed SysTD structure is provided below (adopted and tailored from MIL-STD-498).

Proposed Structure: Qualification/Acceptance System Test Description

Section 1: Scope. This section shall be divided into the following paragraphs:

1.1: *Identification.* A full identification of the system and the software to which this document applies.

1.2: *System Overview.* A brief statement of the purpose of the system to which this document applies. In addition it shall describe the general nature of the system, operation and maintenance and identify the project stakeholders (e.g., system's sponsor, acquirer, user, developer, and support agencies).

1.3: *Document Overview.* A summary of the purpose and contents of this document.

Section 2: Referenced Documents. This section shall list all the documents referenced in this document.

Section 3: Qualification/Acceptante Test Descriptions. This section shall be divided into paragraphs, each describing a unique test case.

3.x *(1, 2, … ,* N*):* Test Identifier. These subsections shall identify system qualification/acceptance test cases by a unique identifier, state the test's purpose and provide a brief description of the test. In addition, each test case paragraph shall provide the following relevant information:

a. *Hardware, Software and Other Preparations.* The procedures necessary to prepare the hardware, software and other elements for the system qualification/acceptance test.
b. *Requirements Addressed.* The system requirements addressed by the qualification/acceptance test case.

c. *Prerequisite Conditions.* Any prerequisite conditions that must be established prior to performing the qualification/acceptance test case.

d. *Qualification/Acceptance Test Inputs.* The test inputs necessary for the test case.

e. *Expected Test Results.* All expected test results for the test case. Both intermediate test results as well as final test results should be provided, as applicable.

f. *Criteria for Evaluating Results.* The criteria to be used for evaluating the intermediate and final results of the test case.

g. *Test Procedure.* The test procedure for the test case. The test procedure should be defined as a series of individual steps listed sequentially in the order in which the steps are to be executed.

h. *Assumptions and Constraints.* Any assumptions made and constraints or limitations imposed in the description of the test case due to system or test conditions, such as limitations on timing, interfaces, equipment, personnel and database/data.

Section 4: Requirements Traceability. This section shall contain traceability from each test case in this qualification/acceptance SysTD to the system requirements and vice versa.

Methods and Further Literature

Section 5.7.6, Qualification testing
Section 5.7.7, Acceptance testing
Section 5.7.8 Certification and accreditation testing
Section 5.7.10, Production testing
Section 5.7.11, Installation testing

- MIL-STD-498 (1994)

2.6.3 Perform Virtual System Testing by Means of Simulation

Objective The objective of this activity is to test a virtual system (rather than the physical system) in a simulated manner in order to reduce lead time and decrease overall testing costs as well as reduce the number of required physical prototypes.

Description Assessment of a developed system often requires many test sequences on physical prototypes. Sometimes, simulating the behavior of the system and its environment rather than physical testing of prototypes can be effective in order to reduce lead time and decrease overall testing costs as well as reduce the number of required physical prototypes.

For example, virtually all passenger cars are produced to individual buyers' specifications. In fact, the same make and model of a modern car may be produced in many thousands of permutations, depending on specific purchase orders. It is often significantly cheaper and faster to test all these types of car products in a simulated manner. Likewise, simulating crash tests in the automotive industry depict an instance where using quantitative information to simulate system behavior reduces the time and cost of a very long and expensive suite of physical tests on fully equipped system prototypes. Along the same line, studying the consequences of a car crash on humans is only possible by simulations of the entire process or conducting real crash tests using dummies to represent human beings.

The ability of modern simulation tools to perform probabilistic design studies may increase the capabilities in the qualification area even further allowing the construction of probability density functions for system responses in different conditions. This is of course very difficult to achieve by any other test/qualification methods.

Methods and Further Literature

Section 4.2.6, Design of experiments (DOE)
Section 4.3.4, System test simulation
Section 4.3.7, Model-based testing
Section 4.3.8, Robust design analysis
Section 5.3.1, Boundary value testing
Section 5.3.2, Decision table testing
Section 5.3.3, Finite-state machine testing
Section 5.3.4, Human-system interface testing
Section 5.4.1, Automatic random testing
Section 5.5.1, Usability testing
Section 5.5.2, Security vulnerability testing
Section 5.5.3, Reliability testing
Section 5.5.4, Search-based testing
Section 5.5.5, Mutation testing
Section 5.6.3, Destructive testing
Section 5.6.4, Reactive testing
Section 5.6.5, Temporal testing
Section 5.7.3, Regression testing

- Karnopp et al. (1990)
- Matko et al. (1992)
- Ogata (2003)
- Zienkiewicz and Morgan (2006)

2.6.4 Perform Qualification Testing/Acceptance Test Procedure (ATP)—System

Objective The objective of this activity is to perform either qualification testing or ATP at the system level in order to assure that the system performs according to documented requirements and the customer's expectations. There are slight differences between system qualification testing and system acceptance testing. The objective of the former is to assure the developer's satisfaction, whereas the objective of the latter is to assure the customer's satisfaction.

Description This activity encompasses the validation of a system composed of components, subsystems and enabling products and their interrelated functions. The qualification of a system can be performed by comparing it with a previous version of the system, a similar legacy system or, most commonly, the specifications and system requirements. The validation of a complete system may be performed by mixing a complementary set of VVT test methods.

Enabling products are a necessary complement to the integrated system. They support the Qualification, Production and Use/Maintenance phases by providing simulation, tools, testers and so on. Examples of enabling products are dedicated test facilities, laboratories, full-scale or scaled-down test facilities, simulation setups, on-board and external instrumentation and sample factories having reduced production capabilities. The enabling products must be qualified separately before system integration in order to be available and to support the qualification process.

The qualification of the system together with its enabling products can be achieved either within the real intended environment or by employing a simulation of the real environment. As this may involve lengthy testing, this activity has direct impact on the risks related to time-to-market and budget of the project.

The reader should note that we refer to "system qualification testing" to indicate a developer-internal system testing performed after the component, subsystem and enabling product integration testing was completed. In contrast, we refer to "system acceptance testing" to indicate a process of validating the system with acquirer participation or, sometimes, acquirer supervision. The following is a generic procedure to perform system acceptance testing (adopted and tailored from MIL-STD-498).

Proposed Procedure: System Qualification Testing/Acceptance Test Procedure

The developer shall perform system acceptance testing in order to demonstrate to the enquirer that the system requirements have been met. It shall cover the system requirements, as defined, for example, in the SSS.

If a system is developed in multiple builds, final acceptance testing of the completed system will not occur until the final build. System acceptance testing in each build should be interpreted to mean planning and performing tests of the current build of the system to ensure that the system requirements to be implemented in that build have been met. The following rules should be met:

a. *Independence in System Acceptance Testing.* The person or persons responsible for the acceptance testing should not be the person or persons who actually developed the system. This does not preclude those who developed the system from contributing their expertise to the process.

b. *Testing on Target System.* The developer's system acceptance testing shall include testing on the target system or an alternative system approved by the acquirer.

c. *Preparing for System Acceptance Testing.* The system developer shall participate in preparing the test data and procedures needed to carry out the test cases, as described in the SysTD. In addition, the system developer shall provide the acquirer advance notice of the time and location of system acceptance testing.

d. *Dry Run of System Acceptance Testing.* If system acceptance testing is to be witnessed by the acquirer, the system developer shall participate in dry running the system test cases and procedures to ensure that they are complete and accurate and that the system is ready for witnessed testing. The developer shall record the results of this activity and shall participate in updating the system test cases and procedures as appropriate.[12]

e. *Performing System Acceptance Testing.* The system acceptance testing shall be conducted in accordance with the system test cases and procedures. It is recommended that the system developer also participate in the system acceptance testing.

f. *Revision and Retesting.* The developer shall make necessary revisions to the system, provide the acquirer advance notice of retesting, participate in all necessary retesting and update the relevant documents as needed, based on the results of system acceptance testing.

g. *Analyzing and Recording System Acceptance Test Results.* The developer shall participate in analyzing and recording the results of the system acceptance testing and sum it up in the SysTR.

Methods and Further Literature

Section 4.2.5, Classification tree method

Section 4.2.6, Design of experiments (DOE)

Section 4.3.4, System test simulation

Section 4.3.5, Failure mode effect analysis

[12]This paragraph refers, in fact, to internal system qualification tests, which are often much broader than normal acceptance test procedures.

Section 4.3.6, Anticipatory failure determination
Section 4.3.7, Model-based testing
Section 4.3.8, Robust design analysis
Section 5.3.1, Boundary value testing
Section 5.3.2, Decision table testing
Section 5.3.3, Finite state machine testing
Section 5.3.4, Human-system interface testing
Section 5.4.1, Automatic random testing
Section 5.4.2, Performance testing
Section 5.4.3, Recovery testing
Section 5.4.4, Stress testing
Section 5.5.1, Usability testing
Section 5.5.2, Security vulnerability testing
Section 5.5.3, Reliability testing
Section 5.5.4, Search-based testing
Section 5.5.5, Mutation testing
Section 5.6.1, Environmental Stress Screening (ESS) testing
Section 5.6.2, EMI/EMC testing
Section 5.6.3, Destructive testing
Section 5.6.4, Reactive testing
Section 5.6.5, Temporal testing
Section 5.7.1, Sanity testing
Section 5.7.2, Exploratory testing
Section 5.7.3, Regression testing
Section 5.7.6, Qualification testing
Section 5.7.7, Acceptance testing
Section 5.7.8, Certification and accreditation testing
Section 5.7.11, Installation testing

- Karnopp et al. (1990)
- Matko et al. (1992)
- MIL-STD-498 (1994)
- Ogata (2003)
- Zienkiewicz and Morgan (2006)

2.6.5 Generate Qualification/Acceptance System Test Report (SysTR)

Objective The objective of this activity is to document and publish the results of the system qualification/acceptance testing process. These tests verify that the qualification/acceptance of the system and enabling products were successful and applications function correctly in end-to-end testing.

Description The qualification or acceptance SysTR records the results of verifying the operation of the system. It should include a purpose, an introduction, test objectives, a description of how the test was conducted and a summary of the test results. In addition, the report should describe any follow-on testing that may be required as a result of problems found during the qualification/acceptance testing.

Each requirement identified in the SSS must be tested during qualification/acceptance testing. This ensures that the product will satisfy all of the requirements and will not include inappropriate or extraneous functionality. A proposed SysTR structure is provided below (adopted and tailored from MIL-STD-498).

Proposed Structure: Qualification/Acceptance System Test Report

Section 1: Scope. This section shall be divided into the following subsections:

1.1: *Identification.* A full identification of the system to which this document applies.

1.2: *System Overview.* A brief statement of the purpose of the system to which this document applies. It shall describe the general nature of the system; summarize the operations and maintenance and identify the project stakeholders (e.g., sponsor, acquirer, user, developer, support agencies).

1.3: *Document Overview.* A summary of the purpose and contents of this document.

Section 2: Referenced Documents. This section shall list all the documents referenced in this report.

Section 3: Overview of Test Results. This section shall be divided into the following subsections to provide an overview of test results:

3.1: *Overall Assessment of System Tested*

a. An overall assessment of the system should be provided based on the test results indicated in this report.
b. Any remaining deficiencies, constraints or limitations which were detected by the testing performed should be identified.
c. For each remaining deficiency, the following should be described: (1) its impact on the system and system performance, including identification of requirements not met, (2) the impact on system and system design and (3) a recommended solution/approach for correcting the deficiency.

3.2: *Impact of Test Environment.* An assessment of the manner in which the test environment may be different from the operational environment and the effect of this difference on the test results.

3.3: *Recommended Improvements.* Any recommended improvements in the design, operation or testing of the system.

Section 4: Detailed Test Results. This section shall be divided into the following paragraphs to describe the detailed results for each test, often composed of a collection of test cases:

4.x *(x = 1, 2, ..., N):* Project-Unique Identifier of a Test. These subsections shall describe each individual test. Each test shall be assigned a project-unique identifier and its corresponding paragraph shall summarize the results of the test. This summary shall include the completion status of each test. When the completion status indicates a failure, its paragraph shall be expanded to include the following information related to the problem(s) that occurred:

a. A description of the problem(s) that occurred
b. The deviation(s) if any, from the original test case/procedure (e.g., substitution of required equipment, procedural steps not followed, different input parameters) and the rationale for the deviation(s)
c. An assessment of the testing deviations and their impact on the validity of each given test.

Methods and Further Literature

Section 5.7.6, Qualification testing

Section 5.7.7, Acceptance testing

Section 5.7.8, Certification and accreditation testing

Section 5.7.10, Production testing

- MIL-STD-498 (1994)

2.6.6 Assess System Testability, Maintainability and Availability

Objective The objective of this activity is to assess the testability, maintainability and availability of the system. Meeting these objectives is not simple because the concepts themselves are often not agreed upon and quantitatively measuring or calculating their value is often a problematic task.

Assessing Testability At an intuitive level, the word *testability* is used to indicate how easy (or difficult) it might be to test a given system. A better description for testability is the degree to which a system facilitates testing in a given "test context." The test context typically includes the intended use of the system (e.g., life critical, financial), the test criteria applied, the test tools used and the test constraints (e.g., available budget and time, required quality). This definition of testability is similar to the IEEE definition,[13] but it emphasizes that testability is a context-dependent attribute of the system.

Complex systems and software contain a large number of components but have only a limited number of inputs and outputs. This causes problems, as it

[13]The degree to which a system or component facilitates the establishment of test criteria and the performance of tests to determine whether those criteria have been met (IEEE Std. 610.12-1990).

is difficult to control individual components and to observe their behavior, because their inputs and outputs have to pass through many intermediate elements. This phenomenon is illustrated in Figure 2.16 depicting a SUT: Input 1 to component A and input 2 to component B are fully controllable, but as we move to other components, the control of inputs is more and more tenuous. Similarly, output 1 generated by component C and outputs 2, 3 and 4 generated by component G are fully observable; however, outputs from other components are less and less observable.

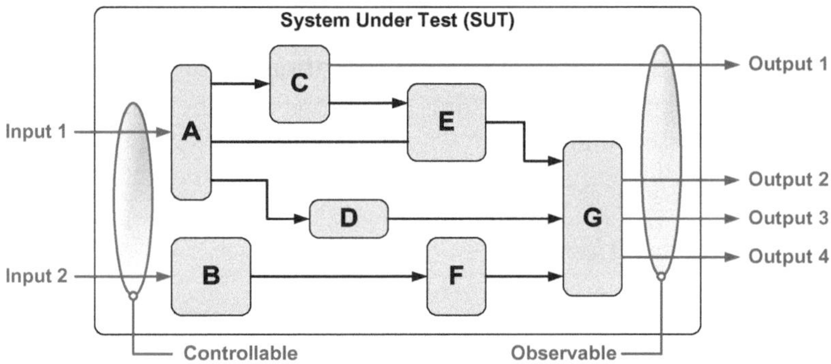

Figure 2.16 Controllable Inputs and Observable Outputs of an SUT.

Testability of distributed real-time systems is a major challenge. First, the behavior of such systems is often nonreproducible so it is difficult to perform regression testing. Second, the observation itself may cause undesired effects on the timing behavior of the system (i.e., the probe effect).

One approach to improve system testability is to increase the controllability and observeability of the SUT. This includes adding internal test points that allow monitoring the status of intermediate components or to bypass intermediate components and directly control particular system elements.

Quantitative measuring of system testability is quite difficult and often uneconomical. Nevertheless there are several approaches for estimating testability in a rather qualitative way, for example, testability assessment by "mutation testing." This concept, also called "mutation analysis," was first introduced as a software testing concept. The original idea was to mutate the code by introducing small errors.

The system then is tested, and if the errors do not damage the performance of the code, then there are two possibilities: (1) either the original code had no effect on performance (i.e., it is not observable) or (2) the test is not effective (i.e., it has no controllability upon the damaged code). More recently, this concept has been extended to hardware testing by adding a step in the testing regime, namely verifying that a checker in the test bench will actually detect the difference in an output when one tampers with a hardware component. This added step serves to give assurance that the system is testable.

The likelihood that faults are hiding from a particular testing scheme is a function of (1) the likelihood that a particular system element is, in fact, activated, (2) the likelihood of a fault at that location causing a wrong behavior and (3) the likelihood of this wrong behavior propagating to the output of the system.

As far as hardware systems, sometimes, faults can physically be inserted into the system (e.g., components, boards or cables may be removed from their place; switches may be set into the wrong position) and the system is tested (preferably by a person or a team unaware of the existence or details of the faults). The fault detection ability of the test suite provides a rough estimate of the system's testability. Similarly, for software systems, tools which automatically generate mutant programs are readily available in the market. Such tools can create "mutant software programs," run the test suite and calculate the testability of programs. In addition, this approach is able to highlight hardware and software areas that require more elaborate testing in order to flash out potential hidden faults. If this solution is not possible, then the next best thing is to increase either the controllability or the observability of the relevant SUT.

Assessing Maintainability Maintainability is broadly understood as the ease with which a system can be modified in order to correct defects and meet new requirements, including coping with a changed environment. Good maintainability means low average duration of all preventive and corrective maintenance activities during a certain period of time.

Researchers have pointed out that the cost of failing to build maintainability into a system is very high and designing for ease of maintenance should already begin when the system is originally conceived. For example, Figure 2.17, adopted (and slightly modified) from the National Aeronautics and Space Administration (NASA) Handbook (NHB 5300.4 1E, 1987), depicts the effect of implementing the maintainability program versus the system lifecycle. The *X* axis shows, broadly, system lifecycle stages and the *Y* axis represents both cost and the amount of design flexibility for application of maintainability.

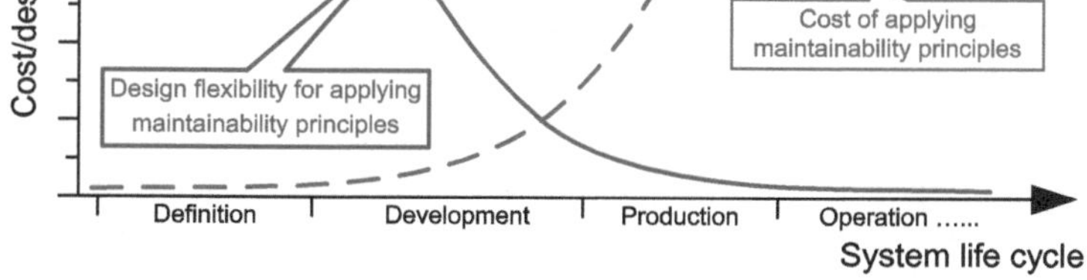

Figure 2.17 Cost versus design flexibility over system lifecycle.

Two plots are shown in the diagram. The first plot, representing the amount of flexibility associated with the application of maintainability, begins at a maximum value, drops nonlinearly and levels off at its minimum value once the operation phase is reached. The second plot, depicting the cost of applying maintainability principles, begins at a minimum value at the start of the definition phase and increases nonlinearly and continues to increase even during the operational phase.

The VVT team should verify system maintainability by assessing the following criteria:

- *Visibility.* Verify that the system is designed for maintenance visibility so that maintainers have maximum visual access to system components. In general, inspecting a component blocked from view will increase a system's downtime.
- *Accessibility.* Verify that the system is designed for maintenance accessibility so that a component can be easily accessed during maintenance, which will greatly reduce maintenance times. When accessibility is poor, other failures are often caused by removal of components or subsystems followed by an incorrect reinstallation.
- *Simplicity.* Verify that the system is designed for simplicity of maintenance. For example, verify that, within reason, the system is composed of a small number of subsystems, the number of components in any given subsystem is small and, whenever possible, these components are standard rather than special purpose. System simplification reduces spares investment, enhances the effectiveness of maintenance troubleshooting and reduces the overall cost of the system while increasing its reliability. Systems designed for simplicity of maintenance will also reduce maintenance training costs as maintenance requires skilled personnel in quantities and skill levels commensurate with the complexity of the maintenance characteristics of the system. An easily maintainable system can often be quickly restored to service by maintenance personnel, thus increasing the availability of the system.
- *Interchangeability.* Verify that the system is designed for maintenance interchangeability, that is, similar components are used within different parts of the system and can be replaced with a similar component if needed. This flexibility in system design usually reduces the extent of the maintenance process and therefore reduces maintenance costs. Interchangeability also allows for system growth with minimum associated costs due to the use of standard components.
- *Human Factors.* Verify that the design takes into account relevant human factors needed during system's maintenance. Verify that the system designers identify requirements necessary to provide an efficient workspace for maintainers and the design does not contain structures and equipment features that impede or prohibit maintainer body movement.

The benefits of this assessment include less time to perform repairs, lower maintenance costs, improved supportability and improved safety.

Unfortunately, today we do not have any useful commonly defined standard for measuring maintainability. Current definitions are too general and do not offer any detailed specification of maintainability. The most detailed quality standard today is ISO 9126 (2007) which defines a set of six (software) quality attributes; one of them is maintainability, defined on a very abstract and general level.

The IEEE Standard Computer Dictionary (1991)) defines maintainability as "the ease with which a (software) system or component can be modified to correct faults, improve performance, or other attributes, or adapt to a changed environment." This vague and incomplete definition is crucially lacking in two respects. First, it does not consider the critical role of the specific context of the system at hand. Second, it fails to provide a precise quantitative definition of maintainability, one that could be used for actual measuring.

Assessing Availability As a practical approach, one can calculate "maintainability of a system" as a function of (1) how frequently, on average, the system fails and (2) how long, on average, it takes to repair it. The first element is measured by the Mean Time Between Failures (MTBF), which represents the average time between failures of a system during its useful life. Calculations of MTBF are made on the assumption that the system is completely repaired after each failure and returns to service immediately.

The second element is measured by the Mean Time To Repair (MTTR). This is the average time required to repair a failure and return the equipment to a condition in which it can perform its intended function. The MTTR takes into account the time it takes for the fault to be correctly identified as well as the time required for maintenance personnel and spare parts to become available. A more rigorous and useful measure is the Mean Down Time (MDT), which is the average time that a system is nonoperational. This includes the amount of time devoted to repair, corrective and preventive maintenance as well as any additional logistics or administrative delays.

As the exact quantitative definition of maintainability is not agreed upon by many researchers, we can adopt a quantitative system availability definition as the ratio of system operating time to total time, where the denominator, total time, can be divided into operating time ("uptime") and "downtime." Underpinning system availability, then, are the reliability and maintainability attributes of the system design, but other logistic support factors also play significant roles. If these attributes, support factors and the operating environment of the system are unchanging, then several measures of steady-state availability can be readily calculated. The equations below depict three concepts of steady-state availability calculations for systems that can be repaired:

1. *Inherent Availability.* System availability assuming corrective maintenance is only undertaken when the system fails:

$$\text{Inherent availability} = \frac{\text{MTBF}}{\text{MTBF} + \text{MTTR}}$$

2. *Achieved Availability.* System availability assuming maintenance is undertaken for both corrective and preventive actions and all logistics (e.g., spare parts, manpower resources, and technical knowledge) is available on location:

$$\text{Achieved availability} = \frac{\text{MTBMA}}{\text{MTBMA} + \text{MMT}}$$

3. *Operational Availability.* System availability assuming maintenance is undertaken for both corrective and preventive actions and average logistic delays are encountered:

$$\text{Operational availability} = \frac{\text{MTBMA}}{\text{MTBMA} + \text{MDT}}$$

The meanings of the relevant system lifecycle and maintenance acronyms are given in Table 2.2. MTBF values may be obtained from similar fielded systems or through system reliability analysis. MTTR or MDT values may also be obtained from similar fielded systems or by inserting various hardware faults and then executing operational scenarios designed to measure the required repair time. Furthermore, it is possible to use stochastic simulation models to assess probabilities of system failures and consequently estimate the variable described above.

TABLE 2.2 Meaning of System Lifecycle and Maintenance Times

Terms	Meaning
MTBF	Mean Time Between Failures
MTTR	Mean Time To Repair (corrective maintenance only)
MTBMA	Mean Time Between Maintenance Actions (corrective and preventive maintenance)
MMT	Mean Maintenance Time (corrective and preventative maintenance)
MDT	Mean Downtime (includes downtime due to active maintenance and logistics delays)

Methods and Further Literature

Section 4.3.4, System test simulation
Section 5.7.1, Sanity testing
Section 5.7.2, Exploratory testing
Section 5.7.6, Qualification testing

- Friedman and Voas (1995)
- IEEE STD 610.12 (1990)
- ISO/IEC TR 9126 (2007)
- MIL-STD-470B (1989)
- NHB 5300.4 (1E) (1987)
- Pecht and Arinc (1995)
- SAE International (1995)

2.6.7 Perform Environmental System Testing

Objective The objective of this activity is to plan and perform an environmental system testing. Environmental testing is used to determine a system's ability to perform its expected functions during or after exposure to a host of detrimental environmental conditions. The objective of these tests is to prove a product's integrity, verify manufacturer's claims regarding operational limits, determine realistic warranty terms and prepare procedures for proper and safe operation.

Description Virtually all systems are subject to environmental stress during their lifetime and they must be able to operate correctly under these circumstances. Environmental testing involves scientific testing of systems under a variety of stressful environmental conditions. Such tests simulate environments with extreme temperatures, humidity levels, altitude, radiation, wind, bacteria, dust, chemical exposure and the like. Environmental testing checks whether a system meets its environmental requirements and therefore is expected to perform successfully during its useful lifetime.

A broad range of standards and custom-designed environmental test facilities are available worldwide. Environmental test equipment sizes range from small bench-top gear to full walk-in/drive-in facilities with a full range of environmental conditions designed to test systems. For example, Figure 2.18 depicts a thermal vacuum chamber for climatic testing and a mechanical vibration apparatus used in dynamic testing.

(*a*) (*b*)

Figure 2.18 (*a*) Climatic and (*b*) dynamic environmental testing (NASA photos).

Choosing an environmental test strategy requires unique specialization and meticulous research. Most testing programs begin by using a specification that identifies environmental requirements and then the procedures to be used for the testing program. Usually engineers familiar with the system should define its test procedure and tests characteristics. The test procedure focuses on ensuring the functionality of the product and has a main goal of improving the product's reliability.

As mentioned, there are several environmental test standards, for example, MIL-STD-810F, Test Method Standard for Environmental Engineering Considerations and Laboratory Tests, Version-F, 2000. This is, in fact, a series of standards issued by the U.S. Army's Developmental Test Command, specifying various environmental tests to prove that equipment qualified to the standard will survive in the field. For the sake of readers' general knowledge, we discuss briefly some of the more frequently used environmental test activities:

1. *Temperature Variation Testing.* In this test, the external temperature is varied between extreme high and extreme low values in a cyclical manner, stressing the SUT. Another variation of this testing is to expose the SUT to simulated solar radiation in order to verify its ability to properly conduct or transmit heat.

2. *Thermal Shock Testing.* Thermal shock is performed to determine the resistance of the SUT to sudden changes in temperature. In this test the SUT undergoes cycles of very low temperature and, within a short period of time, is exposed to a very high temperature. Such temperature shock may cause a permanent change in electrical performance and can cause sudden overloading of materials.

3. *Altitude Testing.* Equipment used in aircraft or at high altitude is subjected to pressures differing from those at sea level. This can cause problems ranging from (1) an increased corona effect on operating electronic equipment to (2) actual equipment failure due to trapped gases. This test simulates the effects of altitude cycling to check the behavior of an SUT under repeated pressure changes. Often, this test is combined with other stress environment conditions (e.g., temperature, humidity).

4. *Mechanical Shock Testing.* In this test the SUT is subjected to a controlled mechanical shock, for example, simulating SUT drop testing and SUT compression testing. In addition, the SUT may be subjected to high levels of accelerations to verify its mechanical properties.

5. *Vibration Testing.* In this test the SUT is vibrated in multiple ways (e.g., ambient and climatic three-axis, random, sine wave, resonant track and dwell). Such tests simulate expected SUT lifetime experience and verify that a system can withstand the rigorous environment of its intended use.

6. *High- and Low-Humidity Testing.* In this test the SUT is subjected to excess moisture to verify that the SUT is not damaged due to corrosion and oxidation. In addition the SUT is subjected to very low humidity to verify that the SUT is not becoming brittle. Similarly, the SUT is subjected to high humidity to verify that components in close proximity are not vulnerable to high electrostatic discharge conditions.

7. *Wet Environment Testing.* In this test the SUT is subjected to typical wet environments, often found in exposed locations and in vessels at sea. These also include rain or freezing rain, wind, icing conditions, salt fog and salt spray. The purpose of the test is to check that the SUT functions properly without rusting, corroding or breaking.

8. *Mold and Fungus Testing.* Products that are exposed to a warm or humid environment are subject to attack by a variety of fungi. These can cause electrical shorts in electronic components as well as mechanical failures and discoloration of exterior surfaces. Finally, fungi may negatively affect human health. In this test the SUT is exposed to warm, moist air in the presence of fungus to see if it grows on the SUT.

9. *Sand and Dust Testing.* Dust and sand blowing occur anywhere in the world as well as in ordinary industrial environments. Products need to be tested for their ability to endure contaminants or abrasion by exposure to them. In this test the SUT is exposed to such conditions to verify proper working conditions and meeting requirements related to surface protection.

10. *EMI/EMC Compatibility Testing.* The past decade has witnessed a significant increase in computer processing speed. As a consequence, electromagnetic radiation of many electronic systems has increased significantly. This causes increased interference with nearby electronic devices as well as increased electromagnetic hazards to humans. Environmental testing of EMI emission from an SUT implies measuring the level and frequency of the electromagnetic energy radiating from the SUT and evaluating it against existing emission requirements and standards. Testing the EMC of an SUT involves ascertaining its ability to operate within the prevailing electromagnetic spectrum and to perform its desired functions without unacceptable degradation under predefined levels of electromagnetic interference.

11. *Explosion Testing.* An explosion test confirms the ability of a component, subsystem or system to operate safely in the presence of hazardous vapors (e.g., oxygen, hydrogen). These tests are common for motors, lighting systems and many aerospace components. The tests can be combined with temperature and altitude variations. In this test the SUT is placed within an appropriate test chamber containing relevant hazardous vapors and the intent is to verify whether sparks created by the SUT device can trigger an explosion.

12. *Highly Accelerated Life Testing (HALT).* The intent of the HALT process is to subject the SUT to stimuli well beyond the expected field environments to determine its operating and destruct limits. It uses step-by-step cycling of environmental variables such as temperature, shock and vibration, simulating accelerated real-world operating environments. The intent of HALT is to ascertain, within a relatively short time, whether the SUT can endure lifetime environmental stress without failing.

13. *Highly Accelerated Stress Screening (HASS).* The HASS is a rather specialized type of environmental screening procedure. It applies stresses

similar to those used in HALT, but it does not intend to damage the SUT. The objective here is to flash out failing parts (mostly in electronic-based systems) resulting from device defects and manufacturing flaws. HASS exploits the statistical "bathtub phenomenon," which indicates a relatively high level of component failure rate during their early life. Once all the infant mortality failures are exposed, the failure rate diminishes to a low "useful life" rate that is relatively constant (see Figure 2.19).

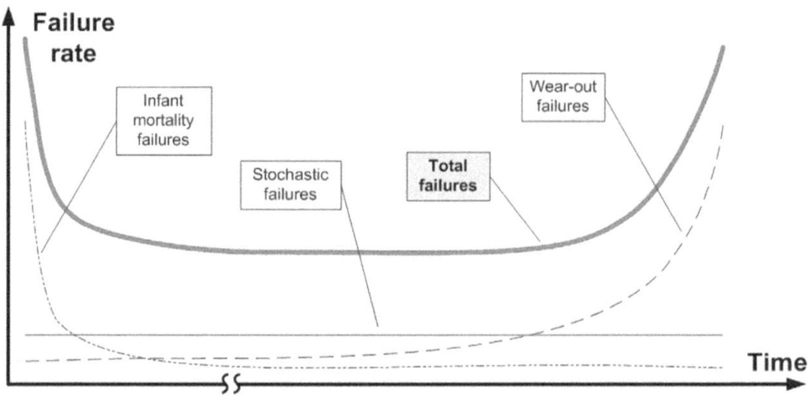

Figure 2.19 Bathtub curve: failure rate versus cumulative operating time.

Methods and Further Literature

Section 4.2.6, Design of experiments
Section 4.3.4, System test simulation
Section 4.3.5, Failure mode effect analysis
Section 4.3.6, Anticipatory failure determination
Section 4.3.8, Robust design analysis
Section 5.7.3, Regression testing
Section 5.7.6, Qualification testing
Section 5.7.8, Certification and accreditation testing

- MIL-STD-810F (2000)

2.6.8 Perform System Certification and Accreditation (C&A)

Objective The objective of this activity is to plan and perform systems Certification and Accreditation (C&A). This is a lifetime, cyclical process involving verification, validation and testing of critical systems in order to insure their proper functionality.

Description Certification and accreditation (C&A) is a process that ensures that systems and major applications adhere to formal and established requirements that are well documented and authorized.

Certification Certification has to do with meeting some criteria. For example U/L certification means that a device or appliance has been successfully tested for safety by Underwriters Laboratories—an independent product safety certification organization that has been testing products and writing standards for safety for more than a century. According to the international standard conformity assessment—vocabulary and general principles (ISO/IEC 17000, 2004), certification is defined as a "third-party attestation related to products, process, systems or persons." In other words, couched in systems VVT terminology, certification is the process in which a third party (e.g., accredited laboratory, the customer) issues a statement indicating that the specified system meets its requirements.

Accreditation Accreditation has the element of permission. Namely, if one is accredited, one is permitted to do certain things legally. For instance, The American Association for Laboratory Accreditation (A2LA) is a nongovernmental, public service membership society which engages in accreditation of a wide range of testing laboratories and industries. According to ISO/IEC 17000, accreditation is defined as a "third-party attestation related to conformity assessment body conveying formal demonstration of competence to carry out specific conformity assessment tasks." In other words, accreditation is a process by which some Designated Approving Authority (DAA) declares, on the basis of some evaluation and review, that a specified organization demonstrated it has the competence to perform specific assessment tasks.

The overall purpose of C&A is therefore to establish uniform standards-based policy for the C&A of systems, provide a disciplined approach to managing the VVT process, use a lifecycle management approach to help program managers implement C&A and identify roles and responsibilities for C&A.

The following is a proposed approach for planning and executing a general system C&A program that is adopted and tailored from the DoD Information Technology Security Certification & Accreditation Process (DITSCAP). We start by adopting the following C&A definitions (from the above source):

- *Certification.* Certification is "a comprehensive assessment of technical and non-technical features associated with the use and environment of a system to establish whether the system meets a set of specified requirements."
- *Accreditation.* Accreditation is "a formal declaration by Designated Accrediting Authority (DAA) that the system is approved for operation, using a prescribed set of safeguards based on residual risks identified during certification."

The two key players that take part in the C&A process should be mutually independent from one another in order to ensure fairness and a biasless process:

- The Desingated Accrediting Authority (DAA) is the person authorized to formally declare the system's accreditation. The DAA assumes the responsibility for operating a system at an acceptable level of risk based on the status of a system, business case and available budget.
- The Program Manager (PM) is the person ultimately responsible for the overall procurement, development, integration, modification, operation and maintenance of the system.

When performing C&A, the entire system is evaluated within the normal operational environment. This includes the systems and all its components (e.g., hardware, software, enabling products). In the normal course of events, a system is certified and then approved by the DAA to become accredited. The C&A is considered a life time process. It must be repeated periodically throughout the entire system's lifecycle, from development to production to maintenance until the system's disposal.

From a top-level view, the C&A process consists of four phases (see Figure 2.20).

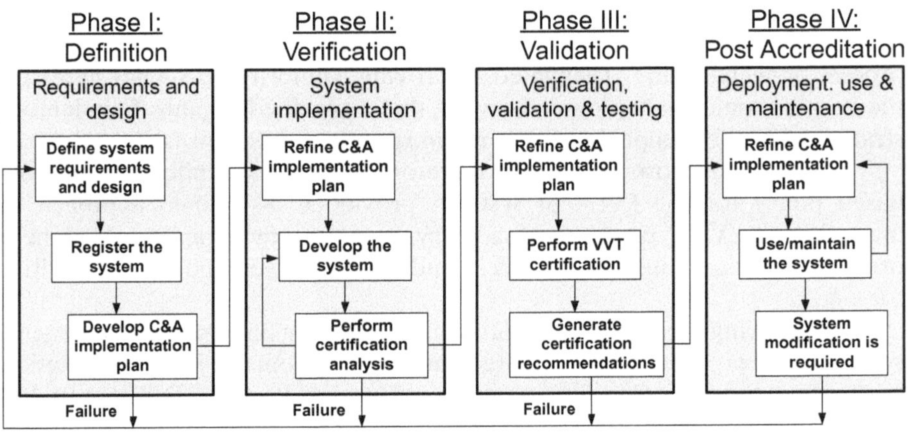

Figure 2.20 Certification & Accreditation Process—four phases.

Phase I: Definition This phase deals with the requirements and the design activities:

1. Define the system requirements and design. This step calls for thorough understanding of the system requirements, capabilities and system architecture as well as potential problems, risks and vulnerabilities. Finally, the operational environment of the system must be understood.
2. Register the system. This step includes identifying the DAA, identifying the organizations involved in the development, operation, maintenance and upgrade of the system. Finally, it involves identifying the system's

C&A scope and estimating funding, schedule and other resources needed for the C&A process.

3. Develop a system Certification and Accreditation Implementation Plan (C&AIP). This step is a formal plan to perform the system C&A. It is used throughout the entire C&A process to guide actions, document decisions, specify requirements, document certification tailoring and level-of-effort, identify potential solutions and maintain operational system functionality.

   The C&AIP must be negotiated and approved by relevant stakeholders and in particular by the DAA and the PM. It is important to note that, if during any phases the system is unable to obtain approval to go on to next stage, it needs to return to the initial phase for redesign.

Phase II: Verification This phase deals with the system implementation activities:

1. Refine the C&AIP to reflect the current state of the system.
2. Develop or modify the system strictly following the C&AIP to ensure that the system is developed correctly. In addition, seek DAA and PM approval to all changes to the system.
3. Perform certification analysis. This step includes system architecture analysis, hardware and software design analysis, integrity analysis, life-cycle management analysis and vulnerability assessment. Sometimes this certification analysis fails and the system must be further developed or modified. At other times, if this certification analysis is passed, check whether the system is ready for certification. If it is ready, then the process moves on to phase III—Validation. Otherwise it goes back to phase I—Definition.

Phase III: Validation This phase deals with the verification validation and testing activities:

1. Refine the C&AIP. This step entails an update to reflect changes and the current state of the system while making sure that all the rules of the C&AIP apply to the developed system. Finally, seek approval of all relevant parties.
2. Perform VVT certification. This step entails system functional verification, validation and testing as well as system management analysis. In addition, this process includes an environment interface accreditation survey, contingency plan evaluation and risk-based management review.
3. Develop certification recommendations based on the above VVT certification results. This step entails creation of a document with all the certification findings for the system as well as recommendations for the system accreditation. If required, the DAA can decide whether to

accredit the system. If not recommended, then the process reverts back to phase I—Definition.

Phase IV: Post Accreditation This phase deals with the deployment, use and maintenance activities:

1. Review C&AIP making sure it is still applicable and maintained up to date. If the plan must be updated, then the DAA and the PM must approve all changes.
2. Use the system and perform ongoing system maintenance and system management operations as well as contingency planning throughout its lifecycle. Whenever appropriate, review the C&AIP to verify its applicability and correctness to any point in time.
3. Whenever a system modification is required, for example, by way of a change request, then first the change request to the system must be reviewed and approved by the DAA and the PM. If approved and it invalidates the system's C&AIP requirement, then the process must go back to phase I for redevelopment. If the change request was not approved, then system operations must be continued without interruption.

Methods and Further Literature

Section 4.2.5, Classification tree method
Section 4.2.6, Design of experiments (DOE)
Section 4.3.4, System test simulation
Section 4.3.5, Failure mode effect analysis
Section 4.3.6, Anticipatory failure determination
Section 4.3.7, Model-based testing
Section 4.3.8, Robust design analysis
Section 5.7.1, Sanity testing
Section 5.7.2, Exploratory testing
Section 5.7.3, Regression testing
Section 5.7.8, Certification and accreditation testing

- DITSCAP (1997)
- Green and Green (1997)
- Hunter (2009)
- ISO/IEC 17000 (2004)
- RTCA/DO-178B (1992)

2.6.9 Conduct Test Readiness Review (TRR)

Objective The objective of this activity is to ensure that the customer or the contracting agency is satisfied that the developer of the system is in fact ready to begin formal system testing. Another objective is to reach technical understanding of the informal system test results and the validity and degree of completeness of the project's key test documents: System Test Plan, System Test Description and System Test Report.

Description The TRR is normally a formal review conducted after the internal system qualification tests have been completed, which take place toward the end of the Qualification phase. The TRR process should determine whether internal testing at the subsystem and integration levels and especially at the system level have been conducted in accordance with the test procedures and that the tests are either complete or problem areas are known and a strategy to resolve them has been established. This review determines whether the system is ready for independent acceptance testing. Reviews of very large systems and certainly Systems Of Systems (SOS) are often broken down into several stages. On the one hand, conducting multiple TRRs has the advantage that each stage is reviewed independently right after the system passes its partial individual qualification tests. On the other hand, if there are multiple TRRs, a final TRR must be conducted in order to assess the overall integrated system.

VVT personnel must either lead or participate in the TRR in order to ensure that during the review the following has been accomplished and verified:

- Changes to the System Requirements Specification (SysRS) that impact the system testing have been carefully reviewed.
- Any changes to the SSDD that impact the system testing have been carefully reviewed.
- Any changes to the SysTP have been carefully reviewed.
- Any changes to the SysTD that was used in conducting the internal system testing, including retest procedures for test anomalies and corrections, have been carefully reviewed.
- Verification that the results acquired during the internal system tests, as depicted in the SysTR, have been carefully reviewed.
- All system test resources, including the status of the development facility, test hardware and software infrastructure and test tools as well as test personnell, have been carefully reviewed.
- The traceability between requirements and their associated system tests has been carefully reviewed.
- All system test limitations (e.g., tests that have not been conducted, tests that failed) and their corresponding unverified system capabilities have been identified and carefully reviewed.
- All known system problems as well as test infrastructure and tool problems have been identified and carefully reviewed.
- The schedules and milestones for the remaining duration of the project have been carefully reviewed.
- The status of all evolving and previously delivered system documentation has been carefully reviewed.

Whereas VVT personnel are expected only to participate in most technical reviews, the TRR is unique in that, often, VVT staff is expected to conduct it. This entails the following responsibilities:

- Gathering all necessary testing information and delivering a "TRR package" on time to the customer and other interested parties
- Attending to the logistics of the TRR, planning it and seeking customer concurrence to an agenda, issue invitations and, finally, leading and controlling the review itself
- After completing the TRR, publishing and distributing copies of "TRR minutes" and seeking the customer's formal approval

Methods and Further Literature

Section 4.4.2, Formal technical reviews

- Faulconbridge and Ryan (2002)
- Horch (2003)

Section 4.4.3, Group evaluation and decision

- MIL-STD-1521B (1995)

2.6.10 Conduct Engineering Peer Review of Development Enabling Products

Objective The objective of this activity is to conduct an engineering peer review related to development of enabling products that were defined, purchased or created during the development period. The intent is to verify that these enabling products appropriately harmonize with the system end products.

Description As mentioned before, engineered systems are, by definition, composed of products that satisfy the operational or mission functions of the system (end products) and products that satisfy the lifecycle support functions of the system (enabling products). Whereas the end products (e.g., hardware, software, databases, communications) provide the desired system capability, the enabling products perform the nonoperational functions of the system. In summary, the enabling products provide lifecycle support to the system that facilitates the progression and use of the operational end product through its lifecycle. Since the end product and its enabling products are interdependent, they are viewed as the engineered system.

The enabling products are assessed to verify their intended functionality vis-à-vis their related end products. Development of an enabling product should be initiated after its requirements have been identified and, often, after the related end product has been defined. Enabling products facilitate the activities of system development (e.g., definition, design, implementation, integration and qualification) as well as production, use/maintenance and, eventually, disposal. Project responsibility therefore, includes the duty of acquiring services from the relevant enabling products in each lifecycle phase.

Engineering peer reviews of development of enabling products generated during the development period (Figure 2.21) should encompass the following three types of products:

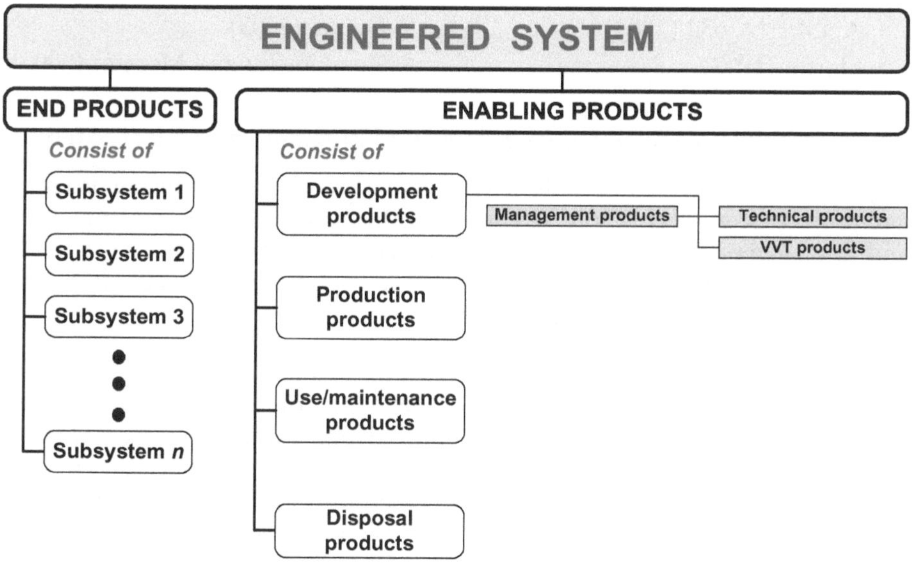

Figure 2.21 Enabling products associated with the development period.

1. *Management Products.* Review the management products including various plans (e.g., SEMP and system integration plan), configuration management audits, program management presentations/summaries/action items, project performance measurements, engineering risk issues, and so on.
2. *Technical Products.* Review the technical products including key technical documentation (e.g., system requirements, system design), COTS tools (e.g., development workstations, laboratory equipment, software compilers, analytical and database tools), in-house development tools (e.g., hardware infrastructure, internally developed software tools and simulators), physical models and system prototypes and presentations from various technical reviews (e.g., SysRR, SysDR).
3. *VVT Products.* Review the VVT products, including VVT plans, policies, procedures and schedules (e.g., RVM, VVT-MP, SysITP, SysITD, qualification/acceptance SysTP/SysTD), special test tools, test facilities and test laboratories (e.g., test-measuring tools, SIL, environmental test facilities, ground, flight and fire test facilities), test demonstrations and test results (e.g., qualification/acceptance SysTR).

Methods and Further Literature

Section 4.3.2, Compare images and documents
Section 4.4.1, Expert team reviews
Section 4.4.3, Group evaluation and decision

- ANSI/EIA-632 (2003)
- Martin (1997)
- Ogata, (2003)
- Zienkiewicz and Morgan (2006)

2.6.11 Conduct Engineering Peer Review of Program and Project Safety

Objective The objective of this activity is to conduct an EPR of the program and project safety, that is, to verify whether the project applies to engineering and management principles, criteria and techniques to achieve acceptable level of mishap risk within the constraints of operational effectiveness and suitability, time and cost throughout all phases of the system lifecycle.

Description This EPR assesses the project for meeting specific system safety requirements. Safety is defined as the "Freedom from those conditions that can cause death, injury, occupational illness, damage to or loss of equipment or property, or damage to the environment." (MIL-STD-882D, 2000). The proposed system safety requirements assessed during the EPR are based on MIL-STD-882D—Standard Practice for System Safety, issued by the U.S. DoD on February 10, 2000. The EPR should examine the following system safety lifecycle requirements:

1. Verification that the system's safety approach has been documented. This should include (1) identification of each hazard analysis and mishap risk assessment process used, (2) information on system safety integrated into the overall program structure and (3) definition of the individual(s) who should be informed of any hazards and the formal mechanism to do so.
2. Verification that hazards have been identified by means of a systematic hazard analysis process encompassing detailed analysis of system hardware and software, the environment and the intended use or application.
3. Verification that a mishap risk assessment of the severity and probability of mishap risks associated with each identified hazard related to potential negative impact on personnel, facilities, equipment, operations, the public and the environment as well as on the system itself has been carried out.
4. Verification that mishap risk mitigation measures have been identified, including alternatives and the expected effectiveness of each mitigation measure. Risk mitigation activity is an iterative process that aims at minimizing any residual mishap risk to a level acceptable to the cognizant authority.

5. Verification that the mishap risk was reduced to an acceptable level and was communicated and agreed to by the developer and other stakeholders of the system.
6. Verification that mishap risk reduction and mitigation have been carried out through appropriate analysis, testing or inspection and the residual mishap risk was appropriately documented.
7. Verification that a hazards and residual mishap risk review is conducted with the appropriate authority, program manager, system users and other stakeholders of the system. The status of the remaining hazards and residual mishap risk should be reviewed and accepted by the appropriate risk acceptance authority.
8. Verification that the status of hazards and residual mishap risks is tracked. Specifically, all hazards, their closure actions and residual mishap risk should be tracked and maintained throughout the system lifecycle.

Methods and Further Literature

Section 4.4.3, Group evaluation and decision

- Leveson (1995)
- MIL-STD-882D (2000)
- Roland and Moriarty (1990)

2.7 REFERENCES

ANSI/ITAA EIA-632, Processes for Engineering a System, American National Standards Institute, Information Technology Association of America, September 2003.

Archbald, W. R., *Built-in test*, Fellows Pub, 1990.

Banks, J., Carson, J., Nelson, L. B., and Nicol, D., *Discrete-Event System Simulation*, 4th ed., Prentice Hall, Upper Saddle River, NJ, 2004.

Barad, M., and Engel, A., Optimizing VVT Strategies—A Decomposition Approach, *J. Oper. Res. Soc.*, 57(8), 965–974. Aug. 2006.

Bardell, H. P., McAnney, H. W., and Savir, J., *Built In Test for VLSI: Pseudorandom Techniques*, Wiley-Interscience, New York, 1987.

Beizer, B., *Software Testing Techniques*, 2nd ed., International Thomson Computer Press, 1990.

Beizer, B., *Black-Box Testing: Techniques for Functional Testing of Software and Systems*, Wiley, New York, 1995.

Booher, R. H., *Handbook of Human Systems Integration*, Wiley-Interscience, HoboKen, NJ, 2003.

Brauer, L. R., *Safety and Health for Engineers*, Wiley-Interscience, HoboKen, NJ, 2005.

Cleland, D., and Ireland, L., *Project Management: Strategic Design and Implementation*, 5th ed., McGraw-Hill Professional, New York, 2006.

Cooper, F. D., Grey, S., Raymond, G., and Walker, P., *Project Risk Management Guidelines: Managing Risk in Large Projects and Complex Procurements*, Wiley, HoboKen, NJ, 2004.

Craig, D. R., and Jaskiel, P. S., *Systematic Software Testing*, Artech House, 2002.

Demillo, A. R., McCracken, M. W., Martin, J. R., and Passafiume, F. J., *Software Testing and Evaluation*, Addison-Wesley, Reading, MA, 1987.

DI-MGMT-81024, Data Item Description, System Engineering Management Plan (SEMP), Draft MIL-STD-499C, Engineering Management, revised March 24, 2005.

DITSCAP, DoD Information Technology Security Certification & Accreditation Process, (DITSCAP), available: http://iase.disa.mil/ditscap/, December 1997.

Engel, A., Requirements Verification Matrix (RVM): A Practical Means for Planning the Systems' Verification Process, paper presented at the 7th International Conference on Software QA and Testing on Embedded Systems, Bilbao, Spain, October, 29–31, 2008.

Engel, A., and Browning, R. T., Designing Systems for Adaptability by Means of Architecture Options, *Systems Eng. J.*, 11(2), 125–146, February 25, 2008.

Engel, A., and Shachar, S., Measuring and Optimizing Systems' Quality Costs and Project Duration, *Systems Eng. J.*, 9(3), 259–280, June 22, 2006.

Faulconbridge, I. R., and Ryan, J. M., *Managing Complex Technical Projects: A Systems Engineering Approach*, Artech House Publishers, 2002.

Friedman, A. M., and Voas, M. J., *Software Assessment: Reliability, Safety, Testability*, Wiley-Interscience, New York, 1995.

Grady, J. (Ed.), *Systems Integration*, CRC Press, Boca Raton, FL, 1994.

Green, D. G., and Green, D., *ISO 9000, Quality Systems Auditing*, Gower Publishing 1997.

Hollnagel, E., Woods, D. D., and Leveson, N. (Ed.), *Resilience Engineering: Concepts and Precepts*, Ashgate, 2006.

Horch, W. J., *Practical Guide to Software Quality Management*, 2nd ed., Artech House, 2003.

Hunter, D. R., *Standards, Conformity Assessment, and Accreditation*, CRC Press, Boca Raton, FL, 2009.

IEEE STD 610.12-1990, IEEE Standard Glossary of Software Engineering Terminology, 1990.

INCOSE-TP-2003-002-03.1, Cecilia Haskins (Ed.), *Systems Engineering Handbook—A Guide for System Lifecycle Processes and Activities*, Version 3.1, International Council on Systems Engineering, August 2007.

ISO/IEC TR 9126, Software Engineering—Product Quality, American National Standards Institute, 2007.

ISO/IEC 17000, International Standard ISO/IEC 17000, Conformity Assessment—Vocabulary and General Principles, 2004.

Juran, J., and Godfrey, B. A., *Juran's Quality Handbook*, McGraw-Hill Professional; 5th ed., 1998.

Kaner, C., *Software Negligence and Testing Coverage*, available: http://www.kaner.com/coverage.htm, 1996.

Karnopp, D., Margolis, L. D., and Rosenberg, C. R., *System Dynamics: A Unified Approach*, 2nd ed., Wiley-Interscience, New York, 1990.

Koomen, T., and Pol, M., *Test Process Improvement: A Step-by-Step Guide to Structured Testing*, Addison-Wesley Professional, 1999.

Law, A., and Kelton, D., *Simulation Modeling and Analysis*, 4th ed., McGraw-Hill, New York, 2006.

Lehtonen, M. (Ed.), *Virtual Prototyping: VTT Research Programme 1998–2000 (VTT Symposium 210)*, Technical Research Centre of Finland, 2001.

Leveson, G. N., *Safeware: System Safety and Computers*, Addison-Wesley Professional, 1995.

Martin, N. J. (Ed.), *Systems Engineering Guidebook: A Process for Developing Systems and Products*, CRC Press, Boca Raton, FL, 1997.

Matko, D., Zupancic, B., and Karba, R., *Simulation and Modelling of Continuous Systems: A Case-Study Approach*, Prentice-Hall, Englewood Cliffs, NJ, 1992.

McCabe, J. T., *Structured testing: A software testing methodology using the cyclomatic complexity metric (Computer science and technology)*, NBS, 1982.

MIL-STD-470B, Maintainability Program for Systems and Equipment, U.S. Department of Defense, May 1989.

MIL-STD-498, Software Development and Documentation, U.S. Department of Defense, December 1994.

MIL-STD-810F, Test Method Standard for Environmental Engineering Considerations and Laboratory Tests, Version F, U.S. Army Developmental Test Command, 2000.

MIL-STD-882D, Standard Practice for System Safety, U.S. Department of Defense, February 2000.

MIL-STD-1521B, Military Standard—Technical Reviews and Audits for Systems, Equipments, and Computer Software, U.S. Department of Defense, 1995.

Monczka, M. R., Handfield, B. R., Giunipero, C. L., and Patterson, L. J., *Purchasing and Supply Chain Management*, 4th ed., South-Western College/West, 2008.

Mooz, H., Forsberg, K., and Cotterman, H., *Communicating Project Management: The Integrated Vocabulary of Project Management and Systems Engineering*, Wiley, HoboKen, NJ, 2003.

Mumford, E., A Socio-Technical Approach to Systems Design, *Requirements Eng.*, 5(2), 125–133, September, 2000.

NHB 5300.4 (1E), Maintainability Program Requirements for Space Systems, NASA Headquarters, March 1987.

Ogata, K., *System Dynamics*, 4th ed., Prentice Hall, Upper Saddle River, NJ, 2003.

Pecht, G. M., and Arinc Inc., *Product Reliability Maintainability Supportability Handbook*, CRC Press, Boca Raton, FL, 1995.

Pennella, R. C., *Managing Contract Quality Requirements*, ASQ Quality Press, 2006.

Pichler, F., Moreno-Diaz, R., and Albrecht, R. (Ed.), *Computer Aided Systems Theory—EUROCAST '95: A Selection of Papers from the Fifth International Workshop on Computer Aided Systems Theory, Innsbruck*, Springer, 1996.

Porter-Roth, B., *Request for Proposal: A Guide to Effective RFP Development*, Addison-Wesley Professional, 2001.

Roetzheim, H. W., *Developing Software to Government Standards*, Prentice Hall, Englewood Cliffs, NJ, 1990.

Roland, E. H., and Moriarty, B., *System Safety Engineering and Management*, Wiley-Interscience, New York, 1990.

RTCA/DO-178B, Software Considerations in Airborne Systems and Equipment Certification, Radio Technical Commission for Aeronautics (RTCA), December 1992.

SAE International, *RMS: Reliability, Maintainability, and Supportability Guidebook*, Society of Automotive Engineers, January 1995.

Sage, P. A., and Rouse, B. W. (Ed.), *Handbook of Systems Engineering and Management*, Wiley-Interscience, New York, 1999.

Schertz, K., and Whitney, T., *Design Tools for Engineering Teams: An Integrated Approach*, Delmar Cengage Learning, 2001.

Spillner, A., Linz, T., and Schaefer, H., *Software Testing Foundations: A Study Guide for the Certified Tester Exam*, 2nd ed., Rocky Nook, 2007.

Suh, P. N., Design and Operation of Large Systems, *J. Manufacturing Systems*, 14(3), 203–213, 1995.

Wiegers, E. K., *Peer Reviews in Software: A Practical Guide*, Addison-Wesley Professional, 2001.

Zienkiewicz, C. O., and Morgan, K., *Finite Elements and Approximation*, Dover Publications, 2006.

Chapter 3

Systems VVT Activities: Post-Development

3.1 STRUCTURE OF CHAPTER

This chapter describes a set of VVT activities that typically occur within the system post development phases (Production, Use/Maintenance, and Disposal). We provide detailed information for each VVT activity in a standard format designed to aid the reader in determining the activity's applicability to a specific system lifecycle phase. As mentioned before, one should (1) tailor the VVT methodology by using the tailoring guidelines presented in the first, introductory chapter and (2) consider using the VVT process model for optimizing the VVT strategy. Subsequently, at the beginning of each system lifecycle phase, one should consider updating the VVT planning document. Typically, each VVT activity may be carried out within one of the following system post development lifecycle phases:

1. *Production.* This produces the completed system in appropriate quantities.
2. *Use/Maintenance.* This operates the system in its intended environment in order to accomplish intended functionality, maintains the system and corrects any defects.
3. *Disposal.* This properly disposes of the system and its elements upon completion of its life.

As mentioned in Chapter 2, each VVT activity is related to one of three aspects: (1) preparing the VVT products, (2) applying VVT to engineered products and (3) participating in or conducting technical reviews. Also, the

Verification, Validation, and Testing of Engineered Systems, Avner Engel

reader should note that we continue to describe each VVT activity in terms of objectives, description and methods and further literature.

3.2 VVT ACTIVITIES DURING PRODUCTION

The purpose of the system Production phase is to reproduce the completed system in appropriate quantities. VVT activities during the system Production phase intend to verify the quality of the incoming material components and subsystems, validate the production process and perform ongoing product quality control (illustrated in Figure 3.1). The following sections define specific VVT activities that are appropriate for the system Production phase.

Figure 3.1 Assembly line testing: comparing products to specifications.

3.2.1 Participate in Functional Configuration Audit (FCA)

Objective The objective of the Functional Configuration Audit (FCA) is to formally validate that the development of Configuration Items (CIs) as well as the completed operation and support documents has been completed satisfactorily and that each CI has achieved the performance and functional characteristics specified in the functional or allocated configuration identification.

Description This description is based on Section 70 (FCA) of MIL-STD-1521 (now withdrawn) and various National Aeronautics and Space Administration (NASA) documents. A FCA verifies that each CI (e.g., component, subsystem or system) meets all the functional requirements, including performance reliability and the like. The FCA embodies a review of the item's performance to ensure it meets the specification without unintended functional characteristics. In addition, the FCA verifies the complete set of operation and support documents. Representatives of the VVT team should verify the availability and quality of the documents needed for the FCA as well as the appropriate execution of the audit itself.

- *FCA Inputs.* Primary inputs for the FCA are the functional requirements for the system and test or operational data showing how it operates. Functional requirement information should include verification methods (analysis, inspection, demonstration, testing or certification) used. FCAs may use, but need not be limited to, data from the following processes and tests:
  a. Functional testing
  b. User trials
  c. Environmental testing
  d. Interface checks and tests
  e. Reliability, availability and maintainability tests and analysis
  f. Software testing, including independent verification and validation (if safety-critical software is involved)
- *FCA Process.* Customarily, the FCA process follows these steps:

  Step 1. Ensure the availability of a verification matrix showing the requirements, verification method and testing procedure name. Ensure that each requirement has a verification method (and procedure) defined.

  Step 2. Add columns to the matrix for test status (i.e., pass, fail and outstanding action items). In addition, add columns to record other details of interest, such as the date the test was conducted and the quality assurance person who witnessed the test as well as any additional information relevant to the FCA process.

  Step 3. Review the test result documentation or inspection/analysis reports that are associated with verifications for each requirement. Record the appropriate information in the expanded verification matrix. When reviewing, ensure that the test was, in fact, sufficient to verify each requirement.

  Step 4. Identify any requirements that are open (i.e., either failed or constitute an outstanding action item).

  Step 5. Write a report which will document the functional configuration audit and its findings.

  Step 6. Resolve any findings and other issues with the project management and, as appropriate, the project stakeholders.
- *FCA Output.* An FCA report, culminating the functional configuration audit, should be generated summarizing the FCA process as well as the

findings, observations and recommendations emanating from the audit. A simple report template is provided below. Tailor the template to fit the needs of the audit.

Functional Configuration Audit Report: Project [...XXX...]

Prepared by: ______________________________
Name, Affiliation

Approved by: ______________________________
Name, Affiliation

Date: ______________________________

Section 1: General

1.1 Reference to relevant document
1.2 List of configuration items
1.3 Test procedures and result versus requirements
1.4 FCA date and list of attendees
1.5 Minutes of FCA
.
.
.

Section 2: Findings. List findings here.

Section 3: Observations. List concerns here.

Section 4: Recommendations. List recommendations here

Methods and Further Literature

Section 4.4.3, Group evaluation and decision

Section 5.7.9, First Article Inspection (FAI)

- MIL-STD-1521B (1995)

3.2.2 Participate in Physical Configuration Audit (PCA)

Objective The objective of the Physical Configuration Audit (PCA) is to technically examine a set of designated CIs and check if each CI "as built" conforms to the technical documentation which defines it.

Description This description is based on Section 80 (PCA) of MIL-STD-1521 (now withdrawn) and various NASA documents.

For complex components, subsystems or systems, the PCA involves comparison of the developed item in its as-built version against its design documentation to ensure that the physical characteristics and interfaces conform to the product specification. In addition, The PCA determines whether the acceptance testing requirements prescribed by the documentation is adequate for acceptance of production units of a CI by the quality assurance activities. The PCA includes a detailed audit of engineering drawings, specifications, technical data and tests utilized in the production of Hardware Configuration Items (HWCIs) and a detailed audit of design documentation, listings and manuals for Software Configuration Items (CSCIs). The review should include an audit of the released engineering documentation and quality control records to verify that the as-built or as-coded configuration is reflected by these documents. For software, the software product specification and software version description must be a part of the PCA review.

Representatives of the VVT team should verify the availability and quality of the documents needed for the PCA as well as the appropriate execution of the audit itself.

- *PCA Inputs.* The PCA may use, but need not be limited to, data from the following processes and tests:
  a. FCA report
  b. Physical HWCIs and CSCIs
  c. Component, subsystem or system specification
  d. Testing and verification reports
  e. Programming process plan
  f. Configuration management records
  g. Deviations and waivers
  h. Problem reports
- *PCA Process.* Customarily, the PCA process follows these steps:

  Step 1. Gather relevant PCA data and documents.

  Step 2. Review FCA reports and verify incorporation (or other appropriate disposition) of action items and findings.

  Step 3. Review the system and its specifications to ensure that (1) the requirements are implemented in the design, (2) the design matches the specifications and (3) the specifications match the actual HWCIs and CSCIs

Step 4. Review all system testing and verification reports and ensure that all design errors that were detected by verification processes were corrected.

Step 5. Review the process plan for component programming or otherwise adjustment to specific configuration. Ensure that the plan has been followed according to the design specifications. Review configuration management records to ensure that the correct design was used.

Step 6. Review problem reports, deviations and waivers to ensure that there are no open issues with the design of the components, subsystems or system.

Step 7. Generate a status report documenting the PCA process and findings of the audit.

Step 8. Resolve any open issues and irregular findings with the project.

- *PCA Output.* A PCA report culminating the physical configuration audit should be generated summarizing the PCA process as well as findings, observations and recommendations emanating from the audit. A simple report template is provided below. Tailor the template to fit the needs of the audit.

Physical Configuration Audit Report: Project [...XXX...]

Prepared by: ______________________________
Name, Affiliation

Approved by: ______________________________
Name, Affiliation

Date: ______________________________

Section 1: General

1.1 Reference to relevant document
1.2 List of configuration items
1.3 Test procedures and result versus requirements
1.4 PCA date and list of attendees
1.5 Minutes of PCA
.
.
.

Section 2: Findings. List findings here.

Section 3: Observations. List concerns here.

Section 4: Recommendations. List recommendations here.

Methods and Further Literature

Section 4.4.3, Group evaluation and decision

Section 5.7.9, First article inspection (FAI)

- MIL-STD-1521B (1995)

3.2.3 Plan System Production VVT Process

Objective The objective of this VVT activity is to plan the system production VVT process at the beginning of the system production cycle.

Description Planning the production VVT process entails formal creation of the production VVT program, including the identification of required production VVT strategy, schedule, management and resources:

- *Production VVT strategy.* Describe the specific VVT strategy for performing VVT activities in support of the manufacturing phase. Table 3.1 depicts a set of VVT activities to be considered as a proposed baseline strategy. The planner of the VVT process is expected to determine an individual level of VVT performance (in the range of 0–100%) for each potential VVT activity.

TABLE 3.1 Proposed Baseline VVT Strategy for Production Phase

Activity Number	VVT Production Activity	Performance Level
Prepare VVT Products		
1	Generate a FAI procedure	
2	Create system Production Test Procedure (PTP)	
3	Validate the production line test equipment	
Apply VVT to Engineering Products		
1	Verify quality of incoming components and subsystems	
2	Perform FAI	

TABLE 3.1 *Continued*

Activity Number	VVT Production Activity	Performance Level
3	Validate preproduction process	
4	Validate ongoing production process	
5	Perform manufacturing quality control	
6	Verify the production operations strategy	
7	Verify marketing and production forecasting	
8	Verify aggregate production planning	
9	Verify inventory control operation	
10	Verify supply chain management	
11	Verify production control systems	
12	Verify production scheduling	
Participate / Conduct Reviews		
1	Participate in FCA	
2	Participate in PCA	
3	Participate in Production Readiness Review (PRR)	

- *Production VVT Schedule.* Plan the production VVT schedule. Production engineering activities and the major milestones shall be identified on Gantt and Program Evaluation Review Technique (PERT) charts, together with the planned production VVT activities as identified above.
- *Production VVT Management.* The VVT organization structure supporting the production phase should be identified and include (1) responsibility of each participating organization involved in the VVT process and (2) identification of subcontractor roles and responsibilities.
- *Production VVT Limitations.* Describe specific limitations that may significantly affect the production VVT plan as well as the expected financial and schedule impact of these limitations. In particular, consider the following issues: (1) resources availability (e.g., manpower, facilities, equipment, funding, schedule) and (2) safety issues (e.g., human health hazards, facilities and equipment protection).
- *Production VVT Personnel and Training.* Identify the required manpower and personnel as well as their training needs for properly carrying out the production VVT plan.
- *Production VVT Sites/Facilities.* Identify the specific sites and facilities needed to carry out the production VVT activities.
- *Production VVT Support Equipment.* Identify the specific test support equipment required to carry out the production VVT plan.
- *Production VVT Expendables.* Identify the type, number and availability requirements for all expendables required to carry out the production VVT plan.

- *Production VVT Budget.* Determine the budget required for performing the identified production VVT activities during the course of the VVT plan.

Methods and Further Literature

Section 4.3.1, VVT process planning

- Bothe (1997)
- Brauer and Cesarone (1991)
- Loch et al. (2003)

3.2.4 Generate a First Article Inspection (FAI) Procedure

Objective The objective of this activity is to create a FAI procedure. FAI provides objective evidence that all engineering design and specification requirements are properly understood, accounted for, verified and documented, so once the inspection has been carried out successfully, system production can commence.

Description The FAI refers to actions that are necessary to maintain high quality and verify the features and characteristics of a material, process, product, service or activity to specified requirements. FAI may be characterized as the analysis of the first item built during the Production phase to confirm correct setup and process configuration. In other words, FAI helps organizations to ensure and review proper documentation of design characteristics, manufacturing parts, referenced exhibits, drawing requirements and product specifications. Having proper documentation helps manufacturers in (1) understanding the appropriate production methods, (2) accounting for all parts of development, (3) verifying the process for reproduction and (4) reporting the findings for management visibility. When complex and critical systems are created, it is of the utmost importance that they are built correctly and repeatedly. Making a mistake in this process could jeopardize people's lives and property. Some of the basic information within an FAI document should includes the following:

- Product name and number
- Specification requirements
- Dimensional measurement
- Detailed statistical analysis
- Design characteristics
- Easy-to-read customer reports

The following proposed FAI procedure is based on: Military Standards DI-NDTI-81307A—Furst Article Qualification Test Plan and DI-QCIC-81110—Inspection and Test Plan, as well as the Society of Automotive Engineers standard SAE-AS9102A—Aerospace First Article Inspection Requirement.

Proposed Procedure: System Level First Article Inspection

1. ***Purpose.*** The following specifies the FAI procedure for verification that a system can be manufactured, assembled and tested in accordance with the prerequisite specifications and drawings with respect to production scheduling, job sheets, production resources and staff skills.
2. ***Field of Application.*** This procedure applies to manufacturing, assembly and inspection of initial production as a basis for subsequent serial production. The FAI must be carried out for new systems, new producers, relocation of production and significant modifications to the design or procedure and after lengthy interruptions of production.
3. ***Definition of "First Article."*** The first article is an assembled system from the pilot production run, first produced with the facilities and processes and under the conditions anticipated for serial production.
4. ***Responsibility.*** The producer shall be responsible for (1) manufacturing and testing of products in accordance with the technical specifications, contractual agreements, approved quality assurance scheduling, approved procedures and manufacture and test scheduling and (2) implementation of the FAIs and issue of appropriate reports.
5. ***Procedure.*** The actual FAI procedure shall be comprised of the following elements: (1) the inspection process itself, (2) documentation of the process and its results, (3) deviation handling, (4) representative witnesses at the inspection, (5) subsequent FAI requirements and (6) final system acceptance.

   5.1. *Inspection.* The first serial-produced system must be fully inspected, ensuring the following:

   a. Accuracy and integrity of manufacture and test scheduling
   b. Configuration conformity
   c. Use of the correct material or parts for production or assembly
   d. Correct heat treatment appropriate to the base material
   e. Conformity of the dimensions of the features to the relevant drawings
   f. Conformity of the surface treatment requested
   g. Implementation of the nondestructive testing requirements
   h. Implementation of the test requirements
   i. Meeting interchangeability/replace-ability requirements
   j. Marking of parts in accordance with the requirements of the specifications

k. Conformity to the specifications in accordance with the drawings
l. Conformity to the procedural specifications and monitoring of procedures
m. Implementation of the procedures by approved personnel using approved facilities
n. Compliance with any additional customer's purchasing requirements
o. Ability of the production machinery to produce acceptable parts
p. Conformity to the specifications regarding serviceability of the test gauges
q. Verification of the manufacturing and testing software used
r. Compliance with the acceptance inspection conditions

5.2. *Documentation.* Documentation needs are:

a. The FAI must be completely documented in the First Article Inspection Report (FAIR).
b. All applicable requirements under Section 5.1 must be formally confirmed.
c. The production and test schedule documents, test specifications and procedural instructions that are subject to approval must be listed.
d. All the main manufacturing and testing resources must be listed.
e. All test figures, measurements and other results obtained during the inspection must be recorded.
f. The first inspected article must be identified in order to enable a subsequent inspection to be carried out.
g. One copy of the fully completed FAI report is to be submitted to the customer with the first article.

5.3. *Deviations.* If any deviation is established during the FAI, preventing conformity to the technical specifications or the purchase requirements, corrective action must be taken. All such deviations must be recorded in a nonconformity report. A corrective action must be specified before acceptance of the FAI report.

5.4. *Representatives.* Selected representatives (e.g., customer's quality assurance, system's licenser or certifier, prospective clients) should be present at the FAI and confirm the orderly conduct of the inspection. The producer must coordinate the FAI procedures and inform the customer's quality assurance in advance of the scheduled FAI.

5.5. *Subsequent FAI Requirements.* Subsequent FAI requirements are:

a. If system's features are modified or added, the customer may request a partial FAI for the first system with the new configuration. The new FAI should cover only the modified or added features.

b. If a change in manufacturing capability of sufficient gravity is reported or established, the customer may demand that the first article manufactured after that change be subjected to a full or partial FAI.

c. Should significant system problems be discovered at the customer site, causing a significant rise in the rate of failures, the customer may instruct a partial or full FAI to guarantee the quality of the supplied systems.

d. The following definitions are used to specify the nature and importance of a production change vis-à-vis the FAI:

- *Change in Facilities.* Change in processing equipment, machinery, tools, adjustment and testing gauges, testing resources or processing facilities.
- *Change in Processes.* Change in the manufacturing and testing methods or process parameters.
- *Change in Personnel.* Change in the staff members that carry out the manufacturing, process, installation or testing operations with special monitoring, so that there is a major change in the group of persons carrying out the work, requiring prior training and teaching of skills courses.
- *Change in Location.* Full or partial relocation of production. A change in location may, but need not, include a change in facilities, procedures or staff.
- *Change in Producer.* Such changes concern the shift of implementation of procedures from the producer to a subcontractor or from the subcontractor to the producer or from one subcontractor to another subcontractor.

5.6. *Final System Acceptance.* Release and acceptance of the system (e.g., pilot production range, serial production batches) shall take place on approval of the FAI report.

Methods and Further Literature

Section 5.7.1, Sanity testing Section
Section 5.7.2, Exploratory testing (FAI)

- Bossert (2004)
- DI-NDTI-81307A (2006)
- DI-QCIC-81110 (1990)

Section 5.7.9, First article inspection

- Geng (2004)
- SAE-AS9102A (2004)

3.2.5 Validate the Production-Line Test Equipment

Objective The objective of this activity is to verify the status of the production line test equipment and to calibrate and test the test equipment, on a regular basis, in order to reduce risk of production line failure.

Description The production line test equipment should be regularly calibrated and validated as part of the production process. The production line test equipment refers to the physical devices that take measurements of products and processes, closing the information loop in order to make decisions about possible modifications in the process. The validation of test equipment can be classified as mitigating strategy risk and must be carefully undertaken in order to optimize this validation process. The main technical characteristics to be considered for the testing equipment are:

- Reliability
- Maintainability (calibration)
- Precision
- Resistance
- Safety

The test equipment must be calibrated and tested under real production conditions. It is recommended that the most critical precision equipment (e.g., gauges) should be calibrated by external laboratories.

Methods and Further Literature

Section 4.3.5, Failure mode effect analysis
Section 4.3.6, Anticipatory failure determination

- Bossert (2004)
- Geng (2004)

Section 5.4.3, Recovery testing
Section 5.7.1, Sanity testing
Section 5.7.2, Exploratory testing
Section 5.7.3, Regression testing

- Jones (1998)

3.2.6 Verify Quality of Incoming Components and Subsystems

Objective The objective of this activity is to verify that incoming materials (i.e., inventory used in the manufacturing process), components or subsystems meet specifications before they are embedded into the produced system.

Description Materials, components and subsystems to be incorporated into a product (i.e., system) must be checked before they are integrated into the system since the system depends strongly on the quality of its parts. The objective of checking the received components and subsystems is to verify that they meet the required specifications. This activity will reduce costs since faulty systems detected further along the production line would lead to expensive corrective action.

Methods and Further Literature

Section 4.3.5, Failure mode effect analysis
Section 4.3.6, Anticipatory failure determination
Section 4.4.1, Expert team reviews
Section 5.4.3, Recovery testing
Section 5.7.1, Sanity testing
Section 5.7.2, Exploratory testing
Section 5.7.3, Regression testing

- Juran and Godfrey (2000)
- Stephens (2001)

3.2.7 Perform First Article Inspection (FAI)

Objective The objective of this activity is to provide objective evidence that all engineering design and specification requirements applicable to a first article manufactured in a production line are properly understood, accounted for, verified and well documented.

Description The FAI should be carried out in accordance with the FAI plan described above. As mentioned, the FAI process consists of a complete, independent and documented physical and functional inspection process to verify that prescribed production methods have produced a fully conforming first article product, as specified.

The first article should be produced on production equipment and using processes which will be utilized on production runs. Subsequent repeated FAIs should be conducted following every major tooling or design change and subsequent to any evident quality degradation for a specific article, component, subsystem or system.

The inspection records and data should identify each characteristic and feature required by design data, the allowable tolerance limits and the actual dimension measured as objective evidence that each characteristic and feature have been inspected and accepted. When testing is required, the parameters and results of the test should also be recorded for the same purpose.

Methods and Further Literature

Section 4.2.5, Classification tree method
Section 4.2.6, Design of experiments (DOE)
Section 4.3.5, Failure mode effect analysis
Section 4.3.6, Anticipatory failure determination

Section 5.4.3, Recovery testing
Section 4.3.8, Robust design analysis
Section 4.4.1, Expert team reviews
Section 5.7.1, Sanity testing
Section 5.7.2, Exploratory testing
Section 5.7.3, Regression testing
Section 5.7.9, First article inspection (FAI)

- Bothe (1997)
- Brauer and Cesarone (1991)
- Loch et al. (2003)
- SAE-AS9102A (2004)

3.2.8 Validate Pre-Production Process

Objective The objective of this activity is to guarantee, to a reasonable extent, the preproduction validation of product and process quality as well as compliance with national and international regulations.

Description The validation of the preproduction product quality and process must follow a set of rules that emanate from the system's specification and, sometimes, from existing national and international regulations. The intent here is to validate the production system before starting full-scale production. Specifically, this entails validating the quality of products and the production process at the earliest possible time after constructing the manufacturing line.

- *Product Quality Validation.* Product quality is intended as conformity to the supply conditions (e.g., geometrical parameters, dimensional tolerances, material characteristics, absence of defects) defined for the system. Usually validation of the product quality is carried out by the customer of the system through a specific "formal review." The customer can be internal (e.g., the manufacturing plant that receives the production system from the development department of the same company) or external (e.g., a car manufacturer plant that receives an engine component from a vendor producer).
- *Process Quality Validation.* Evaluating a mass production process often involves the use of a pilot plant. The pilot plant is equipped with the final production lines, so it is possible to carry out the tests without interference with live production lines. The verification and validation conditions are the same as those in the real plant. This further verification is usually carried out in the presence of all the relevant producers, each of them controlling the correct production/assembly of the component, integrated within the final process.

  Validation of process quality is measured in terms of process performance such as production efficiency (production capacity) and acceptable waste production where the percentage of scrap material must be under a defined threshold. In addition, all the necessary documentation must be generated [e.g., Failure Mode Effect Analysis (FMEA)] in conformance with the quality procedure of the company. At the end of this activity, the production reliability is validated and the production process is certified.

The first set of products built in the pilot plant is used to definitively validate the production process. When the equipment has been tested and the process performance is acceptable, the responsibility for product quality formally passes from the development team to the production team.

- *National and International Regulation Compliance.* Certification requires that a recognized third-party organization (e.g., not the producer or the retailer) attests that a product, a process or a service is in compliance with dispositions, or "essential" requirements, fixed from the technical directives concerning the environment, health, safety and security. Usually a product is compliant if it meets relevant international and national standards. When there is no specific disposition, the conformity is determined from national norms and these dispositions allow the commercialization and circulation of the product.

  In some cases the pioneers in one sector or the most skilled producer define a de facto reference standard that can be recognized by successive producers of the product. Sometimes the market defines a reference product that is universally recognized. It is important to emphasize that regulation conformity appraisal procedures are most often directed to eliminating potential threats to life or well-being.

Methods and Further Literature

Section 4.3.4, System test simulation
Section 4.3.5, Failure mode effect analysis
Section 4.3.6, Anticipatory failure determination
Section 4.3.7, Model-based testing
Section 4.4.3, Group evaluation and decision
Section 5.7.9, First article inspection (FAI)
Section 5.7.10, Production testing

- Bothe (1997)
- Brauer and Cesarone (1991)
- Loch et al. (2003)

3.2.9 Validate Ongoing-Production Process

Objective The objectives of this activity are three-fold: (1) to continuously monitor and validate the production tools and process, (2) to assess ways and means to reduce production cycle time and cost and (3) to evaluate, on an ongoing basis, the manufactured products and systems and to ensure that they fulfill their specified roles.

Description During the Production phase, assessment of the production tools, production processes and resulting products or systems should be undertaken on a continual basis. The intent is to identify faulty products as soon as possible and to improve tooling and processes over time. Continual product modification and improvement requires that the production tools and production processes be updated regularly. In particular, the quality acceptance procedures should be fitted and harmonized before introducing a new version of the product into production.

Throughout the production phase and as a general rule, a sample of each product leaving the assembly line should be tested to verify proper behavior. This activity is required despite the ongoing process control activity as there is still uncertainty about the quality of the produced systems. In addition, failure diagnoses from defective products are useful for process correction planning and improvement. The decision about how much product validation should be performed must be taken after considering other information sources about the product. More specifically and depending upon the situation, one of the following levels of validation may be appropriate:

- *No Validation.* There is sufficient statistical evidence that the product fulfills its specified requirements (i.e., the cost of validation outweighs the risk of no validation).
- *Small-Sample Validation.* There is good historical data on the product that can be confirmed with limited sampling. Without sufficient historical data on the product, small product samples are not enough to draw conclusions, since a batch with as many as 30% defective products may not be detected.
- *Large-Sample Validation.* When there is no substantial previous knowledge of the product, the only way to reliably determine product quality is validation by random sampling. The final decision about how many samples are required depends on economic considerations as well as on the acceptable level of defects in the delivered product. Economic considerations include the cost of validation (which is easy to estimate) and the expected cost resulting from faulty products (which is more difficult to estimate).
- *Complete Validation.* This is the appropriate option for (1) critical system components or subsystems, (2) complex systems or (3) situations when the production process may have difficulty meeting the product specifications. In very critical cases, even more than "complete" validation is attempted as a precaution against the possibility of failure in the validation process itself (This is sometimes called *redundant validation*). When the objective is "zero defects," due to safety, commercial, legal or political reasons, complete validation is attempted (but in reality seldom achieved).

Methods and Further Literature

Section 4.3.4, System test simulation
Section 4.3.5, Failure mode effect analysis
Section 4.3.6, Anticipatory failure determination
Section 4.3.7, Model-based testing
Section 4.4.3, Group evaluation and decision
Section 5.7.9, First article inspection (FAI)
Section 5.7.10, Production testing

- Belytschko et al. (2000)
- Chandra and Mukherjee (1997)
- Ogata (2003)
- Zienkiewicz and Morgan (2006)

3.2.10 Perform Manufacturing Quality Control

Objective The objective of this activity is to perform manufacturing quality control for all the relevant production lines.

Description Manufacturing quality control has traditionally been associated with measuring various products and process parameters and evaluating these parameters for consistency over time. This approach stems from the concept that considers manufacturing quality as "conformance to requirements." Quality pioneers like Walter Shewhart (1986) and Edward Deming (2000) and others established the concept of Statistical Process Control (SPC) and Statistical Quality Control (SQC) as vehicles to follow product quality and ensure conformance throughout the manufacturing process.[14] Several types of control charts (see, e.g., Figure 3.2) are often generated in order to visualize behavioral aspects of the production system.

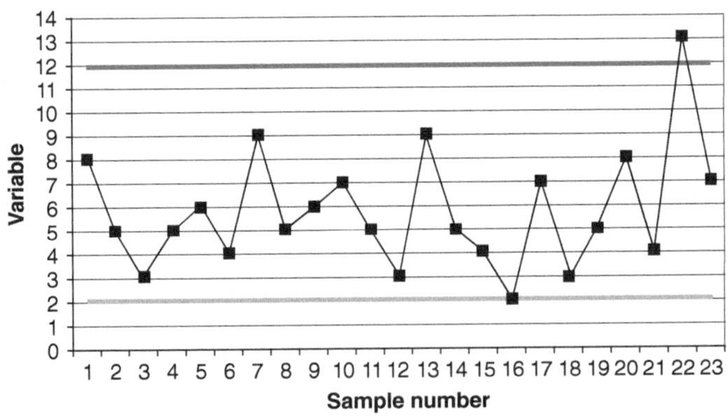

Figure 3.2 Example layout of manufacturing control chart.

[14] Readers should distinguish between statistical process control and statistical quality control. Both methods utilize control charts for evaluating manufacturing; however, SPC is based on process parameters (e.g., measurements of performance such as time, speed and continuity) of production line and equipment, whereas SQC is based on product parameters (e.g., physical dimension, weight, color or other attribute). The basic idea is that a controlled and stable process will produce conforming products.

Typically, the VVT team will perform the following:

Proposed Procedure: Manufacturing Quality Control

Step 1: Planning. The VVT team will define the statistical quality and process control parameters appropriate for the manufacturing plant. These include, among others, the type and size of the product samples as well as the rate of sample collection. In addition, determine (1) which production quality failures would require production intervention (i.e., correcting or adjusting the production process) and (2) the type of control charts the organization would find appropriate for monitoring production. Most common control charts are:

- *X Control Chart.* An X control chart is used to determine the shift in the mean value of a process.
- *R Control Chart.* An R control chart is used to determine the shift in the variance of a process.
- *p Control Chart.* A p control chart is used to determine the shift in a process based on a true proportion of defective elements within a sample. Such charts are appropriate when classifying any given product as either suitable or faulty.
- *c Control Chart.* A c control chart is used to determine the shift in a process based on a number of defects found in individual products. Such charts are appropriate when products can be permitted to have certain levels of minor defects.

Step 2: Sampling. The VVT team will collect appropriate parameters and product samples from the production line and on a regular basis measure the defined relevant parameters. Thereafter, update the various control charts and determine the status of the manufacturing/production line.

Step 3: Optimizing. The intent of quality control in manufacturing is to reduce operating costs by preventing the propagation of defective products through the manufacturing plant and into customer hands. The VVT team must balance between these costs and the cost emanating from performing manufacturing quality control. Here is a summary of this optimization problem:

- *Out-of-Control Cost.* When a manufacturing plant operates without adequate controls, the likelihood of manufacturing defective components and systems increases. The resulting defective products must be repaired or scrapped, which is costly. If, in addition, these defective products were already inserted into larger assemblies, then the cost of extracting/reinserting would add to the cost of product failure. Worst of all, if defective products were used by customers, they might cause harm to people or property, resulting in warranty payments and sometimes lawsuits.

- *Manufacturing Quality Control Costs.* The manufacturing quality control costs may be divided into two categories: (1) sampling and charting cost and (2) failure identification cost:
  a. *Sampling and Charting Cost.* Sampling-and-charting cost involves employing people to extract product samples from the production line, measuring their relevant parameters, inserting the data into a computer and running analyses as needed. Sometimes the sample itself is destroyed in the testing process, which adds to the sampling cost.
  b. *Failure Identification Cost.* When the production process appears to be out of control, the cause for this phenomenon must be determined. Sometimes, the problem stems from an incorrect sampling or charting process. At other times, the production process is indeed out of control, in which case the relevant manufacturing cell or the entire production line must be halted and the specific problem identified and resolved. The cost of VVT personnel involved in the identification process as well as halting production and fixing the problem is obviously quite high.

Methods and Further Literature

Section 4.4.1, Expert team reviews
Section 5.7.9, First article inspection (FAI)
Section 5.7.10, Production testing

- Deming (2000)
- Geng (2004)
- Kalpakjian and Schmid (2005)
- Nahmias (2004)
- Shewhart (1986)
- Tanner (1990)

3.2.11 Verify the Production Operations Strategy

Objective The objective of this activity is to verify the production operation strategy of the manufacturing organization.

Description A production operation strategy is the approach taken by organizations to deploy its resources in order to obtain stated economic and societal goals. The purpose of the VVT actions is to verify the chosen operation strategy in light of the organization's goals. Typically, the VVT team will:

1. Verify that the producer has a clear vision statement elaborated, in a formal (written) way, in its mission statement.
2. Verify that the producer has a clear operation strategy, which includes the following:
   - *Strategy Time Horizon.* Verify that all operation strategies are designed for short-, medium- or long-term implementation, where the strategy time horizon is the length of time required for operation strategy decisions to affect the firm.
   - *Strategy Focus.* Verify that the manufacturing strategy focus is optimally appropriate for the organization and matches the firm's vision statement. In general, this may include (1) adjusting the strategy to market demands (e.g., price levels, required lead time, product reliability), (2) adjusting the production volume at any given period according to projected needs, (3) ensuring an appropriate overall product quality, (4) selecting the appropriate manufacturing mix for each manufacturing location and (5) choosing the optimal manufacturing process technology, that is, balancing technology advantages and risks.
   - *Strategy Consistency.* Normally the term strategy refers to a multitude of company policies, procedures, rules and decisions that affect the entire production organization. This set should be verified for overall consistency. Consistency concerns include (1) clear definition of manufacturing tasks and production capacity, (2) dynamic product proliferation and (3) evolving personnel tasks and responsibilities.
   - *Strategy Evaluation.* Periodical evaluation of the firm's production operation strategy, especially in terms of product cost and quality as well as the overall profitability of the organization and customers' satisfaction.
3. Verify that the firm periodically rejuvenates itself and considers new strategic initiatives. This is in response to new production operation techniques that emerge from industry or academia that would be considered appropriate by the producer's management. Examples of such manufacturing operation strategic initiatives are:
   - *Just-In-Time (JIT).* JIT is a strategy based on establishing close working relationships with suppliers, ensuring a high quality of incoming material components and subsystems and maintaining minimal levels of inventory. The effectiveness of the JIT strategy should be evaluated within organizations adhering to this strategic initiative.
   - *Time-Based Competition (TBC).* TBC is a strategy in which the entire value delivery system is considered. The intent is to minimize the time required for introduction of new features and innovations into the market. The effectiveness of the TBC strategy should be evaluated within organizations adhering to this strategic initiative.
4. Verify that there is an appropriate planning for manufacturing capacity growth. Such planning will determine the ability of the

manufacturing plant to deliver the optimal number of products or systems in the future and thus is critical in ensuring the commercial viability of the organization. Evaluate the capacity growth plan to verify:

- *Planning Factors.* Typical capacity growth planning factors are (1) appropriate prediction of demand patterns, (2) cost of maintaining current plants and construction of new ones, (3) economical ramifications of introducing new technologies and manufacturing processes and (4) information about competing manufacturers.
- *Capacity Change Issues.* If a manufacturing capacity growth plan calls for changes in current manufacturing capacity (either increase or decrease), then it should be further evaluated for an appropriate approach in terms of (1) the specific volume of planned production increase/decrease in capacity, (2) the location where the increase/decrease must take place and (3) the timing of the change.

Methods and Further Literature

Section 4.3.4, System test simulation
Section 4.3.5, Failure mode effect analysis
Section 4.3.6, Anticipatory failure determination
Section 4.4.1, Expert team reviews
Section 4.4.3, Group evaluation and decision

- Geng (2004)
- Kalpakjian and Schmid (2005)
- Nahmias (2004)
- Tanner (1990)

3.2.12 Verify Marketing and Production Forecasting

Objective The objective of this activity is to verify the marketing and production forecasting of the manufacturing organization.

Description Marketing and production forecasting is a mechanism to predict sales of products and systems and to plan future production operations. The purpose of the VVT actions is to verify that these forecasts are performed under a sound process and produce reliable and accurate results. Typically, the VVT team will:

1. Verify that the firm utilizes a well-defined mechanism for marketing and production forecasting which is evaluated periodically and typically includes the following time horizons:
   - *Days/Weeks.* Verify that a short-time forecasting is utilized dealing, typically, with near-term sales, minor manufacturing schedule shifts and immediate resources allocations.

- *Weeks/Months.* Verify that an intermediate-time forecasting is utilized dealing, typically, with forecasting future labor force requirements, overall plant maintenance, intermediate-term resource requirements and the like.
- *Months/Years.* Verify that long-term forecasting is utilized dealing, typically, with long-term capacity needs as well as expected long-term sales pattern and growth trends.

2. Verify that the firm utilizes a well-defined subjective (i.e., based mostly on human judgment) forecasting method; for example:
   - *Customers' Survey.* Verify that formal and informal customers' surveys are conducted regularly in order to determine customers' preferences and expectations.
   - *Sales Force Composites.* Verify that a long-term forecast regarding customers' preferences and expectations is solicited from the organization's sales force.
   - *Management Survey.* Verify that formal and informal management surveys are conducted in order to independently forecast customers' preferences and expectations.
3. Verify that the firm utilizes a well-defined objective (i.e., based on formal data analysis) forecasting method; for example:
   - *Time Series Methods.* These methods predict future behavior based on historical behavior. Verify that short-, intermediate- and long-term forecasts are derived by analyzing time series date to predict (1) behavior trends, (2) cyclical variations, (3) seasonal patterns and (4) no pattern (i.e., only randomness in the time series).
   - *Causal Models.* These methods use data from other sources [e.g., inflation rate, unemployment level, Gross Domestic Product (GDP), exchange rate, consumers' confidence parameters] to forecast future marketing and production parameters.[15] The accuracy of these models and the validity of their input data should be verified.
4. Verify that both the subjective forecasting data sets obtained from the above sources (i.e., customers' surveys, sales force composites and management surveys) and the objective forecasting data sets (obtained through time series methods or causal models or some other method) are correctly aggregated into a single coherent forecast utilizing relevant weights for each set of raw data.

[15] Readers may wonder why the Consumer Price Index (CPI), Gross Domestic Product (GDP) and employment numbers run counter to their personal and business experiences. The problem lies in biased and often manipulated government reporting throughout the western world. Readers should seek ungimmicked parameters with which to base their marketing and production forecasting (see for example: http://www.shadowstats.com/).

5. Verify that the firm formally evaluates the accuracy of the forecasting on an ongoing basis. Two common methods to evaluate the accuracy of forecasting and therefore to improve the forecasting ability of the organization are (1) the Mean Absolute Deviation (MAD) and (2) the Mean Square Error (MSE) between a given forecast and the actual performance data.

Methods and Further Literature

Section 4.3.4, System test simulation
Section 4.3.6, Anticipatory failure determination
Section 4.4.1, Expert team reviews
Section 4.4.3, Group evaluation and decision

- Geng (2004)
- Kalpakjian and Schmid (2005)
- Nahmias (2004)
- Tanner (1990)

3.2.13 Verify Aggregate Production Planning

Objective The objective of this activity is to verify the aggregate production planning of the manufacturing organization.

Description Aggregate production planning is the process of determining how many products or systems are going to be produced and in what mix as well as how many employees are needed at each skill level for a given planning horizon. The purpose of the VVT actions is to verify the aggregate production planning in light of the organization's goals and the marketing and production forecasting. Typically, the VVT team will:

1. Verify the multifaceted handling of the aggregate production-planning problem. This entails the following:
   - *Resource Smoothing.* Verify that the aggregate production planning considers the multitude cost trade-offs associated with changes in production workforce levels.
   - *Production Bottlenecks.* Verify that production bottlenecks are, in fact, eliminated or minimized. Such bottlenecks may result from inadequate production level due to a transitory surge in demand, lack of key resource, machinery failure and so on.
   - *Planned Horizon Determination.* Verify that the planned horizon is determined reasonably and in accordance with market and production plant conditions. In general, rolling schedules are often utilized.
   - *Demand Variation.* Verify that the aggregate production planning considers numerous variations between marketing forecasts and actual sales at any given time. Also verify that the production planning provides an appropriate level of buffer to handle forecast errors.

2. Verify that the aggregate production planning is optimized to minimize typical production waste costs. This entails the following:
   - *Cost of Smoothing.* Verify that the aggregate production plan minimizes the costs emanating from recurring changes in production levels and, in particular, the size and mixture of the workforce.
   - *Cost of Inventory.* Verify that the aggregate production plan minimizes the costs emanating from tying up capital in inventory. At the same time, verify that the planned level of inventory will not lead to undesired cost of shortage, that is, the cost emanating from lack of needed inventory.
   - *Cost of Unit Production.* Verify that the aggregate production plan considers the realistic production cost of each unit, product or system. This cost is composed of direct and indirect personnel cost, material and other manufacturing expenses.
   - *Cost of Plant Underutilization.* Verify that the aggregate production plan considers realistic underproduction costs emanating from occasional delays in deliveries of raw materials, components, subsystems and other supplies, failures of machinery and production lines, underutilization of the workforce and the like.

Methods and Further Literature

Section 4.3.1, VVT process planning
Section 4.3.4, System test simulation
Section 4.3.5, Failure mode effect analysis
Section 4.3.6, Anticipatory failure determination
Section 4.4.1, Expert team reviews
Section 4.4.3, Group evaluation and decision

- Geng (2004)
- Kalpakjian and Schmid (2005)
- Nahmias (2004)
- Tanner (1990)

3.2.14 Verify Inventory Control Operation

Objective The objective of this activity is to verify the inventory control operation of the manufacturing organization.

Description Inventory control is the process of optimizing the quantity of inventory within a manufacturing organization. In general, demand for inventory emanates from customer purchases of end products or systems as well as the demand for raw materials, lower level assemblies and components needed by the various manufacturing entities. The inventory control problem is a complex one, since demand is not constant and not known a priori, whereas filling inventory needs must be undertaken at earlier stages. Therefore, inventory control operation deals primarily with the problem of the type and quantity of inventory needed and when to purchase it. The purpose of the VVT actions with regards to this matter is to verify the inventory control operation, in light of the organization's goals and the marketing and production forecasting. Typically, the VVT team will:

1. Verify that inventory control distinguishes between different types of inventories. This usually entails the following:
   - *Raw Material Inventory.* Verify that all basic materials required for the production process are considered.
   - *Work-In-Process (WIP) Inventory.* Verify that all WIP that is currently in production throughout the manufacturing plant is considered.
   - *Components and Subsystem Inventory.* Verify that all components and subsystems that have been completed and are waiting for further integration into larger systems are considered.
   - *End Products and Systems.* Verify that all completed products and systems which have been accumulated within the manufacturing plant or are in transit (i.e., to distribution centers or to customers) or, in general, are not being delivered to customers are considered.
2. Verify that inventory is being optimally refilled at all times in order to meet the organization's goals as well as marketing forecasts and production plans. This usually entails the following:
   - *Response to Uncertainties.* Verify that an inventory analysis regarding uncertainties (e.g., customer demand, supply availability, inventory lead time) has been conducted and a well-balanced inventory control strategy has been devised and implemented. In particular, smoothing changes in demand patterns due to anticipated factors like seasonality can reduce inventory through comprehensive aggregate production planning.
   - *Economies of Scale.* Verify that the inventory mix and quantity are designed to match production runs. In addition, verify that inventory is optimally obtained (e.g., purchased, transported) due to economies of scale.
   - *Market Considerations.* Verify that inventory control is designed to consider economic market opportunities as they arise. This may be accomplished by, for example, increasing inventory when a price rise is anticipated or decreasing inventory when the cost of capital is expected to increase.
   - *Pipeline Inventories.* Pipeline inventories cover raw material and components that are acquired from outside sources as well as subassemblies and subsystems that are shipped among production cells or sometimes individual manufacturing plants. Pipelined inventory refers also to finished products or systems transported to customers and

markets in general. Verify that the economic effects of inventory transport are carefully considered in the inventory control operation. Sometimes, changing suppliers or reorganizing production distribution configuration may be prudent.

3. Verify that inventory control operations differentiate inventories according to typical characterizations. This usually entails the following:
   - *Demand Inventory.* Verify that inventory control operations identify inventory that is characterized as demand dependent. Such inventory should constitute a response to variations in internal production demand levels as well as the erratic nature of external end products and system demand.
   - *Lead Time Inventory.* Verify that inventory control operations identify inventory requiring explicit lead time to be fulfilled. Such inventory should constitute a response to the elapsed time that takes place from ordering certain items until they are available at the assembly line of the manufacturing plant.
   - *Limited Lifespan Inventory.* Verify that inventory control operations identify inventory items having limited lifespan. For example, drugs, foods, various chemicals and other perishable goods have inherently limited shelf life. Sometimes, machinery spare parts become obsolete once these machines or systems conclude their lifecycle.
   - *Unfulfilled Inventory.* Verify that the inventory control operations recognize the characteristics of unfulfilled or excess demand inventory (i.e., needed inventory which is unavailable at a given time). Unfulfilled inventory may be manifested at supplier chains, at the manufacturing plant or at the end-customer retail level. In general, such unfulfilled inventory will either be satisfied at a later date (back ordered) or be lost (probably fulfilled by other sources).
4. Verify that inventory control operations differentiate inventories according to their cost characteristics. This usually entails the following:
   - *Carrying Cost.* Verify that inventory control operations identify the carrying cost of the inventory. Carrying cost is usually directly proportional to the amount and mix of the inventory and by and large includes storage and insurance as well as certain levels of break and tear typical of any inventory. In addition, the cost of cash tied up in the inventory should also be considered.
   - *Order Cost.* Verify that inventory control operations identify the order cost of the inventory. Order cost depends on the amount or the size of ordered inventory. Often, order cost is composed of a fixed component representing "order set-up cost" and a variable component which is computed on a "per-item cost."
   - *Penalty Cost.* Penalty cost is described as cost emanating from either delivering defective products or lost sales due to reasons such as

product unavailability and late delivery of products, leading in general to customer dissatisfaction. The VVT team should verify that the inventory control operations identify penalty cost and properly estimate its economic effect on the manufacturing operations.

Methods and Further Literature

Section 4.3.4, System test simulation
Section 4.3.5, Failure mode effect analysis
Section 4.3.6, Anticipatory failure determination
Section 4.4.1, Expert team reviews
Section 4.4.3, Group evaluation and decision

- Geng (2004)
- Kalpakjian and Schmid (2005)
- Nahmias (2004)
- Tanner (1990)

3.2.15 Verify Supply Chain Management

Objective The objective of this activity is to verify the supply chain management of the manufacturing organization.

Description Supply chain management may be defined as the management of materials, information and financial flows in networks consisting of producers, manufacturers, distributors and customers. Supply chain management attempts to optimize the flow of raw materials, products and systems as well as information and money between suppliers and manufacturers within the manufacturing entity and between manufacturers and customers. The purpose of the VVT actions is to verify that the supply chain management is optimally efficient in light of the organization's goals and the marketing and production forecasting. Typically, the VVT team will:

1. Verify that all goods in the manufacturing network are transported in an efficient way. This usually entails verifying the optimal scheduling and flow of:
   - Raw materials and required components from suppliers to the manufacturing plants
   - Subassemblies and subsystems among manufacturing cells and production plants
   - Final products and systems from various manufacturing plants into warehouses and final market distributions
2. Verify that the products and systems are designed, among other characteristics, to support efficient supply chain strategy. This usually entails verifying the following two design characteristics:

- That products and systems, especially bulky ones, are designed to permit transportation in parts and then be assembled at the final destination.
- That products and systems are designed to allow postponing, as much as possible, their final configuration. This strategy supports late product variation and modifications due to evolving market conditions or customer requirements.

3. Verify that the supply chain system includes effective electronic commerce capability. Beyond the use of standard commerce enabling tools such as emails and public and privet Web services, verify that the organization uses satisfactory supply chain resources; for example:
   - *Electronic Data Interchange (EDI).* Verify the effective real-time use of regular, computer-to-computer, business transactions both within the organization and between the organization and its suppliers, distributors, customers and other relevant entities.
   - *Web-Based Transaction Systems.* Verify the effective real-time use of Web-based transaction systems for both Business-to-Customers (B2C) and Business-to-Business (B2B) applications.

Methods and Further Literature

Section 4.3.4, System test simulation
Section 4.3.5, Failure mode effect analysis
Section 4.3.6, Anticipatory failure determination
Section 4.4.1, Expert team reviews
Section 4.4.3, Group evaluation and decision

- Geng (2004)
- Kalpakjian and Schmid (2005)
- Nahmias (2004)
- Tanner (1990)

3.2.16 Verify Production Control Systems

Objective The objective of this activity is to verify the production control systems of the manufacturing organization.

Description Production control is the approach used by the organization to obtain raw material and components for the manufacturing process as well as move products and subassemblies within the manufacturing plant. Often, manufacturers select either the Material Requirements Planning (MRP) or Just-In-Time (JIT) approaches.

1. The MRP approach is based on an estimation of the number and mix of end products per unit of time as well as the structure or subassemblies of these products or systems. If the organization is using the MRP approach, then the VVT team should:

- Verify that the Master Production Schedule (MPS), stating the schedule, amounts and mix of all end products needed for a given production horizon, is based on up-to-date, known customer orders and realistic forecasts for future end-product demands, seasonal variations, safety stock considerations and so on.
- Verify that the MRP stating the exact quantity of each individual component needed in the production process, reflects accurately the latest definition of each end-product and system and takes into account yield parameters related to incoming material and components as well as production process yield.
- Verify that the Job Shop Production Schedule (JSPS), stating the scheduling and utilization of each production cell subject to various production line limitations, is sound. The JSPS is a complex problem since, in the real world, there are always various uncertainties and constraints. Verify that the JSPS provides a robust scheduling optimization elucidation that is based on a realistic model using an established optimization technique (e.g., a genetic algorithm) rather than a less accurate heuristic algorithm.

2. The JIT approach is based on the philosophy of reducing the amount of inventory to a minimum and whatever inventory does exist at each production cell is replenished as late as possible. If the organization is using the JIT approach, then the VVT team should:
   - Verify that all production cells are operating at optimal level and the JIT approach is effective at all levels of production (i.e., the JIT approach does not hinder the production process).
   - Verify that quality problems discovered at one production cell are relayed immediately to all relevant outside suppliers and relevant production cells so they may be corrected as soon as possible. Verify that the JIT approach significantly reduces the amount of manufacturing quality problems.
   - Verify that implementation of the JIT approach is based on full management and worker commitment to the success of the JIT approach. Verify that management trusts and empowers workers on the production line. Also verify that employees seek to achieve quality work and are prepared to act in the long-term interests of the producing organization. For example, verify that employees would halt the production process if it were determined that defective parts, components or subassemblies may flow into higher level assemblies.
   - Verify that the JIT approach is extended to each supplier. Verify that management treats suppliers as partners with significant influence on the success of the organization. Also verify that suppliers are, to the extent practical, located in close proximity to the manufacturing plant and, to the extent possible, sharing computerized databases with the manufacturing organization.

3. Verify that the correct production control approach (MRP, JIT or another one) is adopted by the organization on the basis of sound management, economic and social considerations.
   - In general, verify that the MRP approach is adopted when
     (1) the level of uncertainty regarding future demand for the end products is low,
     (2) the level of uncertainty regarding the production capacity and production yield is low and
     (3) it is possible to forecast relatively accurately the level of end-product demand.
   - In general, verify that the JIT approach is strongly considered when
     (1) suppliers are exceptionally reliable, not too numerous and located in close proximity to the manufacturing plant;
     (2) the nature of end-product demand is stable and predictable and
     (3) the working environment enables management and workers to cooperate in setting goals and with mutual respect to achieve a successful JIT operation.

Methods and Further Literature

Section 4.3.4, System test simulation
Section 4.3.5, Failure mode effect analysis
Section 4.3.6, Anticipatory failure determination
Section 4.4.1, Expert team reviews
Section 4.4.3, Group evaluation and decision

- Geng (2004)
- Kalpakjian and Schmid (2005)
- Nahmias (2004)
- Tanner (1990)

3.2.17 Verify Production Scheduling

Objective The objective of this activity is to verify the production scheduling of the manufacturing organization.

Description Production scheduling is concerned with sequencing activities within a plant or a job shop. The purpose of the VVT actions is to verify that the production scheduling is optimally efficient in light of the production forecasting. Typically, the VVT team will:

1. Verify that all production scheduling considers the characteristics of job shop scheduling problems:
   - *Job Arrival Patterns.* Verify that the production scheduling takes into account the stochastically dynamic number and types of jobs waiting to be processed at any given time.

- *Number and Types of Production Units.* Verify that production scheduling takes into account the number, types and locations of machines and production facilities within the plant or job shop.
- *Number of Workers and Their Skills.* Verify that the production scheduling takes into account the number of workers in the plant and their individual skills.

2. Verify that the production scheduling considers balanced multiobjective optimization for job shop management. Typical objectives are:
   - Meeting product target due dates
   - Minimize production cost
   - Maximize machine and worker utilization
   - Maximize product yield level
   - Minimize Work-In-Process (WIP) inventory
3. Verify that the production scheduling considers optimal sequencing rules: The most common ones are:
   - *First-Come, First-Served (FCFS).* Verify that, if this rule is applied in the production scheduling, then jobs are processed in the order they arrive at the machine or production cell.
   - *Shortest Processing Time (SPT).* Verify that, if this rule is applied in the production scheduling, then jobs requiring short processing time are performed before jobs requiring longer time to process.
   - *Earliest Due Date (EDD).* Verify that, if this rule is applied in the production scheduling, then jobs with an early due date are performed before jobs with a late due date.

Methods and Further Literature

Section 4.3.4, System test simulation
Section 4.3.5, Failure mode effect analysis
Section 4.3.6, Anticipatory failure determination
Section 4.4.1, Expert team reviews
Section 4.4.3, Group evaluation and decision

- Geng (2004)
- Kalpakjian and Schmid (2005)
- Nahmias (2004)
- Tanner (1990)

3.2.18 Participate in Production Readiness Review (PRR)

Objective The objective of the PRR is to determine the status of specific actions that must be satisfactorily accomplished prior to undertaking a production go-ahead decision.

Description The PRR is often the last checkpoint before full rate production is initiated. The PRR is concerned with the gross level manufacturing issues, such as the need for identifying high-risk or low-yield manufacturing processes or materials or any specific requirements for manufacturing development efforts to satisfy design requirements. In addition, the review deals with such concerns as production planning, facilities allocation, incorporation of produce-ability oriented changes, identification and fabrication of tools/test equipment and long lead item acquisition. The VVT team should therefore be involved in the PRR process as follows:

- *Installation Qualification.* Review whether the production equipments and machinery are installed correctly within the production plant.
- *Operation Qualification.* Review whether the manufactured products, subsystems and end systems created in early pilot runs meet all their specifications.
- *Process Qualification.* Review whether the production plant meets expected production capabilities within a stable statistical quality control process.

The PRR is usually organized by a project leader associated with the management of the production system to be reviewed. Representatives of the VVT team should verify the availability and quality of the documents needed for the PRR as well as the appropriate execution of the review itself. The project leader organizes the PRR and determines the date and location of the review, invites the participants and assembles and distributes the documentation a reasonable amount of time prior to the PRR. He or she also proposes the list of critical points to be reviewed.

Invariably, the PRR is conducted in a formal manner. The project leader should invite the customer's representatives as well as the key managers from the manufacturing organization. In addition, a few specialists working on the reviewed production system as well as individuals representing the VVT and quality assurance teams will participate in the review.

The VVT team should verify that the list of issues to be addressed during the formal review meeting has been agreed upon in advance along the following typical set of issues:

- *Integration Issues.* Review typical production integration issues, including (but not limited to) subjects such as:
  a. Geometrical compatibility (e.g., dimensions, envelopes)
  b. Interface compatibility (e.g., mechanical, electrical, data flow)
  c. Thermal compatibility (e.g., dissipated power)
- *System Issues.* Review typical production system issues, including subjects such as:
  a. Completeness of design documentation
  b. Performance specifications and test results
  c. Verification of static and dynamic behavior
  d. Choice of materials in terms of compatibility with specifications

- *Logistic Issues.* Review typical production logistic issues, including subjects such as:
  a. Construction sites and logistics
  b. Preassembly, assembly and storage sites and logistics
  c. Transport, delivery and installation logistics
- *Production Engineering Issues.* Review typical production engineering issues, including subjects such as:
  a. Production operation strategy
  b. Marketing and production forecasting
  c. Aggregate production planning
  d. Inventory control operation
  e. Supply chain management
  f. Production control systems
  g. Production scheduling
- *Quality Assurance Issues.* Review typical production quality assurance issues, including subjects such as:
  a. Construction follow-up
  b. Quality control during construction
  c. Acceptance tests
- *Safety Issues.* Review typical production safety issues, including safety measures as a consequence of chosen materials, construction method, operation handling and test and operation procedures.

At the end of the PRR, the project leader is expected to summarize the review in a written conclusion and propose appropriate recommendations.

Methods and Further Literature

Section 4.4.2, Formal technical reviews
Section 4.4.3, Group evaluation and decision
Section 5.7.9, First article inspection (FAI)

- AFSCR 64-2 (1995)
- Webb (2000)

3.3 VVT ACTIVITIES DURING USE/MAINTENANCE

The purpose of the system Use and Maintenance phase is to operate the system in its actual anticipated user environment and to fulfill its intended purposes. During this phase, the system requires a variety of VVT activities as routine operations performed either automatically by the system (e.g., BIT) or manually by operators and maintenance personnel (e.g., daily checking of the assembly line, yearly checking of an automobile). Such activities are conducted as a scheduled preventive maintenance or whenever problems occur. The appropriateness of all such maintenance operations should be verified prior to actually conducting any maintenance activity. In addition, the proper behavior of the systems undergoing maintenance should also be verified.

3.3.1 Develop VVT Plan for System Maintenance

Objective The objective of this activity is to plan the VVT activities during the system Use/Maintenance phase.

Description The longest system lifecycle phase is, normally, Use/Maintenance. During this phase all necessary VVT activities are accomplished to sustain the fielded system in the most cost-effective manner possible. During this phase, modifications and product improvements are usually implemented to update and maintain the required levels of operational capability as technologies and users' desires evolve. The following covers maintenance concepts, maintenance types, maintenance cost, maintenance obstacles and the role of the VVT engineer within this lifecycle phase:

1. *Maintenance Concepts.* The system's maintenance concepts should be developed early by the maintenance stuff, including the VVT team. The maintenance concept should embody such considerations as how the system will be used, its operational availability goals, anticipated useful life and physical environments. The system maintenance concept should first describe the anticipated levels of maintenance, general repair policies regarding both emergency and nonemergency maintenance, assumptions about supply system responsiveness, the availability of new or existing facilities and the maintenance environment.

Initially, the system maintenance concept may be based on experience with similar systems and should use appropriate optimization analysis. In some cases, maintenance and testing operations are so complex that simulation is required in order to design proper maintenance sequences. For example, maintenance and testing of large power plants or nuclear reactors must be meticulously planned as decreasing output power may cause a system's instability. Such a procedure is usually achieved by means of simulation. Another common use of simulation is in assessing the lifetime fatigue characteristics and preventive testing requirements for a wide variety of systems, from an aircraft's outer skin to machine parts. This is usually accomplished by comparing simulation results to given material data after some statistical extrapolations. Simulation methods may be used in order to create an optimal testing and maintenance operation plan where historical data on the lifecycle of system components or specific material data for fatigue analysis are available.

Usually, an Integrated Logistics Support Plan (ILSP) defines the system maintenance concept. In addition the ILSP covers issues such as maintainability and testing principles, a timetable for performing scheduled maintenance and testing, required manpower and other resource, facilities needed for conducting the maintenance and testing as well as spare-parts policies, test and support equipment and the like. All in all, the role of the VVT team is to participate in the development of the system maintenance concept and derive its own VVT planning. This VVT plan should include long-term schedule, budget, manpower and funding needs.

2. *Maintenance Types.* Broadly speaking, system maintenance is the totality of activities required to provide cost-effective support to systems. These activities lay the groundwork for the system maintenance and are performed throughout usage of the engineered systems. Maintenance is needed to ensure that the system continues to satisfy user requirements over a long period of time. Different maintenance activities may be combined into specific groups; however, undertaking major system change (e.g., substantial system modifications or implementation of costly new user requirements) is not considered below as maintenance and should be carried out as a separate new development projects.

- *Emergency Maintenance.* Unscheduled corrective maintenance which may be classified into two categories:
  a. *Production Issues.* Urgent work which halts a system's operations and must be undertaken as soon as possible. Often, such activities are performed without full VVT attention. Often this strategy assumes greater risk due to the reduced levels of quality assurance and testing.
  b. *Pressing Issues.* Urgent work that significantly impacts business operations but can be undertaken while the system is operational. While the corrective work is considered quite critical, there is more room to perform a more thorough VVT process. Often these conditions lead to some risk, which should be weighted in accordance with the functional criticality of the system at hand.
- *Corrective Maintenance.* Identification and removal of noncritical system defects which in general are well documented and operators know how to get around them. Typically, different corrective actions are identified and processed according to a defined maintenance procedure. VVT of a system's corrective maintenance should be rigorous and thorough as it may be accomplished with nominal cost and schedule pressure and no undue risk is necessary.
- *Perfective Maintenance.* Upgrading the system functionality and performance in a rather limited fashion. This may include improvement in performance, dependability, maintainability, safety, reliability, efficiency or cost effectiveness of an operation. Similar conditions suggest VVT thoroughness level should be similar to corrective maintenance.

- *Adaptive Maintenance.* Modifying the system to keep it up to date with its environment. This includes adapting the system to a new or changed environment (e.g., new hardware, interfaces) or a new regulation that impacts the system's operations. Similar conditions suggest that VVT thoroughness level should be similar to corrective maintenance.
- *Preventive Maintenance.* Identification of activities performed in advance of an immediate need for a system's repair or in advance of accumulated deterioration. The purpose of preventive maintenance is therefore to reduce the rate and severity of system failures in the long term. Consequently, emergency maintenance should be eliminated or reduced to an acceptable level. These activities are usually cyclical in nature and planned in advance, so VVT thoroughness is vital.

3. *Maintenance Cost.* System maintenance and, especially, VVT cost and time investment consumes a major share of the system lifecycle financial resources. A common perception of system maintenance is that it merely fixes faults. However, studies over the years have indicated that only 20% of the system maintenance effort is used for emergency and corrective actions. Additional findings indicate that a strategy of frequent cyclical minor maintenance efforts is consistently more cost effective than infrequent major maintenance efforts. The cost effectiveness of reasonably frequent maintenance may be explained by the exponential increase in disruption affecting unmaintained systems. This is illustrated in Figure 3.3 where the dashed lines represent disruptions to normal system operations and each vertical bar represents accumulated system repair cost over a given time period.

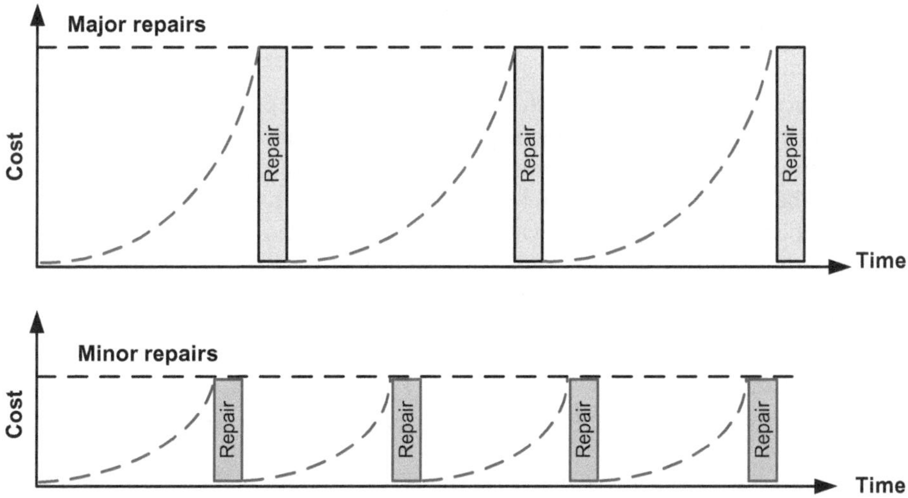

Figure 3.3 Cyclical system maintenance: major/minor repair strategies.

4. *Maintenance Obstacles.* When maintaining a deployed system, the VVT engineers should pay particular attention to the following problems:

- *Planning Maintenance Process.* Often, a system's maintenance tends to be viewed as a simple process that can be done on an ad hoc basis rather than planned carefully in advance. The VVT team should verify that all maintenance activities, including VVT activities, are carefully planned in advance. That planning should include a flexible schedule and an estimate of needed resources. If the resources are insufficient, then the plan should be reformulated to mitigate and control the budget risk.
- *Maintenance and Operational Conflicts.* The VVT team should plan perfective,[16] adaptive or preventive maintenance in a flexible enough manner to accommodate unforeseen schedule changes caused by unanticipated circumstances. The reason is that in most organizations operational obligations determine the availability of the system for maintenance activity and significant schedule conflicts between operational and maintenance needs often end up in favor of postponing maintenance activities.
- *Configuration Management.* The VVT team should be fully cognizant of the three system configurations associated with any deployed system undergoing maintenance: (1) the existing configuration of the system prior to any modification, (2) the temporary modified configuration which is used during the modification and testing of the system and (3) the final system configuration. It is a classical role of VVT to verify that the configuration management of the system is verified properly throughout these stages.
- *Logistics Compatibility.* Modification may change the system's configuration, which in most cases will change the supply, support and maintenance considerations. The VVT team should verify that, if logistics are affected by maintenance activity, then coordination with the logistics community is undertaken.
- *Legacy Systems.* Older systems may not have a producer with a corporate knowledge of the particular system functions and design and the maintenance personnel often do not have complete product baseline data for the system. In addition, legacy systems often use original commercial components that are not available anymore in the market. In such cases, maintaining the system could be a major effort. The VVT team should review maintenance plans of such legacy systems very early in order to identify potential legacy problems.

5. *VVT Engineer's Role.* As was elaborated before, the fundamental role of VVT engineers is to evaluate whether a system behaves in accordance with

[16]Perfective maintenance is a term first coined for software systems. In the context of this book it means maintenance performed to improve the performance, maintainability or other attributes of a system or a product.

its specification as well as evaluate whether a process is carried out in accordance with its approved procedure. This philosophy is also valid during the Use/Maintenance phase. The VVT test engineer's role is therefore confined to testing the system for proper behavior before actual maintenance operations and retesting it after such activities to ensure that maintenance operations did not introduce defects into the system. From a VVT standpoint, the only unique aspects of this lifecycle phase is that various preventive tests are called for before the system actually exhibits visible and concrete failure phenomena.

Methods and Further Literature

Section 4.3.1, VVT process planning

Section 5.7.12, Maintenance testing

- Matko et al. (1992)
- NASA/SP-2007 6105 (2007)
- Ogata (2003)
- SEF DoD (2001)
- Zahavi and Barlam (2000)
- Zienkiewicz and Morgan (2006)

3.3.2 Verify the Integrated Logistics Support Plan (ILSP)

Objective The objective of this activity is to verify the ILSP for the maintenance of the system and associated elements.

Description The ILSP identifies the support elements, management objectives, tasks and events associated with the maintenance of equipment, subsystems and systems. The following verification procedure for system ILSP was created on the basis of U.S. military standard DoD-STD-1702 (1985) and, more specifically, Data Item Description (DID) DI-ILSS 80095 (1985).

Proposed Procedure: System Integrated Logistics Support Plan

Step 1: Verify Integrated Logistic Support Management

1.1: *System Description.* Verify that the ILSP, or "plan" for short, provides a description of the system, including a summary of performance and operational characteristics.

1.2: *List of Equipment.* Verify that the plan identifies all components of the system and equipment addressed in this plan, test equipment or special tools required for maintenance of the system, including (1) equipment logistics data sheets, (2) system block diagrams and (3) documentation of support concepts.

1.3: *Support Transition.* Verify that the plan includes a description for the transition of support responsibility from the producer to the acquirer of the system.

1.4: *Support Validation.* Verify that the plan describes methods to be used to validate and evaluate the support processes established in the ILSP.

1.5: *Points of Contact.* Verify that the plan identifies specific persons within delineated organizations as Points Of Contact (POC) for all significant ILS actions to be implemented.

Step 2: Verify Maintenance

2.1: *General.* Verify that the plan provides a narrative description of the maintenance planning for the system and test equipment and when the planning should be initiated in order to support the system in its operational environment.

2.2: *Maintenance Concept.* Verify that the plan summarizes the general maintenance concept to be used for support of the system and equipment. Also verify that the plan defines how and when effective maintenance can be performed and by whom. This should include:

- *Initial Maintenance.* Summarizing general maintenance procedures for on-site and off-site as well as providing guidance for the return of defective Lowest Replaceable Units (LRUs).
- *Follow-On Maintenance.* Summarizing of general maintenance procedures or other activities on-site and off-site.
- *Contract Maintenance.* Listing hardware, firmware and software end items selected for contract maintenance.
- *Depot Maintenance.* Identifying system or equipment needed at the depot level to test and maintain the fielded system.

2.3: *Maintenance Management.* Verify that the plan identifies applicable maintenance management requirements.

2.4: *Reliability, Availability, Maintainability.* Verify that the plan includes reliability, availability and maintainability requirements.

2.5: *Maintenance, Test and Support Equipment.* Verify that the plan includes specific requirements for Maintenance, Test and Support Equipment (MT&SE), including Built-In Test Equipment (BITE) to the maximum extent practical. In addition, the plan should include requirements and organizational responsibilities for their maintenance and calibration.

2.6: *Maintenance Technical Assistance.* Verify that the plan describes established procedures for obtaining external entities (e.g., original system producer, other government or commercial agencies) as well as technical assistance concerning engineering support problems.

2.7: *Repair/Return Procedures for Faulty Lowest Replaceable Units (LRUs).* Verify that the plan describes established procedures for repair/return of faulty LRUs.

Step 3: Perform Test and Evaluation

3.1: *Test Program.* Verify that the plan identifies applicable regulations, directives, specifications and other documents that describe and define the Test and Evaluation (T&E) requirements.

3.2: *Development Test and Evaluation (DT&E).* Verify that the plan describes and makes reference to the DT&E.

3.3: *Operational Test and Evaluation (OT&E).* Verify that the plan describes and makes reference to the OT&E.

3.4: *Test Support.* Verify that the plan includes:

- *DT&E.* Support material and documentation required for completion of the DT&E phase.
- *OT&E.* Support material and documentation required for completion of the OT&E phase.

3.5: *Emissions Security (EMSEC) testing.* EMSEC is a U.S. military and North Atlantic Treaty Organization (NATO) terminology referring to unintentional intelligence-bearing transmission emanating from computers and other information-processing systems. For such systems, containing sensitive military or commercial information, verify that the plan identifies specific EMSEC testing requirements for each relevant element of the system.

Step 4: Verify Supply Support and Provisioning

4.1: *General.* Verify that the plan describes the supply support concepts and provisioning tasks for the system and a general description of the responsibilities of each organization in this process.

4.2: *Applicable Documents.* Verify that the plan makes reference to applicable documents or contracts for supply support and provisioning.

4.3: *Stock Management/Inventory.* Verify that the plan defines the responsibilities for spares management on-site and identify the organizational responsibilities and method of management to be used.

4.4: *Provisioning.* Verify that the plan provides the scope of provisioning to be accomplished in support of the system or equipment.

4.5: *Support Detail.* Perform the following verification activities:

- Verify that the plan describes the initial support, which begins with end item or system installation and checkout at the customer or user site. The plan should also describe the follow-on support subsequent to the initial support and is normally the responsibility of the lifecycle support authority. The period of follow-on support is usually for the usable life of the system and equipment.
- Verify that the plan defines the following: (1) duration of the initial support period, (2) disposition of installation spares and (3) specific spare/repair parts to be initially provided. In addition, the plan should describe plans and responsibilities for funding and acquiring initial spare/repair parts as well as responsibilities for additional supply support requirements that may develop during the initial support period.
- Verify that the plan defines the duration of the follow-on responsibilities and the date/event/phase they will commence. In addition, verify that the plan should identify organizational responsibilities for providing follow-on supply support.

4.6: *Supply Support during Operation and Maintenance Period.* Verify that the plan identifies the organization responsible for supply support and provides names, addresses and telephone numbers of responsible personnel. In addition, the plan should describe the repair parts/supplies that must be maintained at the site as well as repair parts/supplies that must be maintained off-site. The plan should also describe procedures for the inventory utilization and turnaround requirements for repaired parts.

4.7: *Recording/Storage Media Management.* Verify that the plan identifies requirements for storage of media (e.g., category and size of media, type of media containers, packaging requirements, quantities, shipping address and forwarding instructions, funding method, disposition of used media, magnetic degaussing and reuse procedures, security requirements).

4.8: *Special Tools and Test Equipment.* Verify that the plan defines supply support responsibilities for special tools and test equipment.

4.9: *Depot Test Equipment.* Verify that the plan identifies any special requirement(s) for depot test equipment.

4.10: *Mission Expendable Supplies.* Verify that the plan identifies expendable supplies (e.g., computer and office supplies, fuel) as well as organizational responsibility for providing expendable supplies initially and during the follow-on phase.

4.11: *Disposition of Nonserviceable, Obsolete, Salvaged or Excess Equipment.* Verify that the plan identifies the applicable references for disposition of nonserviceable, obsolete, salvaged or excess equipment and outline any specific directions.

4.12: *Equipment Accountability.* Verify that the plan identifies the applicable references for providing equipment accountability and outline any special directions as well as the organization responsible for equipment accountability once the system is deployed and accepted on-site.

4.13: Cannibalization.[17] Verify that the plan identifies the applicable cannibalization of equipment policy and any special directions toward that end.

Step 5: Verify Packaging, Handling, Storage and Transportation

5.1: *Purpose.* Verify that the plan states the purpose of this chapter and identifies applicable regulations, directives, specifications and other documents that describe and define both domestic and international transportation, packaging, handling and shipping requirements.

5.2: *Organizational Responsibilities.* Verify that the plan describes the organizational responsibilities for ensuring packaging, handling, storage and transportation functions. In addition, verify that the plan identifies any requirements for notifying the affected sites of the shipment of the subject system or equipment and the methods and responsibilities.

5.3: *Material Movement Plans.* Verify that the plan identifies shipping instructions and the shipping coordinator, applicable document reference(s) that provide requirements for material movement, delivery schedules and shipment priorities as well as modes of transportation to be used.

5.4: *Special Handling.* Verify that the plan identifies and describes any special handling requirements for moving, loading, unloading, transporting and storing the system or equipment, such as preservation, temperature control, humidity control, protection from shock or radiation, security requirements and similar information.

5.5: *Preservation and Packaging.* Verify that the plan identifies applicable reference(s) that provide requirements for preservation, packaging and packing of components, subsystems and spare parts.

5.6: *Transportation Requirements.* Verify that the plan provides general planning for transportability requirements related to gross weight and outside dimensions.

5.7: *Technical Data.* Verify that the plan identifies technical data such as documents, drawings and plans that are required to support transportation and handling.

[17] Cannibalization is the process of removing serviceable parts from either a nonfunctioning system or a functioning system (thus making it unusable for its original intended use) with the aim of building or repairing another system of the same kind.

5.8: *Marking.* Verify that the plan identifies applicable requirements for container markings for shipment and storage.

5.9: *Damage or Loss.* Verify that the plan identifies applicable requirements for reporting damaged or lost shipments.

Step 6: Verify Technical Data and Data Management

6.1: *Purpose and Scope.* Verify that the plan provides a summary of and complete information concerning the data deliverables necessary to support the system. In addition, verify that the plan discusses the management techniques and organizational responsibilities to ensure the data are properly specified, obtained in adequate quantities, provided when needed and maintained in an accurate, complete state throughout the system's lifecycle.

6.2: *Data Management.* Verify that the plan describes how the data requirements were established and identifies organizational responsibilities for obtaining it. In addition, verify that the plan describes procedures for reviewing the data for accuracy and completeness, ships it when needed, and monitors and/or revises the data when necessary.

6.3: *Data Deliverables.* Verify that the plan summarizes the data deliverables by category of equipment to be supported and type of support the data will provide, that is, operational maintenance, test specification and so on. The plan should also provide the title of each data product as it appears on the applicable DID and its DID number.

6.4: *Training Documentation.* Verify that the plan describes the types of training and the schedule for development, delivery and validation of training materials and devices.

Step 7: Verify Configuration Management

7.1: *General.* Verify that the plan identifies the objectives of configuration management, the practices to be used and the participating organizations and their respective functional responsibilities.

7.2: *Organization Responsible for Configuration Management.* Verify that the plan identifies the organization responsible for hardware, firmware and software Configuration Management (CM), the function of the CM Configuration Control Board (CCB) and the applicable references that provide guideline for the CCB.

7.3: *Addresses of Configuration Management Organization.* Verify that the plan identifies the CM organization and the POC responsible for system/equipment configuration management.

7.4: *Configuration Items.* Verify that the plan identifies each hardware, firmware and computer program configuration item related so the system and equipment.

7.5: *Configuration Identification.* Verify that the plan identifies the technical data that form the product baseline for the system, equipment, computer software or firmware configuration items.

7.6: *Configuration Control Procedures.* Verify that the plan includes configuration control procedures containing the following general steps:

- *Submission of Engineering Change Proposals (ECPs).* Identification of applicable references that provide guidance for the preparation and processing of ECPs and establishing the chain of review for ECP submittal and provision of a guideline for the preparation of supplementary documentation.
- *Assessment of Impact.* Provisioning criteria for the review of ECPs for determination and assessment of the impact of the change.
- *CM Organizational Review.* Identification of CM organizational CCB responsibilities for reviewing and processing of ECPs.

7.7: *On-Site Configuration Audit.* Verify that the plan identifies requirements and provides a procedure for the conduct of on-site configuration audits leading to system/equipment acceptance.

Step 8: Verify Installation and Facilities

8.1: *General.* Verify that the plan provides a general description of how the system/equipment will be integrated into an existing site or installed in a new site.

8.2: *Site Survey Requirements.* Verify that the plan includes requirements for site surveys which are conducted to determine facility requirement for installation of new systems/equipments. These requirements should include installation of electrical power, heating, cooling, physical space, security and so on. Verify that the plan discusses the purpose of the surveys, organizational responsibilities for their accomplishment and the schedule (plan) for conducting the surveys.

8.3: *Site Preparation and Installation Plan.* Verify that the plan identifies the organizational responsibilities for the preparation of an installation plan with drawings or alternative means.

8.4: *System/Equipment Layout.* Verify that the plan provides a general layout of the equipment comprising the system.

Step 9: Verify Personnel and Training

9.1: *General.* Verify that the plan provides a general description of the personnel and training requirements for the system and organizational responsibilities for the operation and maintenance training of the equipment, subsystem or system.

9.2: *Personnel.* Verify that the plan includes (1) operational personnel, (2) maintenance personnel and (3) software personnel (as needed). In addition, verify that the plan states the maintenance man-hour standards as well as identifies either increases or decreases in all man-power categories caused by the installation and subsequent operation of the system.

9.3: *Training.* Verify the plan as follows:

- *Training Requirements.* Verify that the plan includes training requirements for the initial and follow-on supervisory, operator and maintenance courses and the specific training approach that will be used to satisfy these requirements.
- *Initial Training.* Verify that the plan identifies and describes the initial supervisory, operator and maintenance courses of instruction available to complement the skills identified above as well as funding and contracting responsibilities, organizations responsible for the conduct of the initial training courses and students' prerequisites, load and schedule plans.
- *Follow-On Training.* Verify that the plan identifies and describes follow-on supervisory, operator and maintenance courses of instruction needed to complement the skills identified above.
- *Training Equipment.* Verify that the plan summarizes the training equipment requirements and that their delivery schedule is included in the relevant milestone charts.
- *Training Test and Evaluation.* Verify that the plan identifies the materiel elements of the training subsystem that will be required to be on-hand for DT&E and OT&E.

Step 10: Verify Funding

10.1: *Referenced Documents.* Verify that the plan refers the reader to the appropriate documentation containing information on the funding/budgeting for items of logistic support for the subject project.

Step 11: Verify Computer Resource Support

11.1: *Software Conventions and Standards.* Verify that the plan identifies the source document establishing software design, documentation as well as change authority, convention and standards.

11.2: *Maintenance of Software Programs.* Verify that the plan defines the policies and control requirements for on-site maintenance of software programs, including software lifecycle support responsibility, the method of distribution of programs and updates to the software.

11.3: *Specific Software Configuration Management Requirements.* Verify the following:

- *Software Configuration Management.* Verify that the plan explains unique characteristics of configuration management as it applies to software programs.
- *Software Documentation.* Verify that the plan identifies the organization(s) responsible for ensuring that accurate documentation changes are made and that the documentation is matching the actual software system.
- *Software Change Policy and Authority.* Verify that the plan discusses the policy and authority for making changes to software programs.
- *Preservation of Superseded Program Versions.* Verify that the plan explains or references the procedures for ensuring that superseded software programs are protected until approval for their destruction has been received from the software lifecycle support authority.

11.4: *Software Development, Test and Reviews.* Verify that the plan includes a subplan for developing, testing and reviewing software programs. Such a subplan should identify specific test plans/procedures for testing the operational programs and identify the facilities required to accomplish the test program.

11.5: *Firmware Maintenance.* Verify that the plan assigns firmware maintenance responsibilities by organization/activity. Also verify that the plan describes facilities/resources required for creating replacement Programmable Read-Only Memories (PROMs) and equipment required to embed the program in the Integrated Circuits (ICs) and provides procedures for certifying them.

Methods and Further Literature

Section 4.3.1, VVT process planning
Section 4.4.1, Expert team reviews

- DI-ILSS 80095 (1985)
- DoD-STD-1702 (NS) (1985)

Section 4.4.3, Group evaluation and decision

- Jones (1998)

3.3.3 Perform Ongoing System Maintenance Testing

Objective The objective of this VVT activity is to perform ongoing system maintenance testing seeking to optimize system availability and maintain this availability within technical, performance, legal, commercial and environmental parameters.

Description Maintenance encompasses the set of activities that aim to sustain a system, a condition deemed necessary for it to properly fulfill its functions. Maintenance is focused on testing the target system, repairing failed components or replacing them when cost of repairs exceeds replacement cost. In addition, maintenance requires planning and managing the process in an optimal manner.

As mentioned, maintenance is divided into preventive and corrective activities. From the VVT point of view, preventive maintenance entails inspecting and testing the system to ensure that it performs according to expectations and the day-to-day operations comply with established procedures and regulations. On the other hand, corrective maintenance is conducted when a system malfunctions. The responsibility of the VVT team is to test the system and locate the faulty component or, possibly, operation error leading to the failure.[18]

In general, the objectives of maintenance activities as implemented in everyday practice include (1) minimizing overall system cost by means of preventive maintenance, (2) repairing everything as quickly and cheaply as possible and (3) optimizing the repair/replace strategy to save time or money. Figure 3.4 depicts a qualitative relation between the overall cost of maintenance and the level of preventive maintenance.

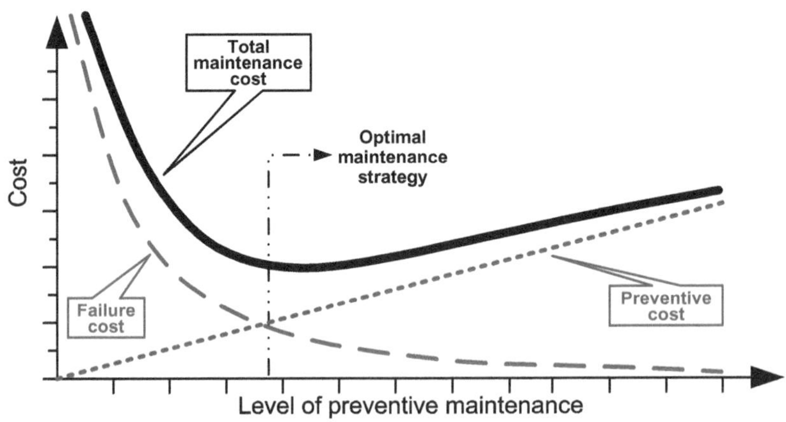

Figure 3.4 Cost of maintenance versus level of preventive maintenance.

[18] As a general rule it is not the role of the VVT team to actually fix the system.

The system failure cost emanating from breakdowns, idle time and extra wear and tear or damage due to late repairs is shown together with the cost of preventive maintenance. Here, the failure cost decreases exponentially with the amount of preventive maintenance, whereas the cost of preventive maintenance is drawn as an increasing linear function. Therefore, the total cost of maintenance is the sum of these two components.

Different engineered systems require different levels of maintenance. We can model this phenomenon and draw some inferences from the following:

1. A hair comb is one the oldest tools (engineered systems); it has been used for over 5000 years. It does not require maintenance, other than cleaning and removing an occasional broken tine.
2. The light bulb, invented by Thomas Alva Edison in 1879, is an engineered system an order-of-magnitude more complex which is fully replaceable and does not require maintenance, other than occasional cleaning.
3. An artificial pacemaker is an engineered medical system that delivers electrical impulses to the heart muscles in order to regulate heartbeat. As a system, it is probably an order-of-magnitude more complex than a light bulb but it is maintained only within the larger system—the human body. Pacemakers are programmable systems containing a BIT mechanism. This is a sophisticated means to test and record automatically the deviations from critical operational parameters, log system failures and the like. Maintenance activities such as charging batteries, evaluating BIT results and adjusting operational parameters every year or a few years are common.
4. Another engineered system, the passenger car, used for transporting passengers and goods, is arguably another order-of-magnitude more complex than a pacemaker. The typical modern automobile contains 20–50 embedded microcomputers, which makes driving safer and relatively comfortable. For example, a modern car will have safety-related systems such as an Anti-lock Braking System (ABS), an Electronic Stability Program (ESP, see Figure 3.5), a Trace Control System (TCS), an airbag control system, a drowsiness monitoring system as well as convenience-related systems such as navigation systems (using GPS), cruise control, automatic parallel parking systems and performance and efficiency systems such as engine fuel injection control. All these systems contain sophisticated BIT mechanisms that inform drivers of any system problem encountered in real time and are used extensively during preventive and corrective maintenance. So automated system testing and driver's advice are performed on a continuous basis during operation. In addition, general maintenance is carried out a few times a year.

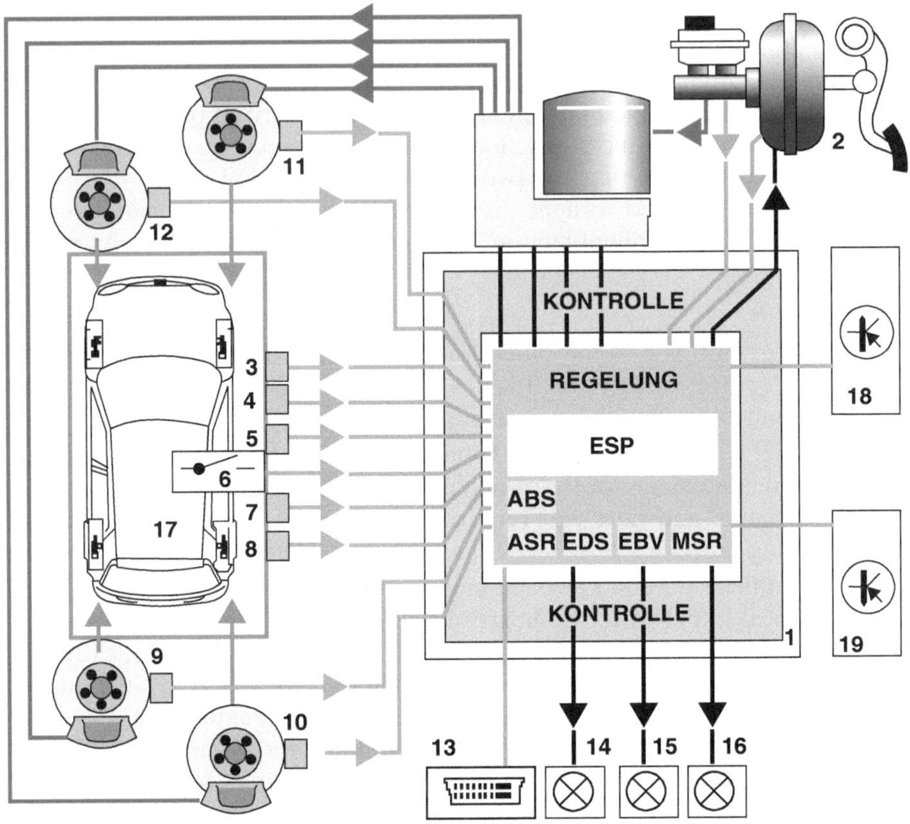

Figure 3.5 Vehicle stability system (Bosch GmbH, Germany).

5. Commercial jet aircrafts are able to fly at altitudes of 10–15 km and speed of up to 900 km/h to a range of 6,000–14,000 km carrying 100–400 passengers or cargo. They are marvelous systems from many engineering perspectives and, again, an order-of-magnitude more complex than an automobile. In addition to continuous automated testing, ongoing system testing is performed several times a day, before, during and after each flight by pilots and ground maintenance crews. Preventive and corrective maintenance is performed on a daily or weekly basis as a matter of necessity and strict international regulations.

Figure 3.6 depicts a positive correlation between complexity and cost of the above engineered systems on a semilog chart. It is the contention of the author that the level of maintenance and, in particular, the testing of engineered systems follow the same pattern (i.e., the more expensive the system, the more funding and other resources must be allocated to system maintenance activities). During systems maintenance VVT includes (1) planning and organizing for a smooth maintenance process and (2) carrying out the actual testing of the system.

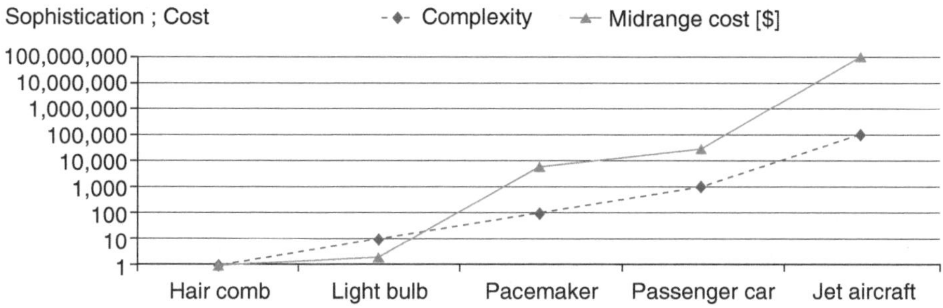

Figure 3.6 Cost and complexity of engineered systems.

Planning/Organizing Maintenance Process The following VVT planning and organizing should be carried out:

- *Maintenance Concept.* Define a general test maintenance concept to be used for testing and validating the system.
- *Test and Support Equipment.* Define specific requirements for Maintenance, Test and Support Equipment (MT&SE), including Built-In Test Equipment (BITE).
- *List of System Elements.* Identify all components and subsystems that may require testing.
- *Personnel.* Provide a general description of the test personnel and training requirements and identify manpower requirements needed to test the system during prevention as well as corrective maintenance. Manpower planning should identify either increases or decreases in manpower categories caused by the installation and subsequent operation of the system.
- *Training.* Identify training requirements for the initial and follow-on system testing activities and the specific training approach and training equipment that will be used.
- *Software, Test and Reviews.* If the system includes software and/or embedded computers, then plan for testing and reviewing software programs. This should include specific test plans/procedures for verifying the operational programs. In addition, the facilities required to accomplish the test program should be identified.

Carrying Out a System's Test and Evaluation The following system testing should be carried out:

- *Preventive Maintenance Testing.* Test the system on a predefined schedule basis and in accordance with the maintenance test plan and identify all failing components that do not meet required specifications.
- *Corrective Maintenance Testing.* Test the system whenever it fails and in accordance with the maintenance test plan and identify the failing component or components causing the system malfunction.

Methods and Further Literature

Section 4.2.5, Classification tree method
Section 4.2.6, Design of experiments
Section 4.3.5, Failure mode effect analysis
Section 4.3.6, Anticipatory failure determination
Section 4.3.8, Robust design analysis
Section 5.7.1, Sanity testing
Section 5.7.2, Exploratory testing
Section 5.7.3, Regression testing
Section 5.7.9, First article inspection
Section 5.7.12, Maintenance testing

- Blanchard et al. (1995)
- Knezevic (1997)

3.3.4 Conduct Engineering Peer Review on System Maintenance Process

Objective The objective of this activity is to conduct an ongoing engineering peer reviews in order to verify the effectiveness of the system maintenance process.

Description Engineering peer reviews are conducted periodically to verify the effectiveness of the system maintenance process. The peer review should be based on a status report summarizing the maintenance activities and the overall condition of the system. In general, the objective of the peer review team is to evaluate, based on available information, whether the system is maintained in a manner acceptable to all stakeholders and in a most cost-effective way. The following provides a list of topics that may be considered for a maintenance peer review. Such peer reviews may be conducted on a cyclical basis covering different topics each time. It was created on the basis of U.S. military standard DoD-STD-1702 (1985) and, more specifically, DI-ILSS 80095 (1985).

Proposed Topics: Engineering Peer Review of System Maintenance

Topic 1: Review Integrated Logistic Support Management

- Review whether all components of the system, test equipment and special tools required for maintenance of the system have been properly identified and updated over time.

Topic 2: Review Maintenance Planning and Concepts

- Review whether there is a clear description of the maintenance planning and maintenance concept to be used for support of the system and the test equipment. In addition, review whether this description is up to date.

- Review whether the requirements for reliability, availability and maintainability are in fact met by the system.
- Review whether the requirements for system MT&SE, including BITE, have been met.
- Review whether the procedures for obtaining outside technical engineering assistance have been exercised successfully.
- Review whether the established procedures for repair/return of faulty Lowest Replaceable Units (LRUs) have been exercised successfully.

Topic 3: Review Test and Evaluation

- Review whether the maintenance testing adheres to applicable regulations, directives, specifications and other documents that define the Development Test and Evaluation (DT&E) and the Operational Test and Evaluation (OT&E).

Topic 4: Review Supply Support and Provisioning

- Review whether the supply support concepts and provisioning tasks for the system/equipment as well as the provisioning responsibilities of each organization are being met.
- Review whether there is a clear definition of responsibilities for on-site spares management and whether the actual level of spare parts provisioning for all system elements as well special tools and test equipment is sufficient. The review should refer to the system's replaceable parts as well as expendables (e.g., computer supplies) located on-site as well as off-site.
- Review whether all nonserviceable, obsolete, salvaged or excess equipment is disposed of in accordance with approved technical, legal, civic and environmental requirements.

Topic 5: Review Packaging, Handling, Storage and Transportation

- Review whether the applicable regulations, directives, specifications and other documents that describe and define both domestic and international transportation, packaging, handling and shipping requirements are in fact adhered to.
- Review whether the organizations responsible for packaging, handling, storage and transportation are performing their duties

effectively and in accordance with shipping instructions, following requirements for material movement, delivery schedules and shipment priorities as well as modes of transportation.

Topic 6: Review Technical Data and Data Management

- Review whether the specified data management techniques and organizational responsibilities to ensure data integrity are properly carried out. That is, review whether data related to the system have been created according to established requirements by organizations responsible for obtaining it.
- Review whether the data are maintained in an accurate, complete state throughout the system's lifecycle. In addition, review whether the procedures for monitoring, analyzing and/or revising the data for accuracy and completeness are, in fact, satisfying the stakeholders.
- Review whether the shipping of a system's related data is carried out when necessary or needed to the full satisfaction of the system's stakeholders.

Topic 7: Review Configuration Management

- Review whether the objectives of Configuration Management (CM), the CM practices used and the participating organizations and their respective functional responsibilities are adequate.
- Review whether the hardware, firmware, and software CM and Configuration Control Board (CCB) are, in fact, performed in accordance with established procedures to the satisfaction of all stakeholders.
- Review whether the configuration control follows, in fact, defined procedures and includes the following general steps:
  a. Submission of Engineering Change Proposals (ECPs)
  b. Assessment of impact on the system by the CCB
  c. Carrying out the engineering change and testing the system according to requirements

Topic 8: Review Installation and Facilities

- Review whether the system was integrated into an existing site or installed in a new site in accordance with prescribed site survey requirements. These requirements should include installation of electrical power, heating, cooling, physical space and security.

Topic 9: Review Personnel and Training

- Review whether the personnel assigned to maintain the equipment, subsystem or system (i.e., operational, maintenance and software personnel) as well as their training met the original planning and actual requirements.
- Review whether the actual maintenance needed to install, maintain and subsequently operate the system was sufficient and met man-hour standards typical to the attributes and character of the system.
- Review whether the training for the initial and follow-on supervisory, operator and maintenance activities was effective and satisfied all system stakeholders. Such training should include supervisory, operator and maintenance courses to complement and enhance staff skills.

Topic 10: Review Maintenance Funding

- Review whether the funding/budgeting for all maintenance activities as well as system logistics is adequate, available on time and meets original planning requirements.

Topic 11: Verify Computer Resource Support

- Review whether the maintained software meets software conventions and standards (e.g., software design, code, documentation).
- Review that the software is maintained according to defined policies and control requirements for on-site maintenance, including software lifecycle support responsibility, identified method for distribution of programs and updates to the software.
- Review whether the software and firmware are developed and tested in a controlled manner, including specific test plans/procedures for verifying the updated operational programs and identifying the facilities required to accomplish the testing process.

Methods and Further Literature

Section 4.4.3, Group evaluation and decision

- DI-ILSS 80095, (1985)
- DoD-STD-1702 (NS) (1985)

Section 5.7.12, Maintenance testing

- Jones (1998)

3.4 VVT ACTIVITIES DURING DISPOSAL

The purpose of the system Disposal phase is to properly dispose of the system and its elements upon completion of its useful life. During this phase, systems should be dismantled, partially or completely recycled and shredded and, finally, toxic materials must be neutralized. The majority of systems have no formal disposal requirements. However, systems with hazardous materials or other safety issues have specific disposal requirements related to environmental protection, levels of materials recovery and methods of disposal.

Let's look, for example, at automobile disposal in the European Union (EU). Every year, End-of-Life Vehicles (ELVs) generate between eight and nine million tons of waste in the EU. In 2000, the EU adopted a legislation (ELV Directive 2000/53/EC) that aims at making vehicle dismantling and recycling more environmentally friendly (see Figure 3.7). Among other elements, the directive sets clear quantified targets for reuse, recycling and recovery of vehicles and their components. In addition, the directive requires producers of cars and their components to manufacture new vehicles with a view to their recycle-ability.

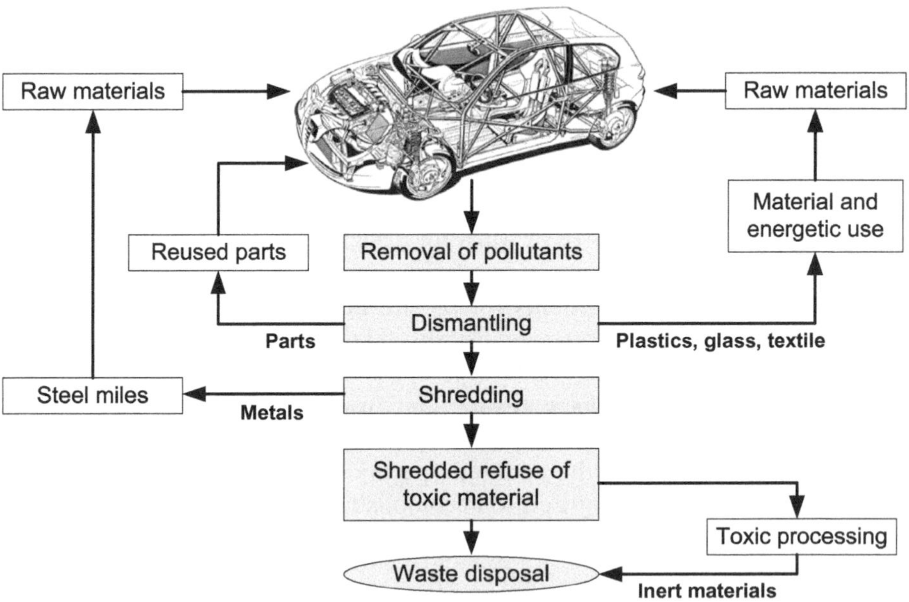

Figure 3.7 Typical vehicle disposal cycle mandated in the EU.

VVT activities during the system Disposal phase include developing a VVT plan for system disposal, assessing the planning of the system disposal process, assessing system disposal strategies by means of simulation as well as assessing the ongoing system disposal process and also conducting engineering peer review to assess the system disposal processes.

3.4.1 Develop VVT Plan for System Disposal

Objective The objective of this activity is to develop a VVT plan for the system Disposal phase.

Description A VVT disposal Program Management Plan (PMP) is a document used to coordinate the VVT activities during the Disposal phase and help guide the program's execution and control from the VVT point of view. The outline of the PMP provided below has been tailored from the US Department of Defense, standard DoD-STD-4160.21-M, Defense Material Disposition Manual as well as the Institute of Electrical and Electronics Engineers standard for software project management plans (IEEE 1058-1998). While the title implies guidance for software projects, the content, scope and flexibility of the IEEE standard facilitate application to a variety of projects that typify wide-ranging system engineering projects.

Proposed Structure: VVT Plan for System Disposal

Section 1: Overview

1.1: *VVT Disposal Program Summary*

- Define the purpose, scope and objectives of the VVT disposal program.
- Describe the assumptions on which the VVT disposal program is based and impose constraints on program factors such as the schedule, budget, resources and components to be reused.
- List the work products that will be delivered, the delivery dates, delivery locations and quantities required.
- Provide a summary of the schedule and budget for the VVT disposal program.

1.2: *Evolution of Plan.* Specify the strategy for generating both scheduled and unscheduled updates to this planning document.

Section 2: References

2.1: *Standards and Documents.* Provide a list of all documents and other sources of information referenced in the document.

2.2: *Deviations and Waivers.* Lists deviations and waivers from the referenced documents.

Section 3: Definitions. Provide references and definitions of acronyms used in the planning document.

Section 4: VVT Disposal Program Organization

4.1: *External Interfaces.* Describe the organizational boundaries between the VVT disposal program and external entities.

4.2: *Internal Structure.* Describe the internal structure of the VVT disposal program organization to include the interfaces among the units of the development team.

4.3: *VVT Disposal Program Roles and Responsibilities.* Identify the nature of each major work activity as well as the supporting process.

Section 5: Management Process

5.1: *Start-Up*

- Specify the cost and schedule for conducting the VVT disposal program as well as methods, tools and techniques used to estimate the program cost, schedule, resource requirements and associated confidence levels.
- Specify the number of VVT staff required by skill level, the VVT disposal program phases in which the numbers of personnel and types of skills are needed and the duration of the need.
- Specify the means for acquiring the resources in addition to personnel needed to successfully complete the VVT disposal program.
- Specify the training needed to ensure that necessary skill levels in sufficient numbers are available to successfully conduct the VVT disposal program.

5.2: *Work Planning*

- Specify the work activities to be performed in the VVT disposal program.
- Specify the scheduling relationships among work activities in a manner that identifies the functional or time-sequencing constraints and illustrates opportunities for concurrent work activities.
- Specify the resources allocated to each major work activity in the VVT disposal program Work Breakdown Structure (WBS).
- List of the necessary resource budgets for each of the major work activities in the WBS.

5.3: *VVT Disposal Program Controls*

- Specify the control mechanisms for measuring, reporting and controlling changes to the VVT product requirements.

- Specify the control mechanisms to be used to measure the progress of work completed at the major and minor VVT disposal program milestones.
- Specify the control means to be used to measure the cost of work completed and compare it to the planned budget.
- Specify the mechanisms to be used to measure and control the quality of the work processes and the resulting VVT work products.
- Specify the methods, tools and techniques to be used in collecting and retaining VVT disposal program metrics.
- Specify the reporting mechanisms and dissemination of VVT disposal program status to entities external to the program. Typical information includes status of requirements, schedule, budget and quality.

5.4: *Risk Management.* Specify the risk management plan for identifying, analyzing and prioritizing VVT disposal program risk factors.

5.5: *VVT Disposal Program Closeout.* Specify plans necessary to ensure orderly closeout of the VVT disposal program.

Section 6: Technical Process

6.1: *Process Model.* Define the relationships among major VVT disposal program work activities and supporting processes by specifying the flow of information and work products among activities and functions, the timing of work products to be generated, reviews to be conducted, major milestones to be achieved, baselines to be established, VVT disposal program deliverables to be completed and required approvals that span the duration of the VVT disposal program.

6.2: *Methods, Tools and Techniques.* Specify the development methodologies, tools and techniques to be used to develop and maintain the VVT disposal program work products.

6.3: *VVT Disposal Program Infrastructure.* Specify the plan for establishing and maintaining the development environment, policies, procedures, standards and facilities required to conduct the VVT disposal program.

6.4: *Product Acceptance.* Specify the acceptance criteria of the deliverable work products generated by the VVT disposal program.

Section 7: Supporting Processes

7.1: *Configuration Management.* Define the configuration management plan for the VVT disposal program.

7.2: *Independent Verification and Validation.* Identify an Independent Verification and Validation (IV&V) mechanism to audit the VVT disposal program and, subsequently, its execution.

7.3: *Documentation.* Define the documentation plan for the VVT disposal program.

7.4: *Quality Assurance.* Submit the VVT disposal PMP to an independent assessor in order to verify that it fulfills its commitments to the process and the product as specified in the requirement specification and any standards, procedures or guidelines to which the process or the product must adhere.

7.5: *Reviews and Audits.* Specify the schedule, resources, methods and procedures to be used in conducting VVT disposal program reviews and audits.

7.6: *Problem Resolution.* Specify the resources, methods, tools, techniques and procedures to be used in reporting, analyzing, prioritizing and processing problem reports generated during the VVT disposal program.

7.7: *Contractor Management.* Specify plans for selecting and managing any subcontractors that may contribute to the VVT disposal program.

7.8: *Process Improvement.* Include plans for periodically assessing the VVT disposal program, determining areas for improvement and implementing improvement plans.

Methods and Further Literature

Section 4.3.1, VVT process planning

- DoD-STD-4160.2-M (1997)
- IEEE 1058-1998 (1998)

Section 5.7.13, Disposal testing

- Spinner (1991)

3.4.2 Assess the System Disposal Plan

Objective The objective of this activity is to assess the system's disposal process plan notwithstanding safety, environmental and economic issues as well as relevant statutory considerations.

Description The majority of fielded systems have few, if any, requirements associated with disposal. Most often, the components are removed, transported to various disposal locations and discarded. In certain circumstances, the system may have materials whose disposal has statutory requirements due to hazard or safety considerations. An example is spent uranium fuel rods from nuclear reactors whose disposal raises both safety and long-term hazard issues.

The system disposal team must identify an appropriate disposal strategy and then develop a disposal plan. This must comply with relevant environmental and economic regulations and current legislation. While the Disposal phase is identified as the final phase of the system lifecycle, the implications for the disposal of components and systems must be considered throughout a system's lifetime. More specifically, the initial disposal planning should be addressed during the system Definition phase and the system Design phase. Disposal of enabling products should also be considered during the system Design and system Production phases when individual system component designs solidify. The planning of the disposal process should be verified in earlier phases, whereas the validation and the verification of actual disposal of the system and the enabling products should take place as part of the Disposal phase.

The disposal plan must be assessed by the VVT team, which should verify that (1) the plan calls for system disposal in accordance with relevant statutory requirements, mainly to avoid hazardous wastes, and (2) the process provides maximum economic benefit as the system comes to its end-of-life stage.

The VVT team should verify, first, that the disposal team is fully satisfied that there is no further practicable use for the system and that it is truly surplus to current requirements before declaring it for disposal. Second, the VVT team should verify that all other, creative system end-use scenarios, which may comprise significant economic value, have been considered. For example:

- Redeploying the system for a different purpose, for example, as a training/instructional or demonstration platform or as a spare system used for parts cannibalization
- Reclamation of the system and expending its lifetime or recycling the usable portions of the system or remanufacturing and upgrading the system
- Reselling the system to other users as potential customers may be interested in deploying the system under a less stringent set of requirements

Third, the VVT team should verify that the system disposal plan clearly defines its goals in realistic and specific terms. The plan should identify all the main issues which need to be addressed as well as the budgetary, manpower requirements and organizational structure with clear responsibilities and accountabilities. In addition, the VVT team should verify that the disposal plan refers to the disposal requirements and how they will be met, the schedule of the plan and all major events and specific strategies for plan implementation.

The VVT team should pay particular attention to the system's disposal requirements since a typical system has a long Use/Maintenance phase and statutory requirements for system disposition may have changed drastically over the life of the system. An example is when the electric generation industry switched from Askarel dielectric and cooling oils to polychlorinated biphenyl (PCB)–based dielectric and cooling oils for large high-voltage transformers. It was discovered that PCBs have serious environmental issues; therefore, disposal of damaged or decommissioned transformers had to be conducted in accordance with new laws and disposal processes had to be developed to meet the new regulations.

Finally, a key VVT activity is to verify that the disposal procedure and infrastructure, as detailed in the system disposal plan, address safety and environmental issues as well as associated statutory obligations. The disposal of a system may require a significant infrastructure, especially if the disposal requirements relate to safety or environmental issues. An example is the shipping industry where, under U.S. and European law, older vessels and especially oil tanker and chemical transport ships must be scrapped under quite stringent regulations.

Verification of the disposal procedures and infrastructure prior to commencement of disposal activities is critical in order to ensure that they meet needed requirements. Often disposal requirements encompass severe economic considerations as well. Therefore, the infrastructure must also be validated against such constraints.

Methods and Further Literature

Section 4.3.1, VVT process planning
Section 4.4.1, Expert team reviews
Section 4.4.3, Group evaluation and decision
Section 5.7.13, Disposal testing

- Blanchard and Fabrycky (2005)
- NASA/SP-2007 6105 (2007)
- Ogata (2003)
- SEF DoD (2001)
- Zahavi and Barlam (2000)
- Zienkiewicz and Morgan (2006)

3.4.3 Assess System Disposal Strategies by Means of Simulation

Objective The objective of system disposal simulation is to assess the environmental impacts and the level of recycle-ability related to different disposal solutions available. Eventually, an optimal disposal strategy is identified and the optimality of this strategy is assessed during this activity

Description Simulation methods may be used in order to assess whether the system disposal strategy is optimal. The advantage of this approach stems from the fact that under simulated conditions the input parameters can be easily adjusted, whereas physical evaluation of different disposal strategies is very complex, time consuming and sometimes hazardous.

A valid assessment of suitable system disposal strategies is not an easy task. As a result, this issue is often ignored or analyzed superficially. Several simulation methods may be used to assess available disposal technologies for the system under study. For example, common techniques such as landfill or incineration may be evaluated. Existing tools provide a general indication regarding the diffusion of harmful substances or the efficiency of the combustion process. Well-established models of this type are commonly used, for example, in the area of nuclear waste storage to assess the risk of contamination due to leaching.[19] Disassembly and recycling activities may also take advantage of simulation results in estimating the amount of salvageable material to be recycled and in the visualization and comprehension of an optimal sequence of disposal operations that are both safer and less expensive.

Usually stochastic simulation techniques are used to define the probability density function needed to assess environmental risk levels and the salvageability level associated with different disposal strategies.

Methods and Further Literature

Section 4.3.4, System test simulation
Section 4.3.7, Model-based testing
Section 4.4.1, Expert team reviews
Section 5.7.13, Disposal testing

- Blanchard and Fabrycky (2005)
- NASA/SP-2007 6105 (2007)
- Ogata (2003)
- SEF DoD (2001)
- Zahavi and Barlam (2000)
- Zienkiewicz and Morgan (2006)

3.4.4 Assess On-Going System Disposal Process

Objective The objective of this activity is to verify that the ongoing system disposal process is performed according to applicable environmental and health regulations and policies. This objective includes verifying that (1) the remains of the system contain no harmful substances to the environment, (2) the disposal process does not constitute any health risk to persons involved in the process and to living organisms in general and (3) the economic maximization of the residual value of obsolete systems by recycling usable components and salvaging exploitable materials.

Description The enormous number of disposed systems every year generate massive amount of hazardous waste. In general, wastes are hazardous if they are toxic to living organisms or ignitable, corrosive and/or reactive or if they appear on a list of about 100 industrial waste streams (Lippitt et al., 2000). Obsolete systems such as electrical and electronic equipment, automobiles, industrial machinery, aircraft, ships and buildings often contain hazardous waste. This may include contaminated sludge, solvents, acids, heavy metals and other chemical wastes. Improper waste disposal is hazardous to human and animal health and the environment and also represents significant economic loss.

[19]This occurs when perched water table conditions exist in the soil profile during rainy seasons. Consequently, after cessation of the rainy seasons, the pollutants are convected downward by the declining perched water table, contaminating large tracts of land as well as the freshwater aquifer.

Today there are many states, national and international regulations mandating safe disposal of systems and waste material. These regulations also direct the salvaging of certain substances for industrial recycling. The role of the VVT team during the disposal phase is to verify that actual disposal processes adhere to existing disposal regulations and policies. Since VVT disposal activities are unique to specific industries, we will give below, by way of example, typical verification activities associated with the disposal of electrical and electronic systems.

Until recently the standard method for disposing of electrical and electronic components, cathode ray tubes (CRTs) and computers was solid-waste landfill disposal. Thousands of tons of such obsolete systems containing vast quantities of toxic materials entered the waste stream annually and caused serious health problems and significant environmental damage near electronic dump sites, notoriously in China India and some parts of Africa. Table 3.2 provides a list of the potential health hazards of materials commonly used in electronic equipment.

TABLE 3.2 Hazardous Materials in Electrical and Electronic Systems

Material	Characteristic Location in Systems and Nature of Hazard
Lead	Lead is a metal used for soldering electronic components onto printed circuit boards and in CRTs. Lead causes damage to blood, kidney systems, central and peripheral nervous systems and the reproductive system in humans.
Cadmium	Cadmium occurs in certain components such as chip resistors, infrared detectors, semiconductor chips and batteries. Cadmium and its compounds are toxic to humans and animals and accumulate in the body, particularly the kidneys.
Mercury	Mercury is used in electrical and electronic equipment. It is used in thermostats, sensors, relays, switches, medical equipment, lamps, mobile phones and batteries. Mercury can cause damage to human organs, especially the brain and kidneys. In addition, fetus development is highly susceptible to mercury exposure.
Chromium	Chromium is used as corrosion protection of untreated and galvanized steel plates and as a decorative or hardener for steel housings. It is easily absorbed into the human body and then produces various toxic effects within the contaminated cells. Chromium can cause damage to DNA and is extremely toxic in the environment.
PVC	Polyvinyl Chloride (PVC) is mainly found in cabling and computer plastic housings, although many computer moldings are now made with the somewhat more benign ABS[20] plastics. As with other chlorine-containing compounds, dioxin can be formed when PVC burns.

[20] ABS (Acrylonitrile, Butadiene and Styrene) is used in the preparation of a wide spectrum of plastics that combine the properties of resins and elastomers, offering toughness, high impact strength and surface hardness.

TABLE 3.2 ***Continued***

Material	Characteristic Location in Systems and Nature of Hazard
BFR	Brominated Flame Retardant (BFR) is used in the plastic housings of electronic equipment and in circuit boards to prevent flammability. Several researchers [e.g., U.S. Environmental Protection Agency (EPA)], suggest that chemical compounds emanating from BFR are toxic and could have harmful effects on humans, animals and water-living organisms.
Beryllium	Beryllium is commonly found on electronic motherboards and "finger clips". Beryllium has been classified as a human carcinogen since exposure to it can cause lung cancer. The primary health concern with respect to this metal is inhalation of beryllium dust, fume or mist.
Phosphor	Phosphor is applied as a coat on the interior of the CRT faceplate. The phosphor is toxic and its coating contains very toxic heavy metals, such as cadmium, zinc and vanadium, as additives.
Toners	Toners are stored in plastic printer cartridges. Ingredients of black toners have been classified as possibly carcinogenic to humans. Some reports indicate that color toners (cyan, yellow and magenta) contain heavy metals, which are hazardous to animals and humans.

There are numerous privacy and environmental protection regulation related to electrical and electronic systems. The EU directives 2002/95/EC[21] on the restriction of the use of certain hazardous substances in electrical and electronic equipment and 2002/96/EC[22] on waste electrical and electronic equipment are designed to tackle the fast increasing waste stream of electrical and electronic equipment and complement EU measures on landfill and incineration of waste. Increased recycle of electrical components will limit the total quantity of waste moving into final disposal. Producers will have to take back and recycle their electrical and electronic equipment. This will also give incentives to design systems in an environmentally efficient way which takes waste management aspects into account. This may include:

1. *Verify Alternative Disposition.* The VVT team should verify that alternative disposition of electronic systems such as computers and peripherals, cell phones and other embedded electronics extracted from household equipment to automobiles, machinery and other engineered systems has been considered prior to actual disposal. This may include:
   - Verification of whether a reasonable effort was made to give the obsolete systems to other units within the organization

[21] Directive 2002/95/EC of the European Parliament and of the Council of January 27, 2003, on the restriction of the use of certain hazardous substances in electrical and electronic equipment.
[22] Directive 2002/96/EC of the European Parliament and of the Council of January 27, 2003, on Waste Electrical and Electronic Equipment (WEEE).

- Verification of whether obsolete systems that have residual value could be sold to outside organizations or donated to charitable or community projects, schools and so on.

2. *Verify Removal of Sensitive Data.* The VVT team should verify that any sensitive or confidential data stored within electronic equipment and any software licensed to the organization have been removed. This includes:
   - Verification of whether all sensitive data held on computers and other equipment containing memory have been irrevocably erased or destroyed before transferring the systems for reuse or disposal. In particular, verify that various privacy legislations be met as such information discovered by a later owner may cause controversy, adverse publicity and lawsuits (see, e.g., in the United States[23] and other countries[24]). Merely deleting the visible files is often not sufficient to achieve irrevocable data erasing since data recovery software could sometimes be used to "undelete" such files.
   - Verification of whether adequate destruction of data was carried out under clear responsibility of the unit that owns the equipment and not delegated to an outside organization without adequate contractual obligations being imposed.
   - Verification of whether data stored in devices which were not in working order were disposed of. Verify that such data were still erased (e.g., by adequately exposing magnetic storage devices to a powerful magnetic field).
   - Verification of whether information-carrying media (e.g., disks, tapes, CD-ROMs) containing extremely sensitive or secret information have been physically destroyed or shredded prior to disposal in accordance with relevant procedures.
3. *Verify the Disposal Process.* The VVT team should verify that if systems cannot be reused in one way or the other, then they should be disposed of in an environmentally friendly manner and appropriate constituents should be recycled to maximize economic benefits and meet existing regulations. This includes:
   - Verification of whether obsolete electronic equipment is completely disassembled and recycled in compliance with rigorous American, European, Japanese or other health and environmental regulations. That is, verify that toxic electronic components have been eliminated

[23] The Gramm-Leach-Bliley Act (GLB), Health Insurance Portability and Accountability Act (HIPAA) and Sarbanes-Oxley Act of 2002.

[24] Canada's Personal Information Protection and Electronic Documents Act (Bill C-6) and the EU's Safe Harbor Accord for the European Commission's Directive on Data Protection.

prior to burial of the remaining material in landfills and the process was accomplished without harming the workers in the industry.

- Verification of whether the disposal process includes harvesting of raw materials such as plastics and heavy metals for reuse.
- Often organizations use external vendors to dispose of their obsolete electronic equipment. The VVT team should verify that the organization has direct and specific knowledge regarding the vendor's disposal practices. A vendors' involvement in offshore dumping or other illegal and environmentally unsound disposal techniques may lead to the vendor's prosecution as well as lawsuits against organizations that used their services.
- Sometimes external disposal vendors give organizations a "certificate of disposal" providing evidence of services performed. The VVT team should verify that the disposal organization maintains such a certificate and demand a full audit trail showing the stage and outcome of each disposal process.

Methods and Further Literature

Section 4.3.4, System test simulation
Section 4.3.5, Failure mode effect analysis
Section 4.3.6, Anticipatory failure determination
Section 4.4.1, Expert team reviews
Section 4.4.3, Group evaluation and decision
Section 5.7.3, Regression testing
Section 5.7.13, Disposal testing

- Lippitt et al. (2000)
- Richard (2002)

3.4.5 Conduct Engineering Peer Review to Assess System Disposal Processes

Objective The objective of this activity is to utilize engineering peer review in order to assess whether the ongoing system disposal process is performed in accordance with the system's disposal process plan and according to applicable environmental and health regulations and policies.

Description Engineering peer review may be used to assess a system disposal process as it is actually performed and should be an ongoing verification process conducted throughout the system Disposal phase. The basis for the peer review should be the system disposal process plan as well as appropriate documents summarizing the ongoing disposal process (e.g., certificates of disposal, disposal audit trail). The following provides a list of topics that may be considered by disposal peer reviews. Such peer reviews may be conducted on a cyclical basis covering different topics each time.

Proposed Topics: Engineering Peer Reviews of System Disposal

Topic 1: Review Alternative System Disposition

- Review whether a reasonable effort was made to give the obsolete systems to other units within the organization.
- Review whether systems that had residual value were in fact sold to outside organizations or donated to charitable or community projects, schools and so on.

Topic 2: Verify Removal of Sensitive Data

- Review whether all sensitive data held on computers and other equipment containing memory have been irrevocably erased before transferring the systems for reuse or disposal.
- Review whether adequate destruction of data was carried out under clear responsibility of the unit that owns the system.
- Review whether data stored in devices which were not in working order were also disposed of.
- Review whether media containing extremely sensitive or secret information have been physically destroyed or shredded prior to disposal in accordance with relevant procedures.

Topic 3: Verify Disposal Process

- Review whether obsolete electronic equipment was, in fact, completely disassembled and recycled in compliance with relevant health and environmental regulation.
- Review whether the disposal process includes the harvesting of raw materials for reuse.
- Review whether the organization has direct and specific knowledge about the disposal process indicating that disposal vendors do not engage in illegal and environmentally harmful disposal activities such as offshore dumping.
- Review whether disposal vendors give the organization certificates of disposal and the organization maintains these documents along with a full audit trail showing the stage and outcome of the disposal process.

Methods and Further Literature

Section 4.4.3, Group evaluation and decision

- Richard (2002)

Section 5.7.13, Disposal testing

3.5 REFERENCES

AFSCR 64-2, Air Force System Command Regulation 64-2, *Production Readiness Rev.*, June 1995.

Belytschko, T., Liu, W. K., and Moran, B., *Nonlinear Finite Elements for Continua and Structures*, Wiley, New York, 2000.

Blanchard, S. B., and Fabrycky W. J., *Systems Engineering and Analysis*, 4th ed., Prentice Hall, Upper Saddle River, NJ, 2005.

Blanchard, S. B., Verma, C. D., and Peterson, E. L., *Maintainability: A Key to Effective Serviceability and Maintenance Management*, Wiley-Interscience, New York, 1995.

Bossert, L. J. (Ed.), *Supplier Management Handbook*, 6th ed., ASQ Quality Press, 2004.

Bothe, R. D., *Measuring Process Capability: Techniques and Calculations for Quality and Manufacturing Engineers*, McGraw-Hill, New York, 1997.

Brauer, C. D., and Cesarone, J., *Total Manufacturing Assurance*, CRC Press, Boca Raton, FL, 1991.

Chandra, A., and Mukherjee, S., *Boundary Element Methods in Manufacturing*, Oxford University Press, 1997.

Deming, E. W., *Out of the Crisis*, MIT Press, Cambridge, MA, 2000.

DI-ILSS 80095, U.S. Department of Defense (DoD), Integrated Logistics Support Plan (ILSP), approved December 17, 1985.

DI-NDTI-81307A, Data Item Description, First Article Qualification Test Plan and Procedures, Nov. 2006.

DI-QCIC-81110, Data Item Description, Inspection and Test Plan, Dec. 1990.

DoD-STD-1702 (NS), U.S. Department of Defense (DoD), Military Standard Integrated Logistics Support Programs for Equipment, Subsystems, and Systems, December 17, 1985.

DoD 4160.21-M, Defense Material Disposition Manual, U.S. Department of Defense, Washington, DC, Aug. 1997.

Geng, H., *Manufacturing Engineering Handbook*, McGraw-Hill Professional, New York, 2004.

IEEE 1058-1998, Standard for Software Project Management Plans, IEEE Computer Society, New York, 1998.

Jones, V. J., *Integrated Logistics Support Handbook,* special reprint ed., McGraw-Hill Professional, 1998.

Juran, M. J., and Godfrey B. A., *Juran's Quality Handbook*, 5th ed., McGraw-Hill Professional, 2000.

Kalpakjian, S., and Schmid, S., *Manufacturing, Engineering & Technology*, Prentice Hall, Upper Saddle River, NJ, 2005.

Knezevic, J., *Systems Maintainability*, Springer, 1997.

Lippitt, J., Webb, P., and Martin, W., *Hazardous Waste Handbook*, 3rd ed., Butterworth-Heinemann, 2000.

Loch, H. C., van der Heyden, L., van Wassenhove, N. L., Huchzermeier, A., and Escalle, C., *Industrial Excellence: Management Quality in Manufacturing*, Springer, 2003.

Matko, D., Zupancic, B., and Karba, R., *Simulation and Modelling of Continuous Systems: A Case-Study Approach*, Prentice-Hall, Englewood Cliffs, NJ, 1992.

MIL-STD-1521B, Military Standard—Technical Reviews and Audits for Systems, Equipments, and Computer Software, U.S. Department of Defense, 1995.

Nahmias, S., *Production and Operations Analysis*, 5th ed., McGraw-Hill Higher Education, 2004.

NASA/SP-2007 6105, *NASA Systems Engineering Handbook*, Revision 1, National Aeronautics and Space Administration, NASA Headquarters, Washington, DC, December 2007.

Ogata, K., *System Dynamics*, 4th ed., Prentice Hall, Upper Saddle River, NJ, 2003.

Richard, C. P., *The Economics of Waste*, RFF Press, 2002.

SAE-AS9102A, Aerospace First Article Inspection Requirement, Society of Automotive Engineers, January 2004.

SEF DoD, Systems Engineering Fundamentals (SEF), Department of Defense, Supplementary Text Prepared by the Defense Acquisition University Press, Fort Belvoir, VA, 2001.

Shewhart, A. W., *Statistical Method from the Viewpoint of Quality Control*, Dover, 1986.

Spinner, P. M., *Elements of Project Management: Plan, Schedule, And Control*, Prentice-Hall, Englewood Cliffs, NJ, 1991.

Stephens, S. K., *The Handbook of Applied Acceptance Sampling: Plans, Procedures & Principles*, ASQ Quality Press, 2001.

Tanner, P. J., *Manufacturing Engineering*, CRC Press, Boca Raton, FL, 1990.

Webb, A., *Project Management for Successful Product Innovation*, 2nd ed., Gower Publishing, 2000.

Zahavi, E., and Barlam, D., *Nonlinear Problems in Machine Design*, CRC Press, Boca Raton, FL, 2000.

Zienkiewicz, C. O., and Morgan, K., *Finite Elements and Approximation*, Dover, 2006.

Chapter **4**

System VVT Methods: Non-Testing

4.1 INTRODUCTION

As discussed in Chapter 1, VVT engineers often use the term "testing" colloquially to mean VVT. But, in a narrower sense, following the VVT definition, testing is a subset of verification and validation, dealing with actively operating the system and verifying or validating it. The term *nontesting* refers to all the VVT activities which are not specifically testing per se. Accordingly, this chapter describes system nontesting VVT methods in the narrow sense. The chapter is divided into three parts: (1) prepare VVT products, (2) perform VVT activities and (3) participate in reviews. Each part describes nontesting VVT methods characteristic of the relevant group.

4.2 PREPARE VVT PRODUCTS

4.2.1 Requirements Verification Matrix (RVM)

A Requirement Verification Matrix (RVM) is usually composed of (1) a requirement identification code, (2) requirement traceability to higher level documents, (3) verification methods to be used, (4) the stage(s) where verification takes place and (5) the verification procedure identification code. Verification methods often listed in the RVM are Analysis, Inspection, Demonstration, Testing and Certification (see typical RVM structure in Figure 4.1).

Verification, Validation, and Testing of Engineered Systems, Avner Engel

Requirement ID	Requirement traceability	Verification method						Verification stage					Procedure ID
		None	Analysis	Inspection	Demonstration	Test	Certification	Definition	Design	Implementation	Integration	Qualification	

Figure 4.1 Typical RVM structure.

The following guidance is proposed in order to assign a specific verification method to a given system requirement:

Verification by Analysis Heuristically, a system analysis method may be used:

- When other verification methods are not possible (e.g., verifying system reliability) or are too expensive (e.g., verifying system behavior in destructive conditions) or endanger humans or property (e.g., test flights outside the normal flight envelope).
- Based on the following means: mathematical models, simulations, algorithms, calculations, charts, graphs and so on.

Verification by Inspection Heuristically, a system inspection method (illustrated in Figure 4.2) typically includes the use of human senses (e.g., sight, hearing, smell and/or touch) or simple physical tools for manipulation or mechanical and electrical gauging and measurements and may be used:

Figure 4.2 Verification by inspection.

- When the intent is to show compliance with very simple requirements (e.g., size, weight, shape and color of a component or a system).
- When it consists of nondestructive examination of items without special laboratory equipment/procedures.
- Typically in component subsystem and system production phase.

Verification by Demonstration A system demonstration method is similar to a system testing method. However, system demonstration is considered a "softer" approach to the verification process. Heuristically, it may be used:

- When the intent is only to generally watch a system accomplishing a certain undertaking within typical operating conditions.
- Quite rarely. For example, Charles Lindbergh "demonstrated" a solo nonstop flight from New York to Paris in a single-seat, single-engine monoplane, the *Spirit of St. Louis*, on May 20–21, 1927. As another example, Richard Rutan and Jeana Yeager piloted the *Voyager* aircraft and "demonstrated" a record-breaking (9 days, 3 minutes, and 44 seconds), nonstop, unrefueled flight around the globe on December 23, 1986 (see Figure 4.3).

(*a*) (*b*)

Figure 4.3 (a) *Spirit of St. Louis* and (b) *Voyager* (NASA photos).

Verification by Testing A system testing method should be considered as the default choice for each entry in the RVM. Naturally, most system requirements will be verified by means of testing. Other verification methods will be selected only under special circumstances. As a general rule it is considered the most rigorous verification method.

Verification by Certification System "certification" may be accepted instead of a test, based on a "verified article" which has been proven under similar operational conditions (e.g., verification of new engine by basing its design on that of a well-performing existing engine). Such certification must indicate the standard/procedure to which the testing was conducted and when, where and by which organization the testing was conducted, state that the testing was successful and state the reason why a certification method is used. Heuristically, a system certification method may be used:

- When a new system is a variant of an existing, tested and proven system.
- When the full verification cycle would be expensive and time consuming.

- When there exists a long-term relationship and trust between the system producers and customers.
- Often during component, subsystem and system manufacturing setting.

Further Literature

- Martin (1997)
- Wasson (2005)

4.2.2 System Integration Laboratory (SIL)

One of the most daunting problems in developing embedded systems[25] is the disconnect that exists between hardware, software and system development loops. As a result most embedded system faults are discovered during integration testing. These faults most often are traceable to misunderstanding requirements or improperly implementing the hardware, software or system interfaces.

One approach to bridge this disconnect is to create a testing environment in which the same tests created to verify the system design are also used to verify the hardware, software and system interfaces. Figure 4.4 depicts such a conceptual environment.

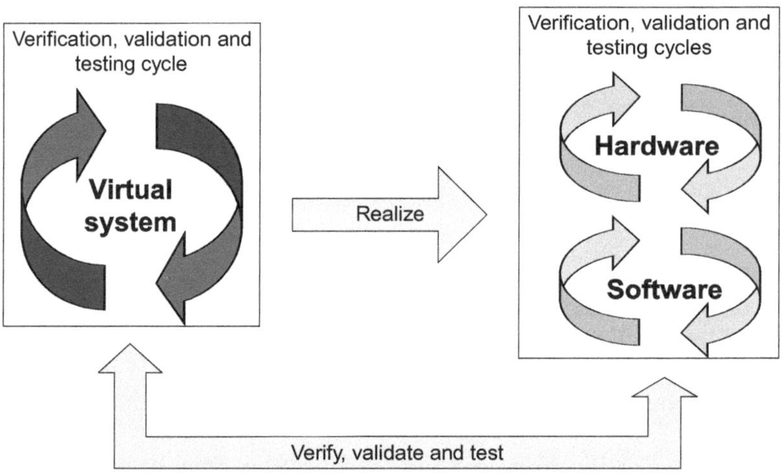

Figure 4.4 SIL concept.

A virtual target system is usually created from Commercial Off-The-Shelf (COTS) hardware and software as well as application software that is developed either manually or by means of executable specifications. In addition, various environmental modules are developed to simulate the external conditions affecting the target system. For example, an aircraft virtual system may

[25] An embedded system is a special-purpose computer-based system designed to perform specific and dedicated functions, often with real-time computing constraints. It is usually embedded within a larger system and may include mechanical and electronic parts such as sensors and actuators.

include simulators to represent flexible body movements, distributed aerodynamics, gravity and fuel slosh. The virtual system may also be interfaced with physical support systems such as hydraulic motion tables and robotic manipulators in order to evaluate certain functionalities such as aircraft's thrust vector control, system actuators and navigation sensors. See, for example, Figure 4.5, where real subsystems A, B and C have already been integrated into the virtual system and others remain to be integrated.

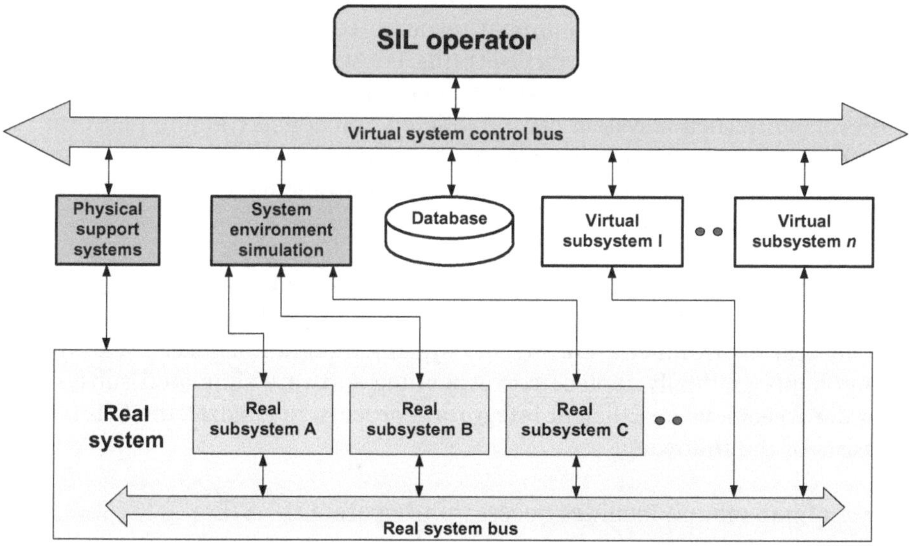

Figure 4.5 Typical System Integration Laboratory facility.

Once the virtual system has been created, a master system level test suite should be generated. This environment-driven test suite is needed to verify the system behavior in realistic nominal and off-nominal scenarios and to gather system performance metrics. Specifically, it will be used to verify and validate the behavior of the virtual system. In parallel, a prototype of the real system is developed and integrated step by step into the virtual system such that simulated elements are eventually replaced with their real prototype counterparts. The beauty of this approach lies in the ability to apply tests from the master system level test suite to the partially real/partially virtual system in stages. The process continues until the entire target system replaces the simulated components and all tests conducted by the master system level pass satisfactorily.

Testing the system by way of virtual SIL provides both test realism of "good" system behavior as well as realistic failure simulation. A typical SIL is capable of exhibiting various levels of "degraded functionality" states. Such ability allows for testing of problematic situations before they occur in the field. For example, one failure condition could be the functional failure (loss) of an individual subsystem. Such loss can be simulated in multiple ways: avoid simulating the subsystem, physical removal of the real subsystem, disconnecting the power cable from the real subsystem and so on.

A properly created SIL offers a unified control structure for the SIL operator, a controlled dynamic environment and a start/stop mechanism. Another important advantage is that individual test engineers may configure the testing environment to interact with their individual test article as well as any desired portion of the system with minimal resource contention issues. In addition, every test engineer may interact with the latest system configuration or any of the earlier versions of the system making regression testing that much easier. Finally, such a system can provide an embedded training platform for mission rehearsal and mission planning evaluation as well as a full environment for testing postrelease fixes, system enhancements and other aspects of system lifecycle considerations.

SIL Description As discussed above, the SIL provides the test engineer or system operator with a real-time dynamic simulation of the target system and its physical environment. During the system Integration phase, real system components gradually replace corresponding software-simulated subsystems in order to achieve an efficient integration process. In general, the SIL facility consists of the following:

- Equipment and facilities necessary to operate the SIL
- Simulation of the elements necessary to operate the system in a real-time environment
- Monitoring and test equipment engaged in the performance of the tests applied to the system and the operational programs
- Facilities to analyze the performed tests
- Real system components [e.g., subsystems, lowest replaceable units (LRUs)]

Typically, the following hardware elements are included in a SIL facility:

- Simulation host computers and peripherals
- Input/output PCs
- Lifelike operational consoles
- Power supplies and a power distribution panel
- Monitoring/test equipment and test point panels
- Operational software development equipment (computers, PCs, etc.)
- Operational subsystems

The SIL software facility provides the capability to test the target system in real time using a simulated system target as well as a simulated environment. In addition the SIL software usually supports saving of simulation data for later analysis. The simulation software is segmented into modules and the modular structure of the software is enhanced by use of the operating system multiprocessing features. Typically, the SIL software is organized within the following packages:

- *Mission Planning.* Software used offline that permits users (i.e., system and test engineers) to interactively define different mission scenarios. This package creates data files for the target system mission initialization.
- *SIL Control.* Software designed to allow users to control physical target subsystems or real-time target simulations or a combination thereof as well as the environment of the target system.
- *SIL Simulation.* Real-time software, which simulates the target systems and their environment and enables the execution of system tests in a realistic, lifelike simulated condition.
- *SIL Monitor.* Software designed to extract relevant data from the real-time target simulation and physical target subsystems as well as the simulated environment, record the data for later analysis and display a subset of the data for users.
- *Post Mission Analysis.* Software designed to read stored simulation data, which was recorded during mission execution, and then analyze it and display the results for users.

Distributed SIL Sometimes, very large systems dispersed over a large geographic area must be tested concurrently. In this case a Distributed System Integration Laboratory (DSIL) may be constructed to provide virtual test systems for multiple test scenarios. Typically, a DSIL comprises multiple simulators, emulators, test beds and control centers interacting with local elements of the target system components and each other over a broadband network (e.g., NASA manned flight missions). A DSIL will be used to perform integration and operational tests (e.g., multielement integration testing, flight element integration testing) as well as system load/stress tests and operational training in much the same way as would a localized SIL.

Distributed system testing presents unique challenges relative to traditional localized testing, especially in terms of system latency, security, timing, data integrity and service availability. At the same time, distributed system testing is sometimes unavoidable and may even yield significant cost benefits in terms of decreased duplication of system hardware, utilization of assets already in place, reduction in maintenance and operations, usage of the most up-to-date system representations, reduction in travel cost and utilization of the more experienced personnel maintaining each system and minimization of system transportation among different facilities.

Distributed system testing may also yield schedule benefits when resources are limited, allowing early testing and yielding less rework due to anomalies in test support equipment. Finally, distributed testing may reduce system development risks by supporting integrated testing throughout the development period. This may be achieved by providing facilities to test prototype system interfaces early to ensure, for example, interface compliance and C^3I interoperability.[26] In addition, risk may be reduced by, de facto, performing early checkout of operational and maintenance procedures.

Generic SIL Sometimes an organization is advised to build and maintain a Generic System Integration Laboratory (GSIL). Such a facility can be instrumental in providing credible proposal data by demonstrating the technical readiness levels of a company's new systems and processes. A functional and technologically up-to-date GSIL may also provide a better starting point from which a program-specific SIL can be tailored, thus reducing the risk of having to start a new SIL design from scratch. A new program's integration and test activity could be performed in a SIL to verify many of the system level requirements using realistic real-time environmental and external stimulus or simulations applied to actual operational hardware and software.

Further Literature

- Braspenning (2008)
- Martinez et al. (2008)
- Obaidat and Papadimitriou (2003)

4.2.3 Hierarchical VVT Optimization

The goal of a hierarchical VVT optimization method is to improve the VVT plans for the complete system, subsystems and its components. Using an iterative process, we can try to reduce or eliminate redundant VVT activities adopting, as much as possible, less costly VVT methods.

Hierarchical VVT optimization may be used when the system development process is underway. At this point, the system has been decomposed into subsystems and components. In addition, it is assumed that the set of requirements at the system level has been prepared and appropriate requirements have been allocated to the various subsystems and components. For example, Figure 4.6 depicts such an allocation of requirements.

[26] C^3I interoperability refers to a Command, Control, Communications and Information architecture that provides interoperability between all elements of such a system.

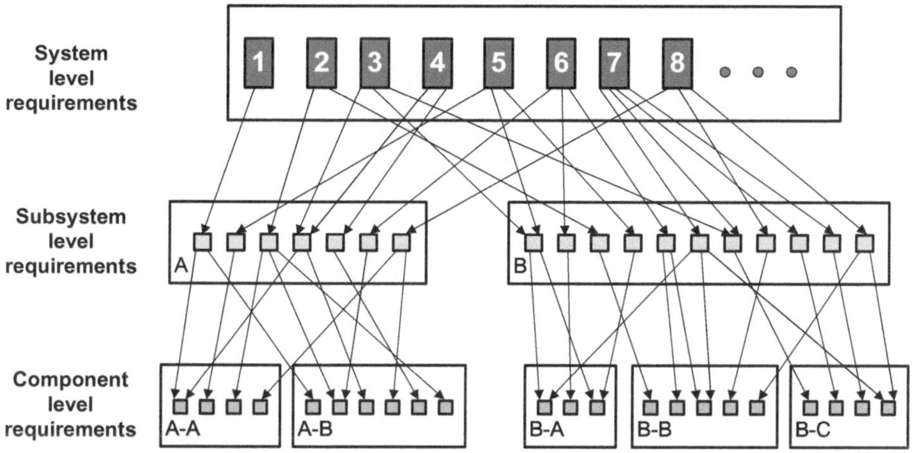

Figure 4.6 System requirements allocated to subsystems and components.

Here, system level requirement 1 is allocated to subsystem A and then further allocated to components A–A and A–B.

In addition, a prerequisite for carrying out the hierarchical VVT method is that the initial versions of the RVMs for the system, subsystems and components are available. Typically each entry in the RVM is composed of requirement identification, requirement traceability to higher level documents, a verification method, a verification stage and verification procedures. Often verification methods consist of analysis, inspection, demonstration, testing and certification. Similarly, the verification stage often follows typical system, subsystem and component development phases: Definition, Design, Implementation, Integration and Qualification.

Hierarchical VVT As mentioned, the intent of hierarchical VVT optimization is to reduce or eliminate, as much as possible, the amount of redundant VVT activities that naturally occur at different levels of the system hierarchy. The inputs to the hierarchical VVT optimization process are the original RVMs associated with the system, subsystem and components as well as a set of constraints applicable to the VVT process. The outputs of the process are updated and hopefully shorter set of RVMs (see Figure 4.7).

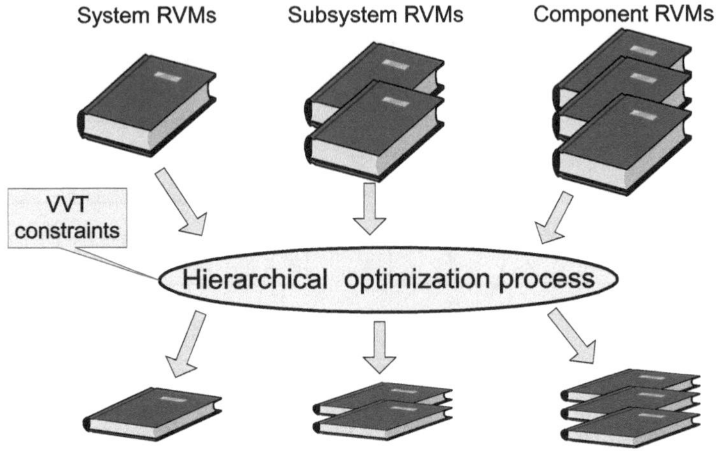

Figure 4.7 Hierarchical optimization of system, subsystem and component RVMs.

At the beginning of the process, all the requirements should be evaluated at the system, subsystem and component levels. Naturally, the first versions of the RVMs may contain many overlapping VVT activities. For instance, requirement 1 in the above example may be tested at the component level (within components A–A and A–B) as well as at the subsystem level (subsystem A) as well as at the system level. Often some testing redundancy could be eliminated based on the nature of the requirements, the test method to be used, the criticality of the function under test and the stakeholders' tolerances for failures.

The optimization process entails reviewing each requirement at each hierarchical level and determining which VVT activity could be eliminated. For example, reviewing the example depicted in Figure 4.6, may suggest that requirement 1 could be tested at the subsystem level and may not require proof at a system level as it must be met in its entirety at the subsystem (A) level.

Hierarchical VVT optimization must be carried out with caution, since individual optimization steps are often subject to various constraints. First and foremost, constraints on funding, schedule or manpower may limit the options here. For example, if VVT funding is only partially available at the Implementation phase (when the subsystem ideally should be tested), then it may be necessary to test it at both the subsystem phase and the system Integration phase. Other constraints may include availability of testing facilities, criticality and safety considerations, geographical distribution as well as stakeholders' involvement in the VVT process. For example, customers wishing to observe the system during acceptance testing may impose an otherwise unnecessary testing activity.

Guidance for Hierarchical Optimization The following guidance can be helpful to someone carrying out hierarchical VVT optimization:

1. Subsystem requirements are derived from the system requirements. Similarly, component requirements are derived from the subsystem requirements. Therefore, requirements at all levels are strongly related and similar validation means may be applied. If such validations are redundant, they should be eliminated if possible.
2. As a general rule, VVT activities should be performed as early as possible. Early corrections of defects are always less expensive than late corrections. As the development progresses from phase to phase, the cost of the correction grows more than linearly.
3. As a general rule, VVT activities should be performed at the component level. Testing components provide better access into the inner recesses of the components (i.e., due to improved controllability). Furthermore, either correct or flawed behavior is more easily observed by testing low-level elements (i.e., due to improved observability).
4. Different verification methods require different investment. Although the testing method may be used most frequently, one should evaluate various verification methods and choose the most effective one.
5. If a given VVT activity is highly critical (e.g., safety- or health-related test) and has a high failure probability, it is recommended that it be performed at the subsystem level and then repeated at the system level.
6. If a given VVT activity has a very low failure probability, it is sometimes recommended that it be performed only at the system level. Savings from this guidance may be realized in terms of both cost and schedule.
7. The hierarchical VVT optimization method often requires negotiations among different system developers, subcontractors and purchasers of the system. This is due to the fact that optimizing the VVT process entails elimination of some VVT activities or transfer of responsibilities among the different organizations involved in system development and validation. For example, if tests to be performed by the suppliers are replaced by tests at the system level, development costs for component or subsystem suppliers may be reduced while cost for the system developer will surely increase. It is then necessary to reach an agreement regarding the development contract.

Advantages of Hierarchical VVT Approach The following advantages are offered by the hierarchical VVT optimization approach:

1. The hierarchical VVT optimization approach can reduce redundancy of the VVT activities by eliminating or scaling down activities that can be made at one level, rather than repeating them at multiple levels.
2. Hierarchical VVT optimization is easy to implement with a limited number of experts.

3. Reducing the number of tests through hierarchical VVT optimization provides both cost savings and time-to-market advantages. In addition, it optimizes the VVT cost of individual verifications by seeking to utilize inexpensive VVT methods whenever possible.
4. This approach fosters a comprehensive and unified visibility of the VVT process at the system, subsystem and component levels and helps to identify the gaps (e.g., missing or inadequate VVT areas) in the overall VVT strategy.

Further Literature

- Siegel (1996)
- Tian (2005)

4.2.4 Defect Management and Tracking

In many organizations the VVT team is tasked not only with the detection of system defects but also with defect management and tracking. The drive for increased system quality demands that developers implement a system to keep track of problems and defects. Customers are increasingly impatient with recurrent system failures. Implementing a system to list and prioritize defects so they are fixed in some logical sequence makes economic sense. This may well be because most of the time spent resolving problems is actually understanding what the fault is and how to eliminate it. In addition, defect tracking helps gain some idea of the amount of work involved in identifying, locating and fixing defects. This knowledge can have quite an impact on resource allocations.

Defect management and tracking may be among the least glamorous aspects of the system development and maintenance process. It lacks appeal, but its importance is at a premium. It is a critical component of a successful quality effort. This laudable practice has mainly been conducted by software developers. We take the liberty of extrapolating and modifying it to the engineered system domain.

Underlying causes of operational failures and defects in products and services are unique in each organization and may be categorized using a Basic Risk Factor (BRF) table. Evaluating the performance of an organization by measuring BRFs provides information about the relative strengths and weaknesses of the organization. Adequately controlling these BRFs will minimize the risk of business disturbances, such as financial losses and diminished reputation. For example, Table 4.1 depicts a list of BRFs associated with an organization engaged in developing and manufacturing large and complex engineered systems.

TABLE 4.1 Example: BRFs for System Development Organization

Category	Description of Basic Risk Factors
Design	Ergonomically poor design of tools, equipment and offices
Hardware	Poor quality, condition, suitability or availability of materials: tools, equipment and components
Maintenance	None or inadequate performance of maintenance tasks and repairs
Housekeeping	None or insufficient attention to keeping the workplace clean and tidy
Error-enforcing conditions	Unsuitable physical conditions and other influences that have a harmful effect on human functioning
Procedures	Inadequate quality, insufficient availability of procedures, instructions and manuals
Training	Insufficient competence or experience among employees
Communication	Ineffective communication between facilities, departments or employees or with other organizations
Incompatible goals	Pursuit of production, financial, political, social or individual goals that conflict with optimal working methods according to established rules
Organization	Shortcomings in the organization's structure, philosophy, processes or management strategies, resulting in reduced revenues
Defenses	Insufficient protection of people, material and environment leading to operational disturbances

Defect Management and Tracking Aims Defect management and tracking aims to:

- Analyze fault history in order to determine the organization's BRFs as well as develop an organization's individual risk profile.
- Identify general weaknesses of an organization in order to improve key development parameters that may improve the organization's quality.
- Define a new strategy to better manage fault and risk.
- Help in defining acceptable quality standard of manufacturing equipment, based on equipment histories, frequency of components failures, and so on.
- Help in managing quality problems during the entire product lifetime, that is, through product Development, production Use/Maintenance and Disposal phases.

Defect Classification Before starting to manage and track any system quality metric, including data about defects, a company or project team should define goals to rationalize such an undertaking. Such goals will directly affect the specific data that are tracked and the complementary analysis effort. With these goals in mind, the team or company can determine the exact data to be collected. For example, the goal of a defect tracking program could be to determine the cause and origin of defects in order to improve the development processes.

Classifying defects is difficult and may result in ambiguous, overlapping or incomplete categories. Yet, the classification of defects into categories can yield important insights, enabling an organization to improve its system development and maintenance process. Figure 4.8, depicts defects categorization scheme tailored for systems. It is based on several publications including Hewlett-Packard concepts published in the late 1990s.

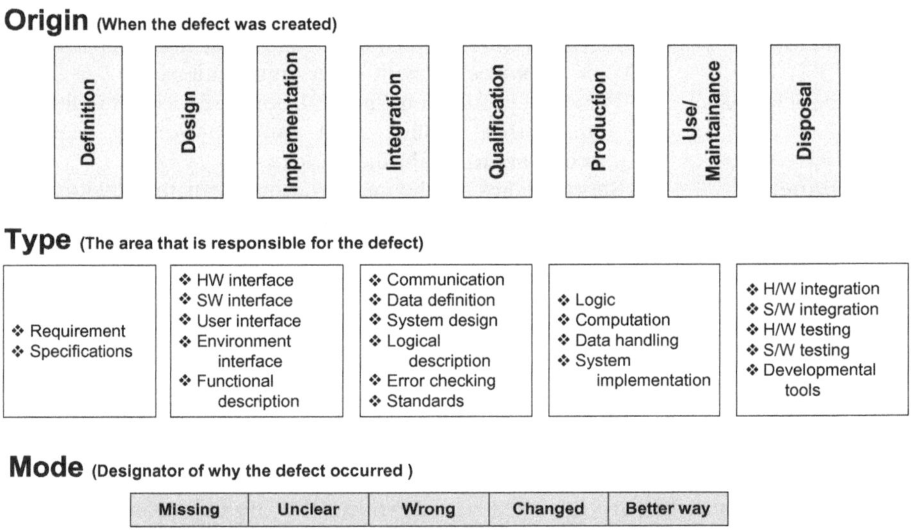

Figure 4.8 Defect classification: origin and type.

As seen in the figure, defects are first categorized by their "origin," that is, the phase in which the defect was introduced into the system. Depending on the phase, each defect is assigned a "type," that is, the area, within a particular origin, that is responsible for the defect as shown in the middle layer of the diagram. All defects, regardless of origin, are further classified based on the defect "mode," that is, a designator of why the defect occurred. For example, a defect which was introduced during the Design phase where a user input control had been omitted would be classified under "missing." An Integration phase defect where a system implementation was incorrect would be classified under "wrong."

Often defects are assigned various attributes. For example, Table 4.2 describe typical defect severity attributes and Table 4.3 describes typical defect priority attributes. In general, defects should be worked on in severity order.

TABLE 4.2 Defect Severity Attributes

Defect Severity	Description
Critical	Application or system shuts down
Major	Errors that prevent continuing system workload
Average	System still functions with a workaround but not as designed
Minor	Minor errors such as user message with spelling or grammar error
Enhancement	System application needs enhancement
Change request	System application functions as designed but not as needed by users
Deferred	Defect will not be fixed immediately or will not be fixed in the current phase

TABLE 4.3 Defect Priority Attributes

Defect Priority	Description
Resolve immediately	Defect requires immediate attention in order to prevent delay in system operations
Give high attention	Defect requires high attention and may delay system operations
Normal queue	Defect requires normal attention and will not delay system operations
Low priority	Defect requires low attention, will not delay system operations and will be addressed after all other defects

When a critical or major error occurs, other VVT activities may be suspended until the defects causing the error have been corrected or a suitable workaround has been identified.

In addition, ancillary information may be collected or computed as part of the defect and management tracking process. For example:

- Number of defects
- Defect discovery rate
- Defect closure rate
- Effort to close a defect
- Elapsed time to close a defect

Defect Management and Tracking Process While not all defects can be avoided, it is possible to minimize their number and impact on a project. One way is to implement a defect management process that focuses on either preventing or identifying defects as early in the process as possible in order to minimize their impacts. A reasonable investment in this process can yield significant returns. The defect management process should be based on the following general principles:

- The process should be risk driven. That is, strategies, priorities and resources should be based on the extent to which risk can be reduced.
- The process should implement defect measurement as an integral part of the development process and be used by the project team to improve the process.

The primary reason for gathering defect information is to improve development processes. When a defect or failure has been detected, a well-designed activity work flow should be followed. Figure 4.9 depicts such a defect management and tracking process. To achieve the aforementioned goals, development teams involved should examine the types of defects that occur most frequently as well as the number and types of defects that occur in each subsystem and component. These latter measures help the VVT team identify system elements that require extra testing or major modification. Additionally, development teams should examine the phases in which defects are encountered. The data gathered could be plotted to identify defect trends.

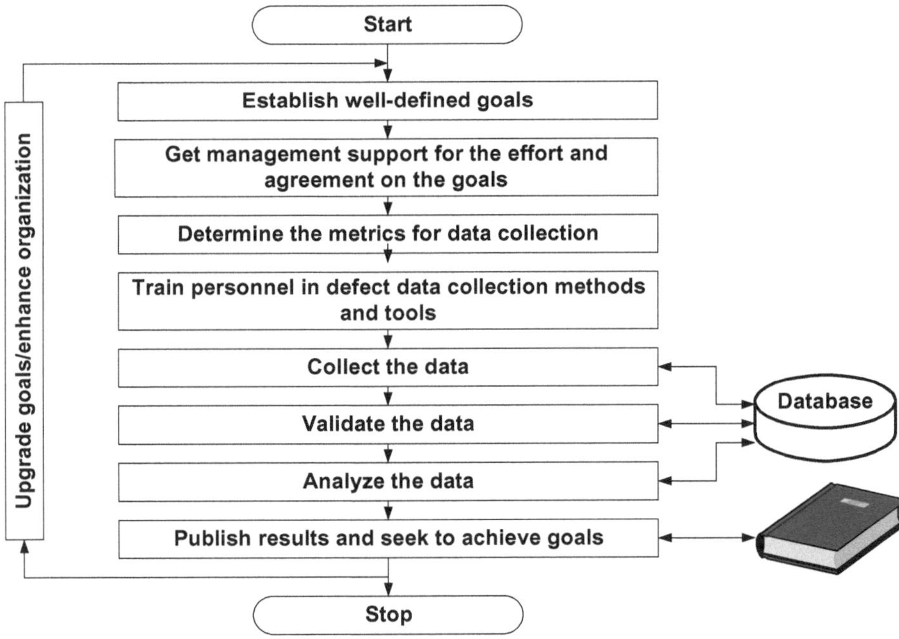

Figure 4.9 Defect management and tracking process.

Defect analysis efforts should focus on the circumstances leading to their introduction as well as the nature of the discovered defects. The intent of using this information is usually to characterize or analyze the environment or a specific development process and then to improve the process in order to eliminate the causes of defects.

Many suppliers of subsystems or components have made defect tracking a part of their ongoing procedures. Usually, it is part of their VVT management system or configuration management. Supplier organizations thus gain understanding about both the products they develop and their development processes. Once defect data are collected, an organization will be able to build a baseline that will allow the VVT team to run statistical analyses to better understand the product and processes. This level of understanding will allow the various development teams to focus their efforts on improving processes. The organization can then recognize its strengths and weaknesses in order to take concrete measures to improve system quality.

Further Literature

- Pfleeger and Atlee (2009)
- Garvey (2008)

4.2.5 Classification Tree Method

A technique for optimizing the functional testing process of systems is the Classification Tree Method (CTM), introduced by Grochtmann and Grimm (1993). These and other authors referenced in this section assess the *input domain* (i.e., space of potential input or environment values) of a test object (system or subsystem) under various operational circumstances. In such a manner, disjoint and complete classifications for test cases are formed. The stepwise partition of the input domain is accomplished by means of classifications represented graphically as a tree. Although the CTM was originally envisioned to classify test objects based on the input domain, we believe the method is also viable when one constructs classification trees associated with *structural or functional domains* (i.e., systems, subsystems or functional capabilities).

The CTM supports functional test case design by systematically and completely segmenting the test object requirement domain into a finite number of mutually disjoint equivalence classes. This is done according to operational aspects relevant to the testing process. Test cases are then generated through a judicious combination of classes.

One of the attributes of the CTM is its simplicity. For that reason, the method is applicable without extensive and time-consuming training. Therefore, over the past few years the CTM has been successfully applied in many industrial software development projects in fields such as aviation and space technology, rail electronics, defense electronics, car electronics, engine electronics and automation technology as well as commercial data-processing applications (Grochtmann and Wegener, 1995).

The CTM is well suited for tool implementation. This is mainly due to (1) the separation of the test case design process into several steps, (2) the graphical representation of a classification tree and (3) the generation of a combination table. Accordingly a Classification Tree Editor (CTE-XL)[27] tool was developed. It recognizes the syntactic rules of the CTM and can act as a stepwise instruction to select test cases.

Method Description The following steps should be undertaken when using the CTM for real-world applications.

Step 1: Selecting Test Objects. A large, real-world system often cannot be tested reasonably with a single classification tree and, as such, a tree would become too large to handle. Therefore, during this step, either the structure or the functionality of the system under test has to be divided into several separate test objects or subsystems. This has to be done in such a way that each of the resulting subsystems can be tested individually and, by testing the combined set of the subsystems, the complete system is tested thoroughly.

Step 2: Designing a Classification Tree. The classification tree identifies specific and relevant requirements for each subsystem. The most important pieces of information required for this task are the relevant functional specifications or requirement documents. Additionally, in order to define the pertinent and critical areas of concern, creativity and expertise on the part of the test engineer are indispensable. For each operational aspect, the input domain should be divided into disjoint subsets. Division into subsets should allow a precise and clear differentiation of possible testing inputs. The partitioning into classes is done separately for each capability of the system and therefore should be easily carried out.

Normally it is useful to introduce subclassifications that include just one component of an existing classification. This use of subclassifications can be continued recursively over several levels until a precise differentiation of all test relevant operational aspects and their classes are achieved. The result is a tree of classifications and classes (i.e., the classification tree).

Step 3: Combining Classes to Form Test Cases. Next, one must build test cases based on the classes in the classification tree. A test case is defined through the combination of classes from different classifications. For each test case, exactly one class of each classification is considered. For this purpose the classification tree is used as head of a combination table wherein the classes that are to be combined are marked. Each line in the table represents a test case and each column represents a final refined class of the classification tree. The number of test cases depends on the test engineer's choice of combinations.

Step 4: Optimizing Testing Process. First, we define a *minimality criterion* as the minimum number of test cases that is necessary to consider each class of the classification tree in at least one test case. Likewise, the *maximality criterion* is defined as all possible combinations of the classification tree classes.

[27]The CTE is a syntax-directed, graphical editor for test case design. It was originally developed by DaimlerChrysler and is marketed by Berner & Mattner Systemtechnik GmbH, Munich, Germany (www.berner-mattner.com).

Selecting a set of test cases meeting the minimality criterion is a straightforward optimization test strategy. However, readers should note that minimizing the number of test cases is not necessarily an optimized testing strategy. In fact, the effectiveness of a system test depends on additional operational aspects such as the interdependency among system functionalities and the criticality of individual test object. Fundamentally, an optimal test strategy entails the execution of a specific set of test cases, where the size of this optimal set is in between the minimality and maximality criterion. Unfortunately, the CTM is silent about this process and the test engineer must use heuristics and common sense to identify this set.

Classification Tree Example The following is an example[28] of a CTM depicting a simplified mobile telephone as the system under test. The inputs to the system include a high-frequency electromagnetic input stream, touch buttons, audio voice and visual images. Similarly, the outputs from the system are a high-frequency electromagnetic output stream, lights, audio and visual images (see Figure 4.10).

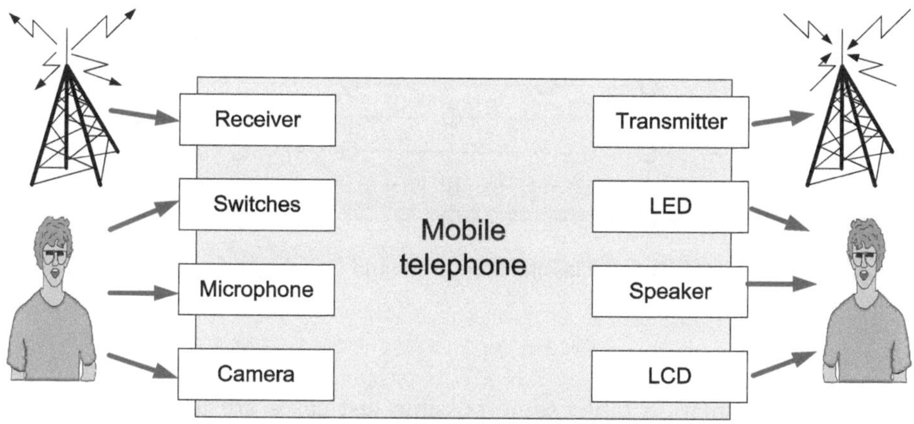

Figure 4.10 Simplified mobile telephone system and its environment.

Appropriate operational aspects for the test in this particular case would be, for example, proper functionality of the various Input/Output (I/O) devices of the system under test, that is, a receiver, switches, a microphone, a camera, a transmitter, an LED (Light-Emitting Diode), a speaker and an LCD (Liquid Crystal Display). The classification based on the radio interface functionality leads to a partition of the I/O domain into a receiver functionality and trans-

[28]This example was inspired by E. Lehmann and J. Wegener, Test Case Design by Means of the CTE XL, in *Proceedings of the 8th European International Conference on Software Testing, Analysis & Review (EuroSTAR 2000)*, Copenhagen, Denmark, December 2000.

mitter functionality and the classification based on the human interface functionality leads to a partition of the I/O domain into a button functionality, audio functionality and visual functionality. Additional operational aspects are introduced for (1) the button class, namely the switches and LEDs, (2) the audio class, namely the microphone and speaker, and (3) the visual class, namely the camera and LCDs. The above classifications and classes are depicted in the classification tree shown in Figure 4.11. Also shown in the figure is the combination table associated with the classification tree.

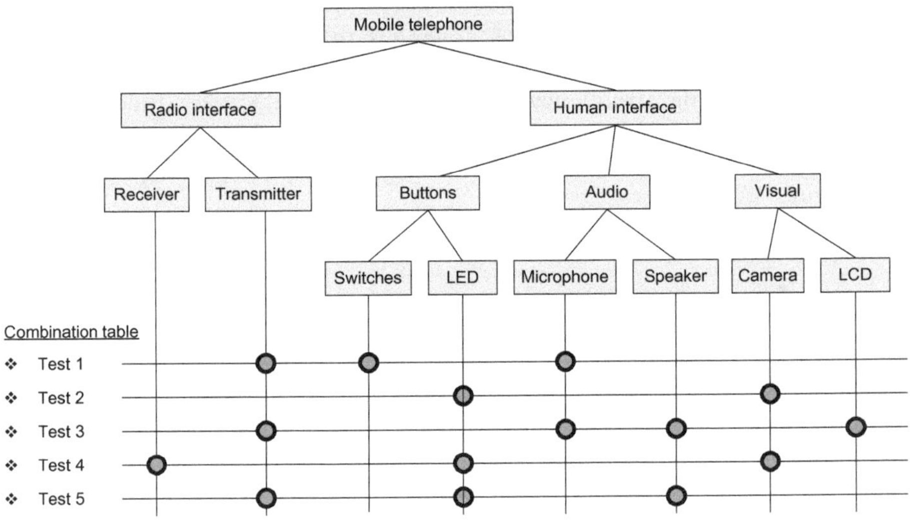

Figure 4.11 Example of a classification tree and combination table for mobile phone testing.

In this combination table, some possible test cases are identified. Test 1, for instance, describes a test involving acquiring microphone audio voice and, under specific switch settings, transmitting it to the external environment (i.e., the relevant cellular antenna tower).

From the minimality criterion, it requires three test cases (i.e., tests 1, 3 and 4) in order to cover all classes of the classification tree in at least one test case. Similarly, in order to compute the maximality criterion (e.g., all possible class combinations), we have to consider the set of all single-system interface tests plus the set of all double-system interface tests and so on up to the set of all system interface tests. This may be computed as follows:

$$
\begin{aligned}
n &= \binom{8}{1}+\binom{8}{2}+\binom{8}{3}+\binom{8}{4}+\binom{8}{5}+\binom{8}{6}+\binom{8}{7}+\binom{8}{8} \\
&= 8+28+56+70+56+28+8+1=255
\end{aligned}
$$

Classification Tool Editor The CTE-XL can be used in a wide range of industry and academic applications since it is independent of specific system functionality. It supports a formal, yet flexible way of specifying and selecting test cases using natural language. In addition, it helps to identify redundant test cases and therefore reduce the overall number of required test cases. The CTE-XL uses a structured graphical representation of test cases. Each test case is specified in a separate line in the combination table. The chronological sequences can be specified by the test engineer in the combination table using an appropriate mechanism. Also, events in the life-cycle of a system could be a classification with corresponding classes in the classification tree. Finally, CTE-XL may be linked with requirement management tools in order to associate requirements with classifications, classes and test cases.

Further Literature

- Alekseev et al. (2007)
- Chen et al. (2000)
- Grochtmann and Grimm (1993)
- Grochtmann and Wegener (1995)
- Lehmann and Wegener (2000)
- Yu et al. (2003)

4.2.6 Design of Experiments (DOE)

Design of Experiments (DOE) encompasses a set of statistical methodologies to efficiently plan and optimize testing processes as well as to analyze their results. The goal of DOE is to maximize the information/cost ratio according to specific objectives. DOE enables the study of complex systems, in particular systems affected by multiple or reciprocal factors. DOE methods are used widely in different disciplines, from social science to economics to engineering. In summary, DOE supports the following three major experimental and testing objectives:

Optimization. DOE helps identify the minimal number of tests necessary to ensure a required level of certainty and robustness.

Screening. DOE helps identify the most influential factors and their interactions affecting responses. As a result, test engineers can determine the necessary investigative direction to achieve optimal testing.

Robustness. DOE helps determine whether the system is robust enough under both controlled and uncontrolled conditions.

According to Montgomery (2004), the DOE encompasses seven steps. The following comparable steps have been elaborated to specifically suit the system testing domain:

1. *Recognition and Statement of Test Problem.* The purpose of this step is to identify the specific system testing problems and the objectives of each individual system test. Focusing on test objectives will lead to an optimal test design and a superior model to extract the maximum information from the VVT to be performed. This step should answer in detail the questions of why and for what purpose the test should be performed and what is the desired result.
2. *Selection of Input and Output Variables.* The purpose of this step is to clearly identify how we want to implement tests, what kind of response is expected from the system and whether or not a given response of a test constitutes a success or failure.
3. *Choice of Factors, Levels and Ranges.* The purpose of this step is to define the metric of the selected factors to be investigated (e.g., controllable, uncontrollable, quantitative, qualitative, multilevel, formulation) as well as their range of interest. Factors, levels and ranges are characterized by the following attributes:
   - Typical factors would be classified into design factors, held-constant factors and allowed-to-vary factors. These factors could further be classified into quantitative and qualitative factors.
   - Typical test levels would be either two levels (high or low) or three levels (high, medium or low) and rarely higher levels
   - Typical test ranges will be based on the previous process knowledge of the test engineer or on a best-guess approach.
4. *Choice of Testing Design.* The purpose of this step is to determine how to organize the experimentation plan. This includes specifics such as test sample size or the choice of test replications as well as the specific order and desired blocking of tests. Available literature can guide testers as to the most appropriate design among the ones available for a given objective. In addition, a fair number of COTS software packages are available to the test engineer for statistical data analysis of various design methods.
5. *Performing Tests.* The purpose of this step is to actually execute the system test according to the established specifications.
6. *Statistical Analysis of Test Results.* The purpose of this step is to analyze the results of the test in accordance with its objectives. For example, a regression analysis is widely used by testers in order to fit raw data to a relevant mathematical model of a system, with the aim of predicting the system behavior. Typically, such models will exhibit linear, quadratic or higher order behavior, depending on the complexity of the system. Another analysis may be aimed at identifying strong interactions between two or more factor inputs, which may imply further testing would be desirable for specific factors.
7. *Conclusion and Recommendations.* First, if the tests revealed any system defects, then in most cases these problems must be corrected and the

system should be progressively tested until it meets its specifications and all requirements have been positively proven. Second, the analysis should identify if there are weak points in the test strategy. If weak points are found to exist, then depending on required resources (funding, schedule, manpower and other resource availability), the test strategy should be amended at those weak points.

Statistical Analysis in Testing Testing systems require the use of multiple tests that replicate the conditions under which the system will actually be used. Clearly, the testing environment is limited in its ability to fully represent actual operating conditions over the life of the tested entity. Thus, tests could a priori be evaluated mathematically to see how close they are to the reality to which the system will be exposed. Statistical analysis is the mathematical set of tools we as engineers depend upon to give us the answers. Engineers involved in system testing need not be mathematicians, but they should be knowledgeable and competent in the use of statistical analysis.

The most important issue in system, subsystem or component testing is the desire to determine if the component, subsystem or system is capable of performing the task for which it is designed. There is never a perfect "yes-or-no" answer to this question. One can only hope to make a yes-or-no decision based on the probabilities determined through statistical analysis. The specific mathematical tool for dealing with this issue is called "hypothesis testing."

A second important issue is to determine the minimum number of test samples required to be reasonably convinced that a given set of system tests will achieve its goal. Namely, does the item being tested fail to meet its stated requirements? This question can be answered with a statistical procedure called "statistical power analysis," which is one of the procedures involved in hypothesis testing. Statistical power is the ability of the statistical analysis of test data to correctly determine that the device or system being tested has failed to meet a requirement. These statistical tools enable the VVT team to efficiently use testing resources, thus making it possible to reduce testing cost.

We summarize here the basics of hypothesis testing and statistic power analysis and then illustrate how these analyses are performed using the free G*Power[29] software package. Additionally, once a set of system, subsystem or component tests have been executed, it is advisable to analyze the test results in order to discover dominating interactions among the various system inputs that affect system behavior. For all these purposes, Analysis-of-Variance (ANOVA) statistical software packages are available free of charge, as are many popular commercial packages, such as SPSS.[30]

[29]A downloadable G*Power software package, as well as various user guides and other relevant materials, is maintained at the Institute for Experimental Psychology, Heinrich-Heine-University, Düsseldorf, Germany. See http://www.psycho.uni-duesseldorf.de/abteilungen/aap/gpower3/.

[30]The SPSS Statistics software package provides a predictive analytic tool for solving scientific, business, engineering and other domain problems. See http://www.spss.com/.

Hypothesis Testing The term *null hypothesis* (labeled H_0) is used by statisticians to indicate a presumed or desired "state of nature." For example, a VVT team receives a newly developed system with the hypothetical but unproven claim that "the system has been constructed in accordance with the required specifications." This might be our null hypothesis. The goal of the VVT team is to determine whether the null hypothesis should be accepted[31] or if it should be rejected in favor of the *alternative hypothesis* (H_1), that is: The system has *not* been constructed in accordance with the required specifications.

A test of some sort is conducted leading to two possibilities: Either the test confirms the null hypothesis or the test rejects the null hypothesis. Because the testing process itself may be flawed, each possibility contains, in fact, two subpossibilities, as depicted in Table 4.4. The test may identify correctly the system as meeting or not meeting its specifications. On the other hand, if the results of the test do not correspond to the actual state of nature, then a testing error has occurred. Broadly, there are two types of testing errors, classified as Type-I and Type-II, depending upon which hypothesis has incorrectly been selected as the true state of nature:

TABLE 4.4 Type-I and Type-II Errors

Real (but Unknown) Situation	Correct Test Results	Incorrect Test Results
System meets specifications	System passes	System fails (Type-I, or alpha error)
System does not meet specifications	System fails	System-passes (Type-II, or beta error)

- *Type-I error*, also known as an **α error**, is the error of rejecting a null hypothesis when the null hypothesis actually is the true state of nature. In VVT parlance, we are finding a defect in a system when in fact the system operates according to its required specifications. Usually Type-I system errors do not constitute a grave problem and are eliminated with relative ease.
- *Type-II error*, also known as a **β error**, is the error of failing to reject a null hypothesis when it is in fact not the true state of nature. Again, in VVT parlance, this is the error of failing to identify a system defect when in truth it exists. Obviously, the consequence of this type of error may be quite severe.

[31] As mentioned, the VVT team cannot positively prove the above null hypothesis but merely assert that the team is unable to disprove it. In other words, the most the testers can say is: "We did not find discrepancies between the systems' behavior and its specifications" (i.e., no defect was found). This is the normal affairs in statistical analysis where the null hypothesis often makes inferences about a universal set based on a limited sample. The null hypothesis may be invalidated but never proved.

There are several approaches to hypothesis testing. The classical *test statistic approach* computes a test statistic from empirical data and then compares it with a critical value. If the test statistic is larger than the critical value or if the test statistic falls into the rejection region, the null hypothesis is rejected. In general, hypothesis testing follows these steps:

- State a null (H_0) and an alternative (H_1) hypothesis.
- Determine significance level (α).
- Compute a test statistic.
- Accept or reject the null hypothesis.

Components of Statistical Power Analysis We can perform statistical power analyses with respect to components, subsystems or the system itself in order to determine the minimum number of test samples required to be reasonably convinced that the system has been adequately tested. For example, a statistical power analysis utilizing the point-biserial correlation[32] model explores relationships among the following four components:

- Sample size (N)
- Population effect size (r)[33]
- Alpha error probability
- Power ($1 - \beta$ error probability)

Sample Size (N) Sample size (N) is the number of observations in a sample. In VVT terminology, this is the number of tests needed to provide reasonable assurance that a system meets a given specification. Often, this is the parameter we seek to determine prior to actually conducting a series of tests. In a priori power analyses,[34] sample size N is computed as a function of the required power level ($1 - \beta$), α and the population effect size. A priori statistical power analyses provide an effective method for minimizing the number of test runs. It is especially desired whenever resources such as the time and money required for the execution of tests are severely limited.

Population Effect Size (r) Effect size (identified as r in the t-test model of the point-biserial correlation) indicates the minimum degree of violation of H_0 a tester would like to detect with a probability not less than $1 - \beta$. Information about a plausible population often comes from previous test runs. However, in the system VVT arena, such data are often not available and we need to derive this value using other means. For example, we can adopt the

[32] The point-biserial correlation is a measure of association between a continuous variable X and a binary variable Y, the latter of which takes on the values 0 and 1. It is assumed that the continuous variables X at $Y = 0$ and $Y = 1$ are normally distributed with means μ_0, μ_1 and equal variance s.

[33] Effect size, in general, is defined as the amount of influence that an independent variable (i.e., the defect being sought) exerts on the dependent variable (the performance of the tested item).

[34] Power analyses prior to actually performing a set of tests.

conventions recommended by Cohen (1988), suggesting that in a t-test of the point-biserial correlation a meaningful set of values of the effect size is

$$r = \begin{Bmatrix} 0.1 & \text{Small} \\ 0.3 & \text{Medium} \\ 0.5 & \text{Large} \end{Bmatrix}$$

Alpha Error Probability Alpha is often called significance level and is the probability of committing a Type-I error. As mentioned above, this error occurs when a null hypothesis is rejected when in fact it is true. The counterpart $(1 - \alpha)$ is called the confidence level, which is used in the form $(1 - \alpha) \times 100\%$ confidence interval of a parameter. Alpha is related to the extent that we are willing to accept a risk of erroneously declaring a system defective when in fact the system functions perfectly. This parameter is chosen subjectively, usually, in the range of 0.01–0.1. A hypothesis test using a lenient α of 0.1 (10%) is more likely to lead to the rejection of the null hypothesis. But if the null hypothesis is concluded on the basis of a lenient α, this conclusion is less convincing than it would be if the same conclusion were reached on the basis of $\alpha = 0.01$. An a of 0.01 identifies significant effects only when the deviation from H_0 is unlikely, leading to a more convincing conclusion.

Power (1 − β Error Probability) As mentioned above, the Type-II error (β) represents the probability of failing to identify a system defect when in truth it is there. The counterpart to this concept is the power of a statistical test $(1 - \beta)$, which is the probability that the test will reject a false null hypothesis. As statistical power increases, the chances of committing a Type-II error decrease.

Component Effects on Statistical Power

Sample Size Generally, a larger sample size increases statistical power. The reason is that when sample size increases, standard error becomes smaller and thus makes the standardized effect size larger. In other words, sample size affects the balance between Type-I errors (α) and Type-II errors (β). In the t-test, for instance, the standard error is a sample standard deviation divided by the square root of the sample size (N):

$$S_{\bar{x}} = \frac{S_x}{\sqrt{N}}$$

Alpha Error Probability Maintaining other parameters constant and increasing α is tantamount to increasing the probability of a Type-I error, which simultaneously decreases the probability of Type-II errors, leading to an increase in the statistical power. Another way to put it is: If a tester changes the significance level from 0.05 to a more lenient value of, say, 0.1, the critical values are shifted to the left, increasing the rejection regions. As a result, β decreases and consequently the statistical power $(1 - \beta)$ increases.

Statistical Power Example We now illustrate how to calculate the required number of independent system tests (sample size) at a given statistical power using the free statistical software package G*POWER 3.

The Problem An Unmanned Air Vehicle (UAV) has been designed for an autolanding capability. The UAV should be able to navigate and fly autonomously from any point within a defined three-dimensional (3D) space to a landing strip (i.e., an airstrip, currently designated for landing) and land there without human intervention (see Figure 4.12).

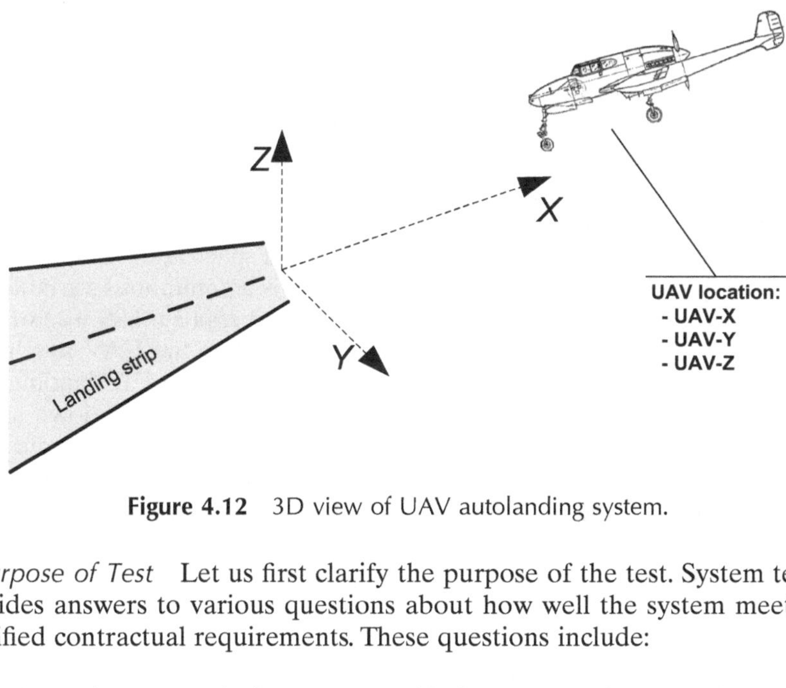

Figure 4.12 3D view of UAV autolanding system.

Purpose of Test Let us first clarify the purpose of the test. System testing provides answers to various questions about how well the system meets the specified contractual requirements. These questions include:

1. Does the system design meet specified system performance?
2. If the system is produced in quantity, what is the percentage of produced systems that fail to perform as specified?
3. Under what conditions will the system continue to perform its function, even when used outside of specified environmental parameters?
4. Will the system meet its specified performance throughout its lifetime?

For our example, the purpose is merely to provide the answer to question 1. Even though we are confining the testing to only one of the four basic questions, we nevertheless have a daunting task ahead of us. System requirements usually specify a range of ambient conditions over which the system must perform well. Also, the system must meet its performance requirement when it has aged as well as when it is brand new. Thus, system performance testing must give appropriate attention to all these issues if the tests are to be unbiased. In our example, we shall assume that due consideration has been given to make the tests realistic and representative of conditions found in the deployed system. In other words, the test shall be planned so that:

- Several different UAV replicas shall be tested.
- The environmental conditions (temperature, wind velocity, precipitation, etc.) shall be varied over the specified ranges.
- Maintenance shall be performed in accordance with specified procedures (no more, no less).
- Selection of which UAV for which test condition shall be entirely random.

This design capability must be tested under simplified but realistic conditions to see whether or not the system meets the requirements. More specifically, the UAV autolanding capability must be tested by bringing the UAV to any location within a 3D space located in front of the landing strip and initiating the automatic landing sequence.

So we can describe the experiment as a system with three inputs factors $\{X, Y, Z\}$ representing the initial location of a UAV in space and an output which indicates a Test Success Score (TSS). Here TSS is a continuous variable representing either total success (TSS = 1), partial success (0 > TSS > 1) or complete failure (TSS = 0). The TSS is computed based on the UAV touchdown rate of descent, UAV angles (i.e., pitch, roll, yaw) relative to landing strip centerline and speed as well as landing strip locations of touchdown as well as completion of UAV rolling. A failed test (i.e., TSS = 0) is declared if (1) the ground operator has to abort the UAV autolanding sequence and manually control it or (2) the UAV either touches down or completes its rolling run outside the confines of the landing strip or (3) the UAV has been damaged in the landing process.

$$\text{TSS} = f(X, Y, Z) \quad \text{TSS} = \{0-1\}$$

Constrained Problem The number of possible tests for this problem is, for all purposes, infinite. However, the cost of each test is considerable, and if we can consider the effect on performance by a given defect within certain intervals to be linear and continuous, we can reduce the number of tests to a reasonable number. For example, we may limit the number of tests by defining a specific set of values for each factor or, similarly, defining a set of rules for determining these values (see Table 4.5). As can be seen, the total number of initial points, and hence tests according to the rules defined in the table is, $3 \times 5 \times 4 = 60$.

TABLE 4.5 Rules of Initial UAV Locations

UAV Initial Location (km)				
Factor Name	Minimum	Maximum	Step Size	Number of Alternatives
UAV-X	3.0	5.0	1.0	3
UAV-Y	–2.0	2.0	1.0	5
UAV-Z	0.5	3.5	1.0	4

According to these rules, the initial UAV positions for this example are depicted as small circles in Figure 4.13.

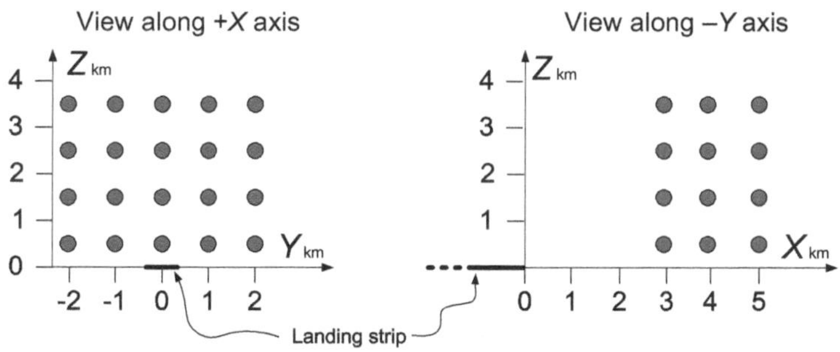

Figure 4.13 Views of UAV initial autoland starting position in space.

Optimizing Number of Tests Due to budget and time constraints, our intent is to further reduce the number of autolanding tests by about 75% (i.e., to execute some 10–15 tests). The problem is how to select the most meaningful tests for actual execution. Usually, in the testing domain, we refer to meaningful tests as the ones that have the highest probability of detecting system failure. Sometimes, selection of such tests can be done intuitively.

For example, let us compare initiating a test from either UAV location {*X*, *Y*, *Z*, = 3, 2, 3.5} or UAV location {*X*, *Y*, *Z* = 4, 0, 1.5}. The first test seems to require a more complex autolanding maneuver; therefore, heuristically, we prefer it as a more meaningful test. However, often the problem does not lend itself to this kind of selection. Furthermore, one facet of this example is the interactions between factors. This is often the case in engineered systems, and therefore the testing problem should be better defined as follows:

$$\text{TSS} = f(X, Y, Z, XY, XZ, YZ, XYZ) \quad \text{TSS} = \{0.0 - 1.0\}$$

Suppose we have an initial set of tests but cannot identify a preferred subset of tests for actual execution. We would like to find a priori (prior to actually performing the set of tests) a reasonable minimum number of system tests of a given statistical power.

We initiate the software package G*POWER 3 and choose the statistical test "*t*-test correlation: point-biserial mode."[35] Next, we select the type of power analysis to be "a priori: compute required sample size" with the intention of performing an upper one-tailed test. We intentionally select a statistical power of 0.8, which introduces relatively large error probability due to the substantial cost of performing each UAV autolanding test. The total cost of the experiments is the dominating factor here.

We proceed by selecting a relatively large effect size of 0.5 as well as a relatively large alpha of 0.1. Finally, we command the software to commence computation and we obtain the results depicted in Figure 4.14 and Table 4.6.

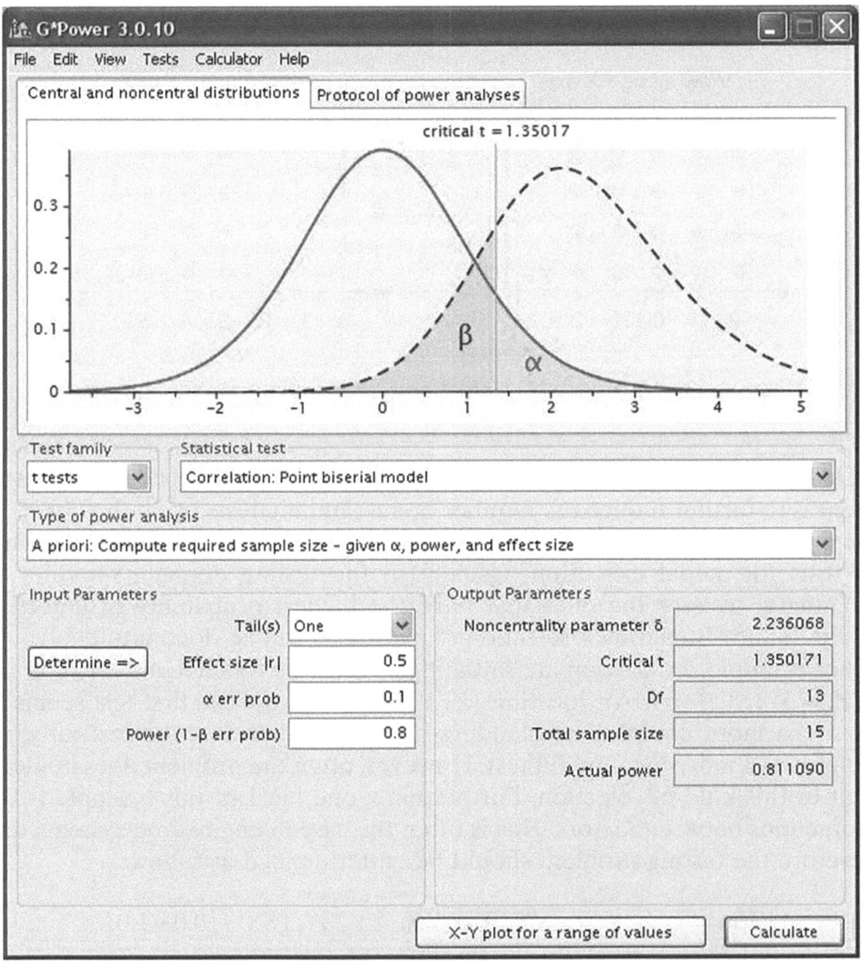

Figure 4.14 Sample size plot for one sample *t*-test.

[35] For a problem as complex as this, the model may be an oversimplification. The use of this model depends heavily on the truth of the assumption that UAV autolanding performance success or failure depends mainly on its initial position in space relative to the landing strip in order to land safely.

TABLE 4.6 Sample Size Computations for One Sample t-Test

I/O	Parameter	Value
Input	Tail(s)	1
	Effect size $\|r\|$	0.5
	Significance level or error probability (α)	0.1
	Power of statistical test (1 – β)	0.8
Output	Non-centrality parameter (δ)	2.236
	Critical t	1.350
	Degrees of freedom	13
	Total sample size	15
	Actual power	0.811

As can be seen, the recommended number of tests (total sample size) is 15 and the actual statistical power is calculated to be 0.811.

Now that we know how many tests should be performed, we can in principle determine the initial locations for starting the 15 or 16 UAV tests. If we do not have any inkling as to more effective locations, we can simply choose these initial locations randomly using an initial set of 60 predefined locations.

We can then plot any selected parameter (α, 1 – β, effect size or sample size) against any other parameter. Of the remaining two parameters, one can choose to draw a family of graphs, whereas the fourth parameter is kept constant. For instance, Figure 4.15 depicts the power (1 – β) against total sample

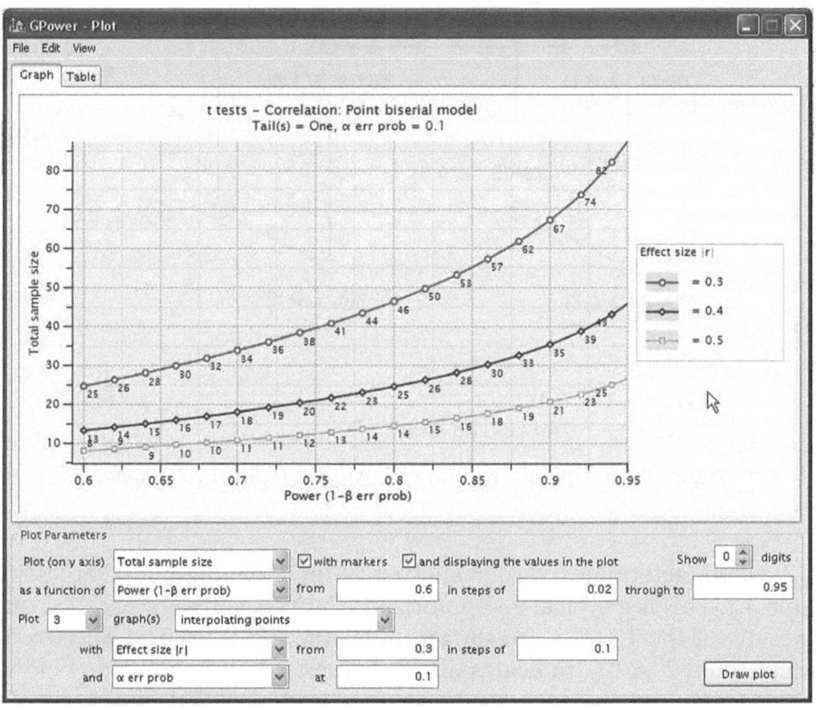

Figure 4.15 Exemplary parameter relationships in statistical power analysis.

size at three levels of effect sizes (0.3, 0.4 and 0.5) while α is kept constant at 0.1. We can observe that, at statistical power level 0.8 and effect size 0.5, the sample size or number of tests (N) is 14.49 (rounded to 15 in the a priori power analysis, in order to guarantee that the test power is at least 0.8). As soon as we select effect sizes of 0.4 and 0.3, the numbers of tests increase dramatically to 25 and 46, respectively.

Post test Analysis Literature on DOE describes many ways of analyzing the results after tests have been performed. In this case, we describe a "2-cubed factorial" test design. In such a design we examine the result of a set of UAV flight tests starting at different initial locations in space. In particular, we like to analyze the results of the tests in order to determine the joint effects of the factors on the success or failure of the flight tests.

In 2-cubed factorial tests one assumes three factors, that is, initial UAV location in three-dimensional space (x, y, z), and limits the test to only two levels, that is, minimum and maximum. In this case we randomly run a set of 8 tests and repeat the process twice, so there are a total of 16 tests. The initial UAV flight configurations are depicted in Figure 4.16.

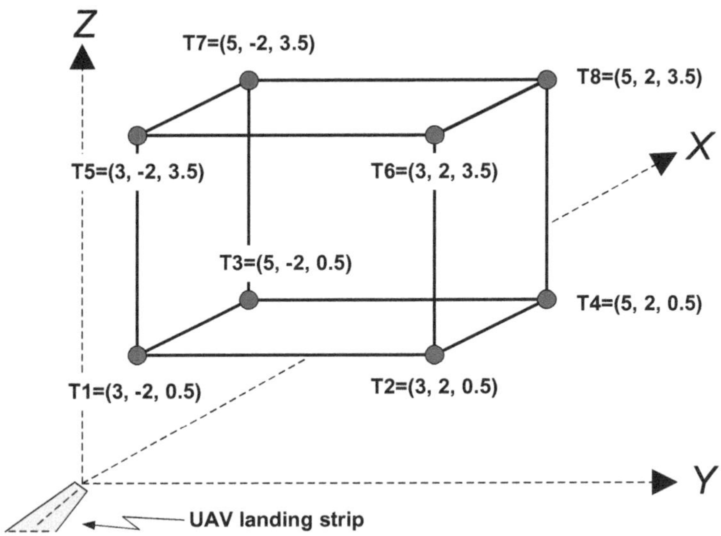

Figure 4.16 Initial location of UAV test flights in 3D space.

A 2-cubed factorial test design analysis of the UAV flight tests are depicted in Table 4.7. This is a typical computerized ANOVA software package output. The results of the 16 tests are shown under the "TSS" or Test Success Score columns 1 and 2. A "1" indicates a fully successful test, any value below one indicates a less and less successful test score and a "0" indicates a failed test.

TABLE 4.7 Results and Analysis of UAV Flight Tests

Run	TSS 1	TSS 2	Variation Source	Sum of Squares	Degrees of Freedom	Mean Square	F	P value
T1 = (3, −2, 0.5)	0.8	0.2	X	0.04	1	0.04	0.48	0.506
T2 = (3, 2, 0.5)	0.0	0.1	Y	0.16	1	0.16	1.94	0.201
T3 = (5, −2, 0.5)	0.9	1.0	Z	0.01	1	0.01	0.12	0.737
T4 = (5, 2, 0.5)	0.3	0.5	XY	0.01	1	0.01	0.12	0.737
T5 = (3, −2, 3.5)	0.8	0.1	XZ	0.64	1	0.64	7.76	0.024
T6 = (3, 2, 3.5)	0.4	0.8	YZ	0.16	1	0.16	1.94	0.201
T7 = (5, −2, 3.5)	0.1	0.5	XYZ	0.04	1	0.04	0.48	0.506
T8 = (5, 2, 3.5)	0.6	0.9	Error	0.66	8	0.08		
			Total	1.72				

As can be seen, one test was fully successful, one test failed, and all the other tests were partially successful.

In this case, the analysis identifies the XZ interaction (i.e. the interaction between the initial X and Z locations of the UAV) as the dominating variation source in this process, accounting for 60% of the total variability. Each of the other factors and interacting factors account for only 16% or less of the total variability. In this example, the P value for the variation emanating from the XZ interaction is 0.024, or 2.4%. ($P < 0.24$ indicates that the probability of observing these data, given that the null hypothesis H_0 is true, is smaller than 0.24.) Customarily, we accept any value below 5% as indicating that the test data are significant and not a result of a random event.

The sum of squares, the mean squares as well as F (the statistic for testing for no differences in treatment means) often provide rough but reliable indicators as to the relative importance of each factor or combination thereof. The identified significant variability of the XZ interaction leads to the conclusion that this area may contain more of a potential for hidden system defects. Therefore, if the VVT team has some extra budget, time and other relevant resources, they should add supplementary UAV flight tests adjusting the X or Z parameters in the initial locations of the UAV rather than modifying the Y parameter.

Further Literature

- Antony (2003)
- Cohen (1988)
- Kenett and Zacks (1998)
- Montgomery (2004)
- Montgomery (2008)
- Murphy et al. (2008)

4.3 PERFORM VVT ACTIVITIES

4.3.1 VVT Process Planning

This section explains how to perform VVT process planning. We briefly discuss (1) project planning (2) key tools for VVT process planning and (3) VVT process planning guidance.

Project Planning VVT process planning at any phase of the system lifecycle should be considered a project planning unto itself. Like any project, it is "the art and science of using the historical data, archived information, personal expertise, institutional memory, organizational knowledge, and project scope statement to predict a project's resource expenditures, total cost, and duration" (Rad and Anantatmula, 2005). From a simple and practical standpoint, VVT process planning may be divided into four steps:

Step 1: Setting Measurable Objectives. A VVT process is successful when the needs of the stakeholders have been met. Here, a stakeholder is anybody directly or indirectly affecting or impacted by the VVT process. Examples of VVT process stakeholders are the project team and management and customers and users of the project deliverables. Once stakeholders have been identified, their needs should be established. One way to do this is by conducting stakeholder interviews. Based on these interviews, a comprehensive list of needs should be drawn up and a set of prioritized measurable goals should be developed and recorded in the VVT process plan.

Step 2: Identifying Deliverables. Using the goals defined in step 1, generate a list of deliverables (reports or products) the VVT process needs to create in order to meet those goals. Identify each deliverable within the VVT process plan together with a rough estimate of delivery date. More accurate delivery dates will be established during step 4.

Step 3: Identifying Needed Resources. For each deliverable identified in step 2, identify the following: (1) the amount of effort (days or weeks) required to complete the task and (2) the specific resource needed to carry out each task. Specifically, the organizations as well as the number and type of individuals needed to carry out the VVT process must be identified together with a description of their roles and responsibilities within the VVT process. Also, a description must be provided of each resource along with an estimated duration of usage and the method for obtaining the resource. More often than not, the required funds or other resources exceed the amount budgeted for the VVT process. The available amelioration options are to (1) renegotiate the budget for VVT process funding, (2) find other resources or (3) reduce the scope of the VVT process.

Step 4: Planning Schedule. Once the amount of effort for each task has been established, one can work out an appropriate completion date for each deliverable. One may use manual means or a software package such as Microsoft Project to generate the VVT process schedule. A common problem discovered at this point is that some VVT activities do not meet required system or project deadlines. Again, the amelioration options available in this situation are similar to the ones mentioned above.

Key Tools for VVT Process Planning Of the many tools supporting project planning, we mention the following:

VVT Process Planning Matrix The VVT Process Planning Matrix (PPM) shows activities and results as well as the conditions necessary for achieving both. These conditions are important assumptions on which rest key process decisions. The PPM usually originated at stakeholder workshops that are scheduled throughout the life of a system.

The PPM is usually a matrix of four columns and four rows, providing 16 squares for a comprehensive description of a VVT process. The PPM lists the links between VVT inputs/activities and VVT objectives to be achieved under certain assumptions. The information in the PPM is organized along two axes in order to show (a) why the VVT process is being undertaken and (b) what are the VVT process outputs.

Objectives or Activities	Objectively Verifiable Indicators	Means of Verification	Assumptions
	Overall Goal		
The broader development impact to which the VVT process contributes	Measures of extent to which a contribution to the goal is made	Sources of information and methods used to collect and report these data	
	Process Purpose		
The development outcome expected at the end of the VVT process	Conditions at the end of the VVT process (used to evaluate the VVT process at completion)	Sources of information and methods used to collect and report these data	Assumptions concerning the purpose or goals of the VVT process

Objectives or Activities	Objectively Verifiable Indicators	Means of Verification	Assumptions
Results or Outputs			
The direct measurable results of the VVT process	Measures of the quantity and quality of outputs and the timing of their delivery	Sources of information and methods used to collect and report these data	Assumptions concerning the output or components objective of the VVT process
Activities or Inputs			
The activities carried out to implement the VVT process and deliver the identified outputs	The resources required for implementation of the VVT process (i.e., funding, manpower, facilities, raw materials, etc.)		Assumptions concerning activities or input requirements

PERT Chart A PERT (Program Evaluation Review Technique) chart is a tool used to schedule, organize and coordinate project tasks. A PERT chart presents a graphic illustration of a VVT process as a network diagram consisting of nodes representing VVT process activities or tasks linked by directional arcs representing the execution sequence of these tasks. A PERT chart can easily indicate task dependencies, but the VVT process status is not immediately apparent on the chart. Figure 4.17 depicts an example of a PERT chart containing five system activities ($S_{1.1}$ through $S_{1.5}$) and eleven VVT activities ($V_{1.1}$ through $V_{1.11}$) and an impact activity (IMP1) representing a system defect correction task.

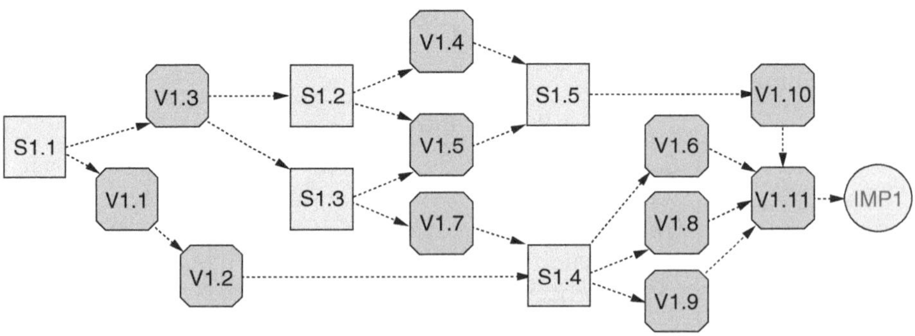

Figure 4.17 Example of a PERT chart.

Gantt Chart A Gantt chart is a horizontal bar chart providing a graphical illustration of a schedule that helps to plan, coordinate and track specific tasks in, for example, a VVT process. The horizontal axis represents the total time span of the VVT process broken down into increments (e.g., days, weeks or months) and the vertical axis represents the tasks that make up the VVT process. Horizontal bars of varying lengths represent the order and time span for each task. A Gantt chart can give a clear illustration of the VVT process status, but indicating task dependencies is rather tricky. Figure 4.18 depicts an example of a Gantt chart containing the same tasks as depicted in the PERT chart of Figure 4.17.

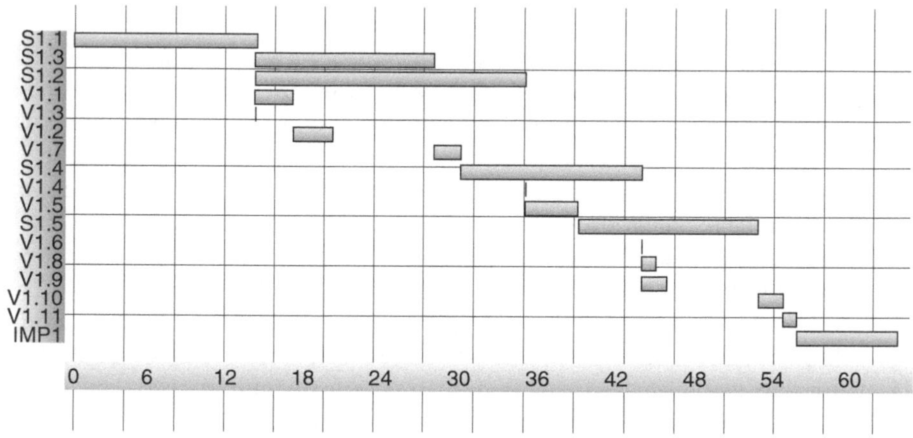

Figure 4.18 Example of a Gantt chart.

Automated PERT as well as Gantt tools may store a great deal of additional information such as cost, dependencies and other resources needed for carrying out each task, number of people and their skill levels as well as names of individuals assigned to specific tasks. Such tools also offer the benefit of being easy to change, which is helpful. Charts may be adjusted frequently to reflect the actual status of the VVT process.

VVT Process Planning Guidance

General Planning Guidance The following are general planning guidances:

- The VVT planner should read and reread the requirement document (or contract). It nearly always contains clauses that impact the VVT process plan.
- An effective way to perform VVT process planning is by way of iterations, regarding the specific VVT tasks, their cost and other resources and their timing and schedule.

- Creating a VVT process plan forces one to think about reducing risk, because various strategies and approaches are considered and the most sensible approach is usually selected during a properly implemented VVT process.
- When planning a given VVT task, it is often prudent to start by first specifying the outputs of the given VVT task and only then considering the inputs needed and the required resources for that task.
- The VVT planner should consider very early on which organization or individuals should perform each VVT task. Similarly the planner should determine who should contribute detailed sections to the VVT process plan itself and at what time these sections are operative in the system lifecycle.
- It is an effective practice, when starting a new project, to copy a previous VVT process plan or import relevant sections from other similar plans and use them as a template in order to retain some of the previous insights and settings.
- Planning assumptions are always made whether one is aware of them or not. Similarly, constraints on resources are always considered by the VVT planner. It is a useful practice to always recognize and document these assumptions and constraints in an organized fashion.
- Although controversial, the VVT planner should always consider adding "hidden slack" into his or her estimates. This strategy is warranted in order to negate a frequent underestimation of time, budget and other resources. Unfortunately, the VVT planner must also participate in the all-too-common, built-in game of negotiated estimation. In this game the planner guesses the required resources in anticipation of a downward negotiation where the project manager forces down all engineering estimates in order to push the schedule and price of the system into alignment with customer expectations.
- Most VVT planners have more experience of a few particular operational aspects and less experience in other areas. Therefore, it is advisable for planners to seek advice from colleagues, experts in areas unfamiliar to the planners.
- VVT planners are advised to make the best use of known benchmarks or other examples to calibrate their own plans.
- The VVT planner should remember to include training as part of the VVT process plan. Training usually occurs at the beginning of a VVT process so that team members can learn the fundamentals of any new skills that they will need. Some training will also be needed throughout the VVT process, particularly for new staff.
- Engel's 5–5–50 law[36] states: "The first 5% and the last 5% of a project takes 50% of the time." Thus, the planner is encouraged to set aside

[36]The author's observation derived over many years of project engineering and management experience.

sufficient and reasonable amount of additional time just for starting and closing out each VVT task.

Estimating Guidance The following are estimating guidances:

- The best estimates are done by (usually experienced) VVT engineers who are doing the actual VVT work. After all, their reputations are at stake and they do learn from experience.
- Cost and time estimates performed by way of "bottom-up" procedures are considered superior to "top-down" estimates, because estimates for small tasks tend to be more accurate than estimates for general tasks.
- When resources are limited, cost and time estimates performed by way of top-down procedures are necessary. Only in this way is it possible to allocate limited resources to vital activities.
- A procedure for achieving minimal over- or underestimations of needed resources calls for conducting both top-down and bottom-up estimates and then negotiating in order to achieve a single and acceptable estimated solution.
- An effective approach to cost and time estimation is to produce a data triplet (minimum, most likely and maximum) range (see Chapter 7). In general, the further into the future that a VVT task is to be conducted, the greater will the range of the estimate need to be.
- It is recommended to update cost and time estimates throughout the VVT process. As actual values are becoming known and the dates of VVT task execution come closer, the planner may have a better idea as to what the estimate parameters will actually be.
- Once cost and time estimates of individual VVT activities are made, one can use optimization methodologies and tools to fine tune the VVT strategy in order to assure delivery of the required product for a reasonable price at a suitable level of quality (see Chapter 7).

Scheduling Guidance The following are schedule guidances:

- Top-level scheduling should be undertaken early on in the project schedule, with the proviso that detailed and accurate planning should be undertaken only for near-future tasks. The recommended approach is to implement a cyclical Just In Time (JIT) planning strategy, that is, when the status and needs of the VVT process are well known.
- The engineers working on the deliverable product should be actively involved in the VVT process scheduling. They are motivated to get it right, they have skills to understand the dependencies and they need to be in agreement with the project work schedule.
- VVT task scheduling should be reviewed and revised iteratively, producing a list of specific deliverables at the end of each scheduled iteration. Only in this way can VVT task progress be validly measured, as these

reviews provide concrete documentation that the VVT process tasks are actually being performed.

- It is highly recommended to schedule demonstrations of the VVT process accomplishments to management, internal and external groups, customers and other stakeholder representatives at the end of each (or some) schedule iterations. This is an opportunity to confirm the approach taken by the VVT team vis-à-vis its ongoing VVT process.

Further Literature

- Rad and Anantatmula (2005)

4.3.2 Compare Images and Documents

Comparing images is the process of observing two images, schemas and so on, usually, in order to verify whether dissimilar details exist between them. Similarly, comparing documents is the process of reading two documents and analyzing them, usually, in order to verify whether both documents contain similar or related text.

A considerable amount of VVT effort involves document comparisons, for example, when assessing completeness and accuracy of a system proposal against a Request For Proposal (RFP), when generating a RVM from a project proposal or an RFP, when assessing a System Requirement Specification (SysRS) against user requirements and when assessing a System/Subsystem Design Description (SSDD) against systems requirements.

Method There are several heuristic methods to compare two objects. Some are more methodical than others, but virtually all of them are based on a "divide-and-conquer" strategy. That is, divide a complex object into smaller and simpler segments and then compare between each relevant pair of segments, rather than attempting to compare the original objects themselves.

One strategy of comparing two rectangular images, illustrated in Figure 4.19, is relatively straightforward. First, each of the two images is divided into $n \times m$ rectangular segments. Thereafter, each individual segment in image A is compared to its corresponding segment in image B (i.e., comparing $A_{1,1}$ and $B_{1,1}$, $A_{1,2}$ and $B_{1,2}$ and so on, until $A_{n,m}$ and $B_{n,m}$). Clearly the number of comparison for a full image is equal to the number of segments, or $n \times m$.

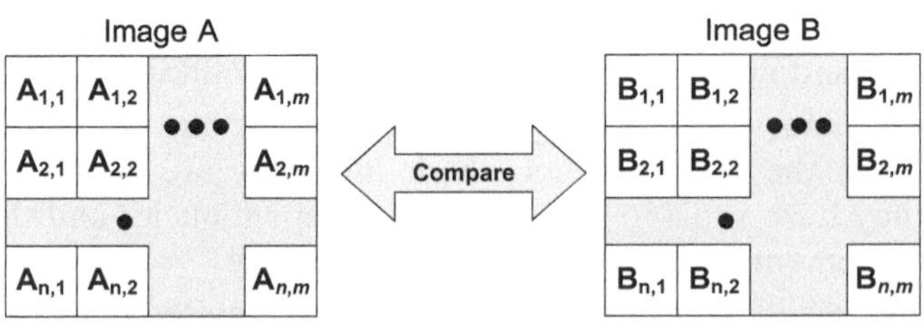

Figure 4.19 Method for comparing two images.

Simple as it may appear, sometimes it still requires considerable human effort to identify differences between two images, especially when the number of different features is unknown (computers, of course, can find such differences easily). Readers are invited to identify the differences between Figures 4.20 and 4.21. (*Hint*: There are five differences between the two images.)

Figure 4.20 Example of an original image for an image comparison exercise.

Figure 4.21 Example of a modified image for an image comparison exercise.

Comparing two documents is quite a challenge that VVT professionals undertake often. Performing this activity manually is a laborious process and is also error prone. Sometimes two documents that have evolved from one another and therefore have similar structures and text must be compared. This problem can be fairly easily solved by using various word processors with a side-by-side comparison feature. Such comparisons are especially relevant for tracking version differences between documents. Microsoft Word as well as several other commercially available tools have a document comparison feature, but this is only applicable if the documents are basically similar.

Comparing any two general structured documents is, by far, more complicated and time consuming. The reason for it is that such documents, in general, may express similar or dissimilar concepts and ideas in quite different wording and manners and, in general, have different structures and sizes. Therefore, after dividing a document into segments we must, in principle, compare each segment from the first document with each segment from the second document. The document comparison process is illustrated in Figure 4.22. The first document is divided into m segments and the second document is divided into n segments. Thereafter, each individual segment in document A is compared with each of the segments in document B (i.e., comparing A_1 and B_1, A_1 and B_2, …, A_m and B_n). Clearly the number of comparisons for a pair of documents is equal to the number of segments in document A multiplied by the number of segments in document B, or $n \times m$.

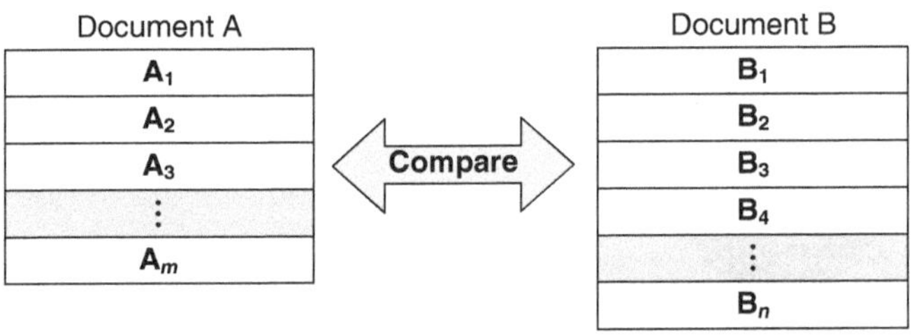

Figure 4.22 Method for comparing two documents.

The subject of automated comparison of documents and texts is a hot topic in computer science and linguistics. One approach among many is indeed document segmentation (i.e., predetermined number of sequential words) and then comparing each segment in one document to all segments in the other document and identifying equal segments. Obviously, the segment size is critical to the effectiveness of the comparison. This size together with the overall size of each document will determine the amount of resources (in particular computer time) needed to perform the process. There are many segmentation methods and we will mention only one of them, called *sentence segmentation*.

Sentence segmentation seems to be the obvious method for segmenting a text, but one must decide how to deal with punctuation such as dots, commas, semicolons, exclamation marks and question marks. A variant of this approach is to use overlapping word segmentation. In this case a segment begins at every word and contains the next predetermined number of words. In total, then, the number of segments per document is equal to the number of words in that text, which makes this method the most reliable in terms of identifying equivalent texts but the worst in terms of resource requirements.

The real problem arises when we must compare documents in terms of ideas or reciprocal concepts, for example, verifying that a system design defined in an SSDD document meets a set of requirements defined in a SysRS document. Here, a manual approach is the only practical method and the VVT engineer must have appropriate skills and comprehensive domain knowledge as a prerequisite.

Further Literature

- Cooper et al. (2002)
- Mitra and Chaudhuri (2000)
- Monostori et al. (2002)

4.3.3 Requirements Testability and Quality

System requirements must be understood by acquirers of the system, users, developers, testers and other stakeholders. Consequently, they are usually written in a natural language. Unfortunately, the use of natural language to describe complex, dynamic systems has severe problems, including ambiguity, inaccuracy and inconsistency. Many words and phrases have multiple meanings which can be interpreted differently by different people. Therefore, it is critical and essential that the VVT team validate all system requirements for both testability and quality.

Evaluating Requirement Testability According to IEEE STD 610.12 (1990), requirement testability is "the degree to which a requirement is stated in terms that permit establishment of test criteria and performance of tests to determine whether those criteria have been met." Requirement testability analysis verifies whether the requirements are indeed testable. The focus of this evaluation is on the system test level and in particular on questions such as "Is it possible to derive test cases from the requirements?" and "Is it possible to define expected system behavior for each test case?"

Requirement testability is performed by checking each requirement individually for testability in order to create the RVM and later to proceed to test planning, design and execution of system testing.

By and large, a testable requirement could be described in terms of (1) the state of the system under test, (2) the inputs to the system under test, (3) the condition or action associated with the requirement and (4) the expected result. This implies that requirements must be stated in a deterministic manner. Determinism means that for a given starting system state, a set of inputs to the system and a set of other conditions specified in the requirement, the results of the test are totally predictable. Testable requirement means that each statement can then be used to prove or disprove whether the behavior of the system is correct. This proof is applicable each time the test is repeated by any tester. For example, the requirement that "the system shall be user friendly" is not testable because the above characteristics are not present.

Evaluating requirements for testability is a tricky business. Researchers suggest that, in combination, the following attributes may be used as a litmus test for this purpose:

1. *Operability.* Operability is an attribute of a system related to its ability to operate satisfactorily under both normal and slightly abnormal conditions which are different from the nominal design conditions. For example, electrical generating power plants rely upon generators with a high degree of operability in order to meet variations in power demand, ambient conditions, fuel supply and so on. A requirement possessing this attribute is more testable because during testing we strive to subject the system not only to normal conditions but also to somewhat abnormal conditions.
2. *Controllability.* Controllability is an attribute of a system related to the ability of an external user to affect system elements (i.e., to compel the system to shift into a desired state or to produce a required output) in its entire configuration space using only external inputs. A requirement possessing this attribute is more testable because performing tests on a system that can be better controlled will allow a more effective testing process.
3. *Observability.* Observability is a measure of how well the internal states of a system can be inferred by knowledge of its external outputs. This means that from the system's outputs it is possible to determine the behavior of the entire system. If a system is not observable, this means the current values of some of its states cannot be determined by observing the output of the system. Obviously, if the requirement possesses this attribute, each operation activity can be easily observed, leading to more effective testing.
4. *Decomposability.* Decomposability is an attribute of a system related to its ability to be broken into components or basic elements. Typically, a simple system has few or weak interactions between its various components. Severing some of these connections usually results in the system behaving more or less as before. On the other hand, complex systems are often irreducible. Sometimes, a complex system cannot be decomposed into isolated subsystems without suffering an irretrievable loss of the essence that makes it a system. Severing any of the connec-

tions linking its parts usually destroys essential aspects of the system's behavior. A requirement possessing this attribute is more testable because such a requirement may be tested within a framework of a subsystem or a component and these tests can, by and large, validate the entire system.

5. *Stability.* In physics, stability is the property of a body that causes it, when disturbed from a condition of equilibrium, to develop forces or moments that restore the original condition. Similarly, in systems engineering, stability refers to the capability of a system to behave in accordance with expected rules. In other words, a stable system is one where, for any given initial state and a specified sequence of inputs, will always behave in the same way and produce the same expected sequence of outputs.

   A requirement possessing this attribute is more testable because testing such a requirement within a stable system will always yield the same result. In this sense the requirement "The display map shall have appealing colors" is not stable since different testers will pronounce different test results for the same system output.
6. *Understandability.* Understandability is an attribute of a requirement where the information provided by it is such that a person with a reasonable knowledge of the subject matter and a willingness to study it with appropriate diligence will be capable of perceiving its significance. An understandable requirement should not leave out anything material but also should not be so comprehensive that the main points of significance are obscured. A requirement possessing this attribute is more testable because testing a requirement which is well understood will usually be carried out in a more effective manner.
7. *Simplicity.* Simplicity is an attribute of a system related to the burden it puts on someone trying to understand it. Something which is easy to understand or explain is simple, in contrast to something complicated. In many uses (e.g., information technology, programming, user interfaces), simplicity often implies beauty, purity or clarity. A requirement possessing this attribute is more testable because testing a requirement which is stated in a simple manner will often entail less testing, which makes the verification process more effective.

Requirement testability may be performed by evaluating each of the requirements individually for testability by means of the attributes defined above. Each requirement should be designated testable only if the test attributes regarding the requirement can be answered positively (e.g., see Table 4.8). Sometimes, under particular circumstances, there might be good reasons for a check not to be fulfilled. In this case, it is appropriate to justify the deviation explicitly.

TABLE 4.8 Requirement Testability Matrix: Example

Requirement ID	Operability	Observability	Controllability	Decomposability	Stability	Understandability	Simplicity	Pass/Fail
System 1	Y	Y	Y	Y	Y	Y	Y	Pass
System 2	Y	Y	Y	Y	Y	Y	Y	Pass
System 3	Y	Y	No	Y	Y	Y	No	Fail

Evaluating Requirements Quality by Attributes The objectives of evaluating requirement quality are to analyze the quality characteristics of each requirement. Good requirements should be organized and written so that information is readily understandable to developers, test engineers, customers as well as all other stakeholders. By and large, system requirements answer the "What" questions, that is, what actions must be carried out by the system under specific conditions. A requirement possesses *quality* when it encompasses all the following attributes:

1. *Traceable.* Each requirement should first have a unique identifier. In addition, it must be traceable to one or more, higher level documents such as user's Request For Proposal (RFP), system and subsystem requirement documents and system and subsystem design documents.
2. *Understandable.* Each requirement must be clearly understood by the implementers and testers of the system as well as by customers, end users and operators of the system. As end users are not engineers, each requirement must be stated in terms that are commonly understood by anyone involved with the system.
3. *Precise.* The bounds of the requirement should be evident and unambiguous. In particular, in the case of numerical bounds, it ought to be evident whether the endpoints are included or not. This may often be achieved by representing requirement bounds in a consistent manner. For example, stating the requirement "The system shall accept valid part numbers from 1 to 1000" raises the question whether the value 03 is a valid input? It is more precise to state "The system shall accept valid integers between 1 and 1000 inclusive, represented without leading zeros."
4. *Succinct.* Requirements should consist of only the necessary information, without additional details and arguments. For example, a requirement may state "Because we feel that this system may be expanded in the future, we require six serial interfaces instead of just four, as asked by the customer." A succinct requirement will state "The system shall have six serial interfaces." One practical approach for maintaining additional information is to create, along with the formal requirement database, a secondary depository or database, where relevant comments, insights, explanations and justifications are maintained.

5. *Clear.* Natural language lends itself to an infinite number of ways to state requirements. Sometimes, specifications are stated in ways that may be unclear to some engineers or end users. For example, the requirement "On a standard day, either rainy or dry, with temperatures between 15 and 25 degrees Celsius, the vehicle will not consume more than 10 liters of gasoline per 100 kilometers on a level road and no more than 15 liters of gasoline per 100 kilometers on a road of 10% upward incline and no more than 8 liters of gasoline per 100 kilometers on a road of 10% downward incline." Perhaps it could be clearer to most people if this requirement was divided into four separate requirements:

   - A "standard day" is defined as either a rainy or a dry day with temperatures between 15 and 25 degrees Celsius.
   - On a standard day, the vehicle will not consume more than 10 liters of gasoline per 100 kilometers on a level road.
   - On a standard day, the vehicle will not consume more than 15 liters of gasoline per 100 kilometers on a road of 10% upward incline
   - On a standard day, the vehicle will not consume more than 8 liters of gasoline per 100 kilometes on a road of 10% downward incline."

6. *Noncompounded.* A compounded requirement is characterized by having multiple subrequirements folded into a single requirement. The example above represents this phenomenon well. Beyond the issue of clarity, the problem with a compounded requirement is twofold. First, several individual tests are needed in order to verify such a requirement. Second, a single failed test may flag the entire requirement as a failure whereas some clearly delineated elements of the requirement meet the specifications. Restructuring a compounded requirement into several unique requirements will again resolve the issue.
7. *Correct.* A correct requirement must reflect the true wishes of the customer. This is not as easy as it sounds. Often different customers (or stakeholders) have different wishes. Sometimes the customer changes his or her perception about the system and so forth. Nevertheless the most common mistake is an incorrect interpretation of customer wishes. For example, the customer requirement was "The system will indicate the length of time associated with each telephone call" and the requirement engineer stated the requirement as "The system shall tag each telephone call with a time-stamp."

   Correct implies "completely correct." That is, the requirement must indicate the fullest possible conditions. For example, a requirement stating "The Radar will be able to track at least 100 targets" may be considered correct, but if the system is expected to eventually expand to track 200 targets, then the requirement should reflect it.

For example, "The Rader system will initially be able to track 100 targets; however, the design should support expending this capability to track 200 targets."

8. *Complete.* A requirement should be complete and give all relevant information on what is required. In other words, the requirement should be considered complete only if it provides all the information that separates an acceptable system behavior from one that is not acceptable. For example, a requirement may be stated as "The system shall provide the operator with safety information needed to shut down the machinery when unsafe conditions occur." The requirement does not specify what type of safety information the system is to provide or the specifics of the machinery to be stopped. A better requirement specification may be "The system shall display a 'High temperature warning' if the temperature inside the boiler will exceed 96.00 degrees Celsius no later than one second after an unsafe condition occurs."
9. *Consistent.* Different requirements should agree with each other. In other words, one requirement should not specify something that is in conflict with other requirements. For example, one requirement may state "The telephone exchange system shall support a maximum of 10,000 users" while another requirement may state "Up to 15,000 subscribers shall be connected to the telephone exchange." In addition, it is always advisable to create requirements in a similar format so their structures also appear consistent to readers.
10. *Unambiguous.* Requirement ambiguity is perhaps one of the greatest problems that affect system development, because the exact meaning of normal human language is notoriously vague and imprecise. An unambiguous requirement must be precise and must have one and only one interpretation. For example, "The aircraft will fly at an altitude of 30,000 feet" is ambiguous since the requirement does not state relative to what this measure is stated. It may be relative to sea level or relative to ground level below the aircraft or any other interpretation.
11. *Feasible.* Feasible means that the requirement has a sound physical and economic basis. That is, there is a known way to accomplish the stated requirement. A requirement stating "Build one more space shuttle for $10,000" is not feasible. Similarly, the requirement "The rocket should be able to fly at two times the speed of light" is probably traceable and also understandable, precise, succinct, clear, noncompounded, correct, complete, consistent and unambiguous, but it is certainly not feasible due to the laws of physics as we know them today.

Each system requirement should be analyzed using the above characteristics and approved if it meets all the above quality attributes (e.g., see Table 4.9).

TABLE 4.9 Requirement Quality Matrix: Example

Requirement ID	Traceable	Understandable	Precise	Succinct	Clear	Noncompounded	Correct	Complete	Consistent	Unambiguous	Feasible	Pass/Fail
System 1	Y	Y	No	Y	Y	No	Y	No	Y	Y	Y	Fail
System 2	Y	Y	Y	Y	Y	Y	Y	Y	Y	Y	Y	Pass
System 3	Y	Y	Y	Y	Y	Y	Y	Y	Y	Y	Y	Pass

Evaluating Requirements by Syntactic and Semantic Means In the late 1990s and early 2000s, several researchers developed tools to automatically evaluate the quality of requirements through their syntactic and semantic attributes. For example, an Automated Requirement Measurement (ARM) tool[37] was developed by the Software Assurance Technology Center (SATC) at the NASA Goddard Space Flight Center as an early lifecycle tool for assessing requirements that are specified in natural language. The objective of the ARM tool was to provide measures that can be used by project managers to assess the quality of a requirement specification document (Wilson et al., 1997).

Similarly, an Italian team from the Istituto di Elaborazione dell'Informazione del CNR in Pisa developed a tool called QuARS (Quality Analyzer of Requirements Specification) for the analysis of natural language requirements.[38] This tool aims at providing a quantitative, corrective and repeatable evaluation of requirement documents. The Italian team defined a set of indicators for automatic syntactic and semantic analysis of requirements; some of these indicators are described below [adapted from Fabbrini et al. (2001) and Gnesi et al. (2005)]:

1. *Optionality.* An optionality indicator exposes a requirement containing an optional part (i.e., a part that may or may not be considered). Typical optionality-revealing words are *possibly, eventually, if case, if possible, if appropriate and if needed.*
2. *Subjectivity.* A subjectivity indicator exposes a requirement containing personal opinions or feelings. Subjectivity-revealing wordings may be

[37]The ARM tool and other supporting materials are available at http://satc.gsfc.nasa.gov/. The tool is accessible to the public at no cost. Unfortunately, it has not been maintained for nearly a decade due to lack of SATC funding and is not functioning properly.

[38]Work on analysis of natural language requirements is alive and well at CNR. A description of the QuARS tool and other supporting materials are available at http://quars.isti.cnr.it.

similar, better, similarly, worse, having in mind, take into account, take into consideration and as [adjective] *as possible.*

3. *Vagueness.* A vagueness indicator exposes a requirement containing words holding inherent vagueness, for example, words having a nonuniquely quantifiable meaning. Typical vagueness-revealing words are *clear, easy, strong, good, bad, efficient, useful, significant, adequate, fast, recent, far, close and in front.*
4. *Weakness.* A weakness indicator exposes a requirement which contains a weak main verb. Typically weak verbs are *can, could and may.*
5. *Implicity.* An implicity indicator exposes a requirement where the subject is generic rather than specific. Typically this appears in demonstrative adjective (e.g., *this, these, that, those*) or pronouns (e.g., *it, they*) or a subject specified by an adjective (e.g., *previous, next, following, last*) or a preposition (e.g., *above, below*).
6. *Multiplicity.* A multiplicity indicator exposes a requirement which has more than one main verb or more than one direct or indirect complement that specifies its subject. Typically multiplicity-revealing words are *and, or, and and/or.*
7. *Unexplanation.* A unexplanation indicator exposes a requirement when it contains an acronym not explicitly defined within the requirement document itself.

Although such tools cannot evaluate requirements in terms of their natural language meaning, it is relatively simple to use the QuARS tool or construct such utilities and use them to reveal syntactic and semantic traps in requirement documents.

Further Literature

- Fabbrini et al. (2001)
- Gause and Weinberg (1989)
- Gnesi et al. (2005)
- IEEE STD 610.12 (1990)
- IEEE STD 830-1998 (1998)
- IEEE STD 1522 (2005)
- MIL-HDBK-2165 (1995)
- Robertson and Robertson (2006)
- Wilson et al. (1997)

4.3.4 System Test Simulation

In the context of this book, simulation means the modeling of engineered systems in an embedded system composed of hardware and computer software. Simulations are useful because they allow us to study phenomena that otherwise are difficult to observe as well as experiment with ideas that otherwise are impossible or quite difficult to implement. In addition, simulations allow us to study advanced systems, subsystems or components that are costly to build.

The concept of simulation is naturally associated with modeling. Modeling and simulation are in fact strictly joined together to include the complex activities needed to construct models representing engineered system behavior and experimentation using these models to obtain required data.

If we loosely define a system as a collection of identifiable interacting parts, called components or subsystems, then the state of the system at a certain time instant is known from the actual conditions of each element at that instant. Not all conditions need to be included in this description, only the ones that are relevant for the study at hand. The time evolution of the system is then described by the time history of the states in their chronological sequence. A model of the system is then a representation of the system itself. This representation can be a physical replica or a symbolic one. In every case the model will not represent all the operational aspects of the system being modeled, and there will be an abstraction level in the model since some properties are omitted or approximated. Given a system and a model, simulation is the use of the model for the chronological production of a history of states of the model, which is considered equivalent to the history of the states of the modeled system. A model once it is used for simulation is called a simulation model.

Based on various definitions available in the literature, we define *test simulation* as the process of designing and creating a computerized model of an engineered system for the purpose of conducting various tests in order to evaluate the behavior of the corresponding real system under a given set of conditions.

Test Simulation Classification There are many kinds of problems that need simulation and one approach of doing simulation cannot satisfy all needs. Different kinds of problems characterize different simulations, for example, (1) when mathematical models of the system exist, (2) when only empirical/statistical data exist or (3) when only words or abstractions exist. Another way of looking at simulations is by classifying them according to the way they are built:

- *Top Down.* In a top-down approach, the simulation is constructed from mathematical models that are known to capture the system's behavior. In this case, the system behavior is known to obey some mathematical model, which is mostly unsolvable, as an analytical solution does not exist. Therefore we use numerical methods for approximation of the original equations. Such simulations are used to simulate the behavior of complex physical systems such as aircraft dynamics, force impacts and fluid dynamics.

- *Bottom Up.* In a bottom-up approach, we build a "virtual" system from the ground up reflecting real behavior of components and subsystems as much as possible and study it instead of the real-world system. In this case, the system behavior is known statistically or empirically. Here, a model of each individual element of the system may be governed by dynamic inputs to the simulated elements as well as a rule-based or probabilistic principle. A computer program integrates this ensemble to reflect the behavior of the system as realistic as possible. Such an approach may be used to simulate a system of production and distribution, information flow within an organization and the like.
- *Indirect.* Sometimes system behavior is not fully known or is too complex to be directly simulated. In an indirect approach, we simulate much simpler models which globally capture the characteristics of the system concerned. Such an approach may be used to understand business growth, crowed behavior under stress and so on.

Another way to classify simulations is on the basis of their construction (see Figure 4.23):

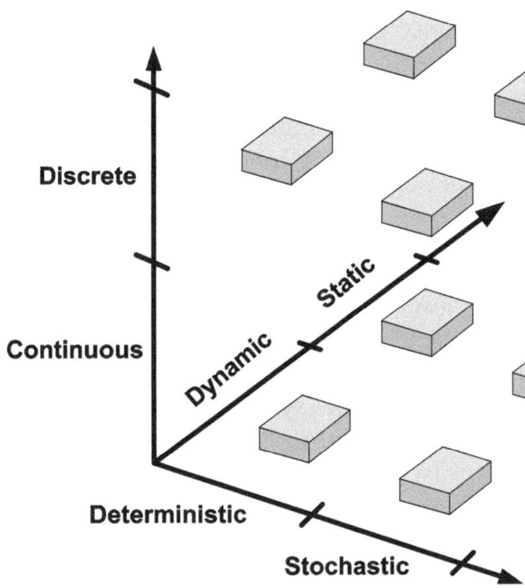

Figure 4.23 Test simulator classification.

- *Dynamic Versus Static.* Dynamic simulation includes the passage of time. It looks at state changes as they occur over time. In contrast, time does not play a role in a static simulation.

- *Continuous Versus Discrete.* In continuous simulations, the state of the system can change continuously over time, while in discrete simulations, change can occur only at separate points in time.
- *Deterministic Versus Stochastic.* Deterministic simulations have no random input, while stochastic simulations operate with at least some inputs being random.

Developing Test Simulations The main objective of test simulation is to evaluate the robustness of a system design with respect to the variation of input parameters. Other objectives may be related to the identification of the functional characteristics of a system and the validation of the design tools by comparing the simulation testing results with a real system being tested under the same initial states and input conditions.

The overall development process of a test simulation is depicted in Figure 4.24. The process alternates between a theoretical phase and an empirical phase. In the theoretical phase the target system (i.e., the system to be tested) is defined in an increasing degree of detail and sophistication. Correspondingly, the models are implemented by means of software and hardware components such that the emerging system can be progressively and iteratively simulated and analyzed. The empirical phase consists of performing manual or automatic tests utilizing the simulated system in place of the real one.

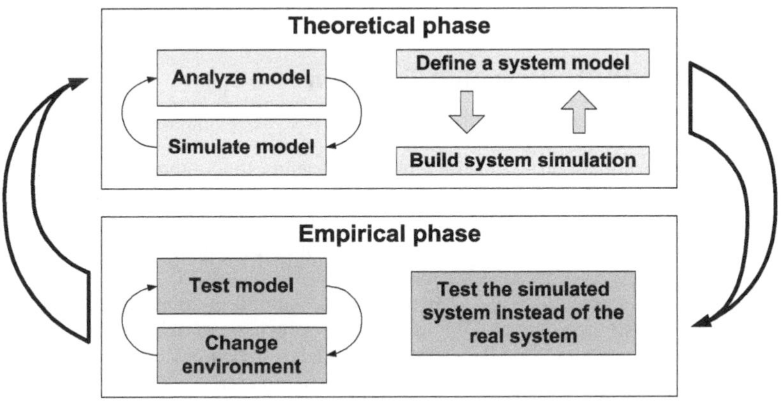

Figure 4.24 Concept of system simulation testing.

Many authors (see Further Literature) offer similar sets of steps to construct and use a simulation process for system verification and validation. Figure 4.25 and Table 4.10 illustrate a derivative procedure considered appropriate for this book.

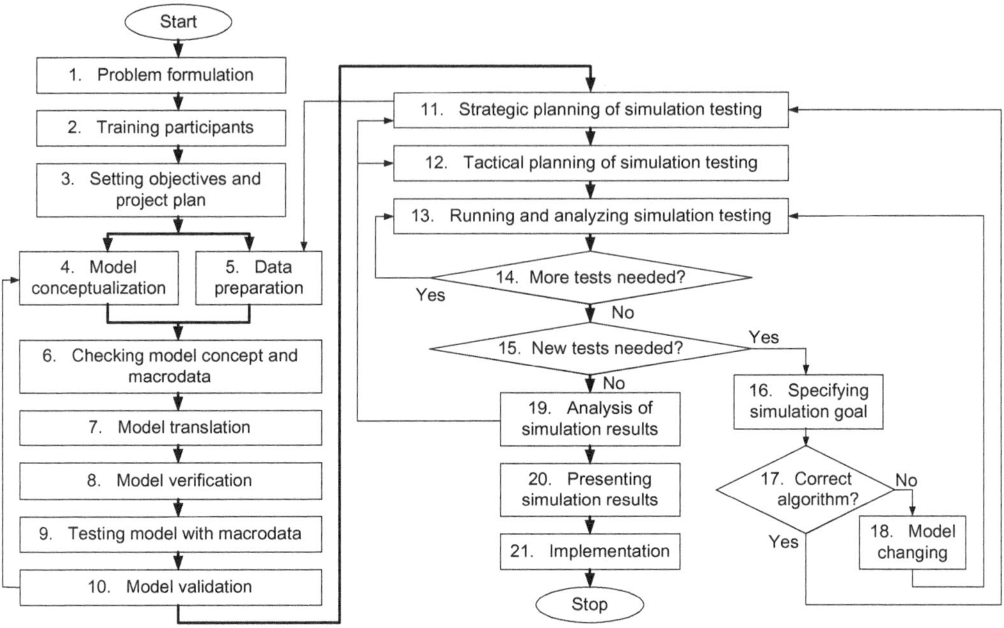

Figure 4.25 System's testing simulation development.

TABLE 4.10 Steps in Developing Test Simulation

Step	Meaning	Comment
1	Problem formulation	Identify and define the system testing problem to be solved.
2	Training project participants	Train relevant involved individuals about test simulation methodologies and how to implement them.
3	Setting objectives and project plan	Specify the simulation objectives and plan the simulation process, including personnel identification, needed resources, schedule and relevant simulation parameter.
4	Model conceptualization	Specify the simulated system and the conceptual model algorithm as well as the important features to be simulated and the expected level of abstraction.
5	Data preparation	Create appropriate data for valid test simulations corresponding with real-life system or its environment. The simulation of random system behavior must be based on realistic statistical considerations.
6	Checking model concepts and macrodata	Evaluate the conceptual model as well as internal and external data elements (e.g., values of key variables at key simulation events).

TABLE 4.10 *Continued*

Step	Meaning	Comment
7	Model translation	Implement the conceptual model by means of the appropriate software and hardware system. Many commercial tools are available to support most simulations, but under special circumstances a simulation environment must be created from the ground up.
8	Model verification	Verify that the realized simulated model accurately reflects the authentic behavior of the real system to be tested.
9	Testing model with macrodata	Evaluate whether the simulated model is sensitive to a particular set of input parameters. If such parameters are identified, then the peculiar behavior of the system should be further investigated and all anomalies must be noted for future retest on the real system.
10	Model validation	Within the defined constraints of the system model, verify that the developed model and the real system operate in an exactly equivalent manner.
11	Strategic planning of simulation testing	Plan the overall (strategic) system testing using the simulation model. The planner should consider testing the simulation model in the same way as it would have been done with a real system.
12	Tactical planning of simulation testing	Develop the test procedure (i.e., the test suite set) to validate the functionality of the simulated model.
13	Running and analyzing simulation testing	Perform the actual simulation tests which have been planned and designed in the previous two steps and record the results.
14	More tests needed	Based on the test results, evaluate whether additional tests are necessary in order to achieve a higher confidence in the simulation results as well as the behavior of the real system to be tested.
15	New tests needed?	If new tests are required, then it is good practice to update the simulation goal specifications. If the model algorithm itself is correct, then the strategic planning of the simulation testing must be updated and the new tests must be run as needed. However, if the model algorithm is incorrect, then it must be fixed and the test simulations must be repeated appropriately.
16	Specifying simulation goal	
17	Correct algorithm?	
18	Model changing	
19	Analysis of simulation results	Analyze the simulation results including both the behavior of the simulation model itself as well as correctness of the simulated system.
20	Presenting simulation results	Share the results of the test simulation with all relevant stakeholders (e.g., development team, management, customer).
21	Implementation	If any defect was discovered, in either the simulation model itself or the real system to be tested, then it is the responsibility of the cognizant system engineers to fix the simulation model or the real system appropriately and submit it for retesting.

Test Simulation Advantages and Disadvantages Modern engineering practice is greatly supported by system modeling and test simulation. Profound insights may be obtained from this technology for many different aspects of system behavior and endurance under severe conditions. In particular, some advantages of using test simulations are listed below:

- *Shortening Schedules.* Modeling and simulation provide means for parallel efforts of developing the target system as well as modeling and testing the simulated system within a virtual environment. The use of simulation can thus result in a substantial time saving.
- *Deeper Knowledge.* Simulated testing can provide very detailed description of system behavior under very different operating conditions. Furthermore, some information available from modeling and simulation may be difficult, if not impossible, to obtain by testing the actual system under stressful conditions.
- *Increasing Flexibility.* Simulation models are often based on parametric architectures which offer inexpensive and rapid means for evaluation systems with alternative solution space.
- *Repeating Tests.* Simulated testing provides possibilities for initializing, recording of internal variables and playback of simulated system and performing repeated tests starting from a precisely known state of the simulated system. Such exact repetition of tests is difficult to achieve in complex systems under realistic conditions.
- *Improving Products and Processes.* The advances in software or hardware technology are useful in offering the means for constructing highly sophisticated testing scenarios. For example, it is now possible to build hierarchies of simulation models that follow a product and related processes in every phase of their lifecycle, thus allowing deeper control of the overall quality and effectiveness. Simulation models are especially useful in diagnosing system problems and reducing risk by testing system potential improvements before attempting to actually implement them.
- *Exploitation of Past Experience.* The use of simulation models increases product knowledge. A simulation model, once validated, can easily be reused for different similar products. Furthermore, the use of hierarchical sets of models can give a detailed description of the product development process, thus highlighting areas of concern.

On the other hand some disadvantages or limits of test simulations are:

- *Return on Investments.* The trend in simulation tools is to evolve in capabilities, complexity and modularity, causing continuous increase in acquisition, maintenance and training costs. The actual return of investments or maintenance expenses is only possible if careful planning and control of simulation activities are exerted.

- *Results Misinterpreted.* A critical aspect of using modeling or simulation techniques in system VVT is the correct assessment and understanding of the results. Interpretation of simulation results is completely under the responsibility of the user and requires great care. For example, sometimes a simulation test fails, not because the underlying system has a defect but, possibly, due to wrong input value or a defect in the model itself. Conversely, a simulation run indicates a valid system under test when in fact the system may contain a defect that is not revealed by a particular test run.
- *Validation Difficulties.* Models used for test simulation reflect the level of knowledge of the system under test. Sometimes, aspects of the modeling process are not known precisely and therefore may be decided upon in quite an arbitrary manner. As a result the validation of the system under test is questionable and subject to interpretations.
- *Capturing Subtleties of Reality.* Model simulations always represent a subset of reality and therefore may obscure some significant problems. Simple analytical models are unable to capture the subtleties of reality whereas complex analytical models may be difficult to construct and fully understand.
- *Overshooting Problems.* Computer simulations offer dramatically improved testing capabilities. They can support complex varieties of testing scenarios unimaginable in the past. However, sometimes, VVT personnel may be caught up in a frenzy of system testing beyond economic justification.
- *People and Organization Commitments.* Technology improvements in the last decades led to the development of user-friendly robust interfaces enabling inexperienced people to use these tools after a short training time. Unfortunately the scientific bases of these tools are usually quite complex so a nontrivial level of knowledge is required for a thorough understanding and correct interpretation of simulation results.

Further Literature

- Banks (1998)
- Kheir (1995)
- Kim (2000)
- Matko et al. (1992)
- SEF DoD (2001)
- Severance (2001)
- Woods and Lawrence (1997)
- Zienkiewicz and Taylor (2006)

4.3.5 Failure Mode Effect Analysis

Failure mode effect analysis (FMEA) is a bottom-up procedure for analysis of potential failure modes within a system or a process and then determining how to eliminate such problems. This is accomplished by identifying the potential types of problems that may occur, their causes and the potential frequency with which they may impact the system or the process at hand. The analysis proceeds with estimating the effects of such failures should they occur. Next a determination is made as to how such events may be detected and/or prevented and, finally, under the FMEA procedure, the priority of handling these corrective actions, whether modifying the system design or the system manufacturing process, is accomplished (see Figure 4.26). FMEA is widely used in various phases of the product lifecycle, especially during the design and manufacturing of systems and their corresponding processes.

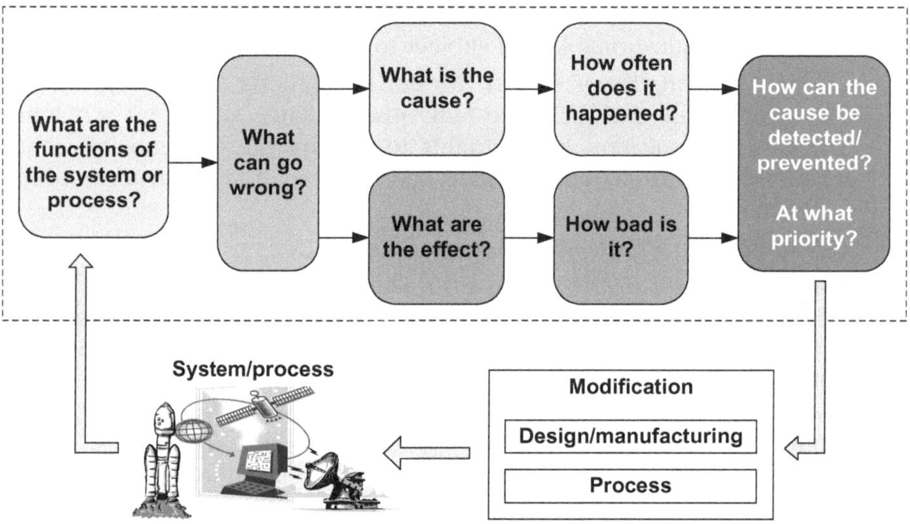

Figure 4.26 Typical FMEA process.

The ultimate purpose of FMEA is to take actions to eliminate or reduce potential future failures. Therefore, a key FMEA practice is to prioritize these potential failures according to how serious their consequences are, how frequently they occur and how easily they can be detected.

Basic FMEA Terms Some of the basic FMEA terms are:

- *Failure Cause.* The underlying cause of the failure or the cause which may initiate a process leading to failure (e.g., defects in design, manufacturing process, quality or part application).
- *Failure Mode.* The characterization of the way a system or process may fail. It refers to a complete description under which the failure

may occur, how the system is being used and the final results of the failure.

- *Failure Effect.* The immediate consequences of a failure on operation, function or functionality or status of the system at hand.
- *Failure Severity.* The consequences of a failure mode, that is, the worst potential consequence of that type of failure, determined by the degree of injury, property damage or system damage that could ensue.

Basic Types of FMEAs There are four basic types of FMEA processes, although most practitioners tend to match and mix them as they see fit:

- *Design FMEA.* This procedure is performed on a system or service during the Design phase. Systems must be analyzed in order to determine how failure modes affect the system operation. This leads to better understanding of design deficiencies which can then be corrected so impact of failure modes is reduced.
- *Functional FMEA.* This FMEA ingredient focuses on the intended function, or use, of a system. For example, the FMEA on an automobile design would investigate the behavior of an automobile of that design without paying much attention to its detailed structure. The FMEA could (1) analyze the potential problem or loss from each potential loss of functionality, (2) estimate the statistical probability of such problem and (3) estimate the potential damage on the automobile, its occupants or the environment of the car. Finally the functional FMEA would attempt to offer remedy to such problems and a priority for implementing each solution.
- *System FMEA.* This "white-box" FMEA can be used to analyze a system at any level, from the piece-part level up to the system level. At the lowest level, it looks at each component in the system to determine the ways in which it can fail and how these failures affect the system. In this procedure the detailed structure of the system takes central stage. The focus shifts from mere system functionality to clear understanding of potential failures and mutual interactions of each individual part of the entire complex system. In the automobile example above, this would mean attention would be given to the intricacies and failure modes of the steering mechanism, the tires and the gas tank as well as every other essential part of the vehicle.
- *Process FMEA.* This procedure is mostly performed on the manufacturing processes, although other engineering processes (e.g., system development, systems VVT) may be examined. The procedure identifies possible failure modes in the process, limitations in resources, equipment, tooling, gauges, operator training or potential sources of error. As in the other FMEA types, this information is used to determine the corrective actions that need to be taken.

FMEA Standards There are several FMEA standards available. Virtually all provide sample inspection forms and instruction documents. They also identify criteria for the quantification of risk associated with potential failures and offer general guidelines on the mechanics of completing FMEA procedures. In addition, most standards describe FMEA procedures encompassing functional, interface, and detailed FMEAs as well as certain preanalysis activities (FMEA planning and functional requirement analysis), postanalysis activities (failure latency analysis, FMEA verification and documentation) and applications to hardware, software and process design. Most FMEA software tools support these standards. The following are a few examples of available FMEA standards:

- *MIL-STD-1629A (1980).* This FMEA standard describes a method used mostly by government, military and commercial organizations worldwide. As found in all FMEA standards, this standard provides formulas for determining criticality and allows rating of failure modes by severity class.
- *SAE J1739 (2002).* This FMEA standard is based on a procedure defined by major international automobile companies and their suppliers. It has been adopted and recommended by the Society of Automotive Engineers (SAE).
- *ARP5580 (2001).* The SAE recommends this FMEA standard for non-automobile applications. It is intended for use by organizations whose product or system development processes use FMEA as a tool for assessing the safety and reliability of system elements within their product improvement processes.

Many organizations use a combination of different standards, modifying them to suit their needs for their particular applications.

Implementing FMEA The FMEA procedure may be divided into four main steps:

Step 0: FMEA Preparation. Before starting with a FMEA, it is important to complete some preliminary work to confirm that robustness and past history are considered in the analysis.

FMEA is initiated by describing the system and its functions or the process that must undergo FMEA evaluation. A good understanding of the FMEA object simplifies the further analysis. This way a test engineer can observe which uses of the system are desirable and which are not. It is important to consider both intended and unintended uses of the system, where unintended use includes improper operation, unexpected environmental effects on the system or perhaps malicious use by a hostile user.

Next, a system block diagram is created depicting an overview of the major components or process steps and how they are related. These are the logical relations around which the FMEA can be developed. Finally, a well-defined set of procedures, forms and worksheets must be created which define important information about the system (e.g., revision dates, names of the components). In addition, all the items or functions of any corresponding element should be listed in a logical manner.

FMEA activities should be supported by appropriate database tools as the procedure tends to be tedious and time consuming. Several techniques can be used to reduce the tedium, time and thus cost of performing a FMEA. For example, failure mode distribution standards can be used to assign common failure modes. Standard reports and input formats may be created to streamline the failure data collection and reporting process. Custom failure mode libraries can also be created and reused for future projects. Several software tools supporting efficient FMEA procedures and standards are available commercially. Such tools can reduce the overall cost of performing and improve the robustness of the FMEA process.

Step 1: FMEA Severity Determination. In this step, we determine all potential failure modes based on the functional requirements of the system and their effects. Examples of failure modes are loss of braking ability in a car and malfunction of a lathe machine in an assembly line. As one failure can lead to another failure mode, it is critical to analyze all the ramifications of each failure type that can occur. A failure effect is defined as the result of a failure mode on the function of the system as perceived by the user, operator or other affected individuals. Examples of failure effects are degraded performance, noisy operation or discomfort by or even injury to a user. Customarily, each potential failure effect is assigned a *severity rating* (*S*) from 1 to 10. For example, Table 4.11–4.13 depict typical FMEA parameters. These systems oriented FMEA elements are derived from Military Standard MIL-STD-1629A and the Society of Automotive Engineers, SAE-J1739.

TABLE 4.11 Design FMEA Severity Evaluation Criteria

Effect	Severity of Effect	Rating
Hazardous, without warning	Very high severity rating when a potential failure mode affects safe system operation or involves noncompliance with government regulation without warning	10
Hazardous, with warning	Very high severity rating when a potential failure mode affects safe system operation or involves noncompliance with government regulation with warning	9
Very high	System inoperable (loss of primary function)	8
High	System operable but at a reduced level of performance; customer very dissatisfied	7
Moderate	System operable but comfort/convenience item(s) inoperable; customer dissatisfied	6

TABLE 4.12 *Continued*

Effect	Severity of Effect	Rating
Low	System operable but comfort/convenience item(s) operable at a reduced level of performance; customer somewhat dissatisfied	5
Very low	Fit and finish/squeak and rattle item does not conform; defect noticed by most customers (greater than 75%)	4
Minor	Fit and finish/squeak and rattle item does not conform; defect noticed by 50% of customers	3
Very minor	Fit and finish/squeak and rattle item does not conform; defect noticed by discriminating customers (less than 25%)	2
None	No discernible effect.	1

These rating numbers help an engineer to prioritize the failure modes and their effects. If the severity of an effect is high (i.e., say 9 or 10), actions must be taken to change the system by either eliminating the failure mode or protecting the user from the effect. A severity rating of 9 or 10 is generally associated with those effects that would cause injury to a user or otherwise result in litigation.

Step 2: FMEA Occurrence Determination. In this step it is necessary to look at the cause of a failure and the frequency with which it may occur. Looking at similar products or processes and the failures that have been documented for them can help in this task. A failure cause may be a design weakness or manufacturing flaws. All potential causes for a failure mode should be identified, analyzed and documented. An *occurrence rating* (*O*), customarily in the range of 1–10 (see Table 4.12), should be assigned to each failure mode.

TABLE 4.12 Design FMEA Occurrence Evaluation Criteria

Probability of Failure	Likely Failure Rates Over Design Life	Rating
Very high: persistent failures	≥100 per thousand items	10
	50 per thousand items	9
High: frequent failures	20 per thousand items	8
	10 per thousand items	7
Moderate: occasional failures	5 per thousand items	6
	2 per thousand items	5
	1 per thousand items	4
Low: relatively few failures	0.5 per thousand items	3
	0.1 per thousand items	2
Remote: failure unlikely	≤0.01 per thousand items	1

Step 3: FMEA Detection Determination by Design Control. A *detection rating* (*D*) represents the general ability to detect a system defect or a failure mode by means of a planned set of tests and inspections. In this step, test engineers look at the system mechanisms that are responsible for detecting potential failures, thus preventing actual failures from occurring. For example, the oil pressure indicator in a car is a mechanism that detects low oil pressure and warns the driver about a potential engine seizure. Test engineers then identify testing, analysis, monitoring and other means that may detect or prevent failures. From these design control efforts, an engineer can learn how likely it is for a failure to be identified or detected. Typical detection ratings are depicted in Table 4.13.

TABLE 4.13 Design FMEA Detection Evaluation Criteria

Detection	Likelihood of Detection by Design Control	Rating
Absolute uncertainty	Design control will not or cannot detect a potential cause or mechanism and subsequent failure mode or there is no design control	10
Very remote	Very remote chance the design control will detect a potential cause or mechanism and subsequent failure mode	9
Remote	Remote chance the design control will detect a potential cause or mechanism and subsequent failure mode	8
Very low	Very low chance the design control will detect a potential cause or mechanism and subsequent failure mode	7
Low	Low chance the design control will detect a potential cause or mechanism and subsequent failure mode	6
Moderate	Moderate chance the design control will detect a potential cause or mechanism and subsequent failure mode	5
Moderately high	Moderately high chance the design control will detect a potential cause or mechanism and subsequent failure mode	4
High	High chance the design control will detect a potential cause or mechanism and subsequent failure mode	3
Very high	Very high chance the design control will detect a potential cause or mechanism and subsequent failure mode	2
Almost certain	Design control will almost certainly detect a potential cause or mechanism and subsequent failure mode	1

Step 4: Computing Risk Priority Numbers. A *risk priority number* (RPN) is a quantitative determination of risk based on multiple factors. Traditionally, RPN is defined as the product of the severity rating (S), occurrence rating (O), and detection rating (D) values of each failure mode:

$$\text{RPN} = S \times O \times D$$

The failure modes that have the highest RPN should be given the highest priority for corrective action. While the above traditional RPN computation is widely used, every project has a unique set of circumstances, and a one-size-fits-all approach to RPN calculation may not produce the most effective results for an analyses. In some situations, such as where human safety is at risk, the RPN could be more meaningful if the severity rating (S) is weighted much more heavily:

$$\text{RPN} = S^2 \times O \times D$$

Further Literature

- ARP5580 (2001)
- Dyadem Press (2003)
- Modarres et al. (1999)
- MIL-STD-1629A (2001)
- SAE J1739 (2002)
- Stamatis (2003)

4.3.6 Anticipatory Failure Determination

As we have seen, traditional risk analysis and prevention methods such as FMEA and Hazards and Operations Analysis (HAZOP) do not offer a systematic procedure for identifying beforehand the dangerous or harmful events that might be associated with a system. The following method, called Anticipatory Failure Determination™ (AFD™),[39] does provide a systematic way for identifying either potential future failures or root causes for already manifesting failures. The following description of the AFD methodology is based mostly on Visnepolschi (2009). However, the reader should note that our presentation is confined only to issues related to a systematic approach to failure prediction. Much wisdom embedded in AFD but not directly related to VVT issues was intentionally left out of this discussion.

AFD methodology offers several strategies to identify failure scenarios. The one that interests us is the concept of finding possible failure initiation events and drawing the resulting failure trees from each. Initiating events are

[39]Research on innovation processes (TRIZ, a precursor to AFD) was conducted in the former USSR over the last half century. These efforts led to the creation of an American company—Ideation International. The company provides consultation and software tools to support AFD process. It is the owner of the trademarks Anticipatory Failure Determination and AFD. See http://www.ideationtriz.com/home.asp.

defined as failures of individual subsystems or components of the system as well as unexpected external events. Thus, in a given system, one would work through each system element, asking, "What would happen if this part failed?" or "What kind of external event can cause this part to behave in an unplanned manner?" This process works because identification of initiating events or failure scenario trees can be carried out at various levels of detail and thoroughness and every failure scenario can be broken down into subscenarios.

Example—Combination of Risk Assessment and AFD Analysis We will present relevant Anticipatory Failure Determination (AFD) ideas by example of a Unmanned Air Vehicle (UAV) mission. Prior to performing a risk assessment for this system, one should be very clear on exactly what that system is. In other words, for a failure scenario to be understood, the "success" (or as-planned) scenario must be clearly specified. Risk assessment denotes this scenario by S_0. In our example, we define five phases of a successful UAV operational scenario (see below and in Figure 4.27:

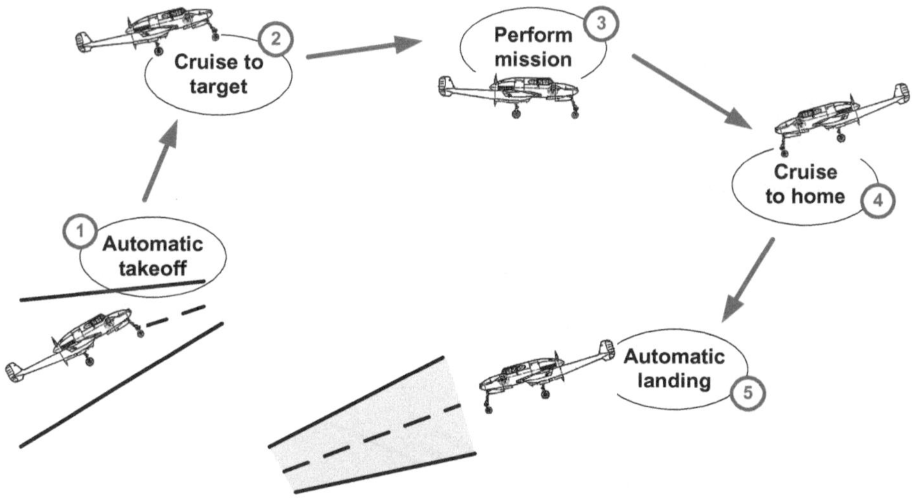

Figure 4.27 Planned UAV operational scenario (S_0).

- *Phase 1: Take Off Automatically.* The UAV performs an automatic takeoff from an airstrip.
- *Phase 2: Cruise to Target.* The UAV flies along a designated route to a designated altitude and location.
- *Phase 3: Perform Mission.* The UAV flies in a predefined flight path and directs its cameras to a certain set of locations.
- *Phase 4: Cruise to Home.* The UAV flies along a designated route back to the original airstrip.
- *Phase 5: Land Automatically.* The UAV performs an automatic landing on the airstrip and comes to a standstill at a designated place on the airstrip.

Risk assessment considers S_0 as a trajectory in the state space of the system, depicting general relations between the system's mission phases and time (see Figure 4.28). Since S_0 is the planned scenario, any failure scenario (S_i) that departs from this plan must have a point of departure from normal system operation.

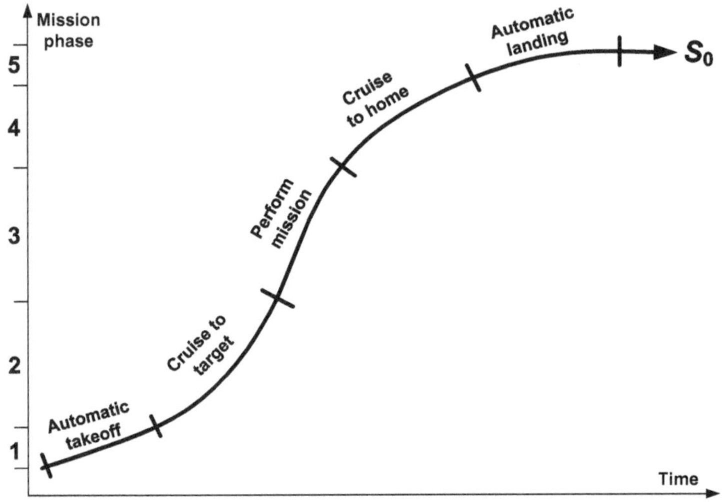

Figure 4.28 UAV system state (system mission phases versus time).

The Initiating Event ($IE_{i,j}$) of S_i may be generated due to internal system failure or due to an unanticipated external disturbance. Two such initiating events are depicted in Figure 4.29.

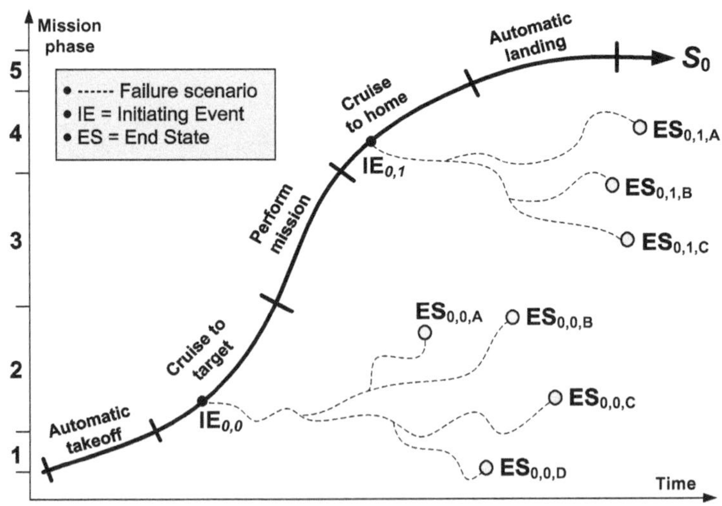

Figure 4.29 UAV system states with several failure scenarios.

From each initiating event, an outgrowth of related failure scenarios emerges, which is referred to as a failure scenario tree. Each path through the tree represents a particular scenario, depending on what happens after the initiating event. Each branch of the tree continues until it reaches some system End State ($ES_{i,j,k}$). For example, Figure 4.29 depicts two failure scenario trees. The first failure tree, occurring during the mission phase "cruise to target," emanates from event $IE_{0,0}$ and ends at one of four system end states $\{ES_{0,0,A}, \ldots, ES_{0,0,D}\}$, and a second failure scenario, occurring during mission state "cruise to home," emanates from event $IE_{0,1}$ and ends at one of three system end states $\{ES_{0,1,A}, \ldots, ES_{0,1,C}\}$.

AFD employs the concept of *resources* to denote all the substances, components, configurations or other factors presented in a situation that can provide means for failure realization. For example, a simplified set of resources in the above-mentioned UAV system example is the six subsystems described below and depicted in Figure 4.30.

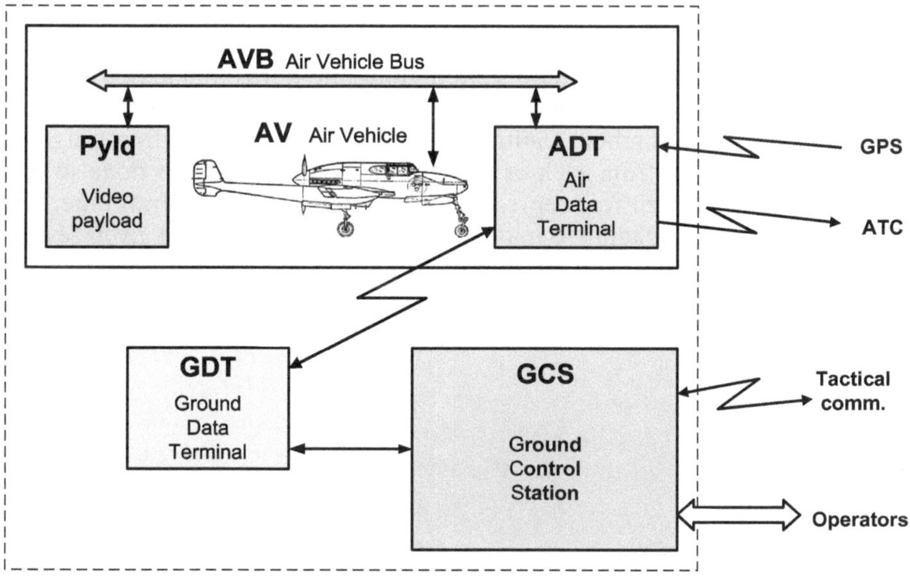

Figure 4.30 A UAV system architecture.

- *Ground Control System (GCS).* The GCS is a small shelter, often mounted on a small truck, housing a UAV pilot and other UAV operators. The UAV team pilots the unmanned aircraft, observes the video and infrared image stream acquired by the UAV and controls the entire UAV system.
- *Ground Data Terminal (GDT).* The GDT is a ground unit containing a powerful transmitter and receiver. It receives commands from the GCS

and transmits them to the UAV and, similarly, it receives UAV telemetry status as well as video and television streams and sends them to the GCS.

- *Air Vehicle (AV).* The AV is an unmanned craft designed to take off, fly and land automatically or manually and carry various payloads and support systems to a desired altitude and location and transmit live video and infrared pictures from that location.
- *Air Data Terminal (ADT).* The ADT is the airborne counterpart of the GDT performing quite similar activities.
- *Payload (PYLD).* The PYLD is a unit containing specialized cameras mounted on a gimbaled platform attached to the AV. It is capable of viewing the external world in visible as well as infrared frequencies and sending the data to the ADT for transmission to the ground.
- *Air Vehicle Bus (AVB).* The AVB is a data bus connecting the ADT, AV and PYLD and allowing the transfer of command, status and other data among these subsystems.

Figure 4.31 depicts the six UAV subsystems along the vertical axis, which we consider a spacelike axis. Similarly, the particular UAV mission S_0 has distinct phases of operation represented along the horizontal axis, forming a timelike axis. For each combination of UAV subsystem and mission phase, we can identify any number of initiating events ($IE_{i,j}$). Next we draw outgoing failure tree (S_i, $i \neq 0$) from each of these initiating events. This is done so that the set of paths in each tree represents a complete set of scenarios emerging from that event and leading to multiple end states ($ES_{i,j,k}$). For a given resolution of system structure and mission phases, the combination of components and phases is finite; therefore, a "complete" set of system failure scenarios may be created.

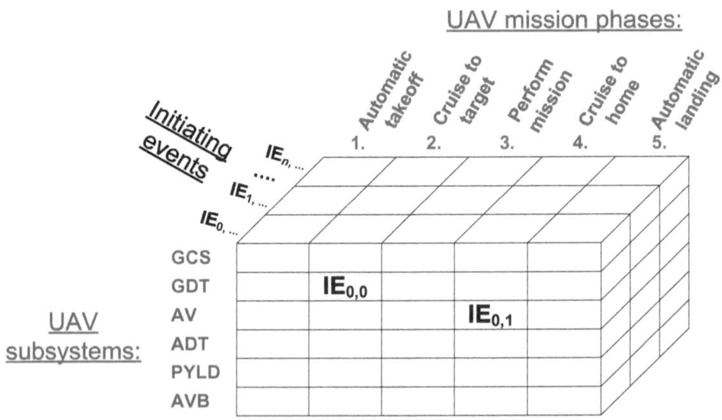

Figure 4.31 Three-dimensional space of initiating failure events in a UAV system.

For example (as seen in Figures 4.29 and 4.31), several potential problems may be caused by the initiating event $IE_{0,0}$—loss of communication between the Ground Data Terminal (GDT) and the UAV which occurs during the cruise-to-target phase of the UAV mission. This situation means that the UAV operators at the Ground Control Center (GCS) are unable to control the UAV or receive any data from it. Four end states have been identified:

- $ES_{0,0,A}$—The UAV is out of control. It flies until it runs out of fuel, at which time it crashes to the ground.
- $ES_{0,0,B}$—The UAV recognizes the loss of transmission condition and initiates its automatic "return-to-home" procedure. The UAV then returns to and automatically lands safely at home base.
- $ES_{0,0,C}$—Similar to $ES_{0,0,B}$ but, unfortunately, the global coordinate address provided to the UAV was pointing to the southern hemisphere instead of the northern hemisphere. The UAV procees to fly away from home base, runs out of fuel and crashes to the ground.
- $ES_{0,0,D}$—The UAV operators initiate a GDT emergency procedure, reestablishing the proper operation of the GDT. The communication between the GDT and the UAV is restored; however, the UAV mission is aborted and the UAV is returned home.

Let us now consider the second initiating event $IE_{0,1}$—UAV fuel runs out—which occurs during the cruise-to-home phase of the UAV mission. This situation means that the UAV engine will stop running within a minute or so. Three end states have been identified:

- $ES_{0,1,A}$—The engine in the UAV stops. Without propulsion the UAV loses its ability to remain airborne. The air vehicle exits its flight envelope and crashes to the ground.
- $ES_{0,1,B}$—The UAV operators recognize the problem and direct the UAV to glide without propulsion and then land at a secondary landing strip located in the vicinity of the stricken UAV. This procedure is successful.
- $ES_{0,1,C}$—Similar to $ES_{0,1,B}$ but the procedure is unsuccessful due to a lack of automatic landing facilities at the secondary landing strip. The UAV hits the landing strip toward its end and crashes against the landing strip perimeter.

Inverted Logic in AFD As mentioned, AFD has two broad applications. The first applies to finding the cause of failures that have already occurred (i.e., failure analysis). The other is concerned with identifying possible failure scenarios that have not yet occurred (i.e., failure prediction). Failure prediction is what interests us in this section. To this end, AFD applies the following philosophy:

- *Changing Attitude Toward Failure.* Instead of asking "What can go wrong with the system?" AFD suggests asking the question "How can we make the system fail in the most effective way?"
- *Adopting Concept of Resources.* For any system failure to occur, all the necessary components must be present within the system or its environment.
- *Eliminating or Reducing Failure.* Any failure, once revealed, can be eliminated or reduced.

Human beings are often subject to a psychological phenomenon called denial, in which they resist thinking about unpleasant things. There is much historical evidence of denial playing a role in disasters and failures. AFD methodologists suggest that inverted questions are useful in counteracting the tendency of humans to deny. So when one asks the inverted question "How can I sabotage the system?" one applies his or her engineering skills and the mind opens up to the full spectrum of failure possibilities.

In addition, there is a plethora of information about the causes of system success. In fact, the literature associated with triumphant war stories like "How we succeeded in building the XXX system" is very rich and hints are often given about avoiding failures. On the other hand, in day-to-day situations, engineers seldom document and publicize failures. Thus, by asking the question "What problems were avoided in building a successful system?" a vast body of useful information becomes available.

AFD Procedure for Failure Prediction Based on the above philosophy, we seek to identify all the possible initiating events ($IE_{i,j}$) as well as all the possible scenarios (S_i, $i \neq 0$) leading to all the failed end states ($ES_{i,j,k}$) using the following procedure:

- *Step 1: Formulating Original Problem.* In this step, the original problem is formulated. For example, considering the UAV system, we can state the following:
  1. There exists a UAV system designed to take off automatically from an airstrip, cruise to a given altitude and location, perform its visual surveillance mission and then cruise back home and land automatically at the home base.
  2. We wish to find all possible undesired effects or failures that can occur within the system or as a result of external events and to identify the ways in which these undesired phenomena can occur.
- *Step 2: Identifying Success Scenario.* In this step, the system success scenario S_0 is described in terms of the phases of the process and the results achieved at the end of each phase.

- *Step 3: Formulating Inverted Problem.* In this step, the problem stated in step 1 is inverted. For example, considering the UAV system, the first sentence remains unchanged and the second sentence becomes "It is necessary to produce all possible undesired effects or failures capable of leading to the system's malfunction or its negative impact on the environment."
- *Step 4: Making System Fail.* In this step, all potentially harmful end states ($ES_{i,j,k}$) and their initiating events ($IE_{i,j}$) generating failure scenarios (S_i) are stated. One may search for failure scenarios by employing the commercially available AFD software package. This software contains a knowledge base consisting of numerous failure checklists. Using these checklists, one can identify categories of harmful end states that might be present, and evaluate initiating events necessary for these end states' spontaneous realization.
- *Step 5: Identifying Available Failure Scenario Resources.* In this step, all the resources (i.e., conditions) available in or around the system that might be instrumental in contributing to a failure are identified. Again, the commercially available AFD software contains a prefabricated template identifying many resources (conditions) that might be present.
- *Step 6: Inventing New Solutions.* In this step, which in fact is not connected with the procedure for failure prediction, one can use the AFD principle that all the resources (conditions) necessary for an initiating event must be present in a situation in order that the event will actually occur. Conversely, if at least one of the necessary resources is not present, then that event will not occur. This principle is most valuable in guiding the search for system failure elimination, namely, remove from the system one of these necessary resources.

Further Literature

- Brue and Launsby (2003)
- Haimes (2009)
- Kaplan et al. (1999)
- Middleton and Sutton (2005)

4.3.7 Model-Based Testing

A model is a description of a system's behavior that is constructed to help us understand and predict its operational behavior. Invariably, models are simpler than the systems they describe. This is so because the model is necessarily an abstraction of the actual system's salient properties. Trying to model every aspect of a system, such as its size, weight, shape or smell, would be both costly and not very useful. Model-based testing is typically achieved using a variety of modeling paradigms such as a finite-state machine, a pre-/postcondition model and a labeled transition model.

Common methods for the quality assurance of systems are simulation, testing and deductive reasoning.[40] These techniques, however, often fail to ensure the high levels of quality required for critical systems, where human life or property may be at risk. Formal methods, on the other hand, provide proof of system correctness based on mathematical models. More specifically, while simulation and testing explore some of the possible behaviors of the systems, model checking conducts an exhaustive exploration of all possible behaviors. Thus, when the model checker verifies a given system property, it implies that all behaviors have been explored, and the question of adequate coverage or a missed behavior becomes irrelevant. Nevertheless, the mathematical formalizations themselves are a possible source of errors and much care and expertise are needed in undertaking these methods.

Model checking, one of several formal system verification methods, may be considered an alternative to simulation and testing. It is a technique for verifying finite-state concurrent and reactive systems such as control systems, sequential circuit designs and communication protocols. Beyond its ability of proving the correctness of system behavior, model checking is highly automatic. Typically the user must provide a high-level representation of the model and the specification to be checked.

Also, if either the system model or its specification contains an error, model checking will produce a counterexample that can be used to pinpoint the source of the error. That is, the model checker will either terminate with the answer true, indicating that the model satisfies the specification, or give a counterexample that shows the conditions under which the specification is not satisfied.

The behavior of reactive systems is usually modeled by transition systems. The inputs to a model checker are finite-state descriptions of the system to be analyzed and properties, often expressed by means of temporal logic, that are expected to hold in the system.

Assume we can create a system model and define a desired set of system properties. Then, a model checker can explore the entire state space of the system model and check whether the system properties are satisfied by the model.

Model-Checking Theory The following are some basic model-checking definitions:

- A *model* (M) of a system can be represented by a Labeled Transition System (LTS) such that

$$\text{LTS} = \langle S, \delta, I, \text{AP}, L \rangle$$

[40]Deductive reasoning is a formal method as well (in fact more general than model checking since it handles parameterized properties) but is difficult to be mechanized.

where S = set of states
$\delta \subseteq (S \times S)$ = transition relation
$I \subseteq S$ = an initial state
AP = finite set of atomic propositions
$L: S \rightarrow 2^{AP}$ = labeling function

a. A *run* of LTS is an ω-sequence $s_0, s_1, \ldots$ s.t. $s_0 \in I$ and $\forall j\ (s_j, s_{j+1}) \in \delta$.
b. A *trace* of LTS is an ω-sequence $\sigma_0, \sigma_1, \ldots$ s.t. there exists a run of LTS $s_0, s_1, \ldots$ for which $\forall j\ L(s_j) = \sigma_j$.
c. The set of all behaviors enabled by a model is the set of all possible traces of the model, denoted by L_M.

- A *property* is a formal description of a requirement. The formalism used to express properties is temporal logic [i.e., Linear Temporal Logic (LTL) or Computation Tree Logic (CTL) or ω-automata]. For instance, LTL consists of atomic propositions, propositional operators such as: $\vee$ (*or*) and $\neg$ (*not*) and special temporal operators such as $\Box$ (*always*), $\Diamond$ (*eventually*) and U (*until*) that are capable of expressing behaviors along the time axis. For instance, the formula $\Box\ (p \vee q)$ means that at every time instant either p or q must hold; the formula p U q means that q necessarily holds at some time instant in the future and p must hold at every time instant until then. Thus, the meaning of an LTL formula φ is the set of behaviors that satisfies φ, denoted by L_P.
- *Model checking* is a technique (algorithm) that, given a model of a system M and a property P, verifies that every behavior of M is indeed a behavior allowed by P. This is stated in formal notation: $L_M \subseteq L_P$. Also, model checking is capable of presenting a counterexample in case of a negative result.

Typical employment of the model-checking procedure is described below:

- *Step 1.* Choose a model-checking tool that appropriately supports the needed type of validation. Different tools have been created to deal with various types of issues (e.g., control, timing).
- *Step 2.* Create a model of the system. Design of a system is usually expressed in a formal form (programming language, VHDL, mechanical design, etc.); hence converting it to a Labeled Transition System (LTS) is carried out automatically by relevant tools.
- *Step 3.* Create the formal specification of the system. Convert the natural language requirements of the system into a formal set of expressions.
- *Step 4.* Activate the model-checking tool and analyze the results. If the property does not hold, examine in detail the countersequence provided to check whether the system model or the specifications are incorrect.

Model-Based Testing in Practice A typical model-based testing process is depicted in Figure 4.32.

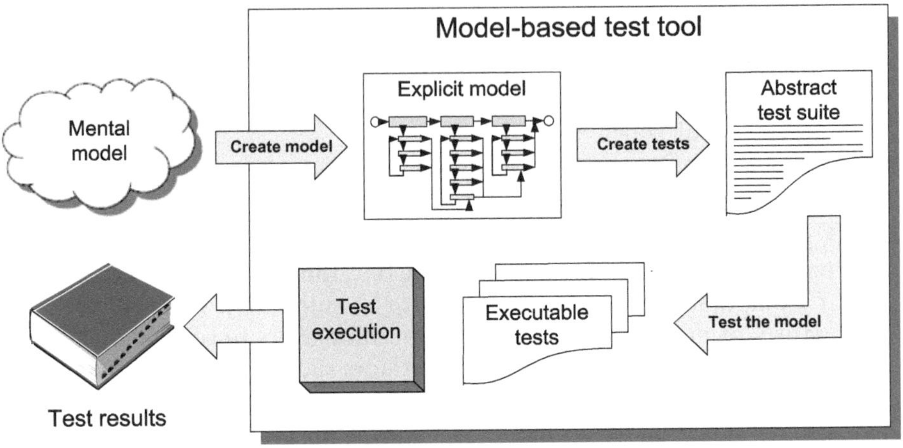

Figure 4.32 Model-based test process.

A mental image of a system is a natural starting point for developing a simplified model of a system. The model is usually an abstract, partial representation of the system's actual behavior. A set of test cases and the test oracle[41] are derived from this model. These are functional tests on the same level of abstraction as the model and are collectively known as the abstract test suite. One of many model-based specification and conformance testing tools is then employed to generate executable tests and these tests are run against the system's model.

The test results indicate whether the system as depicted by its model meets the specifications or not. Discrepancies between actual and expected results are described as *conformance failures*. Such failures may indicate (1) a system failure, (2) a modeling error, that is, a defect in the model definition itself, or (3) a specification error. A specification error may result from a mistake or ambiguity in the system specification (i.e., erroneous representation of the intended system behavior). If the system under test (SUT) has already been built, then this SUT may behave differently than the explicit model embodied within the model-based test tool. The problem then may be located either in the modeling segment or the real system implementation portion.

[41] A test oracle is a mechanism for determining whether a system has passed or failed a test. It is used by comparing the output(s) of the system for a given test case input to the outputs expected by the oracle. Test oracles are always separate from the system under test.

Model Checking—First Example One approach to testing systems that depend heavily on sequences of events or stimuli is to model their behavior using a finite-state machine. Fundamentally, finite-state machines are tested by different "coverage" strategies: (1) state coverage attempts to visit through every state in the model in one or more test cases and (2) transition coverage attempts to traverse through each transition between states in one or more test cases.

However, the problem is more acute when we take into account several additional elements. First, each transition from state to state is dependent on a set of preconditions and postconditions. Should we test separately with respect to all such conditions? Second, we should not automatically assume that states are memoryless. The importance of knowing whether or not the system states have memory is that when they have memory there is a distinction based on what path was taken to reach each given state. In order to achieve all state path coverage when states have memory, the test case should traverse each path that reaches each state.

The number of test case permutations can increase dramatically with the number of states and transitions. Several ways of testing are proposed if it is not feasible to conduct exhaustive testing. For example, we can take a prespecified number of random-walk tests. Another approach is to take a predetermined number of paths of length-N tests.

For example, Figure 4.33 depicts a state machine model of an ordinary digital watch. Superimposed on this model, we can see a single path of a test case traversing from the "time keeping" state through six states and then returning to the original state (i.e., the external stimuli are $\{A, C, C, C, C, C, A\}$).

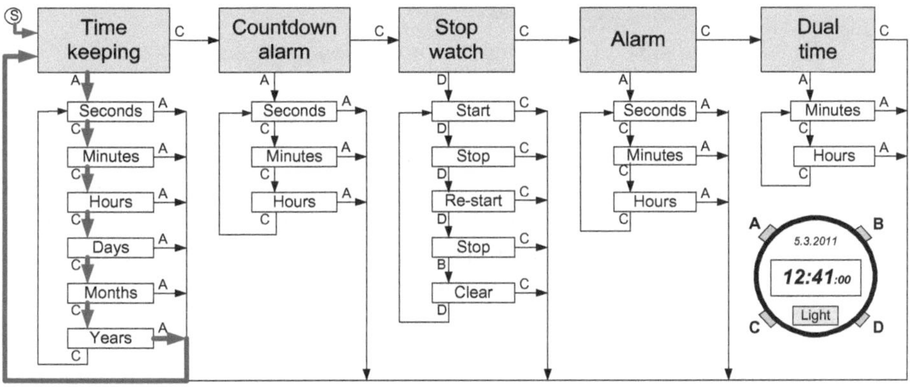

Figure 4.33 Digital watch model tested by means of a state machine.

Model Checking—Second Example Model checking is an effective approach for verifying system requirements or design. A model-checking tool accepts system requirements or design (model) and their properties (specification) that the final system is expected to satisfy. The tool concurs when the given model satisfies the given specifications and generates a counterexample otherwise. By studying such a counterexample, one can identify the source of the error either in the model or in its specifications and correct it. For control systems, an Extended Finite-State Machine (EFSM) is widely used as an ideal abstract notation for defining requirements and design of real-time, embedded systems.

We introduce an example[42] of a steel mill production system, described in Figure 4.34. Molten steel is poured into a vessel of steel and then, when the steel vessel gate is opened, the molten steel flows into a cooling escalator, creating a steel slab. Each steel slab is produced in a predefined width, height and length. The steel production team controls the gate manually. The gate should always be closed, regardless of manual commands, under the following conditions:

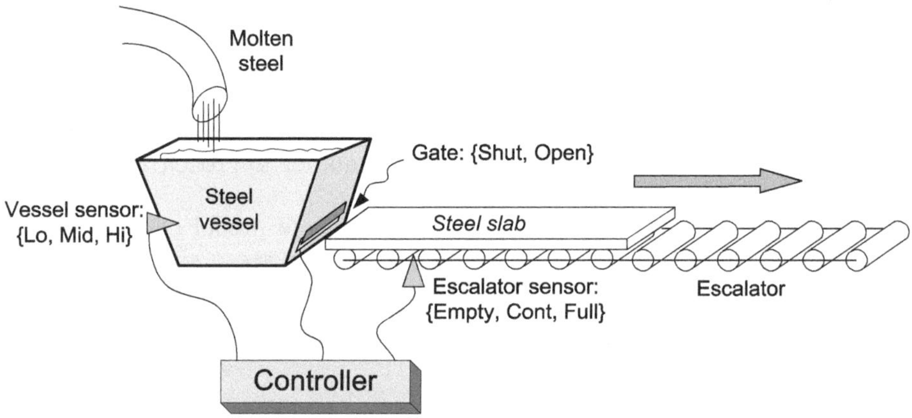

Figure 4.34 Controlling the production of steel slabs.

1. The amount of molten steel in the vessel is low.
2. The steel slab reached the end of the cooling escalator.
3. After manual command to shut the steel gate, it may not open until the current slab completely clears the cooling escalator, at which time a new steel slab may be produced. Thus, an automatic device (controller) makes sure that these rules are obeyed.

The steel vessel has two level sensors to detect whether its molten steel level is low (Lo) or high (Hi). The vessel level is defined as middle (Mid) if the amount of molten steel is between Lo and Hi. The cooling escalator has multilevel sensors to detect whether the cooling escalator is empty (Empty) or full (Full). Similarly, the escalator level is defined as continue (Cont) if it is between empty and full.

[42]This example was inspired by Dr. G. K. Palshikar's paper, An Introduction to Model Checking, published by Embedded Systems Design, February 12, 2004.

Initially, the steel vessel is empty (Lo) and the cooling escalator carries no steel slab (Empty) and the gate is closed (Shut). The production team may open the gate as soon as there is a certain amount (Mid) of molten steel in the vessel. The gate may remain open as long as the steel vessel is not empty, the steel slab does not reach the end of the escalator and the operators did not shut the gate. However, the controller will shut the gate automatically if either the amount of molten steel in the vessel is too low (Lo) or the steel slab reaches the end of the cooling escalator.

Table 4.14 shows a formal model and specification of this system which may use one of several Symbolic Model Verifier (SMV) tools available freely or commercially. First the three system variables {Vessel, Escalator, Gate} are declared, each with its own set of allowable values.

TABLE 4.14 Model and Specifications of System

SMV Portion	SMV Tool Input
Input variable declaration	MODULE main
	VAR
	Vessel: {Lo, Mid, Hi};—Steel vessel (Vessel)
	Escalator: {Empty, Cont, Full};—Slab cooling escalator (Escalator)
	Gate: {Shut, Open};—Steel vessel gate (Gate)
Assignment statements	**ASSIGN**
	next (Vessel): = case
	Vessel = Lo & Gate = Shut: {Lo, Mid};
	Vessel = Lo & Gate = Open: {Lo, Mid, Hi};
(Vessel)	Vessel = Mid & Gate = Shut: {Mid, Hi};
	Vessel = Mid & Gate = Open: {Lo, Mid, Hi};
	Vessel = Hi & Gate = Shut: Hi;
	Vessel = Hi & Gate = Open: {Mid, Hi};
	esac;
	next(Escalator): = case
	Escalator = Empty & Gate = Shut: Empty;
	Escalator = Empty & Gate = Open: {Cont};
(Escalator)	Escalator = Cont & Gate = Shut: {Full};
	Escalator = Cont & Gate = Open: {Full};
	Escalator = Full & Gate = Shut: {Empty};
	esac;
	next(Gate): = case
	Gate = Shut & (Vessel = Mid | Vessel = Hi) &
(Gate)	(Escalator = Empty): Open;
	Gate = Open & (Vessel = Lo | Escalator = Full): Shut;
	esac;
Initialization statement	**INIT**
	(Gate = Shut & Vessel = Lo & Escalator = Empty)
Specifications	**SPEC**
	$\Box$((Vessel = Empty $\vee$ Escalator = full) $\rightarrow$ Gate = Shut)

Next the assignment section defines how the system state changes from one state to another. For visibility they are grouped according to the system variables, but in fact they operate in parallel with each other. In this case the state of the system is defined by a tuple of values for each of these three variables. For example, (Vessel = Lo, Escalator = Full, Gate = Shut) is a system state in which the steel vessel is empty (Lo), the escalator fully occupied with a steel slab (Full) and the gate is shut. Each assignment statement defines how the value of a particular variable changes. For example, the third assignment state, Vessel = Mid & Gate = Shut: {Mid, Hi}, indicates that if molten steel level is in the midpoint and the gate is shut, then the next state will be Mid or Hi (if more molten steel will be poured into the steel vessel). Finally the initialization section defines initial values of the system (the gate is shut, the vessel is low and the escalator is empty).

Specifications usually define rules for system behavior sequence (i.e., state execution trees). In this case, we wish to specify that the controller must shut the gate if either the amount of molten steel in the vessel is too low (Lo) or the steel slab reaches the end of the cooling escalator. In this case, we specify properties of paths and states within the paths by using temporal logic constructs. More specifically, we use Linear Temporal Logic (LTL) consisting of atomic propositions and propositional operators.

As can be seen in Figure 4.35, the depiction of a state diagram of even a relatively simple system generates an almost incomprehensible diagram (see more on the "state explosion" below.

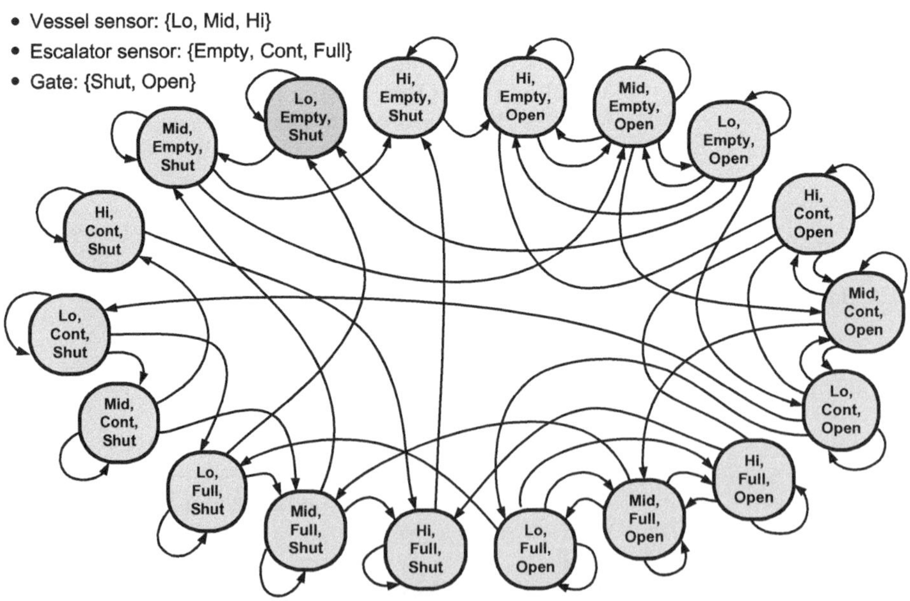

Figure 4.35 State transitions: steel slab production.

Benefits of Model-Based Testing As seen in the above examples, even simple systems exhibit complex behavior. In fact, the number of test cases needed to verify a system is derived from the number of state paths, which tends to be very large. Therefore, the effectiveness of model-based testing is very much dependent on how amenable it is to being automated. Automatic test generation and execution permit running many permutations of test strategies sequentially or in parallel on multiple test stations. Since models are formal entities, their behavior is well defined. Therefore, executing test cases can provide a proof of correctness rather than just evidence that a given set of faults was found. In other words, if full coverage testing can be guaranteed, then the testing process ensures the correctness of the model.

Another benefit of model-based testing is the ability to test early in the system development cycle, perhaps even from the start of the specification stage. This involves the testing early enough to enable detection of engineering design and specification faults.

Weaknesses of Model-Based Testing The model-based testing paradigm encompasses the following major weaknesses:

- *State Explosion.* The main challenge in model checking is dealing with the state space explosion problem which is common in real-life applications. This problem occurs in systems with many interacting components with data structures assuming many different values. In such cases the number of global states can be massive. A widespread approach to deal with this problem is by means of abstraction. This is a process of pruning the system properties by abstracting and simplifying its model. The simplified system may not satisfy exactly the same properties as the original one; therefore, a further process of refinement is often required. Frequently, available resources only permit to analyze a rather coarse model of the system. A positive verdict from the model checker is then of limited value because inconsistencies may well be hidden by the simplifications that had to be applied to the model.
- *Mathematical Limitations.* Whereas model checking for discrete system behavior that can be modeled using a state machine has been successful, such is not the case when dealing with continuous or analogue systems and less so when dealing with heterogeneous systems (i.e., systems that have different properties, depending on what portion of the system is examined). The same limitation applies when dealing with certain data domains. For example, floating point data calculations are not dealt with by most model-checking tools.
- *Nontriviality.* Implementing a model-checking process is not trivial. This method requires experts that understand both the requirements of the model under verification and the technology to implement formal properties.

- *It's a Model, Not the Real System.* The VVT engineer should always keep in mind that the model and the real system are two different physical entities. The implication is that proving the correctness of the model does not necessarily guarantee the correctness of the SUT. Standard procedures such as system testing and formal reviews are necessary to ensure that the abstract model adequately reflects the behavior of the concrete SUT.

Further Literature

- Baier and Katoen (2008)
- Beizer (1995)
- Berard et al. (2001)
- Braspenning (2008)
- Broy et al. (2005)
- Clarke et al. (1999)
- Drusinsky (2006)
- Palshikar (2004)
- Utting and Legeard (2006)

4.3.8 Robust Design Analysis

Robust design is a development philosophy focused on improving system reliability. The method is based on assumptions of scatter, or uncontrollable uncertainties in nature. Scatter in system inputs causes a system to exhibit unexpected behavior and therefore become less predictable. Usually, scatter degrades system performance. From a VVT standpoint, the objective of robust design analysis is to verify that end products or systems are immune, to a reasonable degree, to conditions that could adversely affect their performance. More specifically the intent of robust design evaluation is to ensure minimal product variance with respect to customers' specification or tolerance limits as well as minimized system bias, so that the nominal product operates as would the customer's desired product.

Figure 4.36 depicts a plot of the normal standard deviation identifying how wide the scatter is, or how large the variability is, of a system's response parameter.

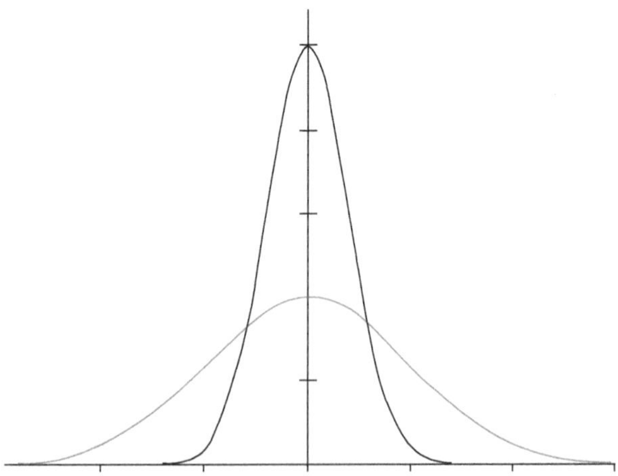

Figure 4.36 Scattering effects of system behavior.

Minimizing the standard deviation will lead to a smaller range of variability; that is, the chance that the response parameters will differ largely from the mean value decreases. So we can state that the goal of a robust design analysis is to minimize the standard deviation of a response parameter.

The importance of this process may be gleaned from Figure 4.37. A product or a system is designed to meet a certain specification (mean) with tolerances $\pm 6\sigma$. This defines the Lower Specification Limit (LSL) and the Upper Specification Limit (USL), respectively. Sometimes, in the presence of noise,[43] the mean is shifting in either direction. If the standard deviation of the system is large, a certain behavior may violate the specification limits, thus producing a system fault. This may be avoided if the system was designed with much narrower required standard deviation.

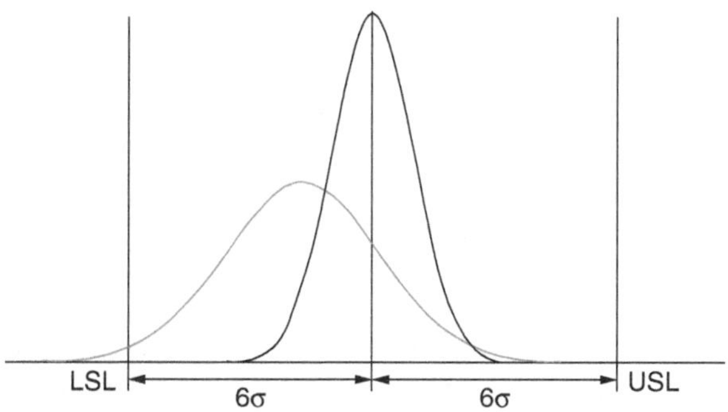

Figure 4.37 System scatter effect due to noise.

From a probabilistic point of view, a system may be considered robust if it is reliable. Therefore, conducting a robust design analysis verifies that the system has been optimized for reliability. Here, reliability is the probability that the product functions as expected, that is, conforms to specifications.

Taguchi's Loss Function According to the traditional view, products and systems are designed and manufactured to meet a specific target value T with allowable tolerance ($\pm t$). So a resistor, for instance, in an electronic circuit may be defined as having a resistance of $50\,\text{k}\Omega$ with tolerance of $\pm 5\%$. Therefore, in Statistical Process Control (SPC), as long as the design or the production is kept within the defined tolerances, we are satisfied.

In the language of Taguchi, one of the quality movement luminaries, according to the traditional view, the quality loss function $L(x)$, is a discontinuous step function: As long as the process or product is within the tolerance limits and quality loss is zero but outside those tolerances, quality loss C becomes unacceptable (see Figure 4.38):

[43]Natural or man-made disturbances (both internal and external to the system) that usually have a deleterious effect on a system's performance.

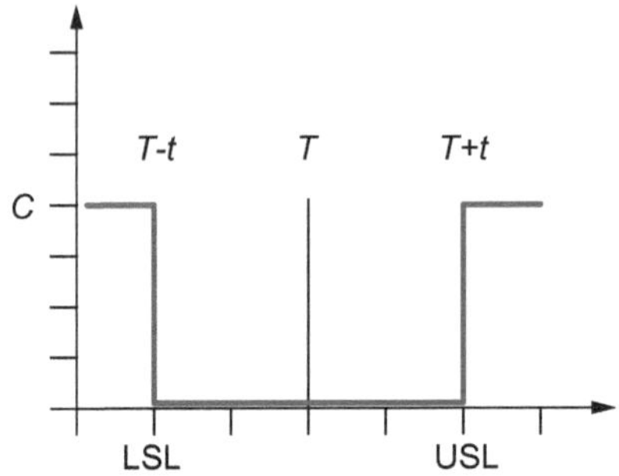

Figure 4.38 Traditional view of loss function.

$$L(x)=\begin{cases} C; & x<T-t \\ 0; & T-t\le x\le T+t \\ C; & x>T+t \end{cases}$$

Taguchi recognized that the traditional view of quality as a step function is not realistic. First, even if a product is manufactured within allowable tolerance, it may not function properly and some added cost will be required to bring it to proper working conditions.

We illustrate this idea in the following example: A box and a cover are produced in an automatic assembly line. Four bolts are welded onto each corner of the box and four holes, fitting perfectly to the bolts, are drilled in each corner of the cover. Each item must be located in a nominal position plus or minus Δ. For simplicity let assume that, for each corner, each bolt or hole is located in one of nine positions. That is, nominal, nominal $\pm\,\Delta_X$, nominal $\pm\,\Delta_Y$ and nominal $\pm\,\Delta_X \pm \Delta_Y$ (see Figure 4.39).

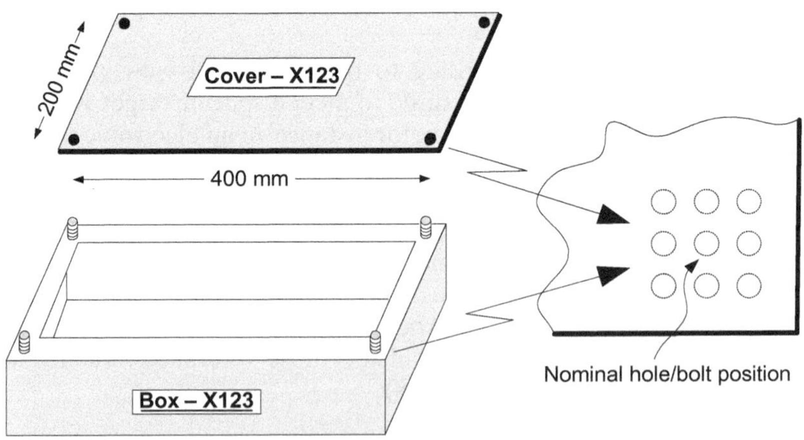

Figure 4.39 Example: cover attached to a box by means of four bolts.

The number of bolt-welding combinations is $9^4 = 6561$. Similarly, the number of hole-drilling combinations is $9^4 = 6561$. Therefore, the number of combinations for the entire box-and-cover system is $9^8 = 43{,}046{,}721$. However, for each bolt combination, there is one and only one fitting hole combination, so there are $9^4 = 6561$ cases where holes in the cover perfectly fit bolts in the box, and therefore the probability of a perfect match is $9^4/9^8 \approx 0.0152\%$. In other words, although all operations were performed within tolerance, virtually every box/cover combination will require some adjustment necessitating extra effort and rendering boxes and covers not exchangeable.

Second, Taguchi suggested that if a system moves away from the nominal specifications outside the tolerance limits, it often still retains some value to its users. A book with a torn page is annoying to a reader but does not render the book worthless. Moreover, in real life, loss of value is often not a linear function of the deviation from nominal specifications. Taguchi suggested a loss function model based on a quadratic function so that gradual deviations from the nominal specifications create squared increments in customer dissatisfaction. Figure 4.40 depicts this model. The loss function $L(x)$ at point x is equal to a loss coefficient C multiplied by the square of the difference between the actual value x and the target value T.

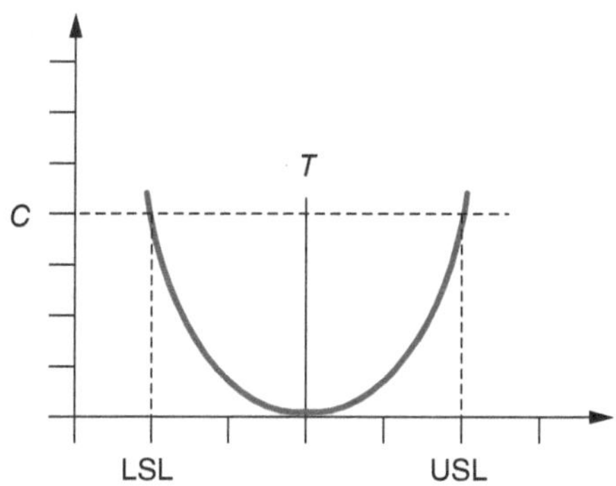

Figure 4.40 Taguchi's view of loss function.

If we accept Taguchi's assertion that quality loss is a quadratic function of the deviation from a nominal value, then the goal of our quality improvement efforts should be to minimize the squared deviations or variance of the product around the nominal specifications rather than the number of units within the tolerance limits (as is done in traditional SPC procedures):

$$L(x) = C(x - T)^2$$

Taguchi's Signal-to-Noise Ratios According to Taguchi and other researchers, all engineered systems should (ideally) always respond in exactly the same manner to the *signals* generated by the user. In other words, ideal systems will only respond to the operator's signals and will be unaffected by random *noise* factors. As a result, we would like to design, manufacture and operate systems having minimum performance variability in the presence of noise.

Taguchi uses the term signal to indicate the inputs users employ to control a given system. For example, we can control a radio receiver by turning it on and off, selecting AM or FM channels and tuning it to different broadcasting frequencies. In contrast, noise is the undesired and usually uncontrolled input affecting our system behavior during design, manufacturing and usage. Noise factors such as manufacturing tolerances, aging, usage patterns and environmental conditions are disturbances that cause system behavior to fluctuate away from the original specifications. They must be identified and quantified so that accurate choices can be made about which effects require compensation. During the system design phase, engineers must therefore compensate for such noise factors that could significantly influence the system away from nominal performance.

Therefore, the goal of a robust design effort is to find the best settings of the controlled factors that are involved in the design, production and operational process in order to maximize the Signal-to-Noise (S/N) ratio of the system. Taguchi (1986) and other researchers suggested several ways to quantify the respective product's response to noise factors and signal factors. Few of them are considered rather controversial while others are more widely accepted. We described some S/N relationships below:

- *Smaller-the-Better.* The following S/N ratio computation may be used in order to measure the occurrences of undesirable product characteristics. In this equation, y_i is the respective characteristic and n is the number of observations on the particular product. For example, the number of errors in a document could be measured as the y variable and analyzed via this S/N ratio:

$$\left(\frac{S}{N}\right)_{(1)} = -10\log_{10}\left\{\frac{1}{n}\sum_{i=1}^{n} y_i^2\right\} \quad i = 1, 2, \ldots, n$$

- *Nominal-the-Best.* Computation of the S/N ratio could be based on a fixed signal (or nominal) value and its production variance around this value, which may be considered the result of noise factors. This equation could be used whenever target quality is equated with a nominal value. For example, the diameter of a bolt must be as close to specification as possible to ensure high fitting to a corresponding nut:

$$\left(\frac{S}{N}\right)_{(2)} = 10\log_{10}\left\{\frac{\mu^2}{\sigma^2}\right\}$$

- *Larger-the-Better.* The following equation should be used when we would like to ascertain the S/N ratio associated with a system's performance, for example, the power of a motorbike engine relative to its fuel consumption:

$$\left(\frac{S}{N}\right)_{(3)} = -10\log_{10}\left\{\frac{1}{n}\sum_{i=1}^{n}\frac{1}{y_i^2}\right\} \quad i = 1, 2, \ldots, n$$

- *Signed Target.* The following equation should be used when we would like to compute the S/N ratio associated with a system where the quality characteristic of interest has a target value of zero and both positive and negative values of the quality characteristic may occur, for example, a pump system that must ensure a zero difference in the pressure of chemicals stored in two tanks within a petrochemical plant. In this equation σ^2 stands for the variance of the quality characteristic across the measurements:

$$\left(\frac{S}{N}\right)_{(4)} = -10\log_{10}\{\sigma^2\}$$

- *Fraction Defective.* The following equation should be used when we would like to compute the S/N ratio associated with efforts to minimize the number of failing elements, scrap and so on. Here, p is the proportion of defective failing elements, for example, of a production batch:

$$\left(\frac{S}{N}\right)_{(5)} = -10\log_{10}\left\{\frac{p}{1-p}\right\}$$

Robust Design Analysis Procedure From a VVT standpoint, the objective of a robust design analysis procedure is to verify, in an organized manner, whether or not the system meets its performance requirements with the highest possible system reliability and within an acceptable systems cost. The process often follows these steps:

- *Step 1: Parameter Identification.* This step entails the identification of the relevant parameters affecting the system. More specifically, it covers (1) the selection of signals for controlling the system, (2) the noise that is always present in the environment of the system and (3) the performance metrics that constitute the response of the system.
- *Step 2: Performance Objective.* This step entails a determination of a set of performance objectives appropriate to the system at hand and other relevant considerations (e.g., available knowhow, resources, budget). Typically, one or more of the following S/N ratios would be selected as the performance objectives:

a. Smaller-the-better
b. Nominal-the-best
c. Larger-the-better
d. Signed target
e. Fraction defective

- *Step 3: Planning the Test.* This step entails the planning of the test runs in the presence of typical environmental noise in order to elicit the desired effects. Depending on economics and other relevant factors, real tests may be conducted or more often than not a set of simulated tests may be performed. The following types of tests are commonly undertaken:
  a. Use of full or fractional factorial designs to identify interactions
  b. Use of an orthogonal array to identify the main effects with minimum of examinations
  c. Use of inner and outer arrays to see the effects of noise factors
- *Step 4: Running the Test.* This step entails the actual conduct of the test(s). In particular, the control and noise factors must represent real-life system usage. The performance metrics should be recorded and the performance objective should be computed.
- *Steps 5: Analyzing Test Results.* In this step the analysis of the test results is performed. In particular, the mean value of the performance objective for each factor setting must be computed and an analysis should reveal which control factors reduce the effects of noise and which ones can be used to scale the response.
- *Step 6: Evaluating Control Factor Points.* This step entails the evaluation of the selected system design settings to maximize or minimize the selected performance objectives while considering existing variations with great care.

Robust Design Analysis The Mean-Squared Deviation (MSD) measures how closely are the dual objectives of (1) achieving average performance close to target and (2) achieving low variation about that target. In the equation below, n is the number of observations, y_i is the measured performance value for observation i, and T is the target value:

$$\text{MSD} = \frac{1}{n}\sum_{i=1}^{n}(y_i - T)^2$$

Minimizing MSD requires meeting both of the following objectives:

- Adjusting the settings of the controllable inputs to center the performance of a system or process at its target value T
- Adjusting the settings of the controllable inputs to minimize the *variation in performance* of a system or process about its average value.

Selection of the appropriate adjustments to achieve both objectives requires that we carry out the following two tasks:

- First we must identify the controllable inputs that influence the *average performance* and generate equations describing the relationship between average performance and those controllable inputs.
- Second, we must identify the controllable inputs that influence the *variation in performance* and generate equations describing the relationship between variation in performance and those controllable inputs.

Robust Design Example The UAV autolanding example given in a previous section can also be used here to demonstrate the Taguchi procedure for robust design and S/N computations. The three controllable inputs are shown in Table 4.15. They are the UAV autolanding starting locations in three-dimensional (3D) space.

TABLE 4.15 System Controllable Inputs (UAV Autolanding Starting Locations)

Factor	Low Setting (−1), km	High Setting (+1), km
UAV-X	3	5
UAV-Y	−2	2
UAV-Z	0.5	3.5

Two uncontrolled variables—wind speed and UAV weight—constitute "noise" factors that affect the behavior of the autolanding system in an unpredictable way. The wind speed may be negligible (denoted Wind = −1) or up to 10 knots per hour (denoted Wind = +1). The UAV may carry a small payload weighing 5 kg and have a near-empty tank of fuel, weighing 1 kg (denoted Weight = −1), or may carry a payload weighing 25 kg and a full tank of fuel, weighing 15 kg (denoted Weight = +1).

The system performance is now calculated on the basis of the following simplified autolanding success model: The UAV landing strip is divided into five zones plus a sixth zone outside the landing strip itself (Figure 4.41). Ideally the UAV should touch down in the front and center of the landing strip but not too close to the beginning of the landing strip (zone A). Similarly, the landing roll of the UAV should end in the center of the landing strip, but not too close to the end of the landing strip (zone A). Each landing performance is calculated based on the sum scores of the UAV touchdown zone and end of the roll zone. For example, an automatic landing with a touchdown at zone D (Score = 1) and end roll at zone B (Score = 2) will produce a total score of 1 + 2 = 3 points for this autolanding test.

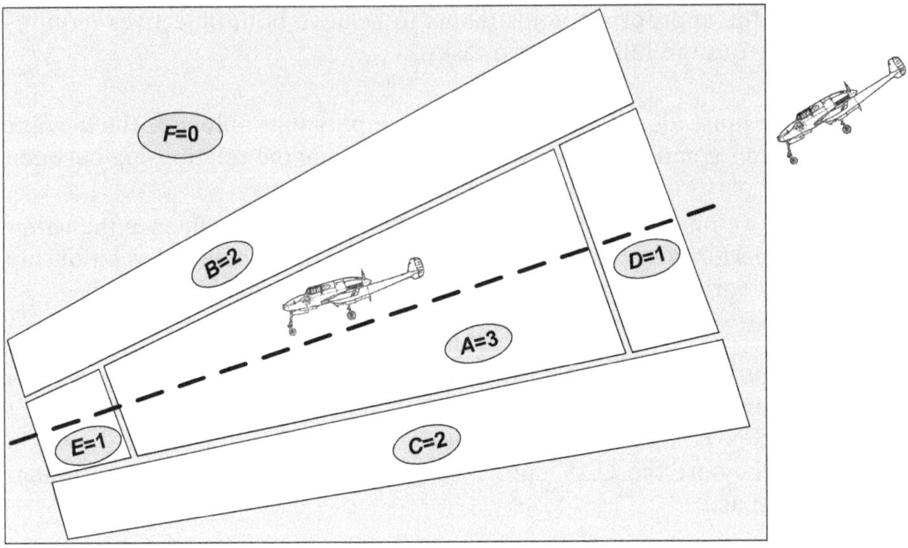

Figure 4.41 UAV landing strip divided into success level zones.

The results from a 32-simulation design run combining inner and outer arrays are shown in Table 4.16.

TABLE 4.16 Autolanding Test Results Under Uncontrolled Wind and Weight Noise

			Wind	−1			−1			1			1					
X	*Y*	*Z*	Weight	−1			1			−1			1			Average	σ	ln(σ)
−1	−1	−1		A	A	**6**	B	A	**5**	C	A	**5**	F	D	**1**	4.25	1.92	0.65
−1	1	−1		A	A	**6**	A	A	**6**	A	A	**6**	A	B	**5**	5.75	0.43	−0.84
1	−1	−1		C	A	**5**	C	E	**3**	C	D	**3**	C	D	**3**	3.50	0.87	−0.14
1	1	−1		A	D	**4**	D	C	**3**	D	A	**4**	D	C	**3**	3.50	0.50	−0.69
−1	−1	1		C	C	**4**	B	C	**4**	D	D	**2**	D	F	**1**	2.75	1.30	0.26
−1	1	1		C	F	**2**	C	E	**3**	A	E	**4**	F	F	**0**	2.25	1.48	0.39
1	−1	1		B	A	**5**	A	E	**4**	B	E	**3**	D	C	**3**	3.75	0.83	−0.19
1	1	1		B	F	**2**	A	B	**5**	B	E	**3**	E	F	**1**	2.75	1.48	0.39

Touchdown zone End roll zone Autolanding score

Figure 4.42 depicts the main effects plots for the average performance of this UAV autolanding example. Such plots, according to Taguchi, identify the controllable inputs that influence the average performance. Accordingly, the initial height of the UAV (Z location) has the largest effect on average performance (autolanding success).

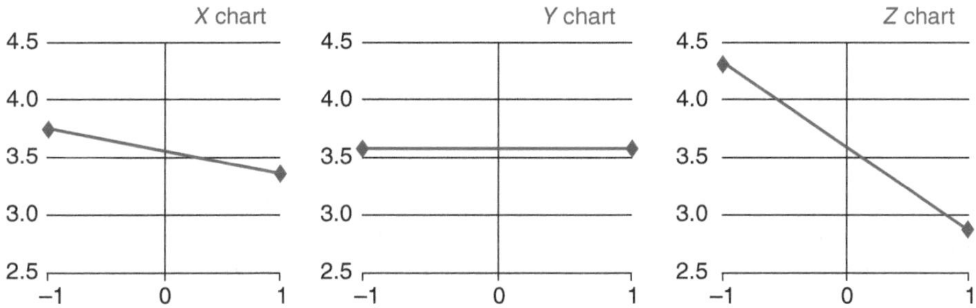

Figure 4.42 Main effects plots for average performance.

A similar analysis performed on the natural log of the standard deviation ($\ln_e \sigma$) produces the results shown in Figure 4.43. These plots suggest that all of the controllable inputs may similarly influence the variation in system performance (autolanding success).

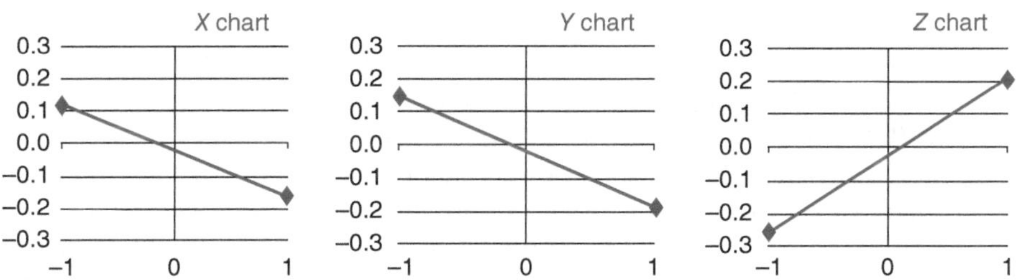

Figure 4.43 Main effects, natural log of standard deviation, autoland performance.

We now compute the relevant S/N ratio, which in our case is larger-the-better. Here, the number of simulated experiments is $n = 32$, y_i (i = 1, 2, …, 32), and represents the autolanding scores of all the landing tests, and the computed S/N ratio is 7.67:

$$\left(\frac{S}{N}\right)_{(3)} = -10\log_{10}\left\{\frac{1}{n}\sum_{i=1}^{n}\frac{1}{y_i^2}\right\} = -10\log_{10}\left\{\frac{1}{32}\sum_{i=1}^{32}\frac{1}{y_i^2}\right\} = 7.67$$

Further Literature

- Park (1996)
- Taguchi (1986)
- Wang (2005)

4.4 PARTICIPATE IN REVIEWS

4.4.1 Expert Team Reviews

We use the phrase *expert team reviews* as a generic term which includes *inspections, walkthroughs, audits and peer reviews*. A systematic description of the first three methods is available from, among other places, Institute of Electrical and Electronics Engineers Standard for Software Reviews (IEEE-STD-1028, 1997). Notionally, there are clear differences among the four types of reviews, but in practice, they often are carried out in a pretty similar ways. The following is a short description of the four types of reviews:

- *Inspections.* Inspections are a class of review processes developed at the International Business Machine (IBM) by Fagan (1976). This process was later improved by Radice (2001) and then Gilb and Graham (1993) and again by Gilb (2008). The process is characterized by examining documents (and computer code in case of software inspections) as well as collecting various metrics about the inspection process itself. This information is used to manage future individual inspections as well as for long-term process improvement. The method of studying documentation is often based on an analysis of a primary document; however, the process is not necessarily sequential. It is characterized by any analysis tactic (e.g., assigning specialized roles to individual inspectors and selecting particular documents or sections of them) that best suits the inspection objectives (e.g., maximizing the effectiveness of inspections, measuring defect density, helping engineers learn specs).
- *Walkthroughs.* Structured walkthroughs are considered descendants of the IBM inspection methodology. Usually, the creator of the evaluated object (most often a document or software code) presents it to a group and they in turn analyze it sequentially and hopefully recognize errors, coding bugs or potential performance problems. IBM carried out research which showed that walkthroughs were less effective than were inspections in identifying software defects. However, the walkthrough format is still favored by many organizations.
- *Audits.* Audits are another variation of team review, which tends to be adversarial in nature. Audits use sampling of actual process performance to determine if an organization is actually following proscribed practices, or the practices they claim to be following. This is quite different from examination of documents, specification and code described above. For example, evaluating an organization to determine its Capability Maturity Model Integration (CMMI) level is typically carried out by means of an audit.
- *Peer Reviews.* Peer reviews are made by people that are normally not the managers of the person whose work is being reviewed, nor are they full-time checkers or inspectors. They are usually peers of the responsible

engineer or author (i.e., individuals of the same type and level doing similar work). The primary idea of a peer review is to achieve open and honest reviews by, among other things, protecting the responsible engineer from being threatened. The implication is that criticism for the person doing the work is confidential and management will neither ask nor expect to hear the criticism. In principle, peers may carry out any inspection, walkthrough or even audit.

Inspections, which we considered most relevant for this book, are perceived in a rather different way by the software community versus the system community. Software inspections are viewed as a disciplined engineering practice to review technical documents as well as software code in order to detect and prevent the leakage of defects into the field. In contrast, system inspections are viewed as a mostly formal process of verifying the condition of existing systems and infrastructures, such as electrical equipment, automobiles, houses, aircraft, buildings, roads, bridges, pipelines and power plants. This section will discuss document inspection methods and system inspections methods leaning toward the software community philosophy.

Document Inspections A document inspection is a disciplined engineering practice for detecting defects in technical documentation and preventing the consequence of their inaccuracies from leaking into production and actual use. Inspection methods are now widely used within various engineering industries so we here describe these topics only briefly. Readers are encouraged to review the existing literature.

Each organization or project must agree on "inspection entry conditions," that is, the quality level of the document or software listing to be inspected (e.g., "at a minimum, the work product is complete and has been signed off by one person besides the author"). Similarly, "inspection exit conditions" must be agreed upon indicating when the inspection process should be terminated (e.g., "no more defects are found and the requirements can go forward to the design phase with little risk").

Document and software listing inspections may be performed with different objectives. But the most important purpose is (1) to identify defects and (2) to reach inspectors' consensus, approving the document for use, once it is considered defect free. Typically, a document inspection process comprises the following steps:

- *Step 1: Inspection Planning.* The inspection leader plans the inspection and selects the inspection team.
- *Step 2: Initial Meeting.* During an initial meeting the author of the work product explains the document or software code to the inspection team.
- *Step 3: Inspection Preparation.* Each inspector on the team examines the document or software listing to identify possible defects.
- *Step 4: Inspection Meeting.* During the inspection meeting the document or software listing is discussed, section by section, and the inspectors

point out the defects for every section. The meeting ends with the writing of an action plan.

- *Step 5: Product Correction.* The author makes changes to the work product in accordance with the action plan from the inspection meeting.
- *Step 6: Inspection Follow-Up.* The inspectors make sure that all problems have been eliminated by checking the changes made by the author.

The following provides guidance for conducting and optimizing the inspection of a system's technical documents. It is an adaptation and generalization of the paper by (Gilb, 1998) on optimizing software inspections for engineered systems. According to Gilb, inspections consist of two main processes: the *defect detection process* and the *defect prevention process*. The defect detection process is expected to find most of the existing defects in a document, whereas the defect prevention process is expected to achieve even greater benefit by teaching engineers how to improve their writing as they go through the defect prevention process. This process will hopefully reduce the number of mistakes made in subsequent work products. The following are some tips about how to conduct and optimize a document's inspection process.

Tips on Optimizing Document Inspection Process[44]

Tip Group 1: Establishing Inspection Purpose

1. Some people seem to think that the only purpose for document inspection is to clean up bad work and defects. More important, inspections should be used to motivate and teach proper document preparation, improve the way we locate the defects remaining in a document, improve document quality as well as improve the document or software preparation processes. In other words, the greatest payback comes when inspection improves future work, that is, reduces the number of documentation defects.
2. Inspections should cover both technical documents and management documents such as contracts, marketing strategies and product development plans.
3. Inspections should be planned to address a set of specific purposes. For example, ensuring document quality, identifying and removing defects, job training and reducing maintenance costs are among the possible purposes. Inspection planning is done by selecting the

[44]Adopted and slightly modified with permission from Gilb (1998).

appropriate document types, choosing an appropriate number of inspectors with relevant skills, assigning them suitable roles and scheduling the timing and duration of inspections in accordance with their purpose.

Tip Group 2: Choosing Work Products Intelligently

1. Resources are always limited in one way or another. Therefore, inspecting upstream work products is more profitable. In particular, inspection of requirements and design documents is rewarding since most system problems tend to reach the implementation phase and beyond.
2. The main purpose of inspections is economic: to reduce lead time and people costs caused by downstream defects. Therefore, we do not like to start document inspection when it is immature and, conversely, we do not like to continue inspecting a document ad infinitum. Document defect sampling is an inexpensive technique to determine entry and exit conditions. Defect sampling is carried out by devoting a short time to inspect a few pages of a document in order to ascertain the amount of major defects in this sample. Such sampling indicates if the document is stable enough to justify a formal inspection process. At a further stage sampling indicates whether the document is mature and is economically safe to release it into the downstream flow.
3. Management inspection is advisable when system development starts with contracts, management and marketing plans.
4. Often organizations waste time checking document features that do not have significant impact on the quality of the final product (e.g., typographical errors in a design document). Defects in such features do not trigger major consequences. One strategy to save inspectors time is to have the author of the document identify important text or graphics that can translate into serious downstream costs in order to distinguish these from less important (commentary or boiler-plate) areas.

Tip Group 3: Focusing on Finding Major Defects

1. Document inspection involves checking each page against several related source documents, checklists and standards. In other words, one must check a single line against many sources. As a result, checking the rate on specific document types may range between 0.2 and 1.8 pages of 300 words per checking hour. This rate range

is seen in the checking carried out both before and during the inspection meeting.

2. A major defect is a document error that, if not dealt with, will probably have an order-of-magnitude or larger cost to find and fix when it reaches the operational stage. It does not matter if a defect is visible or not to a customer. If an error can potentially lead to significant cost were it to escape downstream, classify it as a "major defect" and take care of correcting it as soon as possible.
3. Often inspectors waste time identifying a great deal of minor defects. This "90 percent minor defect" syndrome should be avoided. From an economic standpoint, a clear message must be given to not waste time on minor defects. For example, one should insist only on inspection rules or checklists that emphasize finding major defects or recording only major defects at a meeting. In addition, it is advisable to highlight for management attention all supermajor defects that have been uncovered.

Tip Group 4: Applying Good Inspection Practice

1. Often, organizations do not have the discipline to set up and respect inspection entry conditions. As a result, inspections often start when a given work product is not quite ready, leading to waste of time and money and causing frustration within the inspection team. An important entry condition should be that upstream source documents are available in order to inspect a given document. Another effective entry condition is the assurance that source documents are of high quality. A good step in doing this is to give a numeric quality measure to each source.
2. Inspection necessitates effective work standards, which in turn provides the rules for the authors writing technical documents and then for the inspectors to subsequently check those documents. Standards are built by hard experience. They need to be brief, to the point, monitored for usefulness and, most importantly, respected by the development team.
3. An overall master plan for the entire inspection sequence of a project should be generated early in the project lifecycle. Thereafter, each individual inspection should be specifically planned to include the formal purpose of this specific inspection and the inspected work product, the required supporting documents, the assigned individual inspectors and their roles, the total checking time allocated and any other important issues.
4. Inspection generates a lot of information that is fundamental and useful for managing the process. The inspection team should utilize

commercial or proprietary software tools to capture the data, summarize it and present trends and reports.

5. Because inspection is an imperfect process, one should also focus on defects that may be present in source and kin documents associated with the work product under inspection. For example, if a functional specification is the work product under inspection, there should be a requirement document as one of the source documents and a testing document as one of the kin documents. There is a good chance that these other documents contain defects as well.
6. By and large, an optimum number of people are needed on a specific inspection team. This optimum depends heavily on the purpose of the inspection. Our experience has been that two to four people are needed for an efficient inspection process, four to six people are needed to be effective at finding major defects and larger numbers of people in an inspection team may be justified for teaching purposes.
7. An effective inspection team strategy is to allocate specific defect-searching roles to people on the team such that each person on the inspection team should be looking for different kinds of defects, for example, identification of time and budget risks, checking against corporate standards for engineering documentation and checking security loopholes.
8. Inspection should be performed by professionals committed to making maximum, meaningful progress on the project. Inspectors should avoid suggesting fixes and solutions. The inspection team should not engage in gossip, search for the guilty or malign others on the project team.
9. Exit conditions, if correctly formulated and taken seriously, can be crucial to the success of an inspection. The exit condition "Exit inspection only when the maximum remaining major defects are estimated to be less than 0.2% of the statements in the document" could prove to be very effective. Management must understand the benefits of making clear policy about the levels of major defects that will be allowed.

Tip Group 5: Providing Adequate Training and Follow-Up

1. In order to achieve effective inspections, team leaders must be properly trained. Such training takes about a week (half lectures and half practice). After initial training, they need to be periodically coached by an experienced person and receive a formal inspection certification.

2. An engineering organization should ensure that there are an adequate number of trained people to support inspections. We recommend that at least 20% of all professionals in the organization be qualified to participate in inspections.

Tip Group 6: Publicizing Inspection Results and Statistics

1. Inspections improve the quality of systems and products, prevent embarrassments and save money. Inspection teams should be proud of their contributions to the firm and should publicize their achievements for all to see and follow. The team should place relevant inspection artifacts, standards, statistics, samples of detected problems and experiences on a corporatewide website as soon as possible.

Tip Group 7: Continuously Improving Inspection Process

1. The inspection process should be continuously and systematically improved. Initially, this is required in order to learn the inspection process properly and to tailor it to the needs of the organization. However, over time the inspection process should be more efficient, namely yield detection of more major defects using fewer inspectors devoting less inspection time.

System Inspections System inspections are often portrayed from a maintenance point of view and may be characterized as any task undertaken to determine the condition of a system. Sometimes, people consider the determination of labor, materials, tools and equipment required to repair the system as an organic part of the system inspection process.

Inspection issues are discussed at some length in standard AS-9100, which is derived from standard ISO-9001 (see Myhrberg and Crabtree, 2006). This is a quality management standard specifically written for the aerospace industry. It provides a common set of quality requirements, facilitates development of unified quality systems and enables customers to share results of quality system audits. For example, AS-9100 ensures right of access by the purchaser, the customer and regulatory authorities to all facilities involved in all applicable quality records such as design, test, examination, inspection and customer acceptance requirements and any related instructions and requirements. In addition, it grants access to all requirements for test specimens (production method, number, storage conditions, etc.) for design approval, inspection and investigation or auditing. In fact, AS-9100 is now a family of standards applicable to different areas of the aerospace industry, which include, in particular, AS-9102, the Aerospace First Article Inspection Requirements standard.

The following provides guidance for the inspection of quality systems and processes. It could be used for assessing manufacturer's compliance with quality products and processing. It is an adaptation and generalization of the U.S. Food and Drug Administration's *Guide to Inspections of Quality Systems* (Quality System Inspections Reengineering Team, 1999) for engineered systems. This set of Quality System Inspection Techniques (QSITs) provides ways to conduct an efficient, effective and comprehensive inspection enabling evaluators to focus on key elements of a firm's quality system.

Guide for Inspection of Quality Systems and Processes[45]

This guide concentrates on a "top-down" approach in order to address organizations' quality products and processes from a system point of view. Figure 4.44 shows the seven components of the quality systems and processes. We describe a set of suggested techniques for inspecting each of four key quality system elements which, we think, are the basic foundation of a firm's quality system:

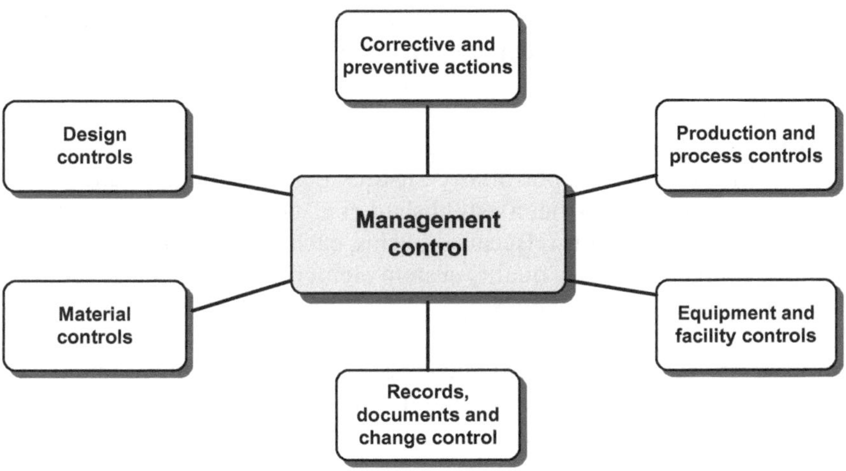

Figure 4.44 Quality system elements (Quality System, 1999).

[45]Based on the document: "Quality System Inspection Techniques (QSIT)", US Food and Drug Administration (FDA, 1999).

1. Management control
2. Design controls
3. Corrective and preventive actions
4. Production and process controls

The QSIT uses the "established approach" in conducting the inspection. In this context, the established approach means assuring a defined and written document implemented routinely. For each quality system element, one first determines if the firm has defined and documented the requirements for that element by looking at procedures and policies. Then, one continues looking at both raw and processed data to determine if the firm is meeting its own procedures and policies and if its program for executing the requirement is adequate.

The duration of inspection is dependent on the depth of the inspection. This guide was designed to accomplish a complete review of all four quality system elements in approximately one week. While the length of inspections vary, following rigorous steps will help assure that one looks at the most important elements of the firm's quality system during the inspection.

Part 1: Management Control

The purpose of management control is to provide adequate resources for system design, manufacturing, quality assurance, distribution, installation and servicing activities; assure the quality system is functioning properly; monitor the quality system; and make necessary adjustments. A quality system that has been implemented effectively and is monitored to identify and address problems is more likely to produce systems that function as intended.

A primary purpose of the inspection is to determine whether management with executive responsibility ensures that an adequate and effective quality system has been established (i.e., defined, documented and implemented) at the firm. Because of this, each inspection should begin with an evaluation of this quality system element. The inspection method should include the following steps:

1. Verify that the following have been defined and documented: (1) quality policy, (2) management review, (3) quality audit procedures, (4) quality plan and (5) quality system procedures and instructions.
2. Verify that quality policies and objectives are in fact implemented.
3. Review the established organizational structure to verify that it includes provisions for responsibilities, authorities and necessary resources.

4. Confirm that a management representative has been appointed and evaluate his or her range of management authority and representative.
5. Verify that management reviews are conducted on a regular basis and include the suitability and effectiveness of the quality system.
6. Verify that quality audits, including repeated audits of previously identified deficient issues of the quality system, are being conducted on a regular basis.
7. Verify that management with executive responsibility ensures that an adequate and effective quality system has been established and maintained.

Part 2: Design Controls

The purpose of the design control quality element is to control the design process to assure that systems meet user needs, intended uses and specified requirements. This should include (1) attention to design and development planning, (2) identifying design inputs, (3) developing design outputs, (4) verifying that design outputs meet design inputs, (5) validating the design, (6) controlling design changes, (7) reviewing design results, (8) transferring the design to production and (9) compiling a design history file in order to assure that resulting designs will meet user needs, intended uses and requirements.

Sometimes, the inspection assignment mandates the inspection of a particular design project. Otherwise, select any project that reflects a good representative of the organization's design control system. This project will be used to inspect the process, the methods and the procedures that the firm has established to implement the requirements for design controls.

If the project selected involves a system that contains software, consider reviewing the software's validation while proceeding through the assessment of the firm's design control system. The inspection method should include the following steps:

1. Select a single design project.
2. Verify that the design control procedures for the selected project meet any regulation requirements if they exist (e.g., aerospace, FDA).
3. Review the design plan for the project at hand to understand the proposed design and development activities, including project assigned responsibilities and interfaces.
4. Confirm that design inputs were established.

5. Verify that the design outputs essential for the proper functioning of the system were identified.
6. Confirm that acceptance criteria were established prior to carrying out the actual verification and validation activities.
7. Determine if design verification actually confirmed that the design outputs met the design input requirements.
8. Confirm that the design validation data prove that the agreed–upon design met the predetermined user needs and intended uses.
9. Confirm that the completed design validation did not leave any unresolved inconsistencies.
10. If the system contains software, confirm that the software was validated.
11. Confirm that risk analysis was performed.
12. Determine if design validation was accomplished using initial production systems or their equivalents.
13. Confirm that all modifications and changes were formally controlled. This includes validation or, where appropriate, verification of such processes.
14. Determine if design reviews were conducted.
15. Determine if the design was correctly transferred into production specifications.

Part 3: Corrective and Preventive Actions

- *General.* The purpose of Corrective And Preventive Action (CAPA) is to collect information, analyze information, identify and investigate product and quality problems and take appropriate and effective corrective or preventive action to prevent their recurrence. Verifying or validating corrective and preventive actions as well as communicating such activities and providing relevant information for management review and documenting these activities are all essential in dealing effectively with product and quality problems, preventing their recurrence and preventing or minimizing system failures. One of the most important quality system elements is the corrective and preventive action. Corrective action taken to address an existing product or quality problem should include action to correct the existing product nonconformity or quality problems and prevent the recurrence of the problem. The inspection method should include the following steps:

1. Determine if the correct reason for product and quality problems has, in fact, been identified. Confirm that data from these sources

have been analyzed to identify existing systems and quality problems that may require corrective action.

2. Determine if sources of systems and quality information that may show unfavorable trends have been identified. Confirm that data from these sources are analyzed regularly to identify potential systems and quality problems that may require preventive action.
3. Challenge the quality data information system. Verify that the data generated by the CAPA system are complete, accurate and timely.
4. Verify that appropriate statistical methods are employed to detect recurring quality problems. Determine if results of analyses are compared across different data sources to identify and develop the degree of product and quality problems.
5. Determine whether failure investigation procedures are followed. Determine if the degree to which a quality problem or nonconforming product is, in fact, investigated in accordance with the level of risk involved. Determine if failure investigations are conducted to determine the root cause of the problem. Verify that preventing distribution of nonconforming product is in fact under control.
6. Determine if appropriate actions have been taken for significant systems and quality problems identified from data sources.
7. Determine if corrective and preventive actions were, in fact, effective and verified or validated prior to implementation. Confirm that corrective and preventive actions do not adversely affect the finished system.
8. Verify that corrective and preventive actions for systems and quality problems were implemented and documented.
9. Determine if information regarding nonconforming systems and quality problems and corrective and preventive actions has been properly disseminated and reviewed by management.

- *Malfunction Product Reporting.* The purpose of malfunction product reporting is to ensure the identification, investigation and reporting of all malfunction information related to a firm's products and systems. This is usually the first step in a process of product corrections and removals as well as product tracking. For example, the medical device reporting regulation mandates that medical device or system manufacturers, device or system use facilities and importers of medically related equipment or substances establish a system that ensures the prompt identification, timely investigation, reporting, documentation and filing of system-related death, serious injury and malfunction information. Such event may require the relevant authority to initiate corrective actions to protect the public health. Therefore, compliance with appropriate device reporting must be verified to ensure that an appropriate surveillance program receives both timely and accurate information. The inspection method should include the following steps:

1. Verify that the firm has defined an appropriate System Reporting Procedure (SRP) and this SRP is indeed established and maintained. In certain industries (e.g., aircraft, health and medicine, nuclear power) such SRPs must address appropriate regulatory requirements.
2. Confirm that the appropriate SRP information is being identified, reviewed, reported, documented and filed.
3. Confirm that the firm follows its SRP and they are effective in identifying reportable malfunctions and their consequences.

- *Systems Corrections and Removals.* The purpose of system corrections and removals is to ensure that manufacturers and importers of products and systems notify the public or appropriate authorities of any product or system correction or removal initiated to reduce a risk to the public. In other words, the inspection should ensure that a system posing known hazards to users, operators or the public be corrected or removed from use. For example, an automobile with a known defect should be recalled for a corrective action. The inspection method should include the following steps:

1. Determine if the manufacturer initiated corrections or removals of a system.
2. Verify that the organization has established and continues to maintain a database for all nonreportable corrections and removals.
3. If formal reporting to government authorities or the public is required by law or appropriate regulation, then confirm that the firm's management has implemented that reporting requirement.

- *System Tracking.* The purpose of system tracking is to ensure that manufacturers or importers of products and systems expeditiously locate and remove defective systems from the market or notify appropriate authorities and the public of significant system problems. The inspection method should include the following steps:

1. Determine if the firm manufactures or imports a tracked system or product.
2. Verify that the firm has established a written Standard Operating Procedure (SOP) for tracking of defective systems and products. In certain industries such SOPs must also comply with appropriate regulatory requirements.

3. Verify that the firm's quality assurance program includes audits of its failed systems, devices and product-tracking system within an appropriate and acceptable timeframe.

Part 4: Production and Process Controls

The purpose of production and process control is to manufacture systems and products that meet specifications. Developing processes that are adequate to produce systems or products that meet specifications, validating those processes and monitoring and controlling the processes are all steps that help assure the result will be systems that meet specifications.

In order to meet the production and process control requirements the firm must understand when deviations from system specifications could occur as a result of the manufacturing process or environment. Determination of such deviations may be accomplished via product and process risk analyses.

For inspection purposes one should select for evaluation a manufacturing process in which deviations from system specifications could occur as a result of the process or its environment. The inspection method should include the following steps:

1. Select a process for review based on the following criteria:
   - CAPA indicators of process problems
   - Use of the process for manufacturing higher risk systems
   - Degree of risk of the process to cause system failures
   - Firm's lack of familiarity and experience with the process
   - Use of the process in manufacturing of multiple systems
   - Variety in process technologies and profile classes
   - Processes not covered during previous inspections
2. Review the specific procedures for the manufacturing process selected and the methods for controlling and monitoring the process. Verify that the process is controlled and monitored.
3. If review of system history records (including process control and monitoring records) reveals that the process is outside the firm's tolerance for operating parameters or rejects or that product nonconformance exists:
   - Determine whether any nonconformance was handled appropriately.
   - Review equipment adjustment, calibration and maintenance.
   - Evaluate the validation study in full to determine whether the process has been adequately validated.

4. If the results of the process reviewed cannot be fully verified, confirm that reviewing the validation study validated the process.
5. If the process is software controlled, verify whether the software was validated.
6. Verify that personnel have been appropriately qualified to implement validated processes or appropriately trained to implement processes that yield results that can be fully verified.

Further Literature

- Fagan (1976)
- Freedman and Weinberg (1990)
- Gilb (1998, 2005, 2008)
- Gilb and Graham (1993)
- IEEE STD 1028 (1997)
- Myhrberg and Crabtree (2006)
- Quality System (1999)
- Radice (2001)

4.4.2 Formal Technical Reviews

A formal technical system review is used to evaluate the quality of a system at various points throughout its lifecycle. The role of a formal technical review is to bring together the most relevant people to criticize the work done, solve open issues and decide on the action items required to pass to the next formal review. These formal reviews often coincide with milestones in the management of a project and carry contractual obligations on both supplier and purchaser. A formal meeting constitutes the peak of the technical review where the most qualified people review the results presented.

Formal system technical reviews are conducted in order to assess the degree of completion of technical efforts related to major milestones before proceeding with further technical effort. More specifically, the objective of reviews is to satisfy all relevant individuals (e.g., system developers and maintainers, management and customer representatives as well as other relevant stakeholders) that the system and its comprising hardware and software satisfy all aspects of the system requirement and mission needs. In addition, the formal technical review assures timely and effective attention to the technical interpretation of contract requirements and monitors program progress and risk. It also evaluates the validity and completeness of technical documentation in order to assess the maturity of the development effort. Finally, the review provides a vehicle for communicating the status of the system to all interested parties.

At the end of a formal review, a decision must be made whether or not to declare the review "passed." Such a declaration is reached if critical action items are fulfilled within a date specified during the review meeting. Otherwise, the team must do some rework and schedule another review. The term "formal" attests that the review is governed by agreed-to written rules. Most commonly, formal reviews are mandated by the Statement Of Work (SOW), usually reflect major system lifecycle milestones and are given well-defined entry and exit criteria. Research studies support the conclusion that formal reviews greatly outperform informal reviews in their cost effectiveness.

Typical Formal Technical Reviews/Audits Formal system technical reviews and audits are performed at different phases of a system's lifecycle. The most common reviews are depicted in Table 4.17.

TABLE 4.17 Typical Technical Reviews and Audits

• Alternative System Review (ASR)	• Integration Readiness Review (IRR)
• Software Requirement Review (SRR)	• System Verification Review (SVR)
• System Requirement Review (SysRR)	• Acceptance Test Review (ATR)
• System Functional Review (SFR)	• Functional Configuration Audit (FCA)
• Preliminary Design Review (PDR)	• Physical Configuration Audit (PCA)
• Critical Design Review (CDR)	• Test Readiness Review (TRR)
• System Design Review (SysDR)	• Production Readiness Review (PRR)

Other Advantages of Formal Technical Reviews The most obvious value of formal technical reviews is that they can identify problematic issues earlier and more economically than they would be through testing or field use. The cost to find and fix a defect by a well-conducted review may be one or two orders of magnitude less than when the same defect is found by testing or in the field.

In addition, formal reviews are a mechanism to make major system decisions. A formal review has a key role in project management because management, quality and financial issues are naturally intertwined with technical considerations. As mentioned, formal reviews facilitate information exchange, as many experts are around the table to give and receive valuable inputs and comments on the work done. Stopping to prepare and evaluate the work completed to date creates an opportunity for reflection on the technical and management issues. Additionally, the documents and presentations prepared for the review are useful not only for the project at hand but also to guide future projects.

Generic Process of Formal Technical Reviews IEEE STD 1028 defines a common set of activities for formal (software) reviews. The following is a variant of this procedure oriented for engineered system formal reviews:

- *Step 0: Entry Evaluation.* The review leader is expected to use a standard checklist of entry criteria to ensure that optimum conditions shall exist for a successful review.

- *Step 1: Management Preparation.* Responsible management ensures that the review will be appropriately resourced with staff, time, materials and tools and will be conducted according to policies, standards or other relevant criteria.
- *Step 2: Planning Review.* The review leader identifies or confirms the objectives of the review, organizes a team of reviewers and ensures that the team is equipped with all necessary resources for conducting the review.
- *Step 3: Overview of Review Procedures.* The review leader ensures that all reviewers understand the review goals and the review procedures. In addition, he or she is responsible for making all necessary material available to the participants and all relevant procedures for conducting the review are well known.
- *Step 4: Individual Preparation.* The reviewers individually prepare for group examination of the work under review by examining it carefully for anomalies, the nature of which will vary with the type of review and its goals.
- *Step 5: Conducting Review.* The reviewers meet at a planned time to pool the results of their preparation activity and arrive at a consensus regarding the status of the system and the activities or documents to be reviewed.
- *Step 6: Rework/Follow-Up.* The persons responsible for the reviewed objects undertake whatever actions are necessary to satisfy the requirements agreed to at the review meeting. The review leader verifies that all action items are closed.
- *Step 7: Exit Evaluation.* The review leader verifies that all activities necessary for successful review have been accomplished and that all outputs appropriate to the type of review have been finalized.

VVT Activities: Pre-Review The VVT team leader should prepare for formal technical reviews along the following steps:

- *Collect Results of Activities.* The VVT team leader must collect all relevant VVT data from subproject leaders before the review and ensure that all VVT documentation has been produced and approved internally.
- *Prepare Material for Review.* The VVT team leader has to prepare, with the help of the project team, all VVT material necessary to the review:
  a. Agenda for VVT issues to be discussed during the review
  b. Technical VVT documents
  c. Material for VVT status presentation

- *Analyze Material.* The VVT team leader must analyze all VVT-related data and provide a synthesis to the reviewers that must show both the technical and management status of each VVT activity under review.

- *Create Review Package.* The VVT team leader must provide all VVT-related material for the creation of the review package. Normally such a package includes an agenda and the material to be examined by the review participants.

VVT Activities: During Review The VVT team leader should contribute VVT-related input and be involved in technical reviews along the following lines:

- *Review Meeting Agenda.* The agenda is formally presented at the beginning of the meeting and some adjustments may be proposed and decided during the meeting. The VVT team leader should ensure that key VVT issues are presented and discussed during the review.
- *Review Project and System Status.* The project master plan is presented and actual as well as potential delays are discussed. In addition, a summary of the budget-planned resources versus the actual expenses is presented. The role of the VVT team leader is to ensure that both schedule and budget issues related to VVT are presented and discussed.
- *Review Technical Items.* A technical status is presented to the attendees, including achievements and open issues. All specialists, including VVT domain experts, should make a presentation of their work. They will receive remarks and critics from the review team.
- *Review Open Issues and Action List.* Toward the end of the review meeting, the attendees will usually reconsider all the open issues. An action list is created showing the open issues to be resolved. Each action item is assigned to a person in charge of solving the related issues within a precise completion deadline. Naturally the VVT team leader will attend to any VVT problem discovered during the review.
- *Decisions: Pass or Fail.* The review team together with management, customers and contract specialist' representatives conducts a synthesis of the review meeting. These individuals make a decision of whether the review has passed or not. Generally, if the review is "not passed," critical action items have to be closed first before another partial review can be conducted to address these problems and move ahead in the project. A decision can be taken to "pass" the review, pending the closure of a given set of action items, if it is not a critical one. Again, the role of the VVT team leader is to monitor all open VVT issues and provide professional advice to the rest of the group.

VVT Activities: Post-Review At the end of a formal technical review, the review leader should create minutes of the review, recording decisions and agreements reached along with a list of follow-up action items. The review's final report should be completed and distributed within a reasonable time (e.g., a week or two) and should include meeting minutes (review topics, objectives, participants, agenda, list of materials covered), an action item list, a review of score results and the scoring system used and lessons learned. The VVT team leader should contribute all data and advice related to his or her specialty.

Guidance for Technical Reviews

- Each formal technical system review should have a clear and predefined set of objectives and a clear statement of purpose.
- It is always advisable to conduct a meaningful set of internal reviews first, and they must produce honest criticism. Furthermore, training reviewers in formal technical system review procedures and techniques prior to assigning them to a project is most advisable.
- Scheduling technical reviews too early, before relevant system documentation and work products are available, may lead to decisions based on insufficient information. Conversely, scheduling technical reviews too late can mean that project commitments have already been made which cannot be changed without incurring heavy financial or time losses.
- Within technical reviews, careful attention should be paid to areas that contain new and unfamiliar problems. It is good practice to call in outside experts to provide such advice.
- Selecting proper reviewers is crucial. One should strive to bring tough reviewers and challenge them to find faults in the material presented to them.
- It is recommended that the review team be comprised of (1) representatives of the customer and relevant stakeholders, (2) the program manager, (3) the chief system engineer, (4) one or more quality assurance, configuration control and process improvement representatives and (5) one or more system developers, maintainers, and user domain experts. Keep in mind that too many reviewers may create havoc in the reviewing process.
- Reviews should be encouraged to perform the following: (1) agree on the scope of the review, (2) collect and review data, (3) inspect the review package, (4) assess review readiness, (5) present findings to the review team, (6) assess review completeness and (6) improve the review process.
- Reviewers are not put on for purposes of gaining approval for a project. They should educate the participants and project team as well as emphasize process improvement. Hiding project weaknesses is counterproductive. Asking for advice is the wisest strategy.
- Management support is a prerequisite to a successful review. This should include allocating adequate manpower, facilities and time for the review and encouraging the review team to bring all significant problems into focus.

- In the final analysis, a good review produces constructive criticism and removes confusion. Therefore, all involved in a review should recognize that a success criterion, more important than "passing," is the illumination of validly identified problems.
- Often (but not always) having the customer as an active participant in the review is valuable. It gives the customer visibility as to the level of requirement understanding and progress of the project. Conversely, it gives the producers and maintainers of the system a better understanding of customer expectations.

Further Literature

- Faulconbridge and Ryan (2002)
- IEEE STD 1028 (1997)
- MIL-STD-499B (1993)
- MIL-STD-1521B (1995)

4.4.3 Group Evaluation and Decision

This collection of methods is based on a group's evaluation and decision meetings, attended by technical experts, convened specifically to evaluate engineered systems and make a decision regarding the suitability of the system to meet relevant requirements. Such group meetings may be partially active throughout the entire system lifecycle and are scheduled whenever needed. For example, a group evaluation and decision may verify a system's design, test and qualification process, production of some objects, a maintenance activity or the disposal of the system.

Typically, technical reviews are conducted by means of the group evaluation and decision process. They provide leaders, system designers, builders, test engineers and production engineers with valuable insight into the state of the system with which they are involved.

Evaluation and decision processes carried out within groups have distinct advantages over similar processes performed by individuals due to the following:

- Research shows that the effectiveness of groups as decision makers is generally superior to individual members. Groups can discuss issues and process information and are more likely to identify errors in logic and facts as well as reject incorrect solutions.
- By nature, groups bring to the table a broad representation of opinions and personalities so that more ideas are generated and the option for evaluation increases. In addition, a group represents greater informational resources and possesses a more accurate memory of facts and events than do its individual members.
- Groups generally set standards for conducting evaluations and making decisions. Usually, following formal procedures solidifies the process and

ensures that all aspects of a problem have been addressed. Well-defined decision rules (e.g., majority rule, unanimous decision, quantitative decision procedures) ensure, at least to some extent, that all group members had a chance to air their opinions and open issues were settled in a fair manner.

- By and large, people are more likely to follow through if decisions have been made by means of an accepted group process. This increased commitment for implementation fosters diligence and expedience as well as better cooperation among the members of the group.

Group Evaluation and Decision Process We assume in this discussion that the members of the group of which we are speaking are suited to the task put to the group. For instance, if the task involves reviewing a technical issue, all group members have some expertise and knowledge that apply to the technology involved. Based on this assumption, the basic phases involved in a typical group evaluation and decision process are:

- *Phase 1: Defining Issue at Hand.* The first phase of the group evaluation and decision process starts with a group orientation and development of shared mental model of the issue. More specifically, the group tries to arrive at an accurate understanding of the system to be evaluated by means of discussion as well as exchanging and sharing information. If initial evaluation of the data available to the group identifies a problem, then the nature of the problem, the extent and seriousness of the problem as well as the likely cause of the problem and the possible consequences of not dealing effectively with it are analyzed. Based on this analysis, the group generates a number of appropriate and feasible alternative lines of action among which an acceptable choice of one or more actions should exist.
- *Phase 2: Making a Decision.* During the next phase, the group uses one of several decision schemes to select a single alternative line of action from the various alternatives originally proposed by the group. Typical decision schemes are an individual (usually managers) who makes the decision for the group, voting using a majority rule, consensus rule (where all members of the group must agree to a certain decision), and so on.
- *Phase 3: Implementing and Evaluating the Decision.* During the next phase the group reviews the implementation of the selected solution and evaluates the consequences of this process. In particular, the group needs to be fully cognizant of the relative merits and disadvantages of all available alternatives in order to learn how the group can be more effective in the future. More specifically, postmortem (i.e., after the problem has been solved or after the problem could not be solved) discussions provide valuable learning lessons to the group, facilitating a retrospective look at past decisions and the decision-making process itself.

Factors in Group Processes Research in several disciplines (e.g., economics, business, engineering, psychology) indicates that both individual and group characteristics influence group dynamics and decision-making processes. Current research shows that group process effectiveness in terms of decision-making speed, correctness or accuracy often depends on the following characteristics:

- *Individual and Group Skills.* Individual and group skills, communication skills and problem-solving skills among group members are important components of effective groups. Similarly, group skills such as conflict resolution, group goal setting or egalitarian leadership foster effective group performance.
- *Cognitive Mechanisms.* Cognitive mechanisms include the mental activities involved in processing information and their related dynamic mental models. Cognitive strategies are the formal mechanism controlling the mental processing of information, whereas heuristics are informal mechanisms controlling the mental processing of information.
- *Communication Dynamics.* Beyond the communications skills of individuals within the group, the characteristics of the communication process itself is significant to group dynamics and decision making. Communication patterns among group members expose information power relationships and the social status of group members.
- *Decision Policies.* Decision policies are the agreed-upon rules that cement the required discipline for group decision making. Such decision policies may be formal, for example, Delphi technique or majority vote or nominal group methods. Conversely, decision policies may be informal, for example, discursive group processes. The aim of informal processes is to deliberate openly and democratically in order to obtain reasoned agreement among equally qualified group participants.
- *Task Complexity.* Task complexity significantly affects the behavior and dynamics of the group. Complexity can be measured in many ways, including the amount of information that must be absorbed and processed, the number of possible decision options available to the group or the number of steps required to perform a certain task (e.g., evaluating the behavior of a system's performance).
- *Social Factors.* Social factors determine the nature and dynamics of interpersonal relationships within the group. They often include interpersonal influence and power as group network cohesiveness and role definitions assumed by group members.
- *Environmental Influences.* Environmental factors affect group decision making. Organizational characteristics such as size, formal structure and culture influence the decision-making processes. In addition factors such as working environment and financial or time pressure can produce stress, which affects group behavior.

Group Process Leadership Styles Typically, leaders of evaluation and decision groups may be categorized into the following decision-making styles:

- *Autocratic.* Under the autocratic management style, leaders tend to solve problems on their own based on information available to them at the time. The information or advice provided by group members is utilized only when it coincides with their own ideas or when proof that they are wrong is irrefutable. Otherwise, they seldom seek information or advice from group members.
- *Consultative.* Consultative leaders tend to share problem solving with members of the group. However, they still rely heavily on their own knowledge, experience and opinions.
- *Participative.* Participative leaders discuss the problems with the members of the group and together the leader and members devise an appropriate solution. In this management style the leader acts as a chairperson of a committee and, by and large, accepts a group decision, which typically is arrived at on the basis of decision by majority or consensus.

Group Process Risks Group evaluation and decision processes are not always successful. First, all such group processes are time consuming. If derived solutions and appropriate mitigating solutions are not timely, the group process may be a failure. In addition, sometimes the group makes a bad decision. Among causes that may be to blame for a bad decision are bias in sharing information, cognitive limitations, group polarization and, most notoriously, groupthink phenomena as well as plain old social loafing. The following describes these pitfalls, often found in bad decisions made by groups:

- *Shared Information Bias.* Shared information bias is the tendency for groups to discuss issues familiar to all members and avoid examining information that only a few members know. This leads to poor decisions making due to ignorance of important facts by the group. For example, evaluating system test information where certain failures are known to some members but are not exposed to the rest of the group may cause judgment errors and heuristic biases.
- *Cognitive Limitations.* Poor communication skills as well as biases in an individual's cognition and motivation can often lead to judgment errors on the part of individuals in the group. Another cognitive limitation on the part of individuals is the tendency to seek out information that confirms their inferences rather than disconfirms them. Again, this may lead to errors in judgment and a failed decision process. In addition, individuals tend to overestimate their judgmental accuracy because they remember mostly the times their decisions were confirmed. Finally, some group participants lack inquiry and problem-solving skills or their information processing is limited relative to other persons, affecting their cognitive abilities.

- *Group Polarization.* Research in social comparison theory identifies the phenomenon of group polarization, the tendency to respond in a more extreme way when making a choice as part of a group. Under this condition a group has difficulty assessing the facts rationally and often fails to reach a decision acceptable to all (illustrated in Figure 4.45).

Figure 4.45 Polarization—not an effective group strategy.

There are a number of possible explanations to group polarization incidents: First it is likely that extreme majority alternatives get more group discussion time. Second, often extreme individuals become more extreme in the heat of an argument. More often than not, group polarization manifests itself when the group (1) lacks maturity and heterogeneity, (2) contains persons tending to egocentrism or (3) most commonly is managed by a person lacking conflict resolution skills.

- *Groupthink.* Irving Janis's (1972) groupthink theory states that decision-making groups will sometimes succumb to a groupthink phenomenon. This occurs when group members become so focused on achieving concurrence that the search for consensus overrides any realistic assessment of other views. Groups affected by groupthink ignore alternatives and tend to take irrational actions. A group is especially vulnerable to groupthink when the group is insulated from outside opinions and is highly cohesive.

  Symptoms of groupthink are group pressures toward uniformity, invariably expressed in either overt or covert criticism of any dissenting views. Typically, the group tends to overestimate its power and invulnerability and manifest close-mindedness and stereotype views about the world outside the group. Other typical causes for groupthink are structural failures in the makeup of the group, entrapment in sunk costs, control by an autocratic leader or a domineering member in the group and finally plainly defective decision-making processes.

Groupthink is a particularly vicious phenomenon resulting in a system that either does not meet requirements or contains problems that were not properly addressed. Groupthink can be prevented or their effect can be greatly reduced by taking the following steps:

1. Enhance the group process. This entails assigning the role of devil's advocate to one or a few members of the group. Given this title, a person would more readily voice different or contradictory views in the group discussions. In addition, the enhanced group process should mandate the obligation to always create multiple alternatives for an eventual selection and adoption of a preferred approach. It will also require reexamining advantages, weaknesses and potential risks of each alternative discussed by the group. Finally the enhanced group process should require that a contingency plans be established in case something goes wrong with the current approach.
2. The group should attempt to obtain expert or outside advice. This is important in order to correct group misperceptions and biases.
3. The group should adopt an effective decision-making technique that will eliminate the tendency of the group to get trapped in stereotyped views. One technique that may be effective is to divide the evaluation and decision group into two smaller groups which would discuss the issues separately and then present their findings in a joint session.
4. Finally, autocratic leaders should adopt a more open style of leadership. In addition, domineering members of the group must be persuaded to make their suggestions later, after others members have had their say.

We should hastily add that the groupthink phenomenon is rarely recognized by members of such groups. As a result, the group will not usually take steps to remedy this tendency. Unfortunately, only after a particularly disastrous error in judgment on the part of the group will it be open to corrective action.[46]

- *Social Loafing.* Research shows that, sometimes, people do not work as hard in groups as they work alone. This is especially true on easy tasks in which individual contributions are blended and indistinguishable. For

[46]For example, after the Bay of Pigs invasion fiasco (1961), U.S. President John Kennedy sought to avoid groupthink in his cabinet meetings. He encouraged cabinet members to discuss possible solutions within their own departments and invited outside experts to share their viewpoints. Occasionally, he divided his cabinet into subgroups to break the group cohesion and sometimes he deliberately left the cabinet room for a while in order to avoid pressing his own opinion. Later, in September 1962, the Soviet government placed offensive nuclear missiles in Cuba, precipitating a crisis that came closest to a strategic nuclear war. The same group that blundered into the Bay of Pigs tackled this political and military challenge with notable wisdom and ingenuity.

example, in rope-tugging experiments, Ringelman (1880s) showed that the larger the group, the less effort individual expand (i.e., one person pulled a rope at 100 units, two people at 186, three people at 255, and eight people at 392 units). Researchers suggest the following reasons for social loafing:

a. *Diffusion of Responsibility.* Naturally, in a group setup the responsibility for the final outcome is diffused among members of the group. More specifically, often, members of the group are less exposed to individual responsibility and this may lead to a reduction of efforts.
b. *Free-Rider Effect.* Sometimes members of a group sense the benefit of belonging to a group in terms of prestige and power and yet feel that their individual contribution is not essential. As a result, they are likely to offer little in return and often practice decisional avoidance tendencies (e.g., avoiding responsibility, ignoring alternatives, procrastination).
c. *Sucker Effect.* In a group situation, everyone is benefiting and getting credit. Often individual members do not want to be ones who do all the work without specific recognition. As a result, members are willing to do what they conceive as their fair share but not more than that. In other words, contribute as little as possible.

Based on this phenomenon, it is fair to conclude that quite often some of the participants in an evaluation-and-decision group do not contribute to the full extent of their capabilities. However, research shows that individuals contribute their best when they think their efforts will help them achieve outcomes they personally value. Therefore, it is possible to identify several social factors that may eliminate or at least reduce social loafing tendencies.

From a positive standpoint, group work should include public acknowledgment of each individual's personal efforts and contributions. Social research shows that people rise to the occasion when the task is challenging and appealing. Therefore, group leaders should instill within the group the notion that evaluating the system and making the correct decisions is a most meaningful and important task. Another factor affecting social loafing is group size as well as familiarity among the group members and cohesiveness within the group. In general, people prefer to work with friends rather than strangers, within a smaller and neatly tied group where they can speak their minds freely.

From a negative standpoint, individuals within a group tend to work hard and contribute to the limit of their abilities if they expect the entire group to be punished for poor performance. Within the well-motivated environment of the VVT engineering community, this latter approach is certainly not a good choice.

Group Decision Methods This section describes specific group evaluation and decision methods (see Figure 4.46).

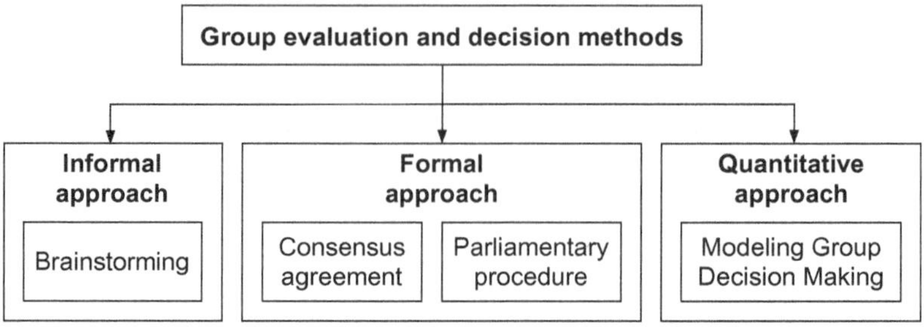

Figure 4.46 Group evaluation and decision methods.

Informal Approach: Brainstorming Brainstorming is an informal but useful method that can help a team or group of people generate creative ideas for evaluating technical problems. Often, brainstorming provides several alternatives for potential solutions to seemingly intractable problems. It also lets everyone in the group know how an idea has evolved and the level of ownership each one has on the outcome, thus setting the stage for consensus and action. Usually one person, perhaps the leader of the team or another experienced person (the facilitator), leads the brainstorming session. Within the expected chaos and confusion of such meetings, the facilitator should enforce the following typical rules:

- *No Egos.* As much as possible people should leave their egos outside the brainstorming process.
- *Anything Goes.* Bizarre and sometimes offbeat ideas are bound to come up. All ideas, however unusual, should be encouraged. Participants in brainstorming should not criticize or propose to modify an idea no matter how wild it is.
- *Quantity over Quality.* The more ideas, the better the chance of finding a desired solution to the problem at hand. It may go against commonly held beliefs, but research shows that, at the early stage of brainstorming, generating lots of ideas should take precedence over generating good ideas.
- *Evolving Ideas.* One advantage of brainstorming is that one person's idea may trigger a derivative inspiration in someone else's mind. Within the context of brainstorming, the facilitator should encourage the evolving generation of ideas based on the ideas of others.

Typical brainstorming may follow these steps:

- *Step 1.* Brainstorming is often most productive if it is preceded by a preliminary discussion that allows people to share their understanding of the problem, its root causes, the barriers to change, the specifics of the present situation and a vision of the ideal solution. Once the problem or issue is clearly defined, brainstorming usually starts as an inventory or listing of old, familiar ideas. Brainstorming works best when the group starts adapting or combining old solutions creatively into new ones.

- *Step 2.* The group is allocated some interval of time in order to brainstorm privately, that is, write their ideas regarding the problem on a piece of paper. This is an effective way to captures one's own ideas. This technique is also helpful in avoiding the syndrome of "group thinking" whereby the entire group goes off in one direction without exploring the full range of possibilities.
- *Step 3.* Each member of the group shares his or her ideas with the other members of the group. As mentioned, the facilitator ensures that no criticism or cynical comments will be expressed. However, a reasonable amount of questioning for better understanding of the ideas should be allowed. At the same time, the facilitator should discourage full-fledged discussion of these ideas. Usually one person (the recorder) notes the group's ideas on the board or on a laptop connected to a projector.
- *Step 4.* Next, the set of ideas generated by the group must be narrowed, focused and combined if any are redundant. This activity should extract a reasonable number of ideas on which the group can work. This may be achieved by means of group discussion as to the practicality and desirability of each idea. Some ideas will be considered outright unacceptable by the entire group and so be eliminated. The remaining ideas should be prioritized. One effective approach to prioritizing is based on a scheme whereby each member of the team rates each idea on a scale of 1–10. A few ideas with the highest combined score will be discussed, further leading to a final decision on the optimal solution.

Formal Approach: General Formal group evaluation and decision represent a process diametrically inverse to obtaining ideas and reaching conclusions by way of brainstorming. Often, a formal approach seems advantageous since evaluating complex technical problems is extremely difficult. First, such difficulty stems from the complexity of the technical issues associated with modern systems facing the VVT team as well as the organization at large. Second, the diversity of agendas and people who are involved in the evaluations, reviews and decisions make the entire process that much more difficult.

Conducting an effective meeting requires the active participation of every person in the group. In general, all group members are expected to actively engage in the group's work, share their views and pay attention to the flow of the meeting. There are various schemes to manage the group evaluation and decision process, but the two basic roles needed are the team leader and the recorder.

Fundamentally, the team leader is responsible for initiating and organizing group meetings as well as guiding the discussions and supporting all who want to participate. Often, the team leader tracks the passage of time and enforces the time limits established in the agenda, although any member can perform that task. The role of the recorder is to capture all relevant information that comes up during the group evaluation and decision process. Sometimes it is a good idea to have these notes taken on a laptop and projected on the wall so people can respond to these summaries in real time. Sometimes, though, this approach causes too much disruption to the ongoing flow of the meeting and the projection may have to be suspended. It is wise to never assign the role of recorder to the team leader.

Chronologically, formal group evaluation and decision will follow these three stages:

- *Step 1: Preliminaries.* The team leader has to prepare the evaluation and decision process. He or she must collect all necessary data needed for the evaluation and prepare it for the review. Once the supporting information package is available, the team leader must prepare an agenda, schedule a group meeting and send invitations along with the information packages.
- *Step 2: Evaluation and Decision.* During the group evaluation and decision meeting, the team leader will start by presenting the team members and the agenda. The main objective of such an evaluation meeting is to check whether the technical solutions that are presented are correct relative to the system requirements.

  Therefore, during the meeting, individuals may present their work to the evaluation group with all relevant information. For example, design activity information can be an analysis of several alternative designs. Similarly, information may be related to a system's test strategy and results or measurements of production performance versus expected target data.

  The evaluation group will examine the presented material based on their knowledge and previous experiences and make a decision regarding the outcome of the evaluation process. Any open issue, especially questions that raise a substantial risk for the project, shall be postponed to a future meeting.
- *Step 3: Closure and Implementation.* The team leader has to prepare a summary of the group findings as well as the decisions made by the group. In addition, the team leader must prepare a list of open actions together with planned closure dates and the details of people responsible for rectifying these problems.

Formal Approach: Consensus Agreement Consensus agreement is a process of coming to an agreement on a particular technical issue. A group evaluation and decision meeting conducted by consensus is usually less formal and the team leader must be willing to share control and allow more leeway in the group discussions.

As a rule, an issue brought up for discussion will be debated until the group reaches an agreement that all sides can accept. In other words, the group cannot take action that is not agreeable to each and every member in the group. Consensus does not necessarily mean unanimity, nor does it mean that all sides are satisfied with the solution but, at least, everyone must agree that they can "live with" and support the decision since it is the best solution acceptable to the group. Depending on national culture, personalities and the specific technical issues, reaching consensus takes considerable time, but the outcome is often worth it.

First, consensus agreement fosters open communication. People talk with one another regarding the technical issues at hand and their ideas about possible solutions. This exchange provides the basis for designing workable and acceptable alternatives.

Second, consensus agreement encourages more informed decisions. It is based on diverse opinions delivered in an open atmosphere and it encourages greater creativity and a larger number of options leading to more satisfactory decisions.

Third, people who interact together to understand the issues and who have developed solutions using consensus will see the reasoning behind a specific decision and, once consensus is reached, members tend to accept it. As a result, all members of the group will cooperate in the implementation and give the proposed decision ample opportunity to succeed.

There are situations where consensus agreement does not seem to be the most prudent way to conduct group evaluation and decision. For example, sometimes the issues are simply not so important or the alternative solutions are not significantly different in their effect on the problem. A one-sided management decision can be taken with minimal risk. Sometimes the extreme opposite occurs where the group is so polarized and emotionally charged that productive face-to-face discussions are not possible. Another example presents itself occasionally where an immediate decision is needed. In such situations, a wrong choice is better than a late choice and no time to convene the group, let alone debate the issue, is available.

Formal Approach: Parliamentary Procedure Parliamentary procedure is also a process of coming to an agreement on a particular technical issue, and its purpose is also to help a group evaluate technical subjects efficiently while preserving a spirit of harmony. It is based on democratic principles as practiced at national levels. Namely, the decisions of the majority are upheld, but voices of dissenting opinions are heard. Parliamentary procedure is simple to implement. First every member of the evaluation and decision group has equal rights. (This precludes the team leader from having unilateral decision power.) Second, each issue presented to the group is entitled to discussion time.

Using parliamentary procedure, the dynamic within evaluation and decision groups is usually quite accommodating and informal. Sometimes, however, this is not the case. For instance, when the technical issues are complex or when they are controversial, disagreements can cause an impasse. Another example is when the evaluation group is rather large or representing different organizations subscribing to different agendas. In such situations, the conflict resolution skills of the team leader and the careful managing of the evaluation and decision process are paramount.

We can sum up by stating that the key difference between consensus agreement and parliamentary procedure is that in parliamentary procedure voting results tend to create a "win–lose situation." As a result, the losers often are unwilling to support the winning position, which hampers implementation of the decision. In contrast, under consensus agreement, usually synthesis of values and ideas manifest itself rather than one side wins and the other loses. By and large, such a result brings about more harmony and individual willingness to participate in implementing the decision.

Quantitative Approach: Modeling Group Decision Group Decision Making (GDM) is a formal quantitative method of making a judgment based on the opinion of different people. Proper decision making is crucial to the functioning of organizations. GDM is an active area of research within MultiCriteria Decision Making (MCDM) studies. Often, we are mostly interested in the aggregation of multiple opinions within a group containing individuals who may be considered not equally influential within the group (i.e., one individual's opinion may be considered more/less valued relative to another individual).

In a group, every person has individual preferences so he or she may choose between a given set of alternatives. More precisely, each individual may choose his or her favorite alternative from each pair of alternatives. For example, given three alternatives a_1, a_2 and a_3, each person could choose between each pair of these alternatives, for instance the combination $\{a_1>a_2,\ a_1>a_3$ and $a_3>a_2\}$ could be the preference set of an individual in the group.

Social choice or, more appropriate for our domain, *Engineering Choice* (EC), is the collection of all possibilities in conjunction with their respective choice sets, and the aggregation of individual preferences. That is, given that each individual has a certain profile of preferences, the engineering choice is a ***function*** that transforms the aggregate set into the level of the collective. For example, in a dictatorship the social choice function that aggregates the preferences of the citizens is, in fact, the preference of just one particular individual, the dictator. We can express this concept formally as follows. For a given set of alternatives $X = \{a_1, \ldots, a_n\}$, we define the tuple of alternatives and preferences (Y, D), where Y denotes the subset of all the pairs in X and D denotes individual preference information. Thus, we can define an Engineering Choice (EC) function:

$$\mathrm{F}: \boldsymbol{X} \times \boldsymbol{D} \rightarrow \boldsymbol{P}(\boldsymbol{X})$$

where $\boldsymbol{X}$ = set of all possible Ys
$\boldsymbol{D}$ = set of all possible preference sets
$\boldsymbol{P(X)}$ = set of all subsets of X

For example, assume X contains two engineering alternatives: a_1 (Test subsystem-A) and a_2 (Test subsystem-B). Suppose the group is composed of only two persons. Each one either prefers the first alternative (+1) or the second alternative (−1) or is indifferent to the two alternatives (0). Here D specifies the preferences (or indifferences) and therefore $\boldsymbol{D}$ for the two individuals has $3 \times 3 = 9$ elements (see Figure 4.47). However for each of the 9 D, $F(X, D)$ can take three output values, i.e. {+1, 0, −1} thus there are a total of 3^9=19,683 engineering choice functions that could be defined.

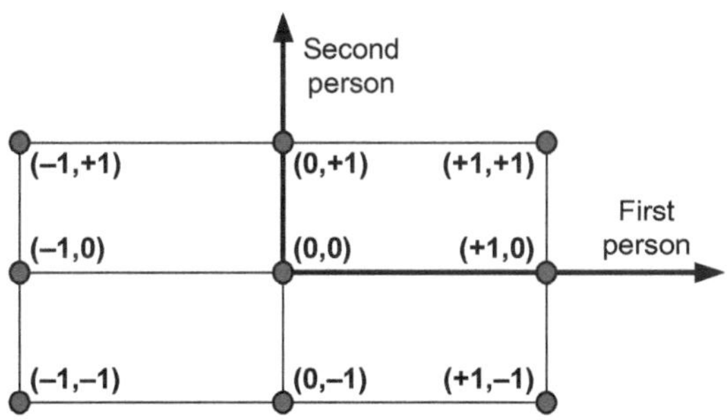

Figure 4.47 Example: universe of engineering preferences for the group.

There are many mathematical ways to obtain data from individuals in a group and then aggregate it into a unified group decision. Let us visualize one simple method of making a group decision by the following example: A technical committee is convened to decide how to deal with a serious budget overrun and a significant schedule delay in a development project. The committee comprises 13 members. It must rank four alternative actions:

- *Action A.* Replace the main contractor.
- *Action B.* Redesign and rebuild one problematic subsystem.
- *Action C.* Develop and produce the system in two builds, postponing problematic capabilities by a year.
- *Action D.* Terminate the entire project.

Each member has equal voting weight within the committee. He or she ranks the four alternatives (A, B, C, D) in order of importance.

This is done by assigning four points to the most attractive action, three points to the next alternative and so forth. The result of the committee members' voting is depicted in Table 4.18.

TABLE 4.18 First Example: Committee Member Vote

		Alternatives			
Support	Member	A	B	C	D
A supporters	1	4	2	1	3
	2	4	2	1	3
	3	4	2	1	3
	4	4	2	1	3
C supporters	5	3	1	4	2
	6	3	1	4	2
	7	3	1	4	2
B supporters	8	2	4	3	1
	9	2	4	3	1
	10	2	4	3	1
	11	2	4	3	1
	12	2	4	3	1
	13	2	4	3	1
Total		37	35	34	24

As can be seen, alternative A is the most valued choice. Nevertheless, it is quite puzzling to see these results (i.e., four members selected one ranking set, three members selected a second ranking set and six members selected a third ranking set). Typically, one would expect that independent individuals with integrity would exhibit much greater variance in their alternative action rankings.

Let us examine the results. First, we might ask, what is the probability that such results would have occurred if each ranking set had equal probability? (Unrealistic but still an interesting yardstick.) We start by noting that each committee member has a total of $4! = 24$ possible ranking combinations. So 13 members have a total of $S = 24^{13}$ ranking set combinations. We select 3 combinations out of 24 and then further select 1 combination out of the 3 and assign it to the first group of 4 out of 13 individuals. We then select 1 combination out of the remaining 2 and assign it to the second group of three individuals out of the remaining 9. Last, we select 1 combination out of the remaining 1 and assign it to the last 6 committee members:

$$N_1 = \binom{24}{3}\binom{3}{1}\binom{13}{4}\binom{2}{1}\binom{9}{3}\binom{1}{1}\binom{6}{6} = 2024 \times 3 \times 715 \times 2 \times 84 \times 1 \times 1 = 729,368,640$$

As can be seen, the probability of this result (based on our yardstick as our sampling space) is extremely low:

$$P_1 = \frac{N_1}{S} = \frac{729,368,640}{24^{13}} = 8.32 \times 10^{-10}$$

The above result may be contrasted with a hypothetical case where each committee member selects a unique ranking solution. In this case we select 13 combinations out of 24 and assign it to 13 committee members:

$$N_2 = \binom{24}{13} \times 13! = 2,496,144 \times 6,227,020,800 = 1.554 \times 10^{16}$$

As can be seen, the probability of this result seems "within an expectable range":

$$P_2 = \frac{N_2}{S} = \frac{1.554 \times 10^{16}}{24^{13}} = 0.0177$$

So we observe P_1 is about seven or eight orders of magnitude smaller than P_2, a very significant difference. One way to explain this puzzling situation is to speculate that the committee members did not vote as free agents with total dedication to the interest of the project but, possibly, were aware of what decision would be acceptable to their respective bosses.[47]

Further analysis of the voting patterns brings another possible "deceptive" strategy common in group decision making, that is, adding a nonrealistic alternative in order to distort the voting results.[48] Let us look at the voting patterns if we eliminate the fourth alternative. Now, each committee member will assign three points to the most attractive alternative, two points to the next alternative and so forth. The result of the committee members' voting is depicted in Table 4.19. Now, alternative B scored the highest and, remarkably, alternative A got the lowest score.

[47]Some readers may disagree with the validity of this example. Is it reasonable to use the above yardstick? Is the resulting speculation valid? Nineteenth-century British Prime Minister Benjamin Disraeli characterized three kinds of lies: "Lies, damned lies, and statistics." We are aware that mathematicians may exercise professional caution about the applicability of statistical inference, knowing that sometimes reality may not conform to assumptions on which these inferential models are constructed. Nevertheless, we think that within engineering this example is telling. As observed by Laplace (Théorie analytique des probabilités, 1820), "The theory of probabilities is at bottom nothing but common sense reduced to calculus."

[48]Kenneth Joseph Arrow was a joint winner of the Nobel Prize in Economics in 1972. He is mostly known for contributions to social choice theory, notably, *Arrow's impossibility theorem*. The condition of *Independence of Irrelevant Alternatives* (IIA) was first proposed by Arrow in 1951.

TABLE 4.19 Second Example: Committee Member Vote

		Alternatives		
Group	Member	A	B	C
A supporters	1	3	2	1
	2	3	2	1
	3	3	2	1
	4	3	2	1
C supporters	5	2	1	3
	6	2	1	3
	7	2	1	3
B supporters	8	1	3	2
	9	1	3	2
	10	1	3	2
	11	1	3	2
	12	1	3	2
	13	1	3	2
Total		24	29	25

As this area is under intensive research, readers are encouraged to further study the existing multifaceted literature dealing with GDM.

Further Literature

- Arrow et al. (2002)
- Best (2001)
- Gallagher (2008)
- Hirokawa and Poole (1996)
- Janis (1972)
- Lu et al. (2007)
- Torrence (1991)
- Vroom and Yetton (1976)

4.5 REFERENCES

Alekseev, S., Tiede, R., and Tollkühn, P., Systematic Approach for Using the Classification Tree Method for Testing Complex Software-Systems, in *Proceedings of the 25th Conference on IASTED International Multi-Conference: Software Engineering*, Innsbruck, Austria, 2007, pp. 261–266.

Antony, J., *Design of Experiments for Engineers and Scientists*, Butterworth-Heinemann, 2003.

ARP5580, Recommended Failure Modes and Effects Analysis (FMEA) Practices for Non-Automobile Applications, July 2001.

Arrow, J. K., Sen, K. A. K., and Suzumura, K. (Eds.), *Handbook of Social Choice and Welfare*, Vol. 1, North Holland, 2002.

Baier, C., and Katoen, J. P., *Principles of Model Checking*, MIT Press, Cambridge, MA, 2008.

Banks, J. (Ed.), *Handbook of Simulation: Principles, Methodology, Advances, Applications, and Practice*, Wiley-Interscience, New York, 1998.

Beizer, B., *Black-Box Testing: Techniques for Functional Testing of Software and Systems*, Wiley, New York, 1995.

Berard, B., Bidoit, M., Finkel, A., Laroussinie, F., Petit, A., Petrucci, L., and Schnoebelen, P., *Systems and Software Verification: Model-Checking Techniques and Tools*, Springer, 2001.

Best, J., *Damned Lies and Statistics: Untangling Numbers from the Media, Politicians, and Activists*, University of California Press, 2001.

Braspenning, N., *Model-Based Integration and Testing: Bridging the Gap between Academic Theory and Industrial Practice*, VDM Verlag, 2008.

Broy, M., Bengt, J., Katoen, J.-P., Leucker, M., and Pretschner, A. (Eds.), *Model-Based Testing of Reactive Systems: Advanced Lectures*, Springer, 2005.

Brue, G., and Launsby, R., *Design for Six Sigma*, McGraw-Hill Professional, 2003.

Chen, Y. T., Poon, L. P., and Tse, H. T., An integrated Classification-Tree Methodology for Test Case Generation, *Int. J. Software Eng. Knowledge Eng.*, 10(6), 647–679, December 2000.

Clarke, M. E., Grumberg, O., and Peled, A. D., *Model Checking*, MIT Press, Cambridge, MA, 1999.

Cohen, J., *Statistical Power Analysis for the Behavioral Sciences*, 2nd ed., Lawrence Erlbaum, 1988.

Cooper, W. J., Coden, R. A., and Brown, W. E., Detecting Similar Documents Using Salient Terms, in *Proceedings of the Eleventh International Conference on Information and Knowledge Management*, McLean, VA, 2002.

Drusinsky, D., *Modeling and Verification Using UML Statecharts: A Working Guide to Reactive System Design, Runtime Monitoring and Execution-based Model Checking*, Newnes, 2006.

Dyadem Press, *Guidelines for Failure Mode and Effects Analysis (FMEA), for Automotive, Aerospace, and General Manufacturing Industries*, CRC Press, Boca Raton, FL, 2003.

Fabbrini, F., Fusani, M., Gnesi, S., and Lami, G., An Automatic Quality Evaluation for Natural Language Requirements, in *Proceedings of the Seventh International Workshop on RE: Foundation for Software Quality*, 2001.

Fagan, M. E., Design and Code Inspections to Reduce Errors in Program Development, *IBM Systems Journal*, Vol. 15, No. 3, 1976.

Faul, F., Erdfelder, E., Lang, A. G., and Buchner, A., G*Power 3: A Flexible Statistical Power Analysis Program for the Social, Behavioral, and Biomedical Sciences, *Behav. Res. Methods*, 39, 175–191, 2007.

Faulconbridge, I., and Ryan, M., *Managing Complex Technical Projects: A Systems Engineering Approach*, Artech House, 2002.

Freedman, P. D., and Weinberg, M. G., *Handbook of Walkthroughs, Inspections, and Technical Reviews: Evaluating Programs, Projects, and Products*, Dorset House, 1990.

Gallagher, S., *Brainstorming: Views and Interviews on the Mind*, Academic, New York, 2008.

Garvey, R. P., *Analytical Methods for Risk Management: A Systems Engineering Perspective*, Chapman & Hall/CRC, Boca Raton, FL, 2008.

Gause, C. D., and Weinberg, M. G., *Exploring Requirements: Quality Before Design*, Dorset House, 1989.

Gilb, T., Optimizing Software Inspections, *Crosstalk*, 11(3), 16–18, March 1998.

Gilb, T., *Competitive Engineering: A Handbook for Systems Engineering, Requirements Engineering, and Software Engineering Using Planguage*, Butterworth-Heinemann, 2005.

Gilb, T., Engineer Your Review Process: Some Guidelines for Engineering Your Engineering Review Processes for Maximum Efficiency, available: http://www.gilb.com/tiki-download_file.php?fileId=143, 2008.

Gilb, T., and Graham, D., *Software Inspection*, Addison-Wesley Professional, Reading, MA, 1993.

Gnesi, S., Lami, G., Trentanni, G., Fabbrini, F., and Fusani, M., An Automatic Tool for the Analysis of Natural Language Requirements, *Int. J. Comput. Syst. Sci. Eng.* (IJCSSE), Special Issue, 20(1), January 2005.

Grochtmann, M., and Grimm, K., Classification-Trees for Partition Testing, *J. Software Test. Verif. Reliabil.*, 3(2), 63–82, 1993.

Grochtmann, M., and Wegener, J., Test Case Design Using Classification Trees and the Classification-Tree Editor CTE, in *Proceedings of Quality Week '95*, May 30–June 2, 1995, San Francisco, CA.

Haimes, Y. Y., *Risk Modeling, Assessment, and Management*, 3rd ed., Wiley Blackwell, 2009.

Hirokawa, Y. R., and Poole, S. M., (Eds.), *Communication and Group Decision Making*, 2nd ed., Sage Publications, 1996.

IEEE STD 610.12-1990, IEEE Standard Glossary of Software Engineering Terminology, 1990.

IEEE STD 830-1998, IEEE Recommended Practice for Software Requirements Specification, October 1998.

IEEE STD 1028-1997, IEEE Standard for Software Reviews, IEEE Computer Society, December 1997.

IEEE STD 1522, IEEE Standard for Testability and Diagnosability Characteristics and Metrics, IEEE (Trial-Use), 2005.

Janis, L. I., *Victims of Groupthink: A Psychological Study of Foreign-Policy Decisions and Fiascoes*, Houghton Mifflin, 1972.

Kaplan, S., Visnepolshi, S., Zlotin, B., and Zusman, A., *Tools for Failure & Risk Analysis: Anticipatory Failure Determination (AFD) & the Theory of Scenario Structuring*, Ideation International, 1999.

Kenett, R., and Zacks, S., *Modern Industrial Statistics: The Design and Control of Quality and Reliability*, Duxbury, 1998.

Kheir, N. (Ed.), *Systems Modeling and Computer Simulation (Electrical and Computer Engineering)*, 2nd ed., CRC, Boca Raton, FL, 1995.

Kim, G. T., *Theory of Modeling and Simulation*, 2nd ed., Academic, San Diego, CA, 2000.

Lehmann, E., and Wegener, J., Test Case Design by Means of the CTE XL, in *Proceedings of the 8th European International Conference on Software Testing, Analysis & Review (EuroSTAR 2000)*, Copenhagen, Denmark, December 2000.

Lu, J., Zhang, G., and Ruan, D., *Multi-Objective Group Decision Making: Methods, Software and Applications with Fuzzy Set Techniques*, Imperial College Press, 2007.

Martin, N. J., *Systems Engineering Guidebook: A Process for Developing Systems and Products*, CRC, Boca Raton, FL, 1997.

Martinez, R. D., Bond, A. R., and Vai, M. M., (Eds.), *High Performance Embedded Computing Handbook: A Systems Perspective*, CRC, Boca Raton, FL, 2008.

Matko, D., Zupancic, B., and Karba, R., *Simulation and Modeling of Continuous Systems: A Case-Study Approach*, Prentice-Hall, Englewood Cliffs, NJ, 1992.

Middleton, P., and Sutton, J., *Lean Software Strategies: Proven Techniques for Managers and Developers*, Productivity, 2005.

MIL-HDBK-2165, Testability Program for Systems and Equipments, in Department of Defense Handbook, July 1995.

MIL-STD-499B, Draft, Military Standard Systems Engineering, Joint OSD/Services/ Industry Working Group, September 1993.

MIL-STD-1521B, Military Standard, Technical Reviews and Audits for Systems, Equipments, and Computer Software, U.S. Department of Defense, 1995.

MIL-STD-1629A, Military Standard Procedures for Performing a Failure Mode, Effects and Criticality Analysis, U.S. Department of Defense, November 1980.

Mitra, M., and Chaudhuri, B. B., Information Retrieval from Documents: A Survey, *Inform. Retrieval J.*, 2(2/3), 141–163, May 2000.

Modarres, M., Kaminskiy, M., and Krivtsov, V., *Reliability Engineering and Risk Analysis: A Practical Guide*, CRC, Boca Raton, FL, 1999.

Monostori, K., Finkel, R., Zaslavsky, A., Hodasz, G., and Pataki, M., Comparison of Overlap Detection Techniques, paper presented at the 2002 International Conference on Computational Science, Amsterdam, The Netherlands, April 21–24, 2002; (I) pp. 51–60, 2002.

Montgomery, C. D., *Design and Analysis of Experiments*, 6th ed., Wiley, Hoboken, NJ, 2004.

Montgomery, C. D., *Design and Analysis of Experiments, Student Solutions Manual*, 7th ed., Wiley, Hoboken, NJ, 2008.

Murphy, R. K., Myors, B., and Wolach, A., Statistical Power *Analysis: A Simple and General Model for Traditional and Modern Hypothesis Tests*, 3rd ed., Psychology Press, 2008.

Myhrberg, V. E., and Crabtree, H. D., *A Practical Field Guide for AS9100*, ASQ Quality Press, 2006.

Obaidat, S. M., and Papadimitriou, I. G. (Eds.), *Applied System Simulation: Methodologies and Applications*, Springer, 2003.

Palshikar, G. K., An Introduction to Model Checking, *Embedded Syst. Design*, February 12, 2004.

Park, S., *Robust Design and Analysis for Quality Engineering*, Springer, 1996.

Pfleeger, L. S., and Atlee, M. J., *Software Engineering*, 4th ed., Prentice Hall, Upper Saddle River, NJ, 2009.

Quality System Inspections Reengineering Team, *Guide to Inspections of Quality Systems*, U.S. Food and Drug Administration, Offices of Regulatory Affairs and Center for Systems and Radiological Health, Washington, DC, August 1999.

Rad, F. P., and Anantatmula, S. V., *VVT Process Planning Techniques*, Management Concepts, 2005.

Radice, A. R., *High Quality Low Cost Software Inspections*, Paradoxicon Publishing, 2001.

Robertson, S., and Robertson, C. J., Mastering the Requirements Process, Addison-Wesley Professional, 2006.

SAE J1739SAE J1739, Potential Failure Mode and Effects Analysis in Design (Design FMEA) and Potential Failure Mode and Effects Analysis in Manufacturing and Assembly Processes (Process FMEA) and Effects Analysis for Machinery (Machinery FMEA), Society for Automotive Engineers, August 2002.

SEF DoD, Systems Engineering Fundamentals (SEF), Department of Defense, Supplementary Text Prepared by the Defense Acquisition University Press, Fort Belvoir, VA, 2001.

Severance, L. F., *System Modeling and Simulation: An Introduction*, Wiley, Hoboken, NJ, 2001.

Siegel, S., *Object-Oriented Software Testing: A Hierarchical Approach*, Wiley, New York, 1996.

Stamatis, H. D., *Failure Mode and Effect Analysis: FMEA from Theory to Execution*, 2nd rev. ed., Quality Press, 2003.

Taguchi, G., *Introduction to Quality Engineering: Designing Quality into Products and Processes*, Quality Resources, 1986.

Tian, J., *Software Quality Engineering: Testing, Quality Assurance and Quantifiable Improvement*, Wiley, Hoboken, NJ, 2005.

Torrence, R. S., *How to Run Scientific and Technical Meetings*, Van Nostrand Reinhold, 1991.

Utting, M., and Legeard, B., *Practical Model-Based Testing: A Tools Approach*, Morgan Kaufmann, 2006.

Visnepolschi, S., and Ramsey, J. D. (Editors), *How to Deal with Failure—Failure Prediction and Analysis Using Anticipatory Failure Determination*, Aptimise-edu, 2009.

Vroom, H. V., and Yetton, W. P., *Leadership and Decision-Making*, University of Pittsburgh Press, Pittsburgh, PA, 1976.

Wang, X. J., *Engineering Robust Designs with Six Sigma*, Prentice Hall, Upper Saddle River, NJ, 2005.

Wasson, S. C., *System Analysis, Design, and Development: Concepts, Principles, and Practices*, Wiley-Interscience, Hoboken, NJ, 2005.

Wilson, M. W., Rosenberg, H. L., and Hyatt, E. L., Automated Analysis of Requirement Specifications, in *Proceedings of the 19th International Conference on Software Engineering*, Boston, MA, 1997, pp. 161–171.

Woods, L. R., and Lawrence, L. K., *Modeling and Simulation of Dynamic Systems*, Prentice-Hall, Englewood Cliffs, NJ, 1997.

Yu, T. Y., Ng, P. S., and Chan, K. Y. E., Generating, Selecting and Prioritizing Test Cases from Specifications with Tool Support, paper presented at the Third International Conference on Quality Software, 2003.

Zienkiewicz, C. O., and Morgan, K., *Finite Elements and Approximation*, Dover Publications, 2006.

Chapter 5

Systems VVT Methods: Testing

5.1 INTRODUCTION

As discussed in Chapter 1, VVT engineers often use the term "testing" colloquially to mean VVT. But, in a narrower sense, following the VVT definition, "testing" is a subset of verification and validation, dealing with actively operating the system and verifying or validating it. Accordingly, this chapter describes system VVT testing methods in the narrow sense. After the introduction, this chapter is divided into two main parts: white-box system testing and black-box system testing. The second part is further divided into (1) basic testing, (2) high-volume testing, (3) special testing, (4) environment testing and (5) phase testing. Each section describes relevant VVT methods.

The fundamental system testing process is depicted in Figure 5.1. System specifications, which include a list of system requirements and other important elements, are the very basis for the design and building of the target system. These are the "musts" and "shoulds" that dictate what the system must be and must do and for which the customer is willing to pay. These same system specifications are therefore the measure by which the system must be judged. Thus, system specifications are instrumental in generating the test cases needed to verify and validate the system. A test engineer or a group of test engineers then perform the specification-directed testing process and thus determine whether or not the system succeeds in meeting all of its specifications.

Verification, Validation, and Testing of Engineered Systems, Avner Engel

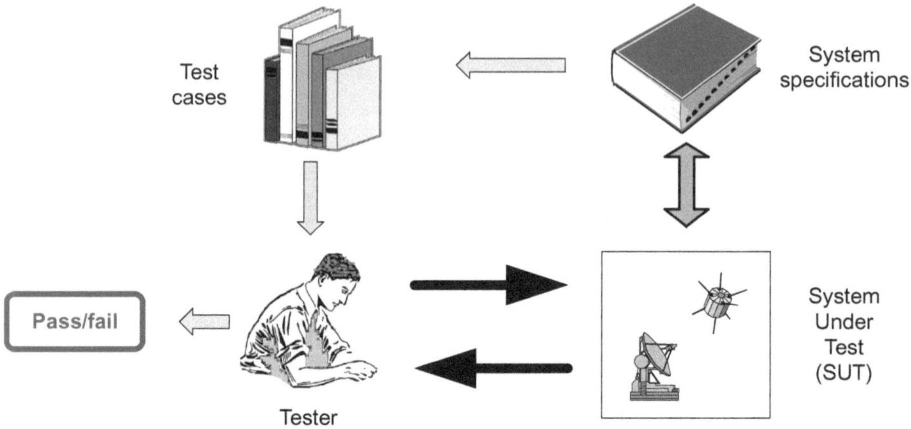

Figure 5.1 Fundamental system testing process.

During any system testing, it must be confirmed that (1) the system is doing what it should be doing (conform to requirements) and (2) the system does *not* do what it should *not* be doing. One could say that this issue is the concern of the writers of the requirements documents. As it turns out, however, one finds few requirements directed toward the avoidance of undesired system behavior. One reason for this is that system engineers and engineers in general tend to concentrate on "what must be done." Less often do they focus on "what should not be done." The more problematic aspect here is that the behavior space of what the system should not do is much greater than the performance space of what the system should do. This can be illustrated in a mortgage approval system shown in Figure 5.2. The requirements for this system are that the principle is permitted to vary between \$100,000 and \$600,000, the fixed interest rate must be in the range of 5–10%, while the inflation rate is expected to fluctuate in the range of 2–6%. In this example, the above variable may take significantly larger values.

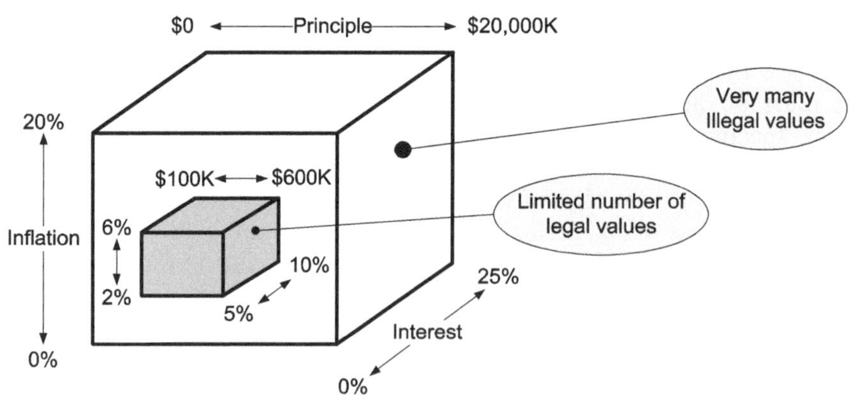

Figure 5.2 A system's legal and illegal behavior space.

For this trivialized example, we assume that the input ranges of the principle, interest and inflation could be \$0–20,000,000, 0–25% and 0–20%, respectively.

In this case, the portion of legal testing space versus illegal testing space is

$$\varphi = 100 \times \frac{(x_2 - x_1)(y_2 - y_1)(z_2 - z_1)}{(X_2 - X_1)(Y_2 - Y_1)(Z_2 - Z_1)} = 100 \times \frac{(600-100)(6-2)(10-5)}{(20{,}000-0)(20-0)(25-0)}$$
$$= 0.10\%$$

The net result of this phenomenon is the following set of empirical testing principles: First, the VVT engineer must select a testing strategy, that is, a compromise between the impossible and the inadequate. On the one hand, an impossible strategy is by definition not achievable, due to limitations in funding, time or other resources. On the other hand, inadequate testing is a fact of life. But, the crucial issue, as discussed in Chapter 7, is to identify a strategy for optimal testing, that is, one that has high potential of uncovering system faults and that costs as little as possible. Second, the VVT engineer should pay close attention and verify that the system requirements contain sufficient references to requirements delineating what the system should not do, especially with regards to safety, security and other important concerns.

This chapter is generally divided into white-box and black-box testing. These terms describe the point of view a test engineer takes when designing the test process. White-box testing is undertaken with an internal or structural view, whereas black-box testing is mainly concerned with a functional or external view of the item being tested. This top-level delineation is important as each type of testing can find different kinds of system faults. More specifically, white-box tests are usually conducted at the unit or component level and tend to discover structural problems, whereas, black-box tests are usually conducted at the subsystem and system levels and typically detect functional defects (see Figure 5.3).

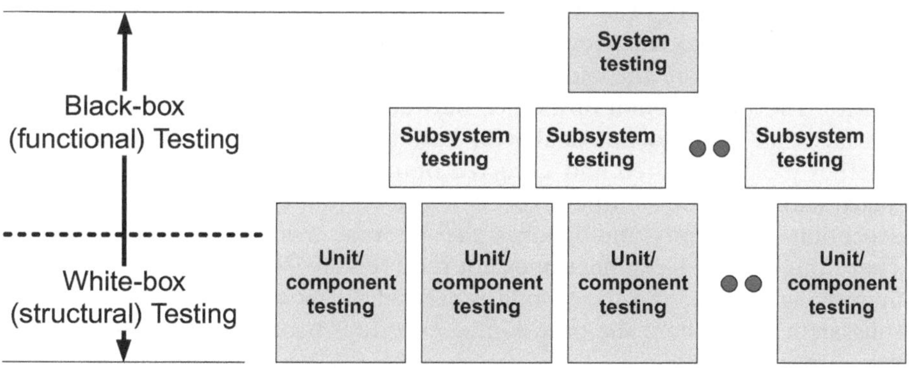

Figure 5.3 Hierarchical testing: white or black-box testing.

1. *White-Box Testing.* White-box testing is sometimes referred to as structural testing. Conducting white-box testing requires an implicit knowledge of the system's inner workings, and testing is generally done by using special features of the development environment. The testing is carried out on individual subsystems or modules which are partitioned on the basis of the system's internal structure.

White-box testing invariably demands that the test engineer select test case inputs that will exercise all paths and determine the appropriate outputs. Therefore, the testing strategy deals with internal logic and structure of the unit under test and seeks to incorporate coverage of each element of the unit under test. In a software unit, tests will incorporate coverage of software code, branches, paths, internal logic of code and so on.

The advantages of white-box testing are derived from the intimate knowledge the VVT engineer has relative to the internal structure of the System Under Test (SUT). In such a case, it is easy to generate input data for testing the application effectively, that is, attacking potential week design points. White-box testing has the added benefit that such testing encourages the test engineers to reason carefully about implementation of the testing process. We should also add that, in case of white-box testing of software, there are many tools available to identify software test coverage as well as measure the complexity of the code.

The disadvantages of white-box testing stem from the fact that the VVT engineer must have skills in the subject matter domains (e.g., hardware, software), as well as having intimate and specific knowledge about the internal structure of the system under test. Another drawback of white-box testing is the limitations to performing exhaustive tests. Modern hardware makes it impossible to reach large portions of the electronic circuitry, and even short pieces of code are so intractable that fully covering all aspects of their structure is difficult. In addition white-box testing will often not detect missing or incorrect functionalities in the system under test.

2. *Black-Box Testing.* Black-box testing is referred to as functional or behavioral testing. The intent here is to validate whether or not a given system conforms to its specifications. The tests present a series of inputs to a system and compare the outputs to a predefined test specification (i.e., test oracle). The fundamental difference between black- and white-box testing is the fact that tests do not deal with how a given output is produced, only whether it is the desired and expected output. The VVT engineer, therefore, focuses solely on the outputs generated in response to selected inputs and execution conditions and ignores the internal mechanism of the system. Therefore, the VVT engineer does not required any specific knowledge of the underlying system, and the testing is carried out at the system or individual subsystem level where the partitioning criteria is based on the system's functional specifications.

Another advantage of black-box testing is that it is appropriate at all levels of development (i.e., component, subsystems and system) and throughout the system's lifecycle (i.e., development, production maintenance, etc.). In fact black-box testing gradually becomes more suitable at higher levels of integration. Finally, black-box testing is perfectly suited, indeed it is designed, to uncover system functionality faults.

VVT engineers must have deep understanding of system specifications as well as stakeholders' expectations. They must be capable of judiciously hypothesizing undesired system responses that have not been specified, even those that have not shown up in previously engineered systems. The very nature of black-box testing (i.e., not having to know the internal structure of the system) generally precludes test engineers from applying extra test efforts in verifying fragile elements of the system design. In fact, in black-box testing, test engineers are naturally oblivious to the internal workings of the unit being tested. The structure of this chapter and a proposed system testing taxonomy is depicted in Figure 5.4.

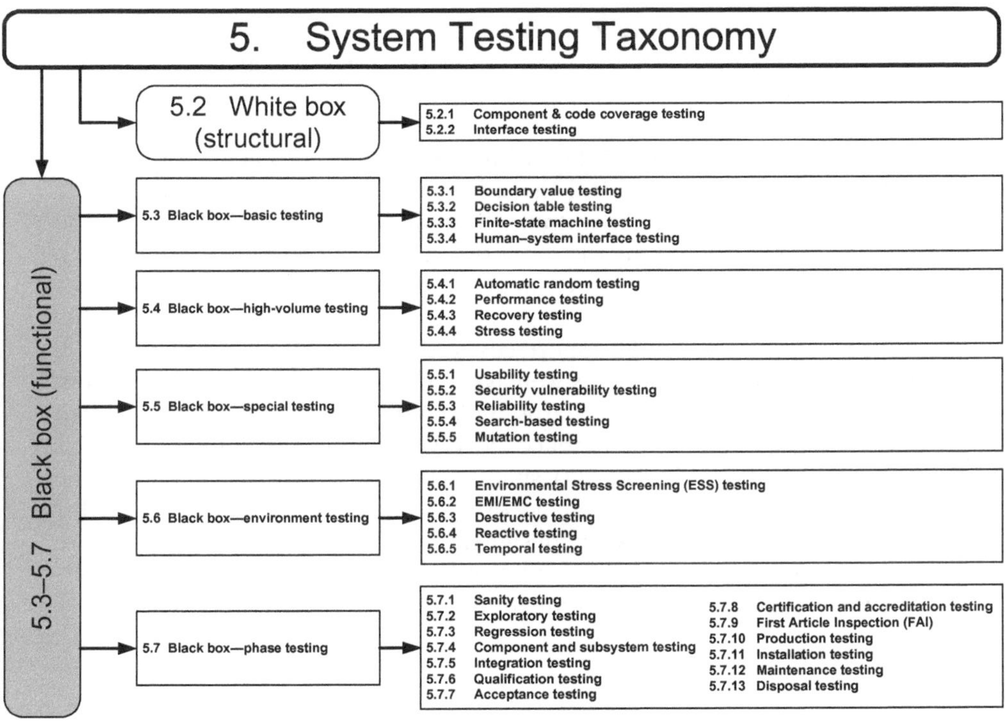

Figure 5.4 Chapter structure and system testing taxonomy.

5.2 WHITE BOX TESTING

5.2.1 Component and Code Coverage Testing

Coverage Testing of Hardware Components or Software Code The emphasis in hardware component or software code testing is on verifying that as large a portion as possible of the Unit Under Test (UUT) has been *covered* by a given set of individual tests. The goal here is to determine input test patterns that will expose existing faults in a UUT by triggering the fault and making its impact visible at the output of the unit. Additional testing goals are high detection rate of real defects in short testing time and low testing cost per UUT with high fault diagnosis (i.e., finding what failed).

In hardware, "component coverage testing"[49] refers to the process of verifying that a certain test sequence has covered (i.e., tested) all the components in a circuit or a system. In software, "code coverage testing" refers to the process of verifying that a certain set of input patterns has traversed (i.e., covered) the entire unit.

Rationale We first ask: Why test at the unit level? (e.g., an integrated circuit, an electronic board or a software unit). The answer is that we seek to detect failure at the lowest package level since, as a rough rule, when a test fails to detect an error at a given level of packaging, it will cost an order-of-magnitude more to detect the error at the next higher level of packaging. The reasons for this cost rule are numerous, but the key difficulty relates to the issues of *controllability* and *observability*. Controllability is the ability to control individual inputs to individual subunits within the system. The larger the system, the more difficult it is to control these inputs. Similarly, observability is the ability to observe individual outputs from individual subunits within the system. The larger the system, the more difficult it is to observe these outputs.

Often, "unit test coverage" measures the percentage of the unit's devices or lines of code which a particular test suite covers. This measure is highly depended on what is termed "short coverage," that is, the percentage of a board- or chip-accessible node, as well as the number of software unit outputs. Nowadays, short coverage of boards and chips is extremely small, due to increased density and minute space between conducting lines as well as complex Three-Dimensional (3D) space geometry layouts. In addition the high-frequency signals often demand precise layouts and offer no room for probe targets. Similarly, software designers tend to avoid inserting software probes into already intricate software in order to avoid the *probe effect*, affecting the behavior of a system by embedding extraneous elements into it.

[49] While "software code coverage testing" is commonly found in the literature, "hardware component coverage testing" is not as well known. Nevertheless, the analogy is strong, so that we feel justified in using the analogy from now on.

Method In white-box testing, we discuss separately test methods for hardware systems and for software systems.

1. *Component Coverage Testing in Hardware.* The universe of potential hardware defects is very large. In fact, defects are too numerous and diverse for simple enumeration. The approach commonly taken is based on creating fault models that identify a well-defined, manageable failure space as targets for the generation of test patterns, analysis and validated by means of testing.

Popular fault models called "stuck-at" models (i.e., stuck-at zero, stuck-at-one) typically affect digital components such as electronic gates (And, Or, Not, etc.) as well as higher level components such as shift registers, latches and memories. More sophisticated fault models identify fault characteristics, such as:

- *Variability.* Nonpermanent hardware faults may appear on an intermittent basis or in relation to transient events within the circuit.
- *Multiplicity.* Sometimes, multiple hardware faults affect the behavior of the unit under test in unexpected ways.
- *Effect on Function and Operating Speed.* Faults may affect the overall functional behavior of hardware. Such faults often manifest themselves only after a specific sequence of inputs.

Current research suggests clever ways of generating and validating test patterns (sometimes called test vectors) either manually or automatically. Test pattern generators based on these new techniques determine test vectors for a given fault model that will propagate error all the way to an observable output. Fault simulations are used to determine the degree of test coverage. Such simulations contain a definition of the hardware circuit under test (i.e., analog and digital components and gates), and they simulate the behavior of the system under both correct conditions (good machine) and faulty machines (bad machine), when test vectors are injected into the system. Bad machine simulations must be repeated many times where, usually, each simulation runs under a single fault assumption. As a result, such simulations require considerable amount of execution time and therefore are often restricted to relatively limited size circuits or portions of larger circuits. Hardware test pattern generation techniques include the following:

- *Manual Generation.* Test patterns may be generated manually by test engineers for functional verification of a UUT. A model of the system should be simulated in order to verify the level of fault detection as well as to identify components whose failure has not been detected.
- *Pseudorandom Generation.* Test patterns are generated using a random number generator and then simulated (most commonly within a stuck-at model) at the circuit level. This technique is often used early in the testing process in order to identify easy-to-detect faults from a fault list.

- *Algorithmic Generation Using D-Algorithm.* The D-algorithm uses a single stuck-at fault model and defines the notions of Primitive D Cubes of Failure (PDCFs) and Propagation D Cubes (PDCs). The D-algorithm is essentially a "branch-and-bound" optimization approach where optimal solutions are made in a sequential manner within the algorithm. The main weakness of the D-algorithm is the fact that its complexity grows exponentially with the number of circuit nodes.
- *Algorithmic Generation Using Path-Oriented DEcision Making (PODEM) Algorithm.* This is an improved D-algorithm in the sense that its complexity grows exponentially with the number of UUT inputs and not with the (much larger) number of circuit nodes. In addition, this algorithm is more efficient in the way it searches the failure space. Several commercial tools are available to support various types of hardware verifications. These model-based tools deal with both digital and analog circuits and perform various functions related to design as well as behavioral modeling, formal verification and physical verification and circuit simulation.

2. *Code Coverage Testing in Software.* In software, code coverage testing results can help improve test cases that will increase code coverage over vital functions. Of the many types of software code coverage, three popular ones (i.e., statement coverage, branch coverage and condition coverage) will be explained by means of a simple software example with three inputs (X, Y, Z) as depicted in Figure 5.5.

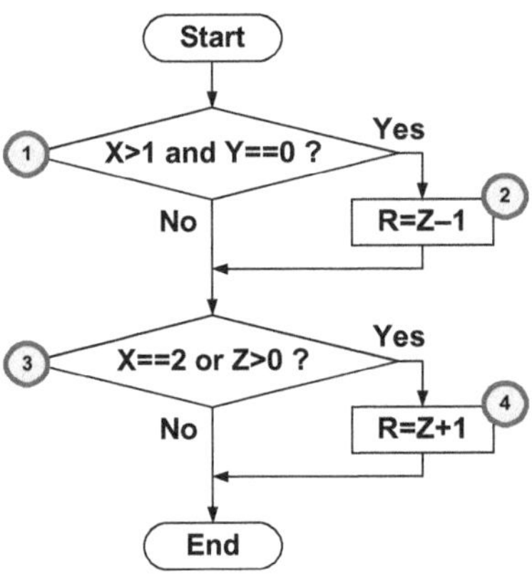

Statement number	Code statements
1	Is X>1 and Y=0 ?
2	R=Z–1
3	Is X=2 or Z>0 ?
4	R=Z+1

Figure 5.5 Software code coverage testing example.

- *Statement Coverage.* In statement coverage testing we verify that one or more test patterns causes the execution of each and every software code statement at least once. In the example depicted in Figure 5.5, a single test pattern where $\{X, Y, Z\} = \{2, 0, 0\}$ will cause the execution of code statement numbers 1, 2, 3 and 4. Therefore, under these conditions the statement coverage is fulfilled.

- *Branch Coverage.* In branch coverage testing we verify that one or more test patterns cause the execution of each and every branch of the control flow at least once. In the example depicted in Figure 5.5, one test pattern where $\{X, Y, Z\} = \{2, 0, 0\}$ will cause the execution of the two YES branches of code statement numbers 1 and 3. Similarly, a second test pattern where $\{X, Y, Z\} = \{0, 0, 0\}$ will cause the execution of the two NO branches of code statement numbers 1 and 3. Therefore, under these conditions the branch coverage is fulfilled.
- *Condition Coverage.* In condition coverage testing we verify that one or more test patterns causes the execution of each and every branch of the control flow and all values of constituents of compound conditions are exercised at least once. So in the example depicted in Figure 5.5, in addition to the test patterns identified in the branch coverage example, we need to create a test pattern where $X > 1$ and $Y \neq 0$, so the NO branch will be selected in code statement number 1. In addition, we need to create a test pattern where $X \neq 2$ and Z is not greater than 0 so the NO branch will be selected in code statement number 3. For example, we can select a test pattern $\{X, Y, Z\} = \{3, 1, 0\}$ which meets the above requirements. Therefore, under these conditions the condition coverage is fulfilled.

Several commercial tools are available to support various types of software verifications. These model-based tools deal with a multitude of software languages and computer types by generating instrumentation at both the source code level as well as the runtime code. In particular, model-based tools support unit testing by enhancing the functionality of unit test case generation, static analysis and regression testing as well as provision for coverage metrics of test cases that execute at various levels, including function, module, class, component and system levels.

Current scientific research seeks to find ways for automatic generation of test vectors that will provide maximum code coverage. Search methods using evolutionary genetic algorithms and similar optimization techniques seem to be a promising research direction. Such an approach yielded high coverage degrees in laboratory experiment and, to a degree, in some advanced industries. Nevertheless, evolutionary testing is not equally well applicable to different items being tested. For example, evolutionary testing of an item being tested with complex predicates might fail. Currently, researchers evaluate the suitability of structure-based complexity measures for the assessment of whether or not evolutionary testing can be performed successfully for a given item being tested (see, e.g., Lammermann et al., 2008).

Further Literature

- Beizer (1990)
- David (1998)
- Kabisatpathy et al. (2005)
- Lammermann et al. (2008)
- Lavagno et al. (2006)

5.2.2 Interface Testing

Purpose Interfaces are agreed-upon mechanisms for interactions and communication between different parts of a system and between different systems. The purpose of interface testing is to evaluate whether systems or components interact properly between them or pass data or control correctly to one another. Usually system testing takes place when modules or subsystems are integrated to create larger systems and interface faults may be detected due to invalid assumptions about the interface requirements.

Rationale Viewing interfaces in a broad manner, we can distinguish among the following categories of interactions:

- *Material.* Material interaction identifies the needs for materials exchange between two elements or systems. For example, a material interface between a pump and a carburetor in a car is the gasoline flowing in a pipe connected between the two system elements.
- *Spatial.* Spatial interaction identifies a need for adjacency, force transfer or orientation between two elements. For example, a dish antenna mounted on a house must have mechanical and spatial interface with the house structure in a prescribed orientation, transferring forces from one system to the other.
- *Energy.* Energy interaction identifies requirements for energy transfer between two elements. For example, a kettle is plugged into a socket mounted on the wall and connected to the electricity grid. The kettle has energy interface with the socket by means of electricity transfer from one system to the other. Similarly, the water in the kettle has energy interface with the kettle heating element by means of heat transfer from one system to the other.
- *Information.* Information interaction identifies requirements for information or signal exchange between two elements. For example, earphones are plugged to a radio transistor via a cable. The earphones have information interface with the radio set by means of electrical signal transfer from one system to the other. The subsequent subsections will concentrate on this type of interface.

Many test engineers will agree that information interface testing is one of the most important types of testing carried out during VVT of complex systems. The following discussion centers primarily on testing of information interfaces. One should keep in mind, however, that proper care and attention should be given to other interface types. Customarily, information interfaces are grouped into the following classes:

- *Hardware/Hardware Interfaces.* This type of interface supports communication between hardware units. For example, a controller in one unit is connected to a relay in another unit. The electrical wires between the two units typify such an interface.
- *Hardware/Software Interfaces.* This type of interface supports interaction between hardware and software. For example, a toggle switch that is monitored by the software and its position affects the behavior of the runtime software typifies such an interface.
- *Software/Software Interfaces.* This type of interface supports communication between software components or subsystems. For example, database software transferring data to display-handling software typifies such an interface.
- *Human/System Interfaces.* This type of interface supports interactions between users and a system. For example, a Graphical User Interface (GUI) used by a programmer developing software code on a console typifies such interface.

Method Normally, interface testing is performed in two phases: During the first phase, each side of an interface is tested using a trusted stub or a "dummy" element representing the other side. This is done in order to mimic the other systems and create a simplified and controlled closed-loop test environment. During the second phase, the two systems are integrated and tested together to verify the proper interaction and communication of the expanded system. In general, the test engineer should be cognizant of the following classes of interface errors:

- *Interface Misuse.* This interface error is generated when one component or system does not follow the prescribed interface rules. For example, one component calls another component and sends more (or fewer) parameters than are required or places the parameters in the wrong order.
- *Interface Erroneous Assumptions.* This interface error is generated when one component or system makes erroneous assumptions about the dynamic behavior of the other system. For example, a calling component assumes at a given time that the called component has sufficient room on the stack, whereas, in fact, the stack is full.
- *Interface Timing Errors.* This interface error is generated when the calling and called component operate at different speeds and obsolete information is used. Another timing problem that may transpire between

two nonsynchronized systems may emanate from the inability of a receiving system to handle incoming information leading to the intermittent loss of data between the two systems.

Hardware Interface Testing Testing hardware related to information interfaces should be conducted at several communication interface layers. Testing of some of the most common ones is described below:

- *Physical Level Interfaces.* Testing the physical connection between different parts of the system, for example, physical layout of electrical harnesses, wiring integrity, correctness, and separation between each conductor as well as isolation from the ground, plugs and sockets compatibility.
- *Electrical Level Interfaces.* Testing the electrical and electronic compatibility of hardware units, that is compatibility of the two systems in terms of signal voltage, current, duration and shape. In other words, test whether an electrical signal created by one system can be accepted by another system.
- *Protocol Level Interfaces.* Testing the internal structure and format of signals between two or more hardware systems. For example, the military standard MIL-STD-1553B (1987) specifies a Mux-Bus communication system that may connect several systems or subsystems. It specifies the physical level and electrical level interfaces as well as a specific protocol level interface; that is, the nature, structure and order of data flow through the interface.

Software Interface Testing Testing of software interfaces should verify the proper interprocesses transfer of control and data among different software components. Testing of some of the more common software interfaces is discussed below:

- *Parameter Interface.* Software parameter interface is based on a protocol whereby a calling procedure or routine transfers control to another procedure together with a predefined set of parameters. Testing a parameter interface entails verifying that both the calling and the called elements agree on the parameters protocol, namely the number and order of the parameters and their exact format and meanings.
- *Message-Passing Interface.* Software message-passing interface is based on a protocol whereby one procedure or routine may pass messages to another procedure. The sender may lock-up, waiting for an acknowledgment or continue execution. All of these operations are usually accomplished by using appropriate operating system services. Testing a message-passing interface entails verifying that both the calling and the called software elements agree on the nature of the message (i.e., number

and order of the parameters as well as their exact format and meanings). In addition, testing must verify that the control hand-shaking dynamics between the two procedures is properly structured so that the receiving procedure is, in fact, able to actually obtain the message and no mutual locking condition can occur under any circumstances.

- *Shared Memory Interfaces.* Software memory interface is based on an agreement between one software element and one or more other software elements whereby one procedure or routine may write predefined information into an agreed memory space and other procedures may read it when they are executed. The advantage here is that usually the operating system is completely oblivious to these transactions. Testing a shared memory interface entails verifying that both the calling and the called elements agree on the number and order of the parameters as well as their exact format and meanings. In addition, testing must verify the appropriate synchronization between the creator of the data and the users of the data. This entailed ensuring that the receiving procedure does not attempt to read data before it has been actually written into memory as well as ensuring that data has not been trampled and updated before the receiving procedure had a chance to acquire it.

Human–System Interface Testing Testing of human interfaces should verify the proper Human–System Interaction (HSI) in terms of controlling the system and receiving appropriate and timely information from it (see Figure 5.6). Testing of some of the most common user interfaces is described below:

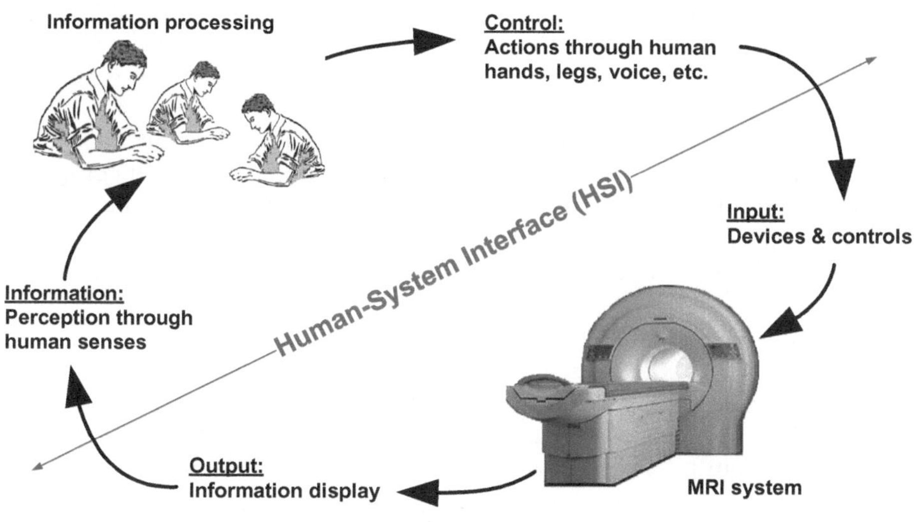

Figure 5.6 Human–system interaction cycle—example.

Human factors engineering is a discipline that applies ergonomic principles to the design and testing of human interactions with a system. Testing of HSIs is critical because good design and implementation of such interfaces can make systems easy to use, that is, better adapted to the person using them and reduce human errors due to misinterpreted information. Testing human–system interfaces is difficult since systems are complex and constantly changing, and information about system operations may also be multifaceted and sometimes inconsistent. Therefore such testing must take into consideration the following:

- *Unpredictability of Users.* Testing must cover the variability among individuals. Often such differences in human behavior are difficult to model. For example, a person's ability to work varies throughout the day, his or her learning abilities and experiences vary and, of course, different individuals hold diverse beliefs systems and cultures. Therefore, test engineers should try to mimic this rich behavioral repertoire during their test processes.
- *System Missions.* Engineered systems are expected to perform large varieties of tasks necessitating enormous range of interactions carried out through HSIs. As a result, testing of users' tasks is influenced by the requirements for interface support as well as the type of information that needs to be available and how it needs to be entered. Testing must take into account what it is that the system end users will be doing and why they will be doing it and design the testing process accordingly.
- *System Technology.* Modern systems tend to evolve fairly rapidly. For example, different generations of passenger cars provide new features, especially in the embedded system area, which changes the total driving experience. Often, the driver's understanding of the interface technology lags behind the technological advances. Therefore, testing of the human–system interface should consider this and attempt to assure a smooth operation of the system at hand.
- *Operational Environment.* Human–system interface testing must also consider the physical layout of the system at hand. An aircraft cockpit is different, of course, from a workstation in an office. Therefore, testing must match factors such as vibration, speed, ambient temperature, noise level, lighting level and ergonomics of the specific system.

The following is a set of HSI testing heuristics:

- *Simple and Natural.* The interface should be tested for a simple and natural dialogue, manifested in aesthetic and minimalist interactions and, to the extent possible, utilizing language familiar to the user.
- *Minimal User Memory Load.* The interface should be tested for minimal user memory load. This may be achieved by verifying that the interface

was designed in a consistent manner, providing adequate user control, flexibility and freedom of actions within appropriate bounds. The interface should also be tested for providing sufficient user feedback and visibility of system status.

- *Handling Users Errors.* The interface should be tested for providing good error messages as well as immediate mechanism to help users recognize, diagnose and recover from errors.

Further Literature

- Reorda et al. (2005)
- Shneiderman et al. (2009)

5.3 BLACK BOX—BASIC TESTING

5.3.1 Boundary Value Testing

Purpose Boundary value testing is a method to verify the behavior of systems at operating boundary areas by selecting test data values that lie at operating extremes. Boundary test values may include maximum or minimum values within the normal operating domain, values just inside and just outside operating domain boundaries, typically encountered operating values or specific error condition operating values.

Rationale The objective of this method is to test systems at boundaries of the operating domain where a substantial number of errors tend to concentrate. Generally, this method is applicable to software, embedded systems and systems that contain some software components. The weakness of boundary value testing is that the testing process is not exhaustive and the method is not appropriate for complete validation of a system.

Method The boundary value testing method is based on selecting test cases within sets of equivalence classes at the "edge" of the class rather than selecting any element at random. As a result, this method facilitates a possible reduction in the number of test cases relative to the number of detected errors. In summary, the system is not fully validated but a high proportion of errors can be found. The method entails two-step operation: (1) defining equivalence partitioning and (2) generating and executing test cases at extreme ends of equivalence classes.

- *Step I: Identifying Equivalent Classes.* This step entails dividing the input domain into "equivalent" classes of data. Under equivalence partitioning we define a test case that uncovers classes of errors, thereby reducing the number of test cases required. In other words, an equivalence class represents a set of valid or invalid states for input conditions. Customarily we can identify either two or three types of equivalent classes:

a. If an input condition specifies a range of values, then one valid and two invalid equivalence classes will be defined, for example, a month in a year:

Valid range:	$1 \leq$ month ≤ 12
Invalid range I:	Month ≤ 0
Invalid range II:	Month ≥ 13

b. If an input condition specifies a specific value, then one valid and one invalid equivalence class will be defined, for example, the height of an aircraft above ground in meters:

Valid range:	$0 \leq$ object height
Invalid range:	Object height < 0

c. If an input condition specifies a set of values, then one valid and one invalid equivalence classes shall be defined, for example, names of family members:

Valid range:	{Tom, Norma, Peter, Amenda}
Invalid range:	{X, 77, Sophia, ...}

- *Step II: Boundary Value Testing.* Applying boundary value testing requires a selection of test cases at each side of the boundary between equivalent classes. That is, for a valid range of values bounded by a minimum (a) and a maximum (b), the test case values should be $\{a - 1, a\}$ and $\{b, b + 1\}$. Therefore, in the above first example, a month specification within a date input stream will entail selecting test data of {0, 1} for the lower boundary as well as a second test data of {12, 13} for the upper boundary. All told, testing will be done by means of four test cases where each of these pairs consists of a "clean" and a "dirty" pair. Clean test cases should result in valid operation, whereas dirty test cases should result in error treatments. More specifically, in case of HSI, the system should issue a warning message and a request to enter the correct data.

  Along the same line, the above second example, a height specification, will entail selecting test data of {–1, 0} for the single boundary. That is, testing will be done by means of two test cases. Similarly, in the above third example, names of family members will entail selecting test data of the entire valid sets. Obviously the invalid range in this case is infinitely large and, therefore, reasonable judgment must prevail as to the appropriate number of required invalid test cases.

Further Literature

- Beizer (1990)

5.3.2 Decision Table Testing

Purpose Decision table testing method focuses on validating responses of a system under specified conditions and constraints.

Rationale System testing is accomplished by means of a decision table, which is a precise and compact way to model complicated logical behavior.

Method Construction of a decision table is accomplished using the following steps:

- *Step 1.* Identify all the possible conditions and their combinations that could affect the behavior of the system.
- *Step 2.* For each and every condition identified in the first step, define all the possible system actions in response to these conditions and their combinations.

Decision tables are typically divided into four quadrants, as depicted in Figure 5.7.

Conditions	Condition alternatives
Actions	Action entries

Figure 5.7 Typical decision table structure.

Each condition corresponds to variables, whose values are listed in the condition alternatives. Each action is an operation preformed by the system under the stated conditions. Typical decision table nomenclature appears below:

C_i denotes ith condition	X identifies action to be taken.
T denotes true	Blank in condition denotes "don't care"
F denotes false	Blank in action denotes "do not take the action"

For example, suppose our system must distinguish among five types of triangles, based on the lengths of the triangle's three sides. Assuming $a \geq b \geq c \geq 0$, the decision table for testing this system may be depicted as shown in Table 5.1.

TABLE 5.1 Decision Table for Triangular Categorization System

Conditions		Condition Alternatives					
Conditions	C_1: $a < b + c$	F	T	T	T	T	T
	C_2: $a = b$		F	T	F	T	
	C_3: $b = c$		F	F	T	T	
	C_4: $a^2 = b^2 + c^2$						T
Actions	Not a triangle	*X*					
	Scalene		*X*				
	Isosceles			*X*	*X*		
	Equilateral					*X*	
	Right triangle						*X*

Finally, for each pair of condition and system action, we must define a test case. In this process, we must ensure that all possible combinations of conditions are covered.

Further Literature

- Beizer (1990)

5.3.3 Finite State Machine Testing

Purpose The purpose of Finite-State Machine (FSM) testing method is mostly to evaluate systems for proper execution of control functions. FSM modeling is based on automata theory, which involves the concepts of system states, events, transitions and activities. Engineered systems that embody FSM philosophy are characterized by a behavior pattern where, under each state or mode, the system behaves (e.g., performs activities and generates outputs) in a specified and unique manner. The system remains in that state until a specific external input or internal event occurs. When that occurs, and certain conditions are fulfilled, the system transitions into another state, under which it may perform an entirely different and unique set of tasks.

Rationale An FSM is a way of thinking about engineered systems and is used to model the dynamic behavior of complex systems. An FSM model has a finite number of states and transitions between those states, which occur in response to specific events within the system or inputs to the system. The state of the system represents a situation during the system life when it performs some activities or waits for some event. More specifically, when the system is in a given state it will perform certain specified activities associated with this state and usually produce specified outputs. A transition is a relationship between two states, indicating that an entity in the first state will perform certain actions and enter the second state when a specified event occurs and specified conditions are satisfied. This is usually shown by a state machine diagram, which shows the behavior of the system in response to external stimuli or internal events and in activity diagrams, which show the behavior of the system in terms of internal processing.

Fundamentally, all engineered systems transition through *superstates*: (1) initial state, where power-up and initialization takes place, (2) operation state, where the system performs its assigned activities and (3) final state, where the system performs closure operations and shuts down.

State machine diagrams describe the states an entity (in this case, engineered system) can have during its lifetime, the behavior in those states and the events that can cause the state to change. States represent the distinct behaviors of a class or system and transitions represent the processes by which the class or subsystem changes behavior. More specifically, transitions must specify the circumstances under which the behavior may change, the paths relating two states, logical conditions necessary to actually perform the transition and any guard conditions which may prevent the transition.

Events are defined as a class, triggering state changes or other system operations. They may occur in response to external events or as a part of a system's operation or may be periodic or be associated with a timer. Furthermore, events may be triggered on entry into a state or exit from a state (i.e., entry events, exit events). They may activate other state machines (i.e., make events happen), generate other events (i.e., call events) or may invoke other system operations (i.e., actions). Also, events may reflect condition changes (i.e., condition events) or times (i.e., time events). Activity diagrams complement the state machine diagrams. They describe the system structure in terms of its subsystems and its work flow, as well as the environment outside the system.

In addition to state charts, an FSM model may be described mathematically using a formal definition. An FSM is described by a 6-tuple (I, S, s_0,, O, SF, OF) where:

- I is a set of inputs $\{i_0, i_1, \ldots, i_m\}$
- S is a set of all states $\{s_0, s_1, \ldots, s_n\}$
- s_0 is the initial state
- O is a set of outputs $\{o_0, o_1, \ldots, o_m\}$
- SF is a next-state function ($S \times I \rightarrow S$)
- OF is an output function ($S \rightarrow O$)

State charts are commonly used to model the behavior of complex, real-time embedded systems and other applications. Several commercial vendors provide tools to support graphical modeling, simulation, dynamic testing and code generation for a rapid development of such systems (e.g., IBM-Telelogic's Statemate tool).

Method From a testing point of view, a system may fail a test if it is exposed to an internal or external event, the guard conditions are appropriate and the system either does not transition to another state or transition to a wrong state. A system may also fail if it does not produce an expected output while in a given state. The following paragraphs discuss the details:

- *State Machine Coverage.* With an FSM model, test coverage criteria can be based on the structure of the state–machine model. This includes testing based on (1) state–event combinations, (2) transition structure and (3) paths specified by the state–machine.
- *Testing Strategies.* There are several coverage criteria for testing an FSM. Transitioning through all the states of an FSM-based system is considered to be the minimum acceptable coverage. Transitioning through all state–event combinations can detect problems when an FSM is not completely specified or there are either missing or extra transitions. Next, transition through all possible one-time transition paths starting from any state can uncover errors stemming from undefined FSM model components or variables.
- *Typical FSM Testing.* Testing for errors in systems based on FSM should include the following:
  a. Test for action fault—the actions on a transition are incorrect, or missing.
  b. Test for guard condition fault—the guard condition on a transition may be incorrect.
  c. Test for an unspecified event or missing transition—there might be no transition specified for a legal event at a particular state.
  d. Test for illegal event failure—an unexpected event may cause a failure.
  e. Test for unintended event failure—the system may accept an event which should not be accepted at any time.
  f. Test for state fault—there might be either extra or missing states.
  g. Test for a next state fault—the system may transfer to illegal or incorrect state.
  h. Test for extra transition—a generally legal event may appear in a particular state, when it was not expected to occur in that state.

Finite-State Machine Example The following depicts a Vehicle Autonomous Driver (VAD) assistant system, described by an activity chart, coupled with a state chart. The purpose of the system is to assist the driver by issuing advice and by controlling the vehicle in an emergency. This system is capable of driving the vehicle autonomously, using various sensors, a computer system and actuators to control the vehicle. In this example, we are interested only in the performance of the VAD controller and assume that the sensors and actuators have already been integrated into the vehicle. From our perspective, the VAD controller is composed of five subsystems and the flow of data and control as well as the operating environment is as depicted in Figure 5.8.

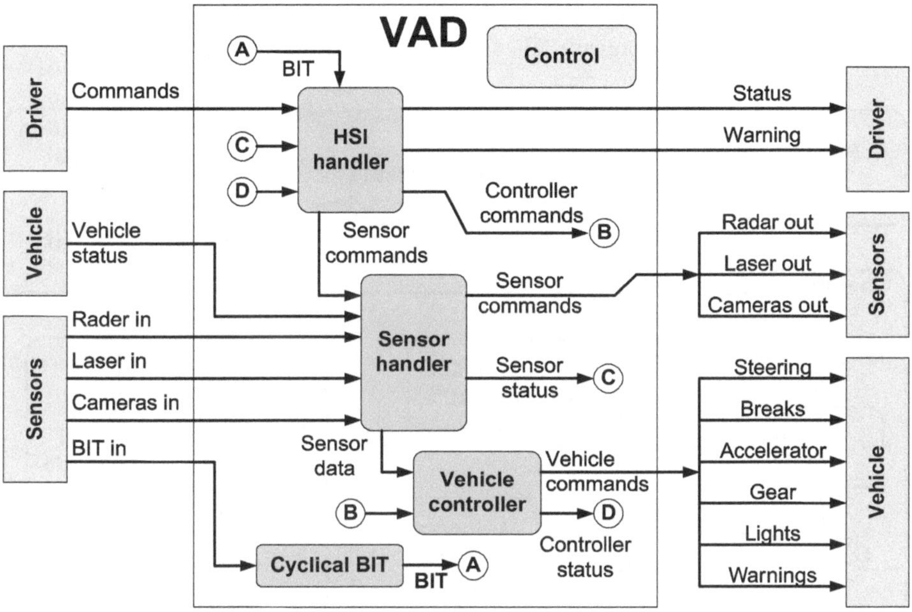

Figure 5.8 Vehicle autonomous driver assistant controller system.

The functionality of each of its subsystems is described separately in Table 5.2.

TABLE 5.2 Functionality of VAD Controller Subsystems

Subsystem	Functionality
Control	Managing VAD's states and transitions
HSI handler	Handling driver inputs and maintenance inputs
	Handling sensors inputs
	Generating Built-In Test (BIT) warning
	Generating VAD status data for driver dashboard
	Generating VAD audio and visual warnings
	Commanding the sensor handler subsystem
	Commanding the vehicle controller subsystem
Sensor handler	Handling vehicle and sensor data
	Handling commands to VAD sensors
	Generating sensor status data for HSI handler
	Generating sensor data for vehicle controller
Vehicle controller	Handling HSI handler data
	Handling sensor handler data
	Analyzing "road picture"
	Generating commands for vehicle control
	Generating vehicle status for HSI display
Cyclical BIT	Obtaining system cyclical BIT data
	Performing system cyclical BIT
	Generating BIT for HSI display

Figure 5.9 depicts the VAD assistant system modes of operations using state chart transition diagram. These modes are described below:

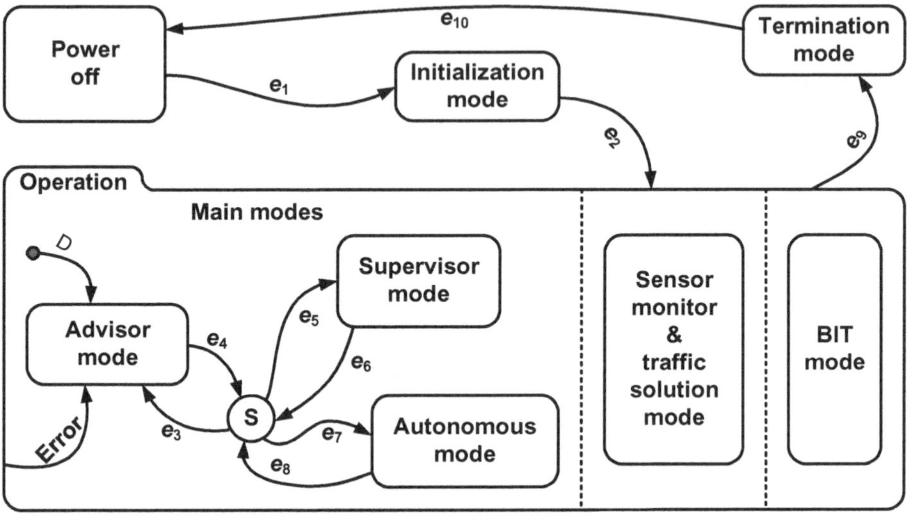

Figure 5.9 VAD assistant modes of operations.

- *Power-Off Mode.* This is the initial mode of the system when the vehicle is not operational.
- *Initialization Mode.* During this mode the VAD assistant system performs the initialization procedure.
- *Operation Mode.* This mode is composed of three parallel submodes: main mode, sensor monitor and traffic solution mode and BIT mode. The main mode is further composed of the following:
- *Adviser Mode.* In this mode the VAD system is passive but provides visual and audio warning to the driver whenever needed.
- *Supervisor Mode.* In this mode the VAD system is semiactive. It provides visual and audio warning to the driver whenever needed. But in case of emergency, it takes control of the vehicle by taking over vehicle steering, braking and acceleration.
- *Autonomous Mode.* In this mode the VAD system is active, fully controlling the vehicle in terms of steering, braking and accelerating, optimizing passenger safety, and adjusting driving speed and maneuvers of the vehicle to meet road and traffic conditions.
- *Termination Mode.* During this mode the VAD system performs termination procedure.

Further Literature

- Harel and Naamad (1996)
- Lavi and Kudish (2004)

5.3.4 Human-System Interface Testing (HSI)

Purpose The purpose of this testing method is to validate that the HSI is functioning properly from both the ergonomic and the functional point of view. HSI testing should consider both the input as well as the output boundaries between people, operators and users of systems and the system itself.

Rationale The discipline of HSI deals with the boundary area between humans and engineered systems. More specifically, HSI deals with input devices, which are the means by which humans control systems, and output devices, which are the means by which humans interpret systems information. HSI performance determines how easily a user may control and comprehend underlying functions of a given system. HSI often is the part of the system that determines the acceptability of the system by end users.

Testing HSIs concentrates on two aspects: the proper functioning of the interfaces and the ergonomics of interface activities. Ergonomics focuses on people's abilities and limitations, as well as what they must do in order to deal with or operate the system. The objectives of ergonomic design activities are to optimize the effectiveness with which work and other human activities are carried out, to maintain important human values such as health, safety and the like and, to the extent possible, stimulate work interest and satisfaction.

Testing for proper HSI ergonomic design will assure easily manipulated operator control interfaces and clear and intuitive representations of system conditions. This will decrease the probability of operator mistakes and misinterpretation of system conditions. Moreover, HSI ergonomic testing increases the likelihood of cost savings in operator training and knowledge retainment. In spite of the importance of this subject, we will not discuss ergonomics in this book, as this is an entire discipline that requires specific specialization.

Testing HSIs is also the process of evaluating user input into the system as well as system output to ensure that the system satisfies the specified requirements. Therefore testing must ensure that systems will not blindly accept any input that the user enters. Conversely, testing must verify whether the system's output is fully comprehensible to users having appropriate capability and training.

Method Humans control systems by issuing appropriate commands. The outputs of these systems, stemming from these commands, are then monitored. Commands may take many forms, including thrown switches, keyboard strokes, mouse moves, screen touches and voice commands. Individual commands or a sequence of commands directs the behavior of the system, provided that the commands are well defined and a complete set of actions is entered by means of the available input devices. Monitors also take on a variety of forms, including computer screens, Liquid Crystal Displays (LCDs) and Light-Emitting Diodes (LEDs), printing on paper and meter dials. Monitors provide humans the information they need to control the system provided that it is unambiguous and easily comprehended.

When one considers the testing task of HSI he or she should be aware of the range of devices[50] used for interfacing with systems (see, e.g., Table 5.3). Test engineers must take into account that each and every interface device connected to the system may introduce a certain problem, distinctive to the given device. For example, switches and buttons may typically introduce timing errors or wrong sequence phenomena, keyboards may introduce text string errors or a display may show incorrect information.

TABLE 5.3 Range of Selected HSI Devices

Input Devices	Output Devices	Combined Input/ Output Devices
• Switches or buttons • Electronic pen or tablet • Joysticks • Mouse • Keyboards • Microphones	• LEDs • Displays or Cathode Ray Tubes (CRTs) • LCDs • Head-Up Display (HUD) • 3D goggles • Earphones or speakers	• Touch displays

As can be seen many, HSIs are uniquely designed to meet the needs of specific applications. For example, Figure 5.10 depicts a ruggedized package of switches, lights, keyboard and display typically used in aircrafts, mobile control centers, and Computer Numerical Controlled (CNC) machine tools.

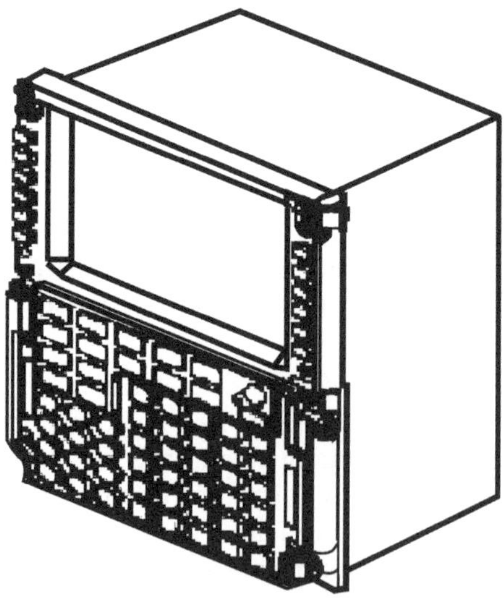

Figure 5.10 Example of an HSI device.

[50]Of course, there is a broad range of other engineered systems with their own specialized Input/ Output (I/O). A shower stall is an engineered system with human input consisting of countless types of faucets and we use our sense of sight and touch in lieu of a system output device.

Human Input Testing To test a human input interface (i.e., a human controlling a system), one must first validate that the system responds correctly to proper commands or sequences thereof. Second one must verify that the system recognizes, tolerates and properly handles operator errors. Here we combine these actions into the requirement that "the system is able to properly process both expected and unexpected input values."

Test cases should be developed to ensure that a system fulfills this latter requirement. In other words, the test engineer must select test data that attempts to show the presence or absence of specific faults pertaining to this input tolerance. In general, we test the input-tolerant properties of systems by verifying that the system is consistently able to (1) detect and handle proper user inputs, (2) detect user input errors, (3) stop input errors from propagating beyond the HSI area, (4) indicate the existence of input error to the user, (5) provide some further suggestion about the nature of the input error and how to correct the error and (6a) permit the user to correct his or her error or (6b) to completely remove the erroneous input from the input interface. When the system is designed to correct specific error inputs, test sequence should be generated to verify the ability of systems to correct those errors automatically and appropriately notify the human operator that the error was corrected.

Essentially, validating human input interfaces encompass generating and executing test sequences composed, first, of proper user commands. In this mode we activate the system under test according to specified procedures and user language definition and validate proper system reaction (e.g., the system meets its specifications and all operational documentations are correct).

Next, we validate suitable system response to improper user commands. This normally includes generating invalid input sequence and illegal text commands while validating proper error handling by the system. Improper user commands may contain invalid syntax, illegal characters and extremely long messages. Such system evaluations may include:

- *Violate Data Type or Size.* Attempting to violate either the data type or size (e.g., entering alphabet characters in a numerical field and vice versa, inserting special characters instead of either alphabet characters or numerical value when not expected, or entering more or less characters than required).
- *Violate User Input Restrictions.* Attempting to violate restrictions on user inputs (e.g., negative or unreasonable high data in age field, unreasonable short or extremely long information in name field or illegal values in date/time fields).
- *Skip Mandatory Fields.* Attempting to skip some mandatory (required) fields in an input form.

- *Inundate System.* Inserting extremely large number of characters (e.g., pressing a key in a keyboard or a button for a long time).
- *Generate Unexpected Sequences.* Activating input switches in unexpected sequence or in a random manner.

Another way to discover system weakness is to study how a system issues an *exception* to a user input. (An exception is an internal system event that signals that an error condition has occurred during the running of the system.) After detecting an exception, the test engineer may utilize this knowledge in order to initiate system failure. Generally, a system may take the following strategies checking for illegal human commands:

- *Real-Time Validation.* After each keystroke, mouse interaction or switch activation, the system checks to see whether the input meets expected value or event. Otherwise the system issues an exception.
- *Committed-Value Validation.* After the user has filled out a given field completely and commits his entry (e.g., by pressing a key to move to the next field). The system checks whether the entire input field meets expected values. Otherwise the system issues an exception.
- *Pass-Through Validation.* After the user has filled out an entire form and commits his entry (e.g., by pressing the carriage return key). The system checks all the fields in the form at once and issues exceptions for the invalid fields.

Readers should note that a substantial gray area exists in the HSI input domain. Should an automobile system check that the driver commands it to travel too fast relative to the road conditions? As it turns out, more and more sophistication is built into engineered systems so they can detect improper and unexpected human inputs. Obviously the test engineer must ensure that such system capabilities are tested.

Human Output Testing In a similar fashion, validating HSI outputs (i.e., human monitoring a system) involves generating and executing test sequences intended first to verify that the system meets its specifications and second that humans react properly to the output information. Since testing human output interfaces is dependent on the specifics of the system and its output devices, we will describe such testing by means of a simple example in the context of a typical Windows operating system. Figure 5.11 depicts a display with an abbreviated flight plan form for which we must first verify the following proper functionality:

- *Text Box.* This field is available for free text insertion. Virtually always, there are limitations on the allowable number of characters, permitted set of characters and the like.

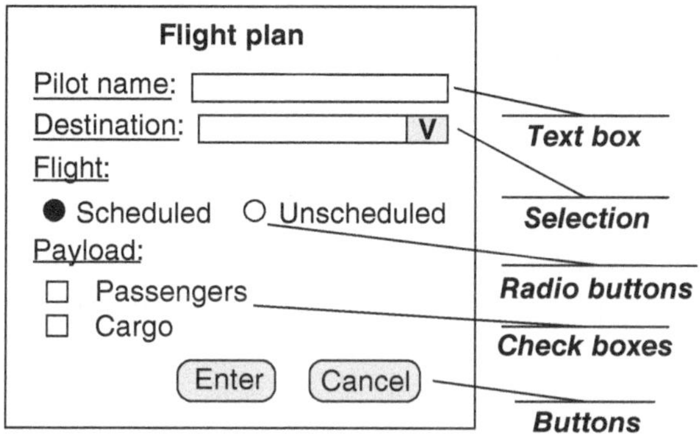

Figure 5.11 Example of an interface display dialogue window.

- *Selection Box.* All relevant destinations are available for selection in the Destination field.
- *Radio Box.* There are only two possible types of flights available (e.g., scheduled and unscheduled), and only one type of flight can be selected, as depicted in the Flight field.
- *Check Box.* There are two possible types of payload available (e.g., passengers and cargo) and either one or both may be selected as depicted in the Payload field.
- *Buttons.* Pressing the Enter button will activate the flight plan and pressing the Cancel button will terminate the request, without activating the flight plan.

In addition, we must validate suitable system outputs to improper or unexpected user requests. Again, using the same example we could invoke the following tests:

- *Inundate System.* Attempt to write extremely long text string into the Pilot Name field and see what appears in this field.
- *Violate Data Type or Size.* Attempts to insert into the Pilot Name field characters which are outside the 26-alphabet character set (numbers, punctuations marks, control characters, etc.).
- *Violate User Input Restrictions.*
  a. Attempt to write a random text string into the Destination field and see what appears in this field.
  b. Attempt to select or unselect *both* radio buttons in the Flight field and see what appears in this field.
- *Skip Mandatory Fields.* Do not select any check box in the Payload field and see what appears in this field.

In general, the system should reject such types of human inputs. This may be done explicitly by the system (e.g., issuing an error message) or implicitly (e.g., not allowing the selection of multiple radio buttons at the same time).

Further Literature

- Charlton and O'Brien (2001)
- Guastello (2006)
- Shneiderman and Plaisant (2004)
- Wise et al. (1993)

5.4 BLACK BOX—HIGH-VOLUME TESTING

5.4.1 Automatic Random Testing

Purpose Automatic random (or statistical or stochastic) testing is based on the concept of automatically injecting very large quantities of random inputs into a system in order to test its behavior. This approach is the opposite of using predetermined and manually selected tests.

Rationale The motivation for conducting random testing stems from the fact that it offers the ability to test the system against a very large and, often unexpected, range of system tests generated automatically and with limited investment. The use of broad test samples assesses the stability and reliability of the system by mimicking, to a large measure, its behavior over a long period of time. On the other hand, random testing in its purest application is somewhat risky, due to the lack of a reliable test oracle (i.e., specifiable failure output values). Without such a test oracle, one could miss finding discrepancies in the specification and can assure finding only obvious faults such as system crashes or certain error conditions. Another concern about this method is that we may need to be careful to restrict the random test data generation to only external conditions that could possibly occur. Otherwise, we would waste valuable time and resources evaluating test results or making system improvements that make no sense.

Method The objective of automatic random testing is to evaluate system performance under unexpected conditions over time. Such high-volume testing, involving a long sequence of tests, where random input values are presented to the system. In this context, we mean random in the mathematical sense, such that a stream of pseudorandom numbers are, in fact, mapped into sequence test cases.

On the one hand, although individual random tests are not very powerful or all that compelling, the generation of a huge number of tests can achieve results beyond the practical abilities of systematic testing. For example, running very large arbitrarily long random sequence of tests can often expose typical long-term software problems such as memory leaks, stack corruption, wild pointers or other garbage that accumulate over time and finally cause system failures. In addition, random testing is inexpensive and the testing environment of the SUT does not require a detailed model of the SUT and is relatively simple to construct and run (see Figure 5.12).

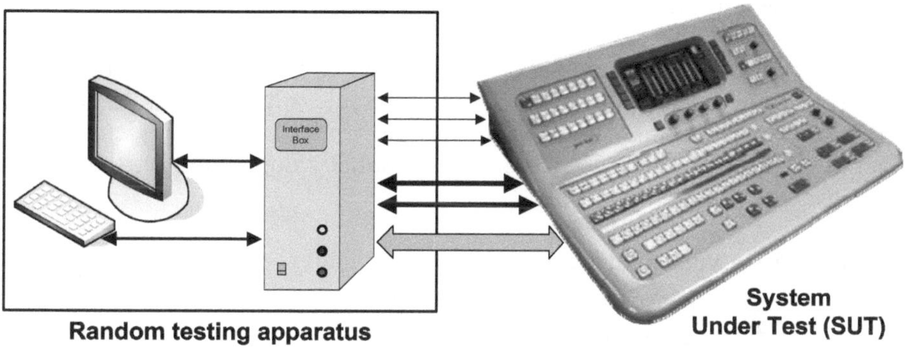

Figure 5.12 Typical environmental setting for random testing.

On the other hand, random testing has severe limitations related to oracle problems. Just figuring out if a random test is functionally allowable is often difficult. Therefore, random testing cannot demonstrate that the system under test meets its specifications. It can only detect SUT failures based on system crashes, error conditions detected by the SUT or improper interactions with other systems. Even then, some test cases yield failures that are very hard to fathom. It is simple to realize that a given failure occurred in a given test, but often the actual trigger may have instigated many tests earlier.

Another problem typifying random testing is that it is not too effective in detecting boundary condition failures. This stems from the fact that boundary conditions are rare in the statistical universe, and this method is not oriented to look for statistically "interesting" places. For example, there is a slim chance that a random algorithm will initiate a flight test at altitude zero.

Similarly, the algorithm is not too effective when an error depends on an unlikely sequence or relationship between inputs. For example, if the system crashes only when a specific input value (A) plus specific input value (B) is equal to, say, specific input value (C), then the probability that this phenomena will be discovered is slim.

Finally the generation of a random input stream is not always trivial. First, often a stream of random numbers is in fact not random at all (i.e., it depends on the random generating algorithm). Second, the random input values must fit reasonably with the operational profile. As random testing invokes a lot of redundant or uninteresting tests, one should consider whether the potential failures hidden in the system are truly important to detect. In this test as in all other VVT activities, test engineers are encouraged to regard Return-On-Investment (ROI) considerations.

Enhanced Random Testing Several ways have been proposed in order to alleviate some of the problems associated with random testing. We will describe some of them briefly here, assuming that interested readers can find further information in the referenced literature and other sources.

Parameterized Random Test Data Generation This method is based on the automatic generation of random data sets, but the data is parameterized in order to control the range and characteristics of those random values. In principle, parameterized random testing allows us to isolate the traits of the data sets. More specifically, it is possible to create a hybrid between equivalence class partitioning and random testing. Under equivalence classes the overall amount of data can decrease substantially depending on the testing strategy:

- Repeating versus nonrepeating attribute values
- Missing versus no missing attribute values
- Categorical versus noncategorical data
- Zero or one label versus nonnegative integer labels
- Predictable versus nonpredictable data sets

Directed Automated Random Testing This method has been used to automate software unit testing. The motivation for this is the recognition that in software practice unit testing is rarely done properly. One reason is that performing manually written tests using specialized harness and driver code is expensive. As a result many software bugs that should have been caught during unit testing remain undetected until late in the development cycle or even into field deployment.

This method proposes to automate unit testing by eliminating or reducing the need for manually written test drivers (i.e., instructions for performing the test) and harness codes. Directed automated random testing using appropriate tools automatically extracts the program interface from source code and generates the test driver for random testing through the interface. The dynamic test generator directs the execution of the software unit along alternative program paths and detects program crashes when they occur.

Specification-Based Random Testing This method combines random testing with formal specification of properties-embedded systems. The method forces people to think in new ways, increasing the understanding of system under test and claims to minimize the difficulty of generating test cases. Specification-based random testing tools (e.g., QuickCheck) accept assertions regarding the properties a given program should satisfy. Then the tool tests whether these properties hold under a large number of randomly generated test cases.

State Model Random Testing This method employs finite-state machine methodology for constructing test cases and test oracles. For any system state, the test engineer can identify the specific actions the user may take and the results of each action in terms of (1) unique system output, as well as (2) the transition to a new state under (3) specific system conditions and guards. Random test cases are executed and the system is evaluated to verify its actual transitions.

Random Testing of Interrupt-Driven Embedded Systems This method has been used to automate testing of interrupt-driven embedded systems by verifying their behavior in the presence of external events impacting the system at random timing. The motivation here is that testing interrupt-driven systems for proper timing behavior typically exercises only a small part of the state space. Random interrupt testing is done by generating interrupts at random times and verifying that the system does not crash or lock up. However, test engineers must be aware of the risk that random interrupts may violate application semantics as interrupts can reenter and overflow the stack of the system. Therefore, the test engineer must restrict interrupt arrivals appropriately.

Regression Random Testing This method is used to enhance and invigorate regression testing. The set of input and output data sequence of previously passed tests are collected and edited so that they don't reset system state. Afterwards, these tests are run in a random sequential order and the results are checked against expected actual outcome. This type of random sequential testing often reveals failures, even though all of the tests have been passed individually.

Random Testing of Integrated Circuits Enhanced random testing is also carried out in numerous types of analog and digital integrated circuit hardware. As this subject is beyond the scope of this book, readers are encouraged to review some of the references mentioned in this section or the many published books and research papers.

Further Literature

- David (1998)
- Dustin et al. (1999)
- Nelson (2004)
- Yarmolik and Demidenko (1988)

5.4.2 Performance Testing

Purpose The purpose of performance testing is to demonstrate that a system meets its defined set of performance requirements. This includes the discovery of performance bottlenecks, verifying that the system contains no discernable faults associated with operating the system at full load and establishing a baseline for future regression testing. Performance testing entails a carefully controlled process of measurement and analysis of the behavior of a system that is being tested which is sufficiently stable so that regular operation can proceed smoothly.

Rationale In general, the motivation for conducting performance testing is to evaluate whether a system can operate at full performance loading within its nominal intended operational environment (e.g., mechanical, thermal, electromagnetic, chemical). In addition, embedded systems should be able to handle external loads given their underlying hardware and software configuration. In nontechnical words, it questions if the system is capable enough to make customers happy.

Some type of system performance testing should be undertaken during different stages of the system lifecycle. Subsystem performance testing should be performed when the the subsystem is implemented in order to verify that the underlying hardware and software supports the application. Nevertheless, significant performance testing should be performed before a system completes its development period so as to verify whether the system meets specifications and is reliable enough to go into production. Finally, during ongoing operations, if the system exhibits performance degradation, performance testing should be repeated to ascertain the cause of this phenomenon.

Since failure of a fielded system can be very costly and embarrassing to a system developer, assuring performance and functionality under real-world conditions and locating potential problems before customers do are paramount to a sensible business strategy. So we can summarize the rationale issue by noting that all testing is risk-driven. Functional testing deals with the risk that the system does not function properly, whereas performance testing deals with the risk that the system will not perform well enough. Ignoring performance risks yields usable systems that may be slow, systems that may be functionally perfect but unusable or systems that are unreliable. Such situations invariably lead to lost business and sometimes may expose companies to costly litigation and payment of damages.

Method A prudent starting point for conducting a system performance test is to develop a Performance Test Plan (PTP) document. This document should cover information related to the entire process of performance testing, including system performance requirements. The PTP should also describe the required resources such as funding, manpower and schedule, as well as needed materials and support infrastructure, which include the target system itself and the testing apparatus setup.

A typical performance testing apparatus setup for evaluating the computational performance of a system under test is depicted in Figure 5.13. The SUT is connected to an environment simulator such that it behaves as if it is performing a nominal mission. The environment simulator can be directed to increase various load parameters, and an observer monitoring the performance of the system can record and analyze appropriate behavior of the system being tested.

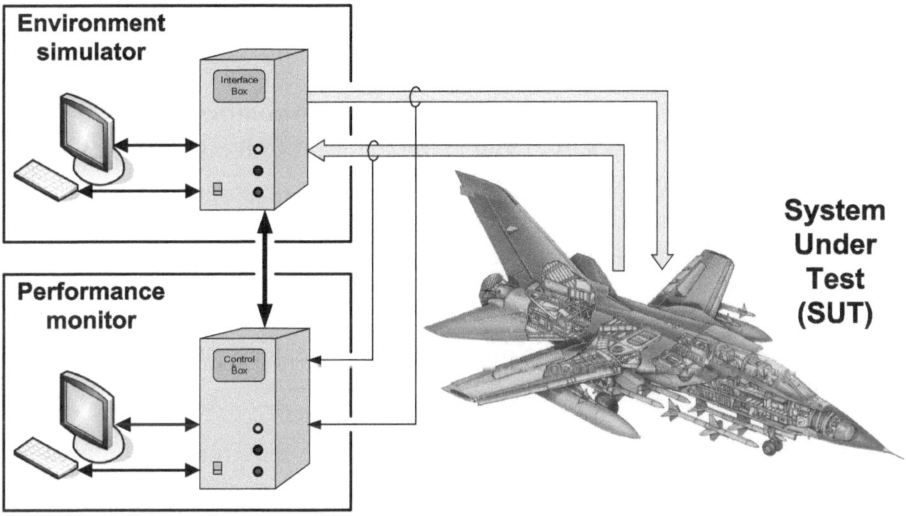

Figure 5.13 Performance testing environment setup.

In summary, the tester must verify whether each system parameter meets its required performance envelope under a required system load. For example, Figure 5.14 depicts such test performance results. In this example a radar system must meet the requirement of acquiring and displaying up to 50 targets using no more than 50% of the CPU (Central Processing Unit) time resource. As can be seen, the system performance varies with load; nevertheless the system does meet its requirement.

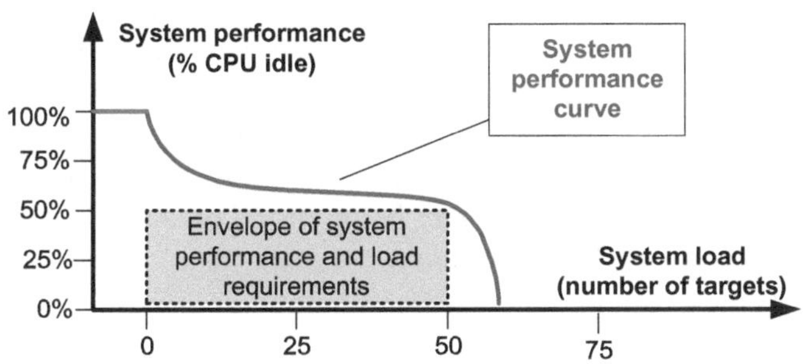

Figure 5.14 Performance and load envelope and actual performance curve.

System performance testing usually includes load and volume testing; that is, testing geared to assess the system's ability to deal with the required I/O throughput as well as maximum utilization of all its other resources. Such tests typically include the following:

- *Task Response Times.* How long does it take to complete a task?
- *System External Capacity.* How many external systems, communication channels or users can the system handle?
- *System Resources.* How many resources are utilized by the system?
- *System Reliability.* How stable is the system under maximum required workload?

A typical procedure for conducting a performance tests usually covers the following steps:

- *Step 1.* Gather and document the performance requirements emanating from the system specifications.
- *Step 2.* Develop a PTP which will include elements like parameters to be tested and their performance/load envelope as well as test resources (e.g., funding, manpower, facilities and test environment set up), test schedule, and so on.
- *Step 3.* Select and purchase performance test tool(s) and then train a number of test engineers in their use. Various automation tools are available commercially (e.g., Mercury—Load-Runner). Although such tools are fairly expensive and complex to operate, they can help test engineers in generating performance test scenarios and test scripts as well as in actual execution and analysis of the performance tests.
- *Step 4.* Develop test scenarios and test scripts for performance testing the system being tested.
- *Step 5.* Develop the performance testing environment setup suit and then install and integrate it with the system being tested.
- *Step 6.* Execute the performance test scenarios using automated test tools iteratively, increasing the SUT load gradually.
- *Step 7.* Collect test results, statistics and graphs and analyze the data to determine whether the system being tested meets the performance specification for each requirement.
- *Step 8.* If the system being tested does not meet specifications, then it is up to the system engineers to carry out appropriate performance tuning or, sometimes, replace hardware or software elements of the system.
- *Step 9.* Generate a performance test report. Such a report will summarize the results of the performance tests and will indicate whether the system meets its performance requirements.

Further Literature

- Jain (1991)
- Molyneaux (2009)
- Musumeci and Loukides (2002)

5.4.3 Recovery Testing

Purpose Many engineered systems, especially real-time, embedded systems as well as computer-based systems and, in particular, distributed systems, are required to have some degree of fault tolerance. That is, certain hardware or network faults, software errors, human errors or loss of data must not cause the system to cease operating or crash. In general, the system must recover from a large variety of faults and resume operating without loss of data and within a specified recovery time.

Recovery testing forces the system to fail in a variety of ways with the intention of verifying that system recovery is properly performed. If recovery is automatic (i.e., performed by the system itself), re-initialization, checkpoint mechanisms, data recovery and restart are examined in terms of process correctness and elapsed time. More specifically, a test engineer should validate that systems with automatic recovery have means for detecting failures and malfunctions, the ability to remove or ignore a failed hardware or software element, perform a switch-over to a standby mode or component and initialize it properly, and, of course, record the system states and all relevant parameters that must be preserved for later corrective action.

If recovery requires human intervention for repair purposes, then recovery testing must examine whether or not the Mean-Time-To-Repair (MTTR) meets specified requirements.

Rationale Error recovery[51] testing is an important part of system testing, especially for safety-critical systems and transactional systems. For example, designers must design various "driver assist" systems (e.g., cruise control system, antilock brake systems, electronic stability system) to meet specific failure behavior requirements or else certain disaster may occur. The rational of testing such mechanism is self-evident.

Similarly, data recovery testing is an extremely important type of evaluation in computer-based transactional systems that contain various data storage devices, databases, distributed client–server architecture and the like. Error detection capabilities that allow an orderly shutdown of a system rather than allow uncontrolled system error propagation should complement data recovery and system restart procedures. If possible however, such mechanism

[51] Error recovery is a preplanned set of procedures for handling system failures in order to minimize disruption and danger to the system itself, the users and the environment.

should record the problem, bypass any damaged data and continue processing as an alternative to a system shutdown.

It is critical therefore that a test engineer evaluate such functionalities and verify that system recovery requirements are indeed being met.

Method Principally, recovery testing is undertaken by injecting some type of fault into the system, observing its behavior and evaluating it against relevant recovery specifications. The technique of fault injection is normally used to induce faults at a hardware level. These type of fault injections involved shorting connections or disconnecting cables and circuit boards and observing the effect on the system. In addition, specialized software may be developed to simulate such processes.

Recovery testing of software-controlled systems can be undertaken by *software mutation* techniques. Under this approach, software tools are used to deliberately modify software code in order to cause system crashes or other abnormal system behavior. The test engineer then observes the resulting behavior of the software-modified system and determines whether or not it meets the required recovery specifications.

Recovery testing also employs more ordinary, albeit aggressive, measures, attempting to sabotage normal system operation, monitoring system failure and examining whether or not the system recovers without loss of data or functionality. For example, such abnormal operation could be achieved by inundating the system with service requests, thus consuming system resources such as memory, disk space, real-time resources, aborting various applications or causing unexpected loss of communication by, for example, disconnect a cable or simply cutting off power.

Beyond validating the proper functional behavior of the recovered system, the test engineer must validate the system data integrity. This involves, among other things, verifying that the last transactions were consistent and robust and that the database and other memory elements remain consistent and integrated.

Further Literature

- Burnstein (2003)
- Myers et al. (2004)
- von Mayrhauser et al. (2000)

5.4.4 Stress Testing

Purpose Stress testing is similar in many ways to performance testing. However, the purpose of stress testing is to operate the system beyond normal operating conditions and observe the results. In stress testing we try to break the system under test by (1) exposing the system to the external environment (e.g., mechanical, thermal, electromagnetic, chemical) beyond nominal operational specifications or (2) overwhelming its resources. The system should be designed with sufficient elasticity so that, when it is overloaded, the system should degrade gracefully rather than fail catastrophically. Furthermore, the system, under certain classes of loads, should fully recover when the unrealistic load is removed. For example, we expect a telephone exchange system to possibly deny some services if the number of callers increase beyond a nominal specifications, but we do not expect the system to crash.

Rationale Stress testing is required to validate robustness and elasticity requirements of the system under test. Robustness is a property of a system to withstand stresses, pressures or changes in procedure or circumstance. In other words it is the degree to which a system can still function in the presence of external adverse or abnormal conditions. Elasticity is the ability of a system to return to its performance parameters after it has been stressed and the stress is removed. Additionally, stress testing often exposes design and implementation flaws that may have remained hidden under traditional testing.

Method As mentioned, the method of performing stress tests is quite similar to the method of carrying out performance tests except that in stress tests (1) we continue to stress the system beyond nominal system specifications and (2) we then decrease the stress all the way to nominal levels while tracking system behavior, as depicted in Figure 5.15.

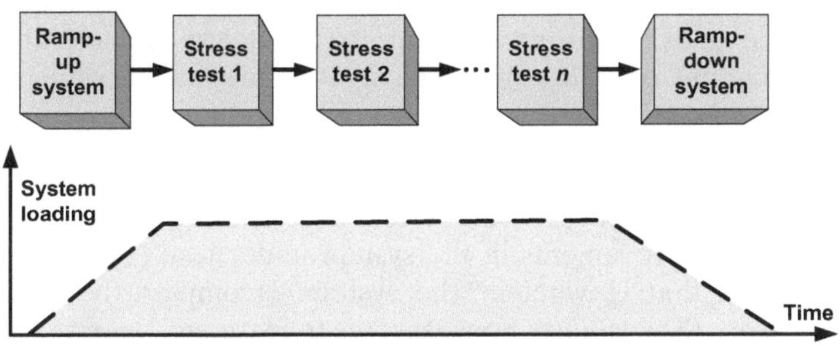

Figure 5.15 General procedure for performing stress tests.

Typically, a number of subtests are performed when the system is stressed beyond its nominal specifications. Most common tests for embedded systems include: (1) testing at maximum input/output data rates, (2) testing at maximum communication channel and data bus usage, (3) exhausting available internal resources such as memory, CPU time and stack level and (4) executing processes that cause transient resource loads.

Typically, performing stress tests is characteristically scripted and generally automated, allowing tests to be repeatable. Post test analysis is performed to identify unexpected anomalies occurring during test and, of course, all problems must be corrected in order to meet system specifications.

Further Literature

- Chan (2001)
- Porter (2004)
- Stamatis (2002)

5.5 BLACK BOX—SPECIAL TESTING

5.5.1 Usability Testing

Purpose *Usability* is the ability of a specific group of users to perform a specific set of activities within a specific environment with effectiveness, efficiency and satisfaction. The purpose of usability testing is to find out practical information about how users actually use a system. Ultimately, usability testing ensures that the design of engineered systems will meet the needs of a representative group of users and, very likely, meet the business needs of the company.

Usability testing involves the observation of typical users performing real system tasks, recording what they do, analyzing the results and recommending appropriate changes if needed. Such user feedback on specific features is of particular interest to the developers of systems. In particular, developers are interested in (1) the level of satisfaction typical users may derive from the system, (2) the efficiency with which users can operate the system, (3) the degree to which users can successfully learn and use the system and (4) the amount of errors that typical users may make while operating the system.

Rationale Usability testing reveals system defects and therefore contributes to the following improvements in the system under test: (1) evaluates functional suitability, that is, whether the system encompass the functionality required by users, (2) evaluates how easy it is to learn and operate the system, that is, whether the users of the system are able to understand how to operate the system accurately (i.e., without errors) and efficiently (i.e., producing the intended result without wasting time, energy or materials) and (3) evaluates the *memorability* of the system, that is, whether users can easily maintain knowledge of a system's operation over time.

Method Usability testing involves recording the performance of typical users doing typical tasks in a controlled environmental setting. The data is used to calculate performance times and to identify and explain users' operational errors. In addition, user satisfaction is evaluated using questionnaires and interviews where the goals and questions focus on how well users operate the system under test. Quite often, when the design of the system has not been solidified, the users are provided with two or more variants of the system embodying different Human-System Interfaces (HSIs) or concepts of operations. In this situation the performance measurements as well as users preferences provide comparison among various system prototypes.

Generally speaking, usability testing is conducted during the system design phase. In the early stages an organization is likely to utilize low-fidelity prototypes and at first employ experts from various disciplines as well as focus groups. Later on in the design process, full usability testing is more likely to be undertaken. Normally usability testing will be conducted by a cross-functional team, as people from different disciplines within the organization bring varying expertise to the team. In addition human factors experts and user interface designers can provide helpful principles about users and design. Typically, usability testing will yield measurements on how well test subjects respond in four areas:

- *Emotional Response (Satisfaction).* A system should be pleasant to use; therefore we try to measure how users feel about each completed task (e.g., confident, confused, stressed).
- *Time on Task (Learnability).* A system should be easy to learn so users can get started quickly; therefore, we measure how long it takes for users to complete basic tasks (e.g., completing a local calling sequence on a mobile phone).
- *Accuracy.* A system should be easy to use, resulting in high productivity. In addition, it should have low error rate and allow error recovery; therefore, we measure how many mistakes users made, what type of error it was (e.g., fatal or recoverable) and how long it took users to recover from their mistakes.
- *Recall (Memorability).* Operating an engineered system should be easy to remember; therefore, we measure how much a user remembers after a period away from operating the system.

Usability testing consists of three broad phases: (1) preparing for the testing, (2) running the actual test and (3) analyzing test results. This process is described in details below.

Preparing for Testing First, the objectives of the usability testing must be defined. This is usually done during the user/task analysis and product scoping. Objectives must be measurable and should indicate the type of user, the task to be performed and the specific performance criteria.

Next, a test plan should be created which will explain what must be tested and how the testing process will be conducted. A usability test plan should not necessary be a very long or detailed document, but rather it should provide a platform for thinking about and organizing the test process. Typically, a usability plan should cover the following topics:

- *Objectives.* These will identify the usability objectives of the testing.
- *Method.* This will detail how the tests will be conducted.
- *Measurements.* These will define the exact test data which will be collected throughout the testing.
- *Analysis.* This will define the nature of the required analysis of the test data.
- *User Profile.* This will describe who are the users and their defining characteristics.
- *Test Environment.* This will define the specific environment where testing will occur (e.g., a laboratory, early system prototype, a system simulation, a fielded system).
- *Test Team.* This will define the roles of individuals supporting the usability test.
- *Resources and Schedule.* This will include all required resources, tasks to be completed, projected schedule, required state of system to be tested and the like.
- *Conclusion.* This will include a list of the expected posttesting activities (e.g., generation of reports, corrective actions).

The last activity prior to actually conducting the usability tests are to create a users questionnaire and to select the test participants. The questionnaire should contain a section for users to provide very general and relevant information about themselves and another section for users to provide their subjective impression about the system under test. Before selecting the test participants, one should gather additional details about the participants' knowledge and experience and ensure that each participant meets the user profile. The number of participants depends on the number of user groups where the target could be two to three individuals from each user area group.

Running Actual Usability Test Usability test sessions should start by participants filling up their personal data in the questionnaire. The facilitator, who is the main contact person with the test engineers, is expected to conduct an appropriate briefing for the user team. As part of the briefing, he or she should assure users that they are helping in evaluating the system. He should also describe to them what will happen during the test process.

The users will then conduct the actual tests one by one while the data recorder, the person assigned to log the usability testing results, will record the results of the usability tests, as well as relevant user comments. Normally, after each task is completed, the users are asked to fill out a questionnaire in order to capture their subjective feelings about the system while the experience is fresh. In the meantime, the facilitator should look and listen for the unexpected. He or she should be ready to handle unplanned situations and should avoid intervening in the normal flow of the test, unless it is necessary.

Analyzing Test Results Normally the analysis phase starts with a debriefing session with the users. The users are asked to elaborate on significant testing events or make general comments. Sometimes they may be asked to watch a video recording of the test and explain what their thoughts were at certain points and the reasons for their specific behavior.

The three areas of test data (i.e., learnability, accuracy and memorability) are then analyzed to verify test performance measurements against required levels. In addition the overall level of users' satisfaction is also considered. The analysis should culminate in a pass or fail decision. If the system does not meet the usability requirements, then further action is required in order to elicit ways to improve the usability of the system.

Example of Usability Test Consider a simulator developed to demonstrate the concept of usability testing. The simulator can evaluate the learnability, accuracy and memorability under normal operations of two different designs of kitchen gas ranges, as depicted in Figures 5.16 and 5.17.

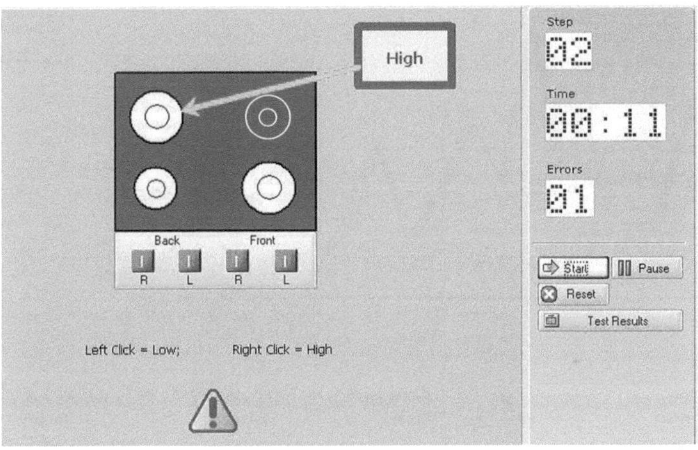

Figure 5.16 Example 1: kitchen gas range first design layout.

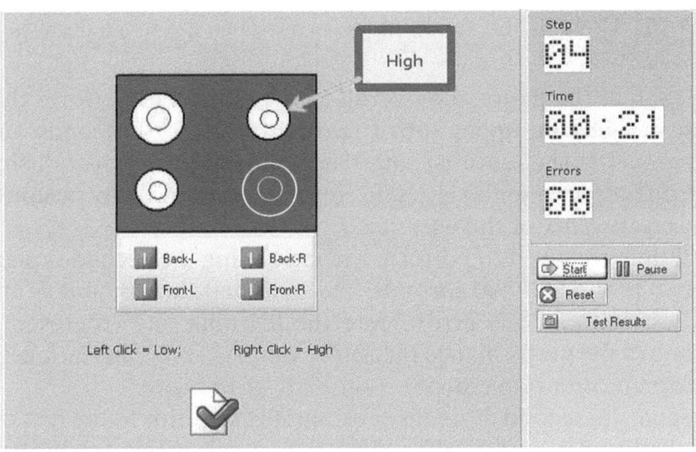

Figure 5.17 Example 2: kitchen gas range second design layout.

A single usability task was defined as an operational test sequence consisting of 20 steps. In each one of these steps the user is asked to turn on either the small or the large burner from a set of four by right or left clicking on the appropriate gas control. The simulator indicates to the user whether a given test step was successful, in which case it moves on to the next step, or in case of a failure it asks the user to try again. The time required to complete each step is recorded as well as the number of errors the user has made throughout the usability test.

The results of two usability tasks (one for the first design and one for the second design) are depicted in Figure 5.18. The *X* axis represents the step number and the *Y* axis represents the amount of time required to complete each step.

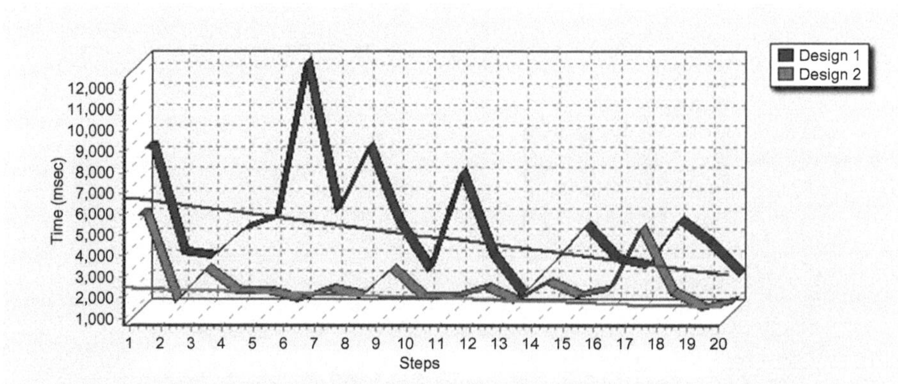

Design	Number of steps	Number of errors	Time total (sec)	Time average (msec)	Learning rate slope (deg)
1	20	3	89.70	4485.00	–10.62
2	20	0	37.90	1895.00	–2.72

Figure 5.18 Usability test results of the two kitchen gas range designs.

As can be seen in the figure, the overall time required to perform the task using the first design (the upper set of two plots) is 89.70 seconds and the number of errors is 3. The learning rate (represented by the overall slope of the performance measurements) is –10.62, indicating noteworthy performance improvement on the part of the user.

The overall time required to perform the task, using the second design (the lower set of two plots), is 37.90 seconds, less than half the amount of time of the previous test, without any error. Here the learning rate (represented by the overall slope of the performance measurements) is –2.72, indicating limited practical performance improvement on the part of the user.

As can be seen, the second design is substantially superior to the first design, as the gas controls are naturally aligned with their respective gas burners.

Further Literature

- Dumas and Redish (1999)
- Rubin and Chisnell (2008)
- Tullis and Albert (2008)

5.5.2 Security Vulnerability Testing

Purpose The purpose of security testing is to identify embedded systems and computer network vulnerabilities in order to protect such computer assets (e.g., servers, applications, Web pages, data). Such attacks, emanating from internal or external sources, may be accomplished through unauthorized access to the system in order to corrupt existing information, carry out financial fraud, steal classified data or cause a denial of service.

Testing a system for security vulnerabilities as well as malware infection requires a specialized type of knowledge. In general, malware and viruses are self-replicating programs that usually have a malicious intent. Some viruses are harmful, for example, they delete valuable information from a computer's disk or modify the operating system causing the computer to crash every now and then. Other viruses are relatively benign and harmless; for example, they display annoying messages or advertisement to attract user attention. Still others may not overtly effect the system but extract valuable information and transmit it to external users. The purpose of Table 5.4 is to acquaint the test engineer with the rich variety of current malware types.

TABLE 5.4 Prevalent Malware Types

Malware	Description
Virus	A virus is a malicious, self-replicating program that uses the Internet to spread from one computer system to other computer systems in an exponential manner. Due to its construction, a computer virus needs human intervention to replicate, which, relatively speaking, slows down the rate of virus propagation through the Internet.
Worm	A worm is a special type of virus which does not need human intervention in order to replicate. Therefore, worms have the ability to spread throughout the Internet in a very brief period of time.
Trojan horse	A Trojan horse is a seemingly innocent application that contains hidden malicious code. Trojan horses are, most likely, useful programs that often are offered free of charge to users and have unnoticeable purposes such as stealing valuable data.

TABLE 5.4 *Continued*

Malware	Description
Backdoor	A backdoor is malware that creates a covert access channel that the attacker may use at any time for connecting, controlling, spying or otherwise interacting with the target system.
Mobile code	Mobile code is a class of either benign or malicious programs obtained from remote systems and downloaded and executed on a local system without explicit installation or execution by end users. Malicious mobile code is downloaded either into client mobile phones through normal telephone connections or Short Message Service (SMS) messages or may be installed in workstations on opening certain emails or while visiting Web pages on the Internet. Results of mobile code attacks include disclosure of confidential information, damage or modification of internal data and denial of service.
Sticky software	Sticky software implements methods that prevent or deter users from uninstalling it manually, for example, by not offering an uninstall capability. Often, under the Windows operating systems, this code sets up the program registry keys to instruct Windows to always launch the malware as soon as the system is booted. This annoying malware method is sometimes perpetrated by software vendors who sell their products aggressively.
Cryptographic worm	Cryptographic worm is a rather new and less common way of using worms to encrypt important data on victims' computers. Such an encrypted data becomes virtually useless to the owner of the data. The perpetrator's intent is to keep the data hostage, demanding ransom for releasing the key that then can restore the information to its rightful owner in its original form.
Adware	Adware is a program that forces unsolicited advertising on end users. Adware is often bundled with a free, limited capability, trial software used to demonstrate and promote the actual, full capability, software package.
Phishing attack	An email message that urges an unsuspecting recipient to provide personal information including bank account numbers, Social Security number, personal data or user name and passwords to Web sites or business accounts. Usually these messages mimic real messages from a reliable source.

Security can be strengthened by physically limiting the access of computers to trusted users. This may be achieved by means of various hardware mechanisms (e.g., physical locks, biometric sensors) or software mechanisms (e.g., imposing rules on entrusted programs, antivirus software to detect malware, secure coding techniques to make software less vulnerable to security attacks).

Rationale The threat to information technology systems is changing. More and more systems with poorly implemented security measures and running critical missions are vulnerable to the changing landscape. First, more systems support Web applications, which are the primary targets of hackers. Second, open-source hacking tools keep improving while the perpetrator population is shifting from amateur hackers to organized crime figures. Third, the sophistication of viruses, spyware and other malwares are increasing dramatically. In this context, malware (i.e., malicious software) is any program that works against the interest of the system user or owner. Typical purposes of malware are:

- *Backdoor Access.* The intent of the attacker is to gain unlimited access to a target computer system.
- *Denial of Service.* The attacker infects a large number of computer systems with the intent to try simultaneously to attack a target server system in the hope of overwhelming it and making it crash.
- *Vandalism.* The intent of the attacker is to disrupt the operations of a target computer system, for example, erasing its disk or defacing a Web site.
- *Resource and Information Theft.* The intent of the attacker is to steal valuable information such as credit card parameters, business or military classified information, and the like.

The number of yearly malware attacks increases exponentially throughout the industrial world. Different numbers of such attacks are reported but, for example, F-Secure Corporation, a computer security service provider located in Helsinki, Finland, suggests that the recent explosion of malware is a result of an industrialization of malware production by hackers who sell their services to professional criminals, who in turn launch worldwide attacks, issue millions of phishing emails or engage in industrial espionage (see Figure 5.19).

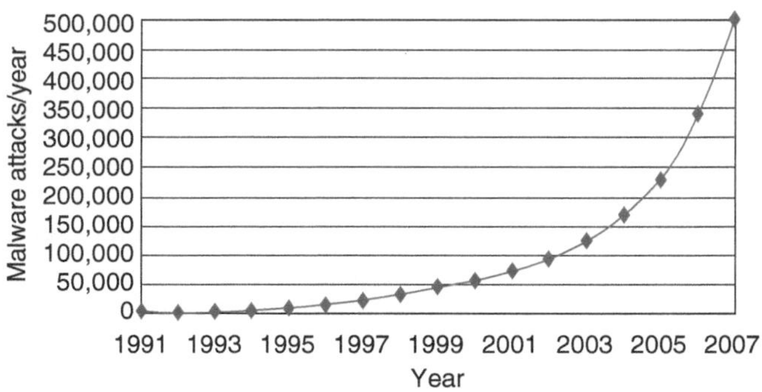

Figure 5.19 Numbers of malware attacks per year (F-Secure Corporation).

The Computer Emergency Response Team (CERT) coordination center, at Carnegie Mellon University (www.cert.org), collects statistics on the total number of vulnerabilities that have been cataloged based on reports from public sources and those submitted to the CERT directly. Here, the term vulnerability is applied to a weakness in a system which allows an attacker to violate the integrity of that system. According to CERT, incident statistics collected between 1995 and 2008, the number of such computer and software vulnerabilities has increased by about two orders of magnitudes during this period (see Figure 5.20).

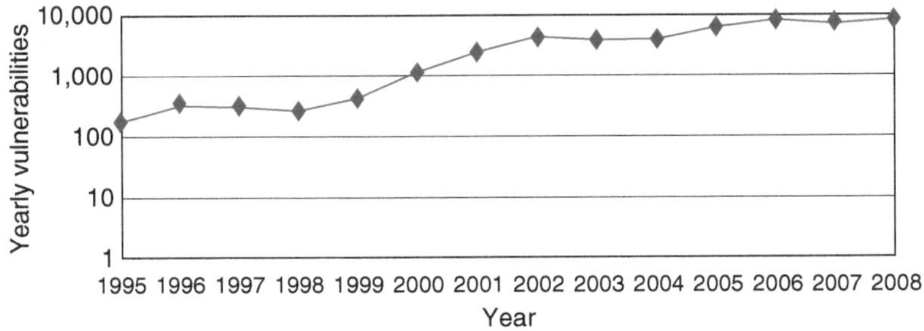

Figure 5.20 Yearly vulnerabilities reported by CERT (Carnegie Mellon University).

The financial impact of a system's security breach is usually manifested in numerous ways. A company may incur financial liabilities due to inappropriate disclosure of sensitive information or it may have to pay fines due to regulatory noncompliance. Or, a company may lose its technical and business edge as competitors access its confidential or proprietary information or, possibly, fail to win new business due to bad press associated with a security breach. Then, if a security breach does occur, the company will incur the cost related to detection, containment, repair and reconstitution of the breached system. Last but not least, after a breach a company must often bear an increase in insurance premiums.

In light of these problems, the rationale for performing security testing may be summarized as follows: finding vulnerabilities in the system before attackers find them and significantly reducing system rehabilitation cost stemming from breaches in system security.

Method A secured system must be able to deal with someone, within or outside the organization, who is intentionally trying to exploit vulnerabilities in the system. Such attack is invariably directed at the system's "attack surface," that is, the interface points where a user may gain access to application resources, that is, Application Programming Interfaces (APIs), network ports, permanent and temporary files and the like. Therefore the objective of security testing is to focus and identify vulnerabilities to unauthorized access or manipulation and thus protect the system.

Security testing can be conducted on a developmental system or an operational system. Test methods depend on the stage of the system's lifecycle and on the security testing process chosen. The U.S. National Institute of Standards and Technology (NIST) has created a recommended set of security requirements. It is defined in Special Publication (NIST 800-53, 2009) Recommended Security Controls for Federal Information Systems. In general, the security requirements needs to be adjusted as a function of the information confidentiality, integrity and the mission criticality of the system undergoing test, as well as the manner in which the system has been implemented. Such matching or augmentation is normally accomplished through a security risk assessment of the system.

The overall objective of the security testing is to ensure that a comprehensive testing activity is identified, covering all appropriate security requirements and involving all necessary individuals. If the system is in operational use, the common approach for testing will use a nonintrusive set of tests. The security testing will include manual as well as mechanized review of critical files from the live system and review of operational procedures. The requirements that will be placed on the operational system should be identified in the system security test plan. This test approach must be designed to avoid any disruption to the ongoing activities. In general, security tests will be conducted in close coordination with individuals familiar with administration of the system to draw on their expertise in system operation and to identify any potential for system disruption.

The testing of an information system's security features starts usually with a series of formal systems tests and operational tests:

- *System Tests (STs).* This group of tests is designed to verify that a system meets its specified requirements. Subsets of the system test are development tests, operational tests, environmental tests and acceptance tests. Each of these elements must verify the fulfillment of all the requirements associated with a system.
- *Operational Tests (OTs).* This group of tests demonstrates that the system is operationally effective and operationally suitable for use. These tests focus on demonstrating that operational requirements have, in fact, been met and that a mitigation plan which resolves known deficiencies has been developed and accepted.

After passing the system tests and the operational tests, the actual security tests are conducted. This includes a vulnerability test and penetration tests:

- *Vulnerability Tests (VTs).* This group of tests are undertaken in order to identify current security vulnerabilities that may compromise the system by using an approved vulnerability scanning method. These may include, but are not limited to, port scans, available services, password checks, system patches and the like.
- *Penetration Tests (PTs).* This group of tests evaluate whether the test team can succeed in gaining access to the system by attempting to

circumvent its security features. Usually, penetration testing on live information systems must have advanced coordination and formal authorization from an appropriate staff officer that owns the system as well as the owners of the information stored on the computer system. Furthermore, if the penetration test could impact one or more related systems, then coordination must include all affected system managers.

A system security test plan should be generated to serve as a tool for developing, implementing and managing the security testing process. A test plane which was derived from NIST 800-53 could contain the following elements:

- *Phase 1: Scope and Rules of Engagement.* In this phase the planner must first determine what elements of the system are to be tested (e.g., applications, databases, servers, interface with other systems or services). In addition, a general security vulnerability test plan must be formulated which should include an estimate of required resources (e.g., funding, equipments, facilities, manpower) as well as schedule for both the testing and the expected corrective process. Finally, the rules of engagement vis-à-vis the conduct of the testing project as well as a list of deliverables must be defined.
- *Phase 2: Develop Evaluation Methods.* In this phase, a detailed, step-by-step test procedure should be developed, identifying the specific test methods (based either on white- or black-box testing categories) applicable for each system element. In addition, the test team should select tools for performing the security tests. Many applicable tools are available commercially, for example, antivirus software Symantec (http://www.symantec.com) and McAfee (http://www.mcafee.com) and a large variety of other hardware and software tools available commercially.
- *Phase 3: Security Testing Execution.* In this phase the actual security evaluation takes place. The system hardware and software architecture should be examined. Similarly the operating procedures are evaluated so that at the end of this phase the overall system vulnerabilities are identified. Next, a security test report should be written which identifies the findings of the test and provides recommendations for corrective action. A possible approach for such a document is to follow the standard FIPS PUB 199 established by the U.S. Computer Security Division, Information Technology Laboratory, NIST.
- *Phase 4: Perform Corrective Measures.* In this phase the corrective actions related to the elimination of system vulnerability must be undertaken. This activity should be based on the general planning undertaken in phase 1, utilizing the resources allocated for that purpose.
- *Phase 5: Retesting.* In this phase the planner should establish expected retesting intervals in order to ensure that the system maintains its secured status on a permanent basis.

Security Architecture Security architecture is a specification that is used as a guide to enforce security constraints. It specifies where security mechanisms (e.g., firewalls, intrusion detection systems, encryption) need to be positioned in the system architecture as well as the individual security level of various applications which constitute key components of the system. Typical security architecture is comprised of the following elements (see Figure 5.21):

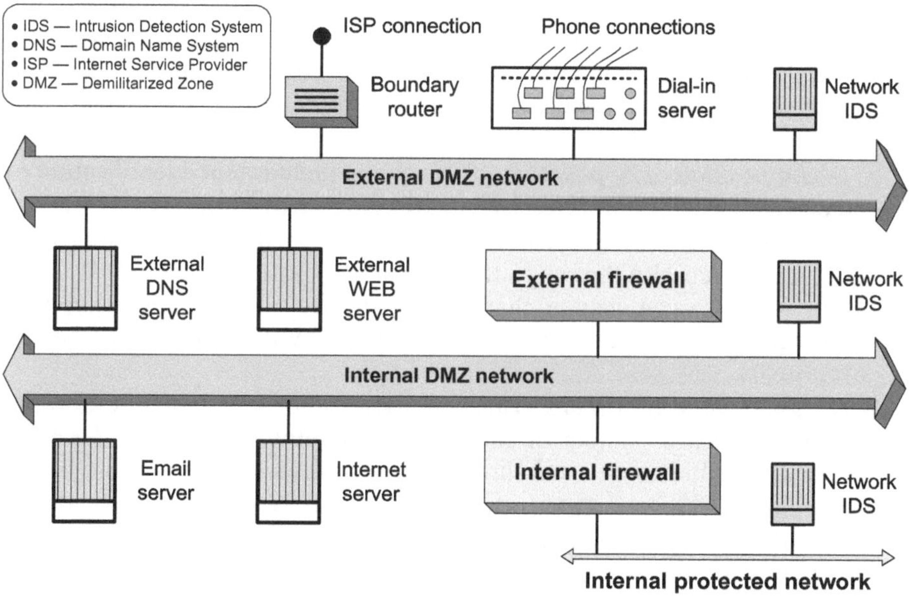

Figure 5.21 Example of two-tiered firewall security architecture.

- *Subsystems.* For example, Web servers, application servers, databases, directories, Web applications and legacy applications.
- *Communication Links between Subsystems.* For example, external and internal networks, local and remote calling facilities and communication protocols.
- *Security Means.* For example, authentication and authorization points, encryption methods, mechanisms for audits, logging, monitoring, intrusion detection, registration, backup and recovery.

There are many security vulnerabilities which arise from poorly designed security architecture, most notable, unauthorized access to data and applications, confidential and restricted data flowing as unencrypted text over unsecured network connections and the like. Accordingly, security architecture is validated using a process called threat modeling. This is usually carried out manually within an inspection process, similar to system requirements or design inspection. Threat modeling is the responsibility of the test team which is commonly composed of systems security experts, test engineers and managers. The test team will typically carry out the following activities:

- *Identification of Assets.* This activity includes identifying valuable information stored within the system which is possibly coveted by intruders. This may include credit card numbers, social Security numbers, computing resources, trade secrets, financial data, and the like.
- *Creation of Architecture Overview.* This activity includes definition of the required system architecture and identification of the trust boundaries and the authentication mechanisms. Trust boundaries define systems and software area limits where users may be admitted depending on their access prerogatives.
- *Decomposition of Application.* This activity includes the identification of data flows, encryption processes, password flows, and the like.
- *Identification of Threats.* This activity includes analysis and identification of existing security threats to the system; for example, verifying if unauthorized users can view or change data, the security limitations imposed on legitimate users and unauthorized access by users to various system resources.
- *Documentation of Threats.* This activity includes the formal description of issues such as system threats, target components, potential forms of attack, possible countermeasures to prevent such attacks, and the like.
- *Ranking of Threats.* This activity includes the ranking of each threat according to its threat area category and level of threat, usually on a scale of low, medium and high (see Table 5.5).

TABLE 5.5 Ranking Security Threats

Category	Description	Rank		
		Low	Medium	High
Damage potential	The damage potential of each security threat (e.g., damage to property, data integrity, financial loss).			
Success probability	The probability that an attempt to compromise the system will, in fact, succeed.			
Exploitability/discoverability	Both the level of difficulty in achieving unauthorized penetration into the system as well as quick discovery of such system breaching by the system's security elements.			
Affected users	The number and the types of users who might be affected by any given security threat.			

Examples of Established Security Tests A proactive approach to security testing will prevent repeated security crises in a computer and embedded computer systems. In general, proactive measures entail the integration of security testing within the system development lifecycle, retesting security elements and recertifying the system if there are significant changes to the system or to the environment and performing recurring architecture review and security gap analysis. The following is a short description of prevailing security tests:

- *Network Scanning.* This security testing involves using a port scanner to identify all hosts connected to an organization's network and the network services operating on those hosts as well as the specific applications running on the identified services. The result of these tests is a comprehensive list of all active hosts and services, printers, switches and routers operating in the scanned address space.
- *Vulnerability Scanning.* This security testing is similar to network scanning but also provides information on various associated vulnerabilities and permits mitigation of the discovered vulnerabilities. This test provides the system and network administrators means by which to identify vulnerabilities before an intruder can find them. Commercially available tools enable relatively efficient ways to quantify an organization's exposure to such vulnerabilities.
- *Password Cracking.* In today's computer systems, virtually all passwords are stored and transmitted in an encrypted form called a hash. When logging on to a computer system, a hash code is generated and compared to stored hash. If entered and stored hashes match, then the user is authenticated. This security testing is used to identify weak passwords by verifying that users select and thus employ sufficiently strong passwords.
- *Log Reviews and Analysis.* This security testing involves automated review of various system logs in order to identify deviations from the organization's security policy. These logs normally collect vast amounts of audit data on the system. Log audits and analysis can provide a dynamic picture of the ongoing system activities that can also be compared with security policy.
- *File Integrity Checkers.* These security testing devices provide tools for the system administrator to recognize unauthorized changes to systems files. Integrity checkers compute the checksum of every protected file and establish an encrypted database of these checksums. The encrypted checksums are regularly compared with current values checksums in order to identify any file that was modified illegally.
- *Malware Detectors.* These security testing devices ascertain whether the system contains malware such as viruses, Trojan horses, worms and the like by having been connected to Internet or via users downloading contaminated software programs or data. The impact of this malware may

be negligible or very serious. It also presents a risk of exposing confidential information to unauthorized individuals.

- *Modem Dialing.* This security testing involves the identification of unauthorized dialup modems that are connected to the computer system surreptitiously. Such modems could provide means to bypass the security measures in place and gain illegal entrance to the system. Several commercially available tools allow network administrators (as well as computer hackers) to dial large blocks of phone numbers in search of such modems.
- *Wireless LAN.* A wireless Local Area Network (LAN) links an external computer to a system by means of radio transmission. This gives users the mobility to move around within a coverage area and still be connected to the network. However, such communication systems are often vulnerable and enable attackers to bypass the security systems. This security testing involves periodic verifications that the organization's wireless connection policy is, in fact, fully maintained and unauthorized users are prevented from entering the system. In addition, the testing involves radio scanning for external incoming signals from neighboring wireless LANs.
- *Penetration Attempts.* This security testing attempts to identify methods of gaining access to the system by using common tools and techniques. The aim here is to identify security weaknesses based on understanding of system design and implementation.

Further Literature

- Basta and Halton (2007)
- Belapurkar et al. (2009)
- DoD 5200.28-STD (1985)
- FIPS PUB 199 (2004)
- NIST 800-53 (2009)
- Solomon and Chapple (2005)

5.5.3 Reliability Testing

Purpose The purpose of reliability testing is to verify that a system meets its reliability requirements. As a general rule, such testing should not occur during the normal defect testing process because testing for defects does not reflect normal system operations. In addition, making reliability inferences about the system should be based on a sample data which is statistically significant.

Rationale Reliability testing measures the quality of systems and predicts the potential for future failures. It provides mechanisms to make management decisions on an impartial basis, for example, in determining when to release a system to its customers and in estimating testing requirements (i.e., to achieve the reliability targets) and costs.

Reliability testing is especially important for safety-related systems, that is, preventing the system from harming users, other individuals, financial interests or the environment. Highly reliable systems are ultimately safer systems, preventing unintended consequences throughout the industrial and service sectors, as well as transportation, space exploration, military operations, and the like.

In the final analysis, the rationale for conducting reliability tests is the simple fact that reliable systems are a prerequisite for satisfied customers, users and the society at large. The ultimate goal here is to adhere to the user's requirements. In addition, a reliable system increases the likelihood of business success for the company, as reliability saves time and money.

Method In order to test the reliability of a system, an operational profile should be generated that reflects as much as possible normal operations of the system. Generating normal test inputs requires significant effort but is a fairly straightforward task. Unfortunately, operational profile includes also "reasonable but unlikely" inputs, and VVT practitioners should be aware that predicting and creating an exhaustive set of such test inputs is a daunting task. The system should then be tested under this operational profile. Failure statistics are gathered and the system reliability is predicted based on appropriate statistical analysis models and tools.

If a system does not meet its specified reliability requirements, then it should be corrected and retested prior to delivery. According to current reliability growth models, system reliability can be improved over time, as the system undergoes this process of testing and defect removal. Nevertheless, reliability does not necessarily increase with such changes, as modifications can introduce new faults. These same mathematical models can also be used to predict future system reliability, by extrapolating from current failure data.

To summarize, reliability validation is usually composed of the following steps:

- *Step 1.* Establish an operational profile for the system. This should include both normal operator inputs as well as reasonable unusual or abnormal inputs.
- *Step 2.* Construct test data reflecting this operational profile.
- *Step 3.* Test the system and observe the number of failures, the time of failure occurrence and their severity.
- *Step 4.* Assess the reliability of the system by means of available reliability tools. This process should take place after a statistically significant number of failures have been observed. This step is accomplished by reviewing the system's failure data, selecting an appropriate statistical model that fits the failure data and estimating the model parameters. Next, verifying the appropriateness of the selected model and parameters

by performing "goodness-of-fit" operation. Finally, make the actual reliability predictions based on the selected models.

System Reliability Models System reliability is the probability that a system will not fail for a specified period of time under specified conditions. Although hardware faults often emanate from material fatigue or heating of components, software does not wear out, and failures are mainly related to design and implementation faults, which are harder to detect, correct and model.

Existing engineered systems tend to fail a fair number of times in the course of their lives. This necessitates correcting inherent problems. Therefore, reliability models show that system reliability tends, in fact, to grow over time. The dynamic of this process is this: We assume that a system fails at times $\{t_1, t_2, t_3, \ldots, t_n\}$, and we ask what is the probability of its failure at time t_{n+1}? In pure hardware we can adopt the uniform model and further assume that the probability of all failures is constant as we simply replace a defective hardware component with an identical one. However, in complex, computer-embedded systems we often correct the problem by treating a core design or production problem (i.e., often fixing the software). This reduces the probability of failure after a repair or increases the expected duration until the next failure at t_{n+2}.

There are two prevalent families of reliability growth models related to our discussion: (1) the basic exponential model which assumes finite failures (ν_0) in infinite time and (2) the logarithmic Poisson model which assumes infinite failures in infinite time. The parameters involved in the above reliability growth models are:

- *Mean Failures Experienced* (μ). This is the mean failures experienced (μ) for a given time period (e.g., one day, week, month, year, of operations). Assuming that P_i is the occurrence probability of failure i and where n is the total number of failures, it is calculated as

$$\mu = \sum_{i=1}^{n} i p_i$$

- *Failure Intensity (λ).* This is the failure rate or the number of failures per unit of time.
- *Execution Time (τ).* This is the duration of time the system is operating.

The relationships between these parameters, mean failures experienced (μ), failure intensity (λ) and execution time (τ) are presented in Table 5.6.

TABLE 5.6 Relationships between Reliability Growth Parameters

Comparison	Basic Exponential Model	Logarithmic Poisson Model
Failure intensity (λ) versus mean failures experienced (μ)	$\lambda(\mu)=\lambda_0\left(1-\frac{\mu}{v_0}\right)$	$\lambda(\mu)=\lambda_0 e^{-\theta\mu}$
Mean failures experienced (μ) versus execution time (τ)	$\mu(\tau)=v_0[1-e^{-(\lambda_0/v_0)\tau}]$	$\mu(\tau)=\left(\frac{1}{\theta}\right)\ln(\lambda_0\theta\tau-1)$
Failure intensity (λ) versus execution time (τ)	$\lambda(\tau)=\lambda_0 e^{(-\lambda_0/v_0)\tau}$	$\lambda(\tau)=\frac{\lambda_0}{\lambda_0\theta\tau-1}$

where: λ_0 is the initial failure intensity, v_0 is the total failures and θ is the decay parameter.

As VVT professionals, we are interested in verifying that a system meets its reliability requirements. As we can see in the above equations, reliability (R) of a system changes over time and follows the general equation:

$$R(\tau)=e^{-\lambda(\tau)\tau}$$

where $\lambda(\tau)$ is a dynamic (time-dependent) failure intensity and τ is a natural unit, usually time in terms of days, weeks, or months. Reliability is a complementary concept to failure so, in order to compute it, we typically, seek failure specification such as (1) time of each failure, (2) time interval between failures and (3) cumulative failures up to a given time. VVT practitioners can use a plethora of system reliability tools. For example, we will demonstrate the computation of system reliability utilizing the Computer-Aided Systems Reliability Estimation (CASRE[52]) tool. Once historical failure data is entered into the tool, CASRE can generate reliability information using a collection of probability models which may be appropriate for different input data and circumstances.

System Reliability Example Toward the end of a software-intensive project, the system was handed over to two test engineers for a comprehensive evaluation which lasted a total of 60 working days. The system developers received defect information on a daily basis, proceeded to correct the system immediately and submitted the fixed system for retesting. Table 5.7 identifies a total of 117 defects found during this period, tabulated on a daily basis.

[52] CASRE is a PC-based tool that was developed by the Jet Propulsion Laboratories in the United States. It is freely available at: http://www.openchannelfoundation.org/orders/index.php?group_id=250.

TABLE 5.7 Results of 60-Day System Evaluation

Day	Defects	Day	Defects	Day	Defects	Day	Defects
1	11	16	3	31	0	46	1
2	7	17	1	32	1	47	0
3	9	18	2	33	0	48	0
4	7	19	0	34	0	49	0
5	5	20	2	35	0	50	0
6	6	21	9	36	0	51	0
7	6	22	1	37	1	52	0
8	9	23	0	38	0	53	0
9	5	24	0	39	0	54	0
10	4	25	0	40	0	55	0
11	5	26	1	41	0	56	0
12	5	27	0	42	0	57	0
13	7	28	1	43	0	58	0
14	7	29	0	44	0	59	0
15	1	30	0	45	0	60	0
						Total	**117**

CASRE provides operations to either display or transform or smooth the failure data. For example, Figure 5.22, created by CASRE, depicts the raw number of detected failures per day during this period.

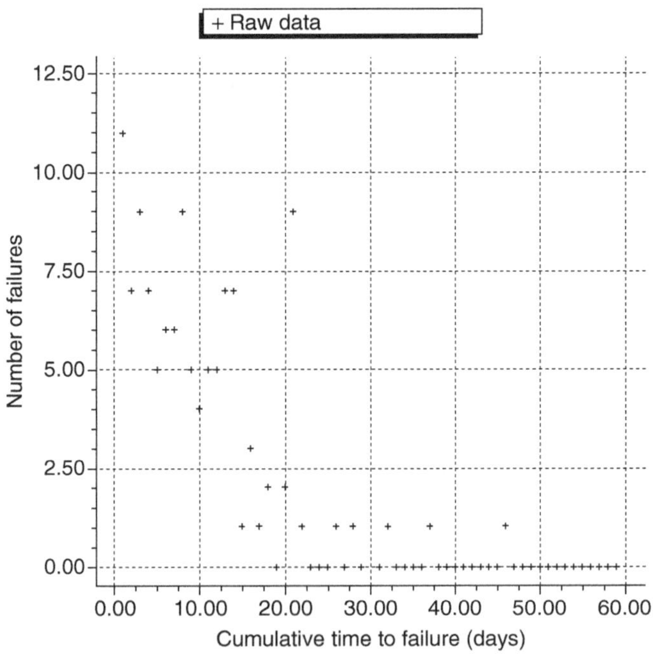

Figure 5.22 Results of 60-day system evaluation.

CASRE also provides a collection of reliability models to capture the dynamics of the failure data. Such results are displayed graphically, in terms of failure counts per test interval, times between successive failures and the cumulative number of errors discovered.

For example, Figure 5.23 depicts failure intensity (number of failures per day) distributed over time, using a Non-Homogeneous Poisson Process (NHPP) model. One reason for the attractiveness of the NHPP model is its assumption that the cumulative number of failures detected at any time follows a Poisson distribution. This distribution is a special case of binomial distribution which (1) takes into account rare events as well as assumes that (2) all events are independent and (3) the average rate of failures does not change over the period of interest.

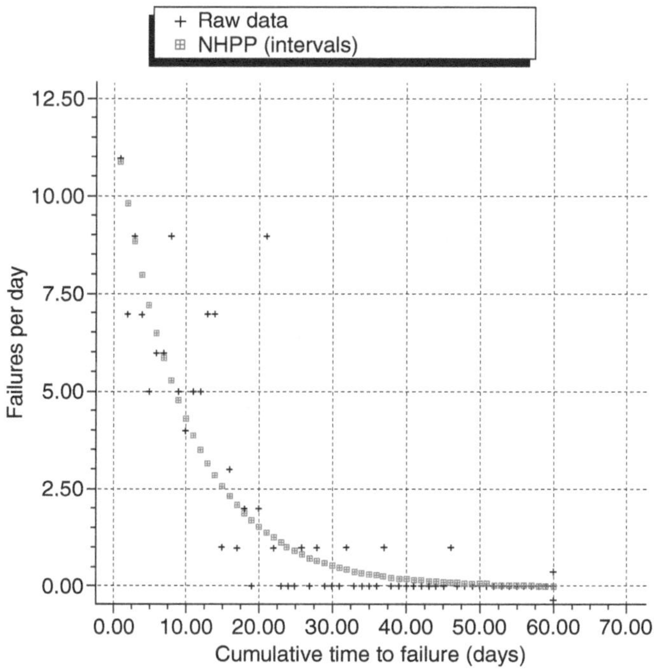

Figure 5.23 Failure intensity plot using NHPP model.

As can be seen in Figure 5.24, the yearly reliability of the system has improved markedly over the 60-day testing and fixing period, but at a yearly reliability level of approximately 0.75 it is just not sufficient for most applications.

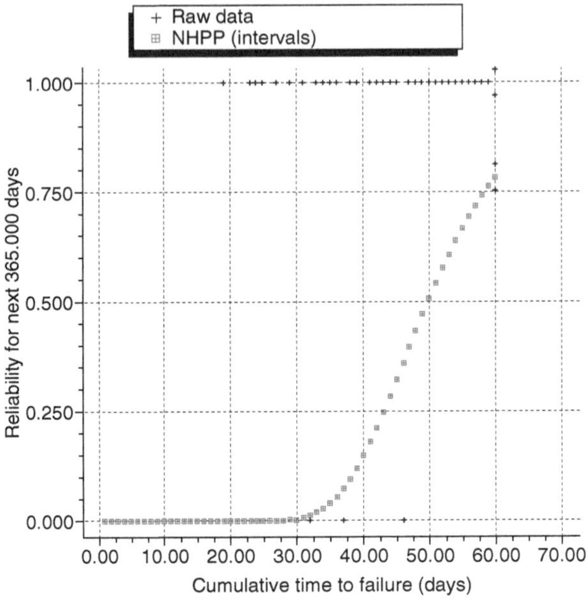

Figure 5.24 System yearly reliability plot using NHPP model.

We can use CASRE to predict system reliability if we continue our testing program and assuming a continuous improvement in the reliability of the systems. For example, if the system testing was to be extended by 30 days, then the yearly reliability prediction of the NHPP model increases to close to 1 (see Figure 5.25).

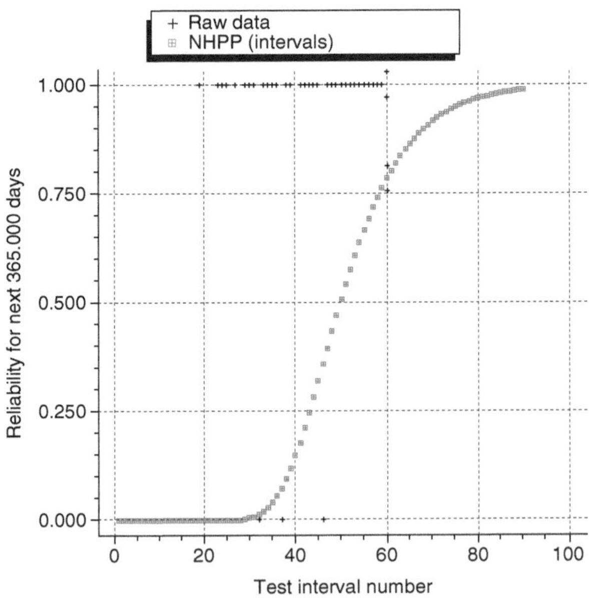

Figure 5.25 System yearly reliability prediction after further testing.

Note of Caution VVT practitioners should view system testing using reliability estimation methods with some skepticism or, more appropriately, follow the dictum: "Suspect the numbers, accept the trend." Reliability estimation is always based on a system's operational profiles, that is, the set of input events that the system will receive during operation, along with the expected behavior of the system. However, reliability estimations are problematic due to the following reasons:

- *Statistical Limitations.* When a population is too large for exhaustive study (e.g., computer-based systems), a statistically correct sample must be drawn as a basis for inferences about the population. In classical statistics we can define the population in quite clear terms. This is not the case in statistical testing where we are unable to specify the reciprocal component, namely the entire system behavior space. Similarly, in classical statistics, we can define a statistical sample and make valid statistical generalizations, whereas in statistical testing we can define a set of test cases but we are unable to make complete and accurate inferences about the behavior of the system resulting from executing each test case.
- *Rare Inputs.* A "rare" input value is one that is unlikely to be selected according to the system's operational profile. Therefore, we must consider rarity as a probability issue, not an abnormality issue. The problem is that, by and large, system testing employs legal inputs applied in a well-organized order. However, during normal operation of a complex system over a long time, the system, being itself in different internal states, is exposed to a deluge of anomalous inputs and their combinations. As a result, systems, and especially digital systems, respond to rare inputs in a quite unpredictable manner. Since any testing cannot be exhaustive, it stands to reason that the domain of system failures is not known a priori and, therefore, true computation of system reliability is, in fact, impossible.
- *Unanticipated Events.* Digital systems are highly susceptible to unanticipated events and drastic changes in a system's state. When the system design or implementation is not robust, such events can corrupt internal data and program states during execution, rendering the behavior of the system unpredictable. It is a fair bet that statistical testing is a limited tool in predicting such events.
- *Cost and Efforts.* Statistical testing is a labor-intensive work requiring sizable resources. In particular, the effort of gathering historical failure data is considerable. In addition, it may be impossible to generate enough credible failure data to draw statistically valid conclusions, especially during the development of new systems.
- *Validity of Statistical Models.* Complex systems, especially digital ones, tend to undergo periodical revisions throughout their lifecycle. Immediately after such revisions systems tend to exhibit a large increase in the number of failures that are then flushed out over time.

Invariably, reliability models are most effective during a single revision period rather than the whole lifecycle. This phenomenon is depicted in Figure 5.26.

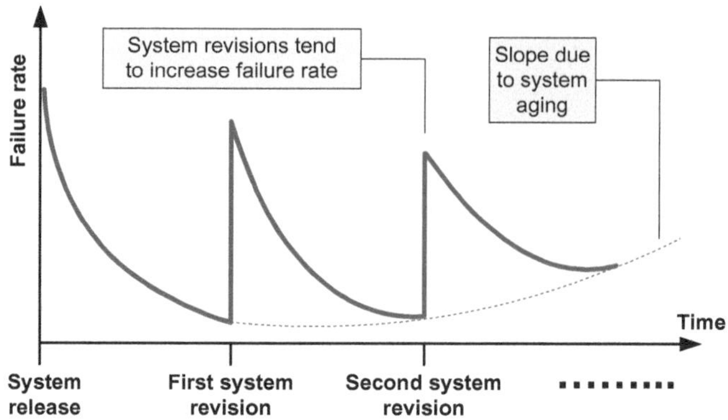

Figure 5.26 Lifecycle failure rate during multiple system revisions.

Further Literature

- Fenton and Pfleeger (1998)
- Kan (2002)
- MIL-HDBK-781A (1996)
- O'Connor (2002)
- Wasserman (2002)

5.5.4 Search-Based Testing

Purpose Search-based testing reformulates testing tasks into optimization problems. The objective becomes that of discovering an optimum set of meaningful test cases from among a huge number of possible test cases, one that is sufficiently good according to an appropriate fitness metric. This reformulation enables automation of previously manually intensive tasks. It solves problems that are intractable by other methods, and often leads to an innovative and insightful view of the system under test.

Genetic algorithms[53] (GAs) are general-purpose, computer-based search procedures patterned after the natural selection mechanisms of biological organisms that have adapted and flourished in a changing and highly competitive earth environment for millions of years. Genetic algorithms have been successfully applied to problems in a variety of engineering and other disci-

[53] In this section, we use genetic algorithms, which are just one of a number of metaheuristic search techniques such as gradient ascent or descent, simulated annealing, taboo search, particle swarm intelligence, ant colony optimization and greedy algorithm.

plines, and their popularity continues to increase because of their effectiveness, applicability and relative ease of use. Examples of test applications using genetic algorithms are depicted in Table 5.8.

TABLE 5.8 Examples of Test Applications Using Genetic Algorithm

GA Test Type	Test Search Objectives
Structural testing	Find test cases which will maximize the white-box coverage of software program constructs.
Functional testing	Find test cases which will seek system operation and logical errors.
Temporal testing	Find test cases which will search for either the longest or shortest system execution time.
Safety testing	Find test cases which will seek violation of system safety constraints.
Robustness testing	Find test cases which will stress the system and overcome fault-tolerance mechanisms.
Mutation testing	Find test cases which will try to detect errors in a mutated system (i.e., a system within which errors have intentionally been injected).
Stress testing	Find test cases which will seek to stress a system beyond its capabilities.

Rationale In certain circumstances, search-based testing can increase the effectiveness and efficiency of the testing process. This type of testing will automatically generate test cases which will evolve and improve over successive iterations of the algorithm. Evolutionary testing is characterized by the use of search techniques for test case generation. The test aim is transformed into an optimization problem. The input domain of the system under test forms the search space in which one searches for test data that fulfill the respective test aim. Due to the naturally nonlinear behavior of computer-based systems, the conversion of test problems into optimization tasks mostly results in complex, discontinuous and nonlinear search spaces. Therefore, various search methods are employed, for example, evolutionary algorithms and simulated annealing. In most cases, evolutionary algorithms are used to generate test data because their robustness and suitability for the solution of different test activities have already been proven in an industrial setting.

In order to transform a test aim into an optimization task, a numeric representation of the test aim is necessary from which a suitable fitness function for the evaluation of the generated test data can be derived. Depending on the specific pursued test aim, different fitness functions may be adopted in order to evaluate the test data.

The advantages of this approach are:

- *Automatability of Test Case Design.* Assuming a test oracle is available (i.e., we know what are the expected results based on a fitness function), a genetic algorithm apparatus can generate automatically test cases. This is not possible with classic test case design procedures.

- *Large Amount of Test Data.* Due to this automatability characteristic, testing can be performed with a large amount of error-sensitive test data and be completed in a relatively short time. This strengthens the confidence in the correct functioning of the system under test.
- *Estimating Test Duration.* It is possible to calculate an optimal time for the completion of the test by analyzing the test's convergence status. For example, if the test has converged, the probability of ascertaining further error-sensitive test cases with the same test run is low.
- *Human Factor Advantages.* Evolutionary tests can be used to process complex test problems that could not be covered by a test engineer with sufficient quality and justifiable time expenditure. In addition, errors, which test engineers can make during test case design, are avoided by evolutionary tests.
- *Numerous Test Aims.* Evolutionary tests are suitable for a great number of test aims. Evolutionary tests can, for instance, be deployed for structural testing, functional testing, temporal testing, safety testing, robustness testing, mutation testing, stress testing and so on.

At the same time, not all the testing needs can be transformed into an optimization problem easily. In many cases, particularly in that of functional tests, the definition of a suitable fitness function for the evaluation of the test data generated can be difficult. In addition, a search-based approach for system testing is relatively new and not many VVT practitioners are familiar with these methods.

Method: Setup Process In comparison with common test activities, evolutionary testing results in an extension of the setup process, particularly within the context of the following activities:

- *Classification of Test Problem.* In order to classify the test problem, the VVT engineer must investigate how the system under test is defined, which interfaces are available or must be installed and how the test problem can be formalized for the item being tested. This includes establishing input data, defining additional parameters if needed, as well as establishing the means for monitoring the output data. In the final stages the system must be encapsulated in such a way that it is controlled entirely by the input data generated and, depending on the dynamics of execution, returns the values required for the fitness calculation.
- *Definition of Fitness Function.* The definition of a particular and unique fitness function for evaluating the test data is always dependent on the test problem addressed. For example, if a temporal or real-time behavior of a system is being tested, the fitness evaluation will typically

be based on the execution times measured during the performance of the test. For safety tests, the fitness values are derived from pre- and post conditions of systems or components; for robustness tests, the number of controlled errors can form the starting point for the fitness evaluation; for functional testing, constraints on the output values will dictate the nature of the fitness evaluation; and for structural testing, the coverage values achieved by a test datum are a suitable basis for the fitness evaluation.

- *Analysis and Visualization of System Behavior.* In order to investigate the behavior and the properties of the system under test, its search space structures must be analyzed and visualized. By and large, systems under test have a large number of input parameters that may result in a complex output behavior. Therefore, producing a comprehensible textual or graphical representation of the fitness landscape is not trivial. If the fitness landscape is limited to one- or two-dimensional domain, then it is possible to illustrate it directly with standard diagrams. If, however, the fitness landscape is composed of more dimensions, only two or at most three dimensions of the fitness search space are suitable for human representation at any given time.
- *Selection of Optimization Process.* The appropriate optimization procedure to be applied and a suitable system configuration are naturally derived from the analysis of the system behavior. Usually, the search spaces of real-life systems are quite complex. Therefore, evolutionary computation is used as a preferred optimization technique. The suitability of such algorithms for optimizing the test process is mainly based on their ability to produce effective solutions and also to do so for complex and little understood search spaces with many dimensions. The reader should also note that the dimensions and the complexity of the search space are directly related to the amount of input parameters of the system under test as well as the complexity and search space discontinuities inherent in the system under test.
- *Configuration of Optimization Procedure.* Next, the specific parameters of the optimization procedure must be determined. The efficiency and effectiveness of evolutionary tests can be increased considerably by an appropriate configuration of the optimization procedure. For evolutionary algorithms the test case population size, the parents selection procedure to be used, the operators for elitism,[54] recombination and mutation[55] and the survival strategy, for example, must be established.

[54] Elitism is a very successful variant of the general process of constructing a new test population that allows some of the better tests from the current generation to carry over to the next, unaltered.

[55] Recombination (also known as crossover) and mutation serve to evolve the population in one generation (i.e., parents) of tests to the next generation (i.e., offspring) of tests.

Method: Testing Process The overall genetic evolutionary and testing cycles are described in Figure 5.27. It is based on a testing cycle (described in the left circle) combined with a genetic algorithm cycle (described in the right circle) working in tandem. The individual activities from which this testing process is composed are described below.

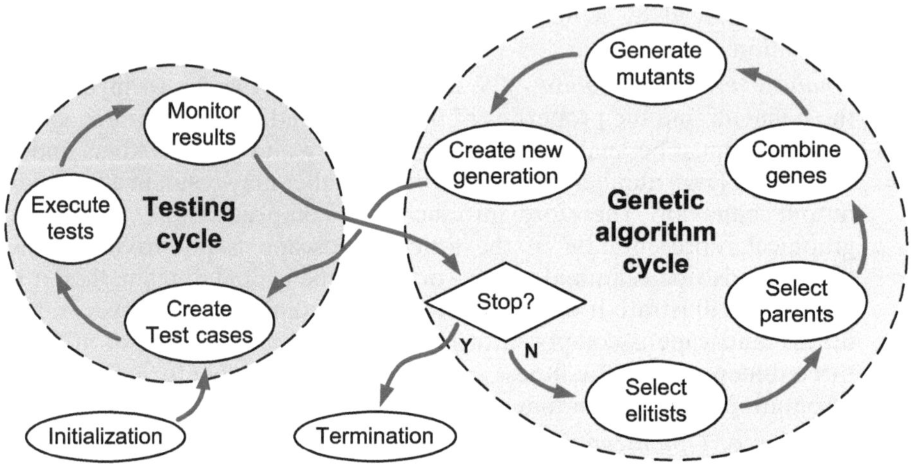

Figure 5.27 Evolutionary and testing cycles.

- *Initialization.* The testing starts at the initialization process. The initial set of test data is usually generated at random. In principle, if test data has been obtained by a previous systematic test, this could also be inserted into the initial set of test data. The evolutionary test could thus benefit from the test engineer's knowledge of the system under test.
- *Creating Test Cases.* At this stage, the input test data is examined to ensure compatibility with the interface definition of the system under test. Thereafter, the actual test cases are created, using first the initial data and then, on successive iterations, offspring test data that has been generated within the genetic algorithm cycle.
- *Executing Tests.* At this stage, the system under test is executed using the current test cases.
- *Monitoring Tests.* At this stage, the results of the test execution are evaluated with respect to the selected test aim. The fitness values for the individual test data are calculated using these monitored results. The aim here is to establish whether an "interesting" datum was encountered (e.g., error, minimum, maximum).
- *Evaluating for Stopping Criteria.* At this stage, the stopping criterion is evaluated against the test results, and a decision is made to either continue and go to the next stage or terminate the testing process. Such termination may be a result of actually achieving the predefined

stopping criteria or a termination request, issued by the person conducting the test.

- *Selecting Elitists.* At this stage, the first genetic algorithm operation (i.e., elitism) is performed. One or a few of the fittest members of the current generation pool are transferred into the pool of the new generation in order to ensure survival of the fittest individuals.
- *Selecting Parents.* At this stage, the next genetic algorithm operation of parents selection is performed. Parents are selected at random with selection chances biased on the fitness measure.
- *Combining Genes.* At this stage, the next genetic algorithm operation of gene combination is performed. Here genetic material from two parents is combined in order to produce the next generation offspring.
- *Generating Mutants.* At this stage, the next genetic algorithm operation of mutant generation is performed. This operation is carried out on a very small portion of the test case population in order to introduce randomness into the population therefore bringing diversity into the test cases.
- *Creating New Generation.* At this stage, a new generation pool is created. The surviving individuals (i.e., test points) are selected on the basis of fitness measure and according to the predefined survival procedure. These individuals constitute the next generation of the test case population.

The above process repeats itself until the test objective is fulfilled or some appropriate stopping condition is reached.

Example: Autonomous Parking System We illustrate the concept of functional search-based testing using an Autonomous Parking System (APS) for passenger cars, a real-life industrial problem, which has been described in the referenced literature.[56] Typical APS sequence evolves as follows: The passenger car drives slowly along a potential parking space, the system measures parking space size, using appropriate sensors. On finding a potentially satisfactory location, it informs the driver of a suitable parking space. If the driver grants an autoparking authorization, the system determines the position of the car with respect to the parking space, plans the trajectory path for the parking maneuver and autonomously drives the car into the parking space.

The aim of testing is to detect errors in the functional behavior of the system. In particular, we are interested in finding out (1) whether or not there exists an initial parking scenario leading to a collision and (2) whether or not there exist parking scenarios leading to an "impossible" attempt to park. In this context, we use the term "scenario" to mean the parameters of the parking space, as well as the initial position of the car relative to this space (see Figure 5.28).

[56]This example was inspired by Wegener and Bühler (2004).

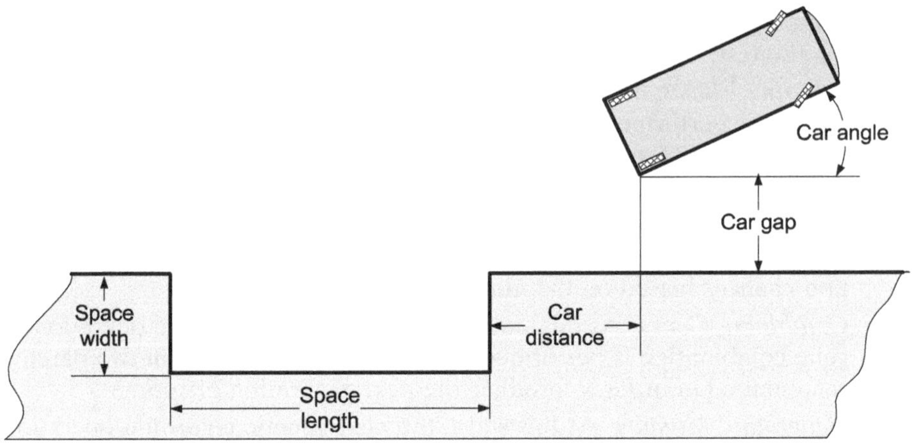

Figure 5.28 Starting parameters for Autonomous parking system.

In this example we can define the smallest (or a negative) distance between the car and any collision surface as the objective value, execute a search-based testing and seek to find a scenario leading to a parking maneuver which generates a negative objective value.

Ideally complex and critical systems of this nature should undergo exhaustive testing of all possible scenarios, but this is not practical. Assuming the system behaves in a linear fashion within a short interval of the input parameters, testing it within 3 centimeters of distance or 3 degrees of car angle could be considered acceptable (see Table 5.9). However, under these assumptions, the number of exhaustive combinations is over 14 million, clearly an unreasonable number of individual tests.

TABLE 5.9 Permutations of Scenarios: Autonomous Parking System

Input Parameters	Units	Minimum	Maximum	Steps	Combinations
Space width	cm	140	200	3	21
Space length	cm	480	600	3	41
Car distance	cm	–20	100	3	41
Car gap	cm	10	80	3	25
Car angle	deg	–15	30	3	16
Total number of combinations					14,120,400

The very real risk here is that the system may not behave in a linear fashion for all scenarios depicted in the table and, therefore, may cause collisions at some obscure combinations of input parameters. The objective of the exercise then is to search for such potential combinations and, if they exist, find them and eliminate any problem within a reasonable time.

Investment in search-based testing seems justified as such systems are critical, safety related and may be installed in hundreds of thousands of vehicles. Any residual defect could result in a high number of failures in the field, accompanied by lawsuits and the necessity of expensive recalls.

Tool Support Three components are required for the technical realization of evolutionary testing: (1) for test data generation, a toolbox is required which will provide efficient evolutionary operators, (2) for the proper execution of the system under test, a test driver for implementing the test sequence has to be implemented and (3) for evaluating the fitness of individual test results, a process monitor is required which is appropriate for the specific test goal.

- *Test Data Generation.* In order to generate appropriate test data, a toolbox of evolutionary algorithms is required. This toolbox could be implemented as a test data generator that produces the appropriate parameters having the required ranges and types of data. The test data generator will then automatically ensure that these constraints are being met when generating individual test data. On the basis of the constraint information, the test data generator will generate the initial test case population with which the test driver will execute the system under test.
- *Test Driver.* The test driver will transform the individual test points into test cases for the system under test. In the simplest case, the variable values of the individual data points may be assigned to the input parameters of the system under test on a one-to-one basis. However, if a variable defines a more complex process, such as an event sequence or a time interval between the occurrences of two events, then the test driver will have to transform it into a suitable test sequence. The test driver has to execute the system under test with a corresponding sequence of events or to maintain the given time interval for the generation of events.
- *Process Monitor.* Process monitoring is a critical element for achieving the testing goals. It determines how to transform the test goal into an optimization task and how to calculate the fitness values for the test data generated. Process monitoring is unique for each test goal and cannot be created in a general manner.

Further Literature

- Bin et al. (2007)
- Karr and Freeman (1998)
- Lammermann et al. (2004)
- Miettinen et al. (1999)
- Wegener and Grochtmann (1998)
- Wegener et al. (2001)
- Wegener and Bühler (2004)

5.5.5 Mutation Testing

Purpose Mutation testing of software is attributed to Richard Lipton in 1971, but the general idea was implemented in engineered systems much earlier, and it is employed in conjunction with traditional testing techniques. The purpose of mutation tests, sometimes called error seeding or fault seeding testing, is to measure the adequacy of test cases and use this measure to estimate the amount of remaining defects in the system as well as get a general notion of the reliability of the system under test.

In mutation testing, defects, usually one at a time, are deliberately introduced into the system design or implementation. This is done either in hardware by disconnecting a cable, removing a component from a socket, or grounding a certain signal or in software by modifying a program either manually or by using automated means. Each temporarily modified system is called a mutant and, of course, many versions of mutants can be created (see, e.g., Figure 5.29).

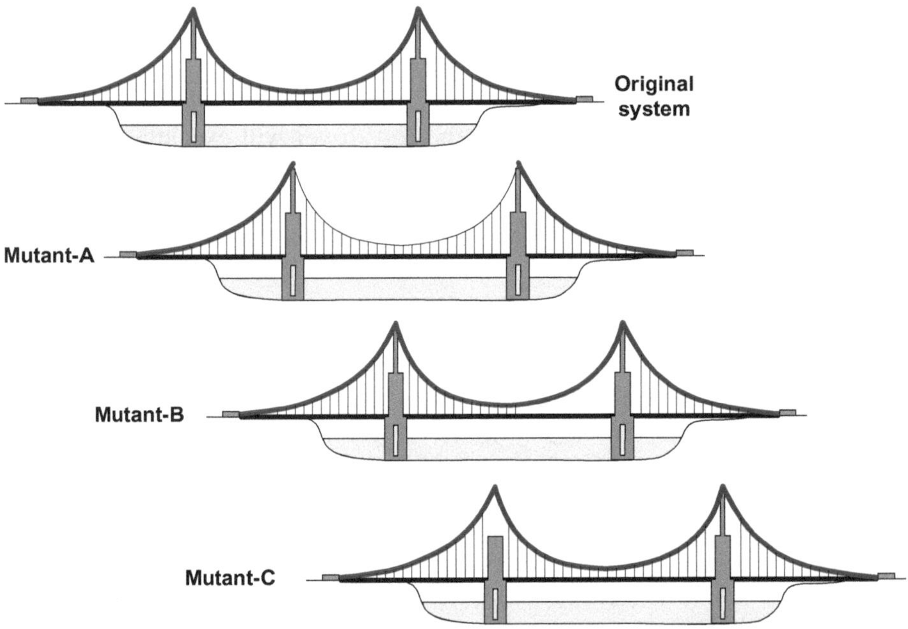

Figure 5.29 Bridge system and three-bridge design mutants.

The test cases are applied to the original system as well as to each version of the mutant system with the expectation that the mutant system will fail but with the real goal of causing the mutant program to succeed, thus exposing weaknesses in the test case suite.

Fault-based testing is widely used in semiconductor manufacturing using models of typical manufacturing faults (e.g., gates stuck-at-one or stuck-at-zero). Several variants of fault-based testing play a role in research of software testing, and some advanced organizations do use this method in critical or safety-related software systems. However, fault-based testing for design errors is more challenging and, in general, is not widely used in industry. The VVT practitioner should be aware that mutation testing rests on some troubling assumptions about seeded faults, which may not be statistically representative of real faults. Nevertheless, a model of typical or important faults is definitely valuable information for designing and assessing test suites.

Rationale The rationale for performing mutant testing is based on the "competent programmer hypothesis" which states, in systems terminology, that engineers are generally very competent and do not design or implement grossly faulty systems. Therefore, an engineer may create a faulty system, but that will be very close to a correct one. Furthermore, an incorrect system (i.e., a mutant) can be created from a nearly correct system by making some minor changes to it.

These facts allow us to evaluate the adequacy of test cases. A test case is adequate if it is able to detect faults in a system containing defects. Therefore, a collection of test cases should prove to be adequate by distinguishing between mutants and the original system. More specifically, adequate collection of test cases will show that each mutant system generates a different output than does the original system. (This demonstration of difference is termed "killing a mutant.") Conversely, if the original system and some mutant systems generate the same output, then the test cases are considered inadequate. The reader should note that it is entirely possible to create mutants which are functionally equivalent to the original system. Obviously, the test suite will not succeed in killing such mutants (see Figure 5.30).

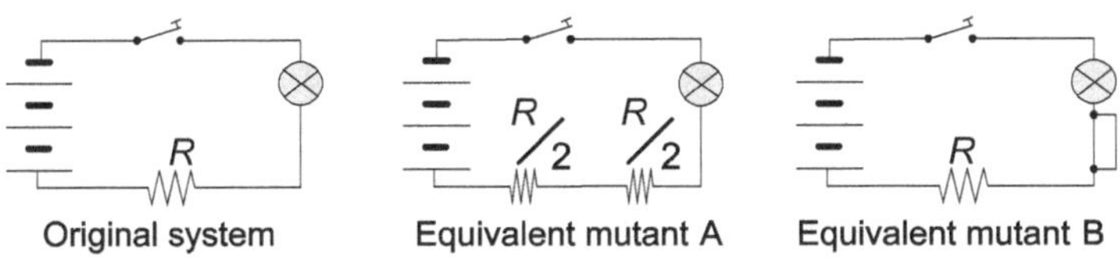

Figure 5.30 Example of original system and functionally equivalent mutants.

If some of the mutants are not killed under the current set of system tests, then we can make a rough calculation in order to estimate the number of remaining faults in the system. Although this approach is simple to implement and useful, the main drawback of mutation testing is the difficulty of establishing that the seeded faults really represent the actual ones.

Method Under mutant-based testing we would like to judge the effectiveness of a test suite in finding real faults, by measuring how well it kills these mutant systems. This approach is valid to the extent that the seeded faults are representative of real system defects. The algorithm of mutation testing follows these steps:

- *Step 1: Generate System Test Cases.* This step entails the creation of a set of test cases needed to verify that the system under test meets its requirements.
- *Step 2: Perform System Testing.* This step entails conducting the system tests. If the output of the system under test is incorrect, then either the system or the test case suite contains one or more defects. Corrections must be undertaken and the system then must be retested.
- *Step 3: Construct Hardware or Software Mutants.* This step entails planning and creating mutant systems either manually or, in case of software systems, by means of one of several commercially available tools.
- *Step 4: Test Mutant Systems.* This step entails executing the set of test cases against each mutant system. If the output of the mutant differs from the output of the original system, the mutant is considered killed. Two kinds of mutants may survive: either not killable or killable.

  As mentioned, nonkillable mutants are ones that are functionally equivalent to the original system. For example, we can create a mutant system by grounding a spare or unused wire. Another example is setting a variable in a software program to an incorrect value. However, as it happens, this same variable is initialized to the desired value by the program prior to its use. In both cases, testing such a mutant system will not identify any problem.

  Killable mutants are ones that are functionally different from the original system. However, if the existing set of test cases is unable to kill individual mutants, then additional test cases must be created to do the job.
- *Step 5: Compute System Fault Statistics.* This step entails the computation of the system's fault statistics. If all mutated systems have been detected, then we may guess that test suit is comprehensible and the system under test is fault free. However, as mentioned, this hypothesis is subject to certain limiting assumptions and in particular it depends on the error-revealing capability of the test set.

Estimating Remaining Faults We can empirically estimate the number of faults remaining after mutant testing by using a method based on statistical maximum-likelihood approximation. This may be done by assuming that the ratio of detected seeded faults to the total seeded faults is the same as the ratio of the detected nonseeded faults to total nonseeded faults. In other words, seeded and nonseeded faults are equally easy or hard to detect, after some period of testing. This may be expressed as:

$$\frac{\text{Detected seeded faults}\,(s)}{\text{Total seeded faults}\,(S)} = \frac{\text{Detected nonseeded faults}\,(x)}{\text{Total nonseeded faults}\,(X)}$$

Therefore, the total number of nonseeded faults is approximately

$$X \cong x\frac{S}{s}$$

Therefore, the remaining faults $\bar{X}$ in the system could be calculated as

$$\bar{X} = X - x = x\frac{S}{s} - x = x\left(\frac{S}{s} - 1\right)$$

For example, a system is seeded with $S = 50$ faults (i.e., 50 system mutants are generated, each with a single defect). The test team performs system testing by executing the test suite against each mutant system and find $s = 40$ seeded faults and $x = 8$ nonseeded (indigenous) faults. Therefore, it is likely that the remaining number of faults is

$$\bar{X} = x\left(\frac{S}{s} - 1\right) = 8\left(\frac{50}{40} - 1\right) = 2$$

Estimating Confidence Level We can also estimate the confidence or the likelihood that the system is fault-free. Suppose we seed a system with S faults and claim that it still has X nonseeded (indigenous) faults. We test the system until we find all S of the seeded faults. If x is the actual number of real faults discovered during testing, then the confidence can be calculated as follows:

$$C = \begin{cases} 1 & \text{if } x > X \\ \dfrac{S}{S - X + 1} & \text{if } x \le X \end{cases}$$

For example, suppose we claim that our system is fault free, that is, to the best of our knowledge, there are no hidden faults and therefore $X = 0$. Suppose we again seed our system with a total of $S = 50$ faults. Thereafter, we find all of these 50 faults without uncovering any indigenous faults. We than proceed to calculate the confidence level that indeed the system is fault free:

$$C \cong \frac{50}{50 - 0 + 1} \cong 98\%$$

Obviously, the level of confidence depends on the number of tested mutant systems. Suppose, in the above example, we generate only $S = 5$ mutant systems. Then our confidence in the assertion that the system is fault free becomes

$$C \cong \frac{5}{5 - 0 + 1} \cong 83\%$$

Further Literature

- Benso and Prinetto (2003)
- Burnstein (2003)
- Voas and McGraw (1998)

5.6 BLACK BOX—ENVIRONMENT TESTING

5.6.1 Environmental Stress Screening (ESS) Testing

Purpose The purpose of Environmental Stress Screening (ESS) is to precipitate and eliminate latent defects in systems which are introduced either during the design of the system or during the manufacturing, assembling and packaging processes. ESS tests, also known as "burn-in," attempt to catch "infant mortality" failures. Such failures rarely emerge during normal testing or visual inspection.

The topic of ESS is highly specialized and we will describe it in a very superficial manner. Interested readers are directed to the references identified in this section for more information. They describe the historical evolution of ESS and its basic concepts as well as statistical and physical quantification of ESS phenomena. By and large, the references concentrate mainly on environmental stress screening of electronic equipment, which typically includes ESS conditions, durations of exposure, procedures, equipment operation, actions taken upon detection of defects and screening documentation.

Rationale The rationale for conducting environmental stress screening is for effectively disclosing manufacturing defects in systems, mainly electronic equipment caused by poor workmanship and faulty or marginal parts. ESS can also identify design problems if the design is inherently fragile or if qualification and reliability growth tests were not effective. The objectives of ESS testing is, therefore, to improve the overall system's economy through fault detection and correction during the product development and manufacturing cycle, to reduce the number of system failures during the warranty period and, in general, to improve product quality.

Undertaking ESS is most appropriate for complex systems that have limitations such as size, weight, and power consumption which are used in critical and safety-related applications. Such system failure could have serious consequences (e.g., avionics, space, medical equipment).

Although the most common elements practiced within ESS are temperature cycling and random vibration, a reasonable ESS program must be dynamic and also be tailored to the particular characteristics of the equipment being tested. In addition, ESS testing should be performed during both the system development phase as well as the manufacturing phase.

Method The environmental stress screening method is based on the technique of applying various types of stresses on systems and components within a controlled manner. The commonly applied stresses are temperature, vibration, humidity and electrical stimuli, and the levels of applied stresses are much greater than the stresses that the product is likely to encounter during normal operation. This is done in order to simulate the expected overall lifecycle stresses within an accelerated manner.

ESS has been proven to find latent defects that would very likely precipitate in end-use applications, causing product failures in the field. As a result, the ESS process can effectively intensify product reliability. ESS tests include the following two variants:

- *HALT (Highly Accelerated Life Testing).* HALT is used during the design phase of a system by applying increased stress to a product in steps and fixing faults, if discovered, to improve the design. This process continues beyond the limits of normal shipping, storage and operational conditions normally encountered in the fields until the destruction limits of the material in the product are reached. Such a procedure is meant to find weak design spots within the system and helps to define the operating limits of a system. It normally consists of the following steps:
    a. Applying environmental stress in steps until the system fails.
    b. Making a temporary change to fix the failure.
    c. Stepping stress further until the system fails again and repeating the stress–fail–fix process.
    d. Finding the fundamental operational and destruct limits of the system beyond which fixing the system is not economic.
- *HASS (Highly Accelerated Stress Screening).* HASS is used after the stresses versus destruction limits from the HALT process are already known. It is performed on manufactured systems in order to identify weak individual products and it helps to verify product performance during the estimated lifetime of the product. HASS is a nondestructive test designed to apply high levels of stress on a system under test in order to reduce test time with the intention of confirming that all reliability improvements made in HALT are maintained. More specifically, it ensures that no defects are introduced due to variations in the manufacturing process and vendor parts. It normally consists of the following steps:
    a. Stress predetermined percentage of the products in order to turn latent defects into exposed defects.
    b. Detect manufacturing defects and perform failure analysis.
    c. Perform corrective actions. This may include fixing failed systems and repeating the stress testing or redesigning appropriate portions of the failed system.

Further Literature

- Chan (2001)
- Kececioglu and Sun (2003)
- MIL-HDBK-2164A (1996)

5.6.2 EMI/EMC Testing

Purpose Electromagnetic Compatibility (EMC) deals with unintentional generation, propagation and reception of electromagnetic energy with specific attention to Electromagnetic Interference (EMI). Electromagnetic interference covers individual electromagnetic pulses, as well as frequencies of tens of hertz to GHz range.

The purpose of EMI/EMC testing is to verify the correct operation of a system in an electromagnetic environment where different equipment may emit or be susceptible to electromagnetic interference effects. EMI/EMC testing must verify the system's susceptibility to both continuous and transient interference.

Continuous interference arises when a source of electromagnetic noise, either within or outside the system, regularly emits a constant range of frequencies. Typical man-made emitters of radio frequencies may be mobile telephone, television and radio receivers as well as industrial, scientific and medical equipment. There are several natural sources of electromagnetic interference, for example, cyclical solar activity and various unstable isotopes that emit interfering frequencies during their natural decay process.

Transient interferences are typically a result of electromagnetic pulses where the source emits a short-duration pulse of energy. Typically, such interference is generated during the operation of electromechanical systems like electric motors as well as bursts of electrical current surge (e.g., switching action of electrical circuitry, power line pulses). The most important natural source of electromagnetic pulse interference is lightning.

Rationale EMI/EMC testing is often carried out when a system is composed of numerous electromagnetic emitting subsystems with potential electromagnetic interference problems. The rationale for performing EMI/EMC testing is twofold: (1) to verify whether the system under test operates, with adequate safety margins and without malfunction or degradation of performance, in the intended electromagnetic environment generated by the system itself and any other system likely to be in its vicinity, and (2) to verify that the system does not emit to the environment electromagnetic radiation above the required threshold, meeting appropriate standards and regulations.

Method EMI/EMC testing verifies that the electromagnetic interference (emission and susceptibility) characteristics of an electronic, electrical and electromechanical system meets its specifications, when it functions in its natural operational and nonoperational environment.

Various U.S., European and other nations military as well as civilian standards establish general testing techniques for use in the measurement and determination of the electromagnetic emission and susceptibility characteristics of such systems. Such test methods are usually divided into the following categories: (1) conducted emissions, (2) radiated emissions, (3) conducted susceptibility and (4) radiated susceptibility.

For example, MIL-STD-461E defines a total of 17 different EMI/EMC areas of testing. Depending on the nature of a given system, appropriate requirements should be selected in order to meet specific electromagnetic compatibilities and resistance to interference. Table 5.10, which contains information from the above military standard, describes a set of verification requirements for the control of the electromagnetic emission and susceptibility characteristics of electronic, electrical and electromechanical systems.

TABLE 5.10 MIL-STD-461E: Emission and Susceptibility Requirements

Requirement	Type of Test	Description	Frequency Range
CE101	Conducted emissions	Power leads	30 Hz–10 kHz
CE102	Conducted emissions	Power leads	10 kHz–10 MHz
CE106	Conducted emissions	Antenna terminal	10 kHz–40 GHz
CS101	Conducted susceptibility	Power leads	30 Hz–150 kHz
CS103	Conducted susceptibility	Antenna port	Intermodulation, 15 kHz–10 GHz
CS104	Conducted susceptibility	Antenna port	Signal rejection, 30 Hz–20 GHz
CS105	Conducted susceptibility	Antenna port	Cross-modulation, 30 Hz–20 GHz
CS109	Conducted susceptibility	Structure current	60 Hz–100 kHz
CS114	Conducted susceptibility	Bulk cable injection	10 kHz–400 MHz
CS115	Conducted susceptibility	Ground–bulk cable injection	Impulse excitation
CS116	Conducted susceptibility	Power & I/O	Damped sinusoid transients, 10 kHz–100 MHz
RE101	Radiated emissions	Magnetic field	30 Hz–100 kHz
RE102	Radiated emissions	Electric field	10 kHz–18 GHz
RE103	Radiated emissions	Antenna spurious & harmonic outputs	10 kHz–40 GHz
RS101	Radiated susceptibility	Magnetic field	30 Hz–100 kHz
RS103	Radiated susceptibility	Electric field	10 kHz–40 GHz
RS105	Radiated susceptibility	Transient electromagnetic field	Pulsed EMI–EMP

This standard also establishes general techniques for use in the measurement and determination of the electromagnetic emission susceptibility characteristic of equipment and systems. These test procedures, test facilities and equipment requirements could be used to determine compliance with the applicable emission and susceptibility requirements of the standard.

By and large, EMI/EMC testing is performed within a shielded enclosure covered internally by a radio-frequency absorbing material in order to reduce the reflected electromagnetic energy. Commonly, each subsystem must pass an individual EMI/ECM test prior to system level tests. Also, all the test and accessory equipment used in conjunction with EMI/EMC measurement must not be affected by electromagnetic noise, nor be degraded during the testing process.

Further Literature

- Mardiguian (1999)
- MIL-STD-461E (1999)
- Montrose and Nakauchi (2004)
- Paul (2006)

5.6.3 Destructive Testing

Purpose Destructive testing is a generic term for all tests, which permanently impair the subsequent usefulness of a component, subsystem or system. We hasten to note that, in the context of this book, we refer to destructive testing of whole engineered systems rather than material or component destruction tests (e.g., a slab of cement, a steel beam). Such system testing combines experimental procedures with numerical simulation typically undertaken by the transportation, aerospace and defense industries. Since the cost of conducting physical destructive testing is quite exorbitant, several analysis and mathematical modeling and simulation tools have been developed in order to compute the behavior of materials and structures under dynamic loading conditions.

The most prevalent and well-known destructive testing is carried out in the automotive industry where passenger safety and care for the environment have become important buzz words in the auto world and all world-class car manufacturers have begun to apply the stringent safety norms in the manufacturing of their vehicles.

In the passenger automobile industry, virtual (i.e., simulated) crash testing is carried out from the earliest stage of developing a new model of vehicle and continues into the systems integration phase. Then, physical tests are undertaken in parallel with simulated destructive tests. By law, passenger cars in most regions of the world must undergo formal certification that involves destructive testing. In addition, automobile manufacturers concerned with vehicle safety rating (i.e., in terms of vehicle safety classification above and beyond the minimums required by law) design their vehicles to withstand such tests in order to enhance the public image of their companies and increase sales as well as avoiding potential lawsuits.

Motorcycles are also crash tested in order to evaluate their safety design parameters, but this type of activity is done rather sparsely as public concerns about motorcycle safety is apparently relatively low. In addition, various road elements like precast concrete barriers or box-beam roadside barriers are subjected to destructive tests.

Destructive testing is not confined to the automobile industry. Several train crash tests have been conducted to understand the resilience of locomotives under extreme impact conditions, as well as to verify the safety sealing mechanisms of nuclear fuel shipping containers. Only a few, fully fledged destructive tests are conducted in the aircraft industry. For example, one or more bird strike tests are conducted on every new type of jet engine. The term bird strike is used in aviation to identify a collision between a bird and an aircraft. It is a common threat to aircraft safety and has caused numerous fatal accidents. Bird strikes happen most often during takeoff or landing or during low-altitude flights. The point of impact is usually any forward-facing edge of the aircraft such as a wing leading edge, nosecone and cockpit windscreen or engine inlet. The impact of such collision depends on the point of impact, weight of the bird and the relative speed of the bird and the aircraft. However, most hazardous bird strike accidents occur when the bird hits the windscreen or is ingested into the engines.

In contrast to automobiles, aircrafts hardly evolve to improve passenger safety. Every year there occur several dozen serious aircraft accidents in which several hundred individuals lose their life so the suffering and economic impact is significant. The reason for this limited proactive action on the part of the industry seems to be the industry's success in convincing the public that air transport is safer than passenger car transport by more than an order of magnitude.[57]

Probably the most spectacular aircraft physical destructive test was conducted in December 1984 by the U.S. National Aeronautics and Space Administration (NASA), Dryden Flight Research Center, and the Federal Aviation Administration (FAA) under the Controlled Impact Demonstration (CID)[58] program. A remotely piloted Boeing 720 aircraft with no crew aboard was deliberately crashed into a barrier intended to rupture its fuel tanks. The

[57]According to the Air Transport Association (ATA) the U.S. yearly fatality rates per 100 million passenger miles between 1989 and 2004 was 0.02 for air travel versus 0.87 for passenger car travel. The fallacy of this statistics is obvious if one realizes that 99% of the commercial air transport accidents occur either in the first few minutes after takeoff or the last few minutes before landing (i.e., the distance covered by each flight is virtually irrelevant). Computing travel safety on the basis of the number of passenger trips taken by either aircrafts or passenger automobiles reveals that the safety record of air travel is, in fact, inferior to that of car travel.

[58]For more details, see http://www1.dfrc.nasa.gov/Gallery/Photo/CID.

aircraft contained 76,000lb of antimisting kerosene designed to inhibit fires and prevent flame propagation of the released fuel in case of an aircraft crash. From the standpoint of antimisting kerosene the test was a major failure, as seconds after the picture depicted in Figure 5.31 was taken, a spectacularly large fireball enveloped and burnt the Boeing 720 aircraft.

Figure 5.31 Controlled impact demonstration preimpact skid (Courtesy of NASA).

Rationale The rationale for either physical or simulated destructive tests is that such tests can reveal hidden system defects that may only be detected under uncommon and very rare events in the life of engineered systems.

However, physical destructive tests are inherently very wasteful, as virtually an entire system must be sacrificed for each individual test. For example, in the automobile industry, at least 10 prototypes of cars must usually be destroyed at test facilities to develop the final safe car that can pass formal certification and be put on the road. Vehicle manufacturers often spend $100 to $150 million on developing a new model of car that is both user-friendly and safe for both passengers and the environment. In a dynamic rollover, one of a battery of destructive tests performed on an actual racing track, a car is sent rolling sideways at a speed of over 50km/h to study the impact of the collision on the vehicle and the passengers. There are also elaborate tests to evaluate the passenger comfort from the seats and head rests as well as their safety aspects in the event of a collision.

In addition the system itself or at least in its prototype form must be available for the test. So, it is not possible to conduct such tests during the early requirements and design phases. Another weakness is the presumption that the destroyed system represents all similar such systems (i.e., the fact is that systems evolve over their lifetime. The system that passed an initial destructive test may not pass it in its upgraded form). Finally, physical destructive tests are very special occasions where test engineers establish a large number of test variables. By definition the test cannot be repeated over and over with different parameter values. As a result, this situation limits the test ability to detect potentially fatal flaws.

Conversely, simulated destructive tests do not require the sacrifice of good parts or systems. Furthermore such virtual tests can verify whether a system meets its safety requirements already during the concept and design phases. It also provides a better understanding of safety dynamic and usually decreases the amount of physical destructive tests substantially. Another important advantage of virtual destructive tests is the potential of studying the biomechanical dynamics of humans within such catastrophic situations using simulated models of human beings rather than dummies.

At the same time, virtual (i.e., simulated) destructive tests necessitate the combined operation of several complex software tools. Typically, such tools may include a tool for numerical simulation, a tool for geometry calculation and more tools to simulate humans occupying the system and their related biomechanical behavior within that environment. Another problem to consider is the potential divergence between a simulated and an actual test. In other words, virtual testing may not represent actual real-life system behavior.

Method Due to the specialty of the subject, we will describe destructive testing within the passenger car industry in lieu of general engineered systems. There are a number of automobile crash testing programs around the world dedicated to providing consumers with a source of comparative information in relation to the safety performance of new and used vehicles. Variants of the New Car Assessment Program (NCAP) include USNCAP, EuroNCAP, JapNCAP and ANCAP. They are practiced in the United States, Europe, Japan, Australia and New Zealand, respectively. For example, Figure 5.32 depicts several collision tests defined by the U.S. National Highway Traffic Safety Administration. Although each program is structured in a slightly different way, the main destructive automobile tests contain the following subtests:

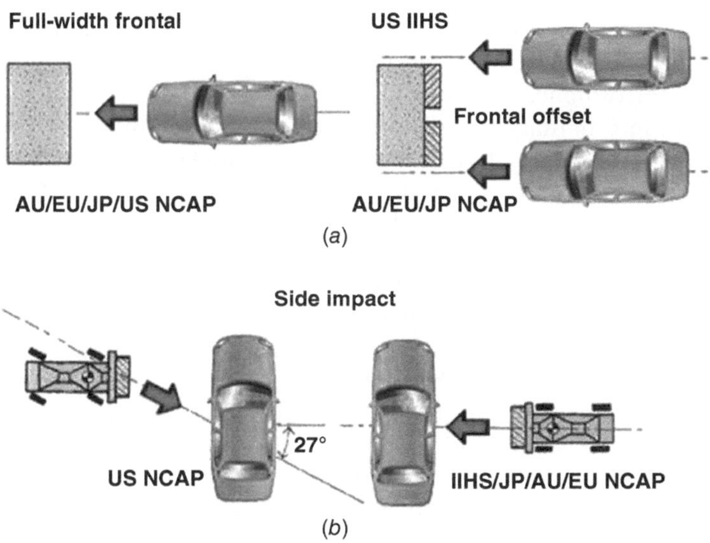

Figure 5.32 (*a*) Front and (*b*) side automobile destructive tests (USNCAP).

- *Front-Impact Tests.* These destructive tests involve a head-on test between a vehicle under test and either a stiff barrier or a relatively soft entity like another vehicle.
- *Offset Tests.* These destructive tests are similar to front-impact tests, but only part of the front of the vehicle under test impacts with a barrier or with another vehicle. Although the collision forces may be less, the smaller fraction of the car which is involved in the collision has to absorb all of the force.
- *Side-Impact Tests.* These destructive tests involve side impact. Although the relative speed between the vehicle under test and the impacting object may not be too high, such tests are very important as cars do not have a significant crumple zone to absorb the impact forces before an occupant is injured.
- *Roll-Over Tests.* These destructive tests evaluate the ability of the vehicle under test to maintain its rigid physical configuration in a dynamic, multidirectional impact, in particular the structure holding the roof.
- *Old versus New Designs.* These destructive tests involve collisions between either an old and big car against a new small vehicle under test or between two different generations of the same car model. These tests are performed to evaluate the improvements in crashworthiness.
- *Roadside Element Crash Tests.* These destructive tests are used to verify whether crash barriers and crash cushions installed on highways will, in fact, protect vehicle occupants from roadside hazards, such as guard rails, sign posts, light poles and similar road-related elements.

The study of a passive emergency situation in the automotive field, leading to the provisions that are designed into the automobile system in order to limit the consequences of accidents, are chiefly derived from destructive tests between two bodies in relative motion. Currently, the level of vehicle passive emergency performance is heavily dependent on the design of new automobiles, and the international safety norms prescribe in fact more and more strict tests in order to obtain the homologation (i.e., formal certification). Moreover, during the past few years, some automotive companies subject new vehicles to tests (rating) even stricter than those required for accreditation, due to the increasing public impact on the image of individual vehicles. The result of this trend requires detailed study of vehicle behavior under collision profiles, and such activity must start in the earliest phases of the product planning.

Computer-Aided Engineering (CAE) tools are used to simulate the behavior of an automobile system. The tools may be divided into three categories: preprocessors, model calculators and postprocessors. Preprocessor tools are used to define the simulation model and the boundary conditions of the system. Often the model is a Finite-Element Model (FEM), and the process starts with a formal description of the system geometry using standard elements such as beams, axles, poles and bolts. The number of simulated elements may vary from a few tens to thousands, depending on the system complexity and the requested detail level. Model calculator tools perform the actual model calculations while the postprocessor tools extract the relevant data and present the results to the users.

The behavior of the automobile system, subjected to various load and stress conditions, could then be investigated with such CAE tools. The FEM calculates a static geometry diagram and takes care of the characteristics of the materials used. During these computations the model takes into consideration both external forces as well as the internal propagation of forces within the material. The user then obtains the stress state, which indicates the probable areas of criticality (e.g., probability of breach in some parts or components).

The CAE tools can also simulate specific collision scenarios between two entities in a relative motion. The fundamental difference regarding the structural analysis is that the calculation refers not to a static condition but to a dynamic one. That is, boundary conditions may vary in time. In addition, while static structural analysis generally considered only the elastic deformation of materials, virtual destructive analysis considers also plastic deformation. This necessitates more sophisticated CAE tools having further knowledge about material behavior as well as an embedded algorithm to compute both elastic and plastic behavior of these mechanical elements.

Further Literature

- Hiermaier (2007)
- Nordhoff et al. (2007)
- Society of Automotive Engineers (SAE) (2005)

5.6.4 Reactive Testing

Purpose Reactive testing is a dynamic approach to systems testing whereby the individual test cases are affected by the behavior of the system under test. In other words, a reactive test is not fully and precisely defined by the test engineer a priori, but rather the test facility itself is able to react and evolve, depending upon the behavior of the SUT itself. This is done by creating mechanisms in the test facility to observe dynamically the output of the SUT during each test execution step. Needless to say, the system under test and the test facility are required to run synchronously, so that test actions can be performed using the same timing framework.

Reactive testing is usually undertaken when a system is either especially complex or exhibits nonlinear or erratic behavior, often necessitating a test strategy of covering a large number of input data combinations. The characteristic behavior of such systems is often not fully predictable. Thus, testing must look for odd behavior in remote niches of the system behavior space.

Automating the test runs in such a way that each test will react to the system's behavior on the previous test constitutes reactive testing.

Rationale Reactive testing is particularly suited for automated testing which, in some manner, depends on the response of the system itself. Advantages of reactive testing are:

- *Automation.* Reactive testing is an automatic process and thus enables the testing of systems with a very large amount of test data. Often, being able to fully automate continuous test cases becomes possible only by using reactive testing. However, when testing complex systems, the input test data must match the exact temporal behavior of the system under test.
- *Reusability.* Reactive testing lends itself to easy reusability during system development stages. This stems from the fact that the temporal behavior of the system, a relatively straightforward issue in reactive testing, is the dominant factor that changes in the course of development. In addition reactive testing also lends itself to easy reusability of test specification across several systems with similar functions.
- *Effectiveness.* Reactive testing of complex systems is significantly more effective at finding defects than are scripted tests. In addition, since the tests evolve from mechanized observations of actual system behavior, reactive testing is effective even when the system is poorly documented and the testing process is under severe time pressure.
- *Robustness.* Scripted tests tend to lose their effectiveness over time, since the faults that they are designed to detect have already been detected. In contrast, reactive tests are more dynamic because of their natural variance over time. Therefore, they tend to be effective indefinitely.
- *Efficiency.* Reactive testing requires less paperwork than other forms of testing and is easier to maintain for the long run because individual test runs are generated automatically. In this respect, some scientists claim that reactive testing is cheaper, when measured in terms of cost per defect found.

Reactive testing has several significant disadvantages. Here are some of the more pronounced ones:

- *Coverage Gaps.* Purely reactive testing approach can lead to coverage gaps in the testing space as the automated testing process may, unintentionally, ignore problem spots. However, in a predesigned testing approach, the test strategy is consciously considered, planned and carried out in an orderly fashion. In reactive testing the specific tests are generated on the fly and often in an opaque manner to the test engineer.
- *Repeatability Limitation.* The nature of reactive tests is twofold. First, they evolve over time depending on the behavior of the system under

test. Second, such tests are executed automatically, following one another at a very rapid rate. Under real test conditions, it is often impractical to repeat a test run sequence right after it is run. That is, if the system under test has no memory and its behavior depends only on the test input data, then we can easily repeat any individual test. However, if the system under test does have memory and its behavior depends on past states, then it is quite difficult to repeat the sequence of test runs leading to the same failure.

- *Test Oracle Problems.* With predesigned tests, there is typically a defined expected result or some other way of determining whether the test is passed. In some reactive testing cases, the only test oracle is the judgment of the test engineer. Therefore, the unbiased evaluation of test results is more difficult, as compared to the difficulty of evaluating results under a scripted testing methodology.

Method By and large, traditional functional testing is carried out by intuition. The selection of test data is usually ad hoc and is based on a few typical cases of system use as well as extreme use scenarios and cases with high probability of producing system errors.

Reactive testing facilitates test automation by interacting intelligently with the output of the SUT in order to generate new dynamic tests and derive conclusions (pass/fail) about the behavior of the system. Obviously, a reactive testing procedure requires that the system under test be executable so that a dynamic test can be performed. In addition, in order to guarantee the creation of legal test data, the input and output interface of the system to be tested must be defined explicitly since the system to be tested and the test facility must interact in a closed loop. This is done through channels that transmit data throughout the testing process. The test facility generates test cases that (1) stimulate the system's input channels with appropriate signals and (2) observe the system's output channels, in order to react to the system behavior, as required.

In reactive testing, the test facility acts as the environment of the system under test. In general, behavior within a real-world environment is subject to temporal constraints (e.g., residual magnetism or hysteresis phenomenon occurring in ferromagnetic materials); therefore, functionalities are usually also subject to timing constraints. For embedded systems this means each test criterion needs to account for a temporal sequence in order to validate the proper functioning of the system.

The following is a typical procedure for implementing reactive testing:

- *Step 1: Define SUT.* The specific SUT, its boundary and its environment must first be specified. For example, in the example below, the system under test comprises a controller and a variable-speed electric induction motor.

- *Step 2: Define Test Requirements.* The specific system requirements to be tested must be specified in a formal way. This includes the specific elements to be tested and their test oracle, that is, the constraints on their values. In the example below, the following must be verified: (1) change time, (2) settling time and (3) the maximum surge speed of the motor.
- *Step 3: Define Test Suite.* The structure and capabilities of the test suite must be specified in details. In the example below, the test manager element and the next test generator element must be specified. In addition, means for dynamically measuring the motor speed as well as a method (e.g., genetic algorithm) for the automatic computation of the next test case based on current system parameters must be defined.
- *Step 4: Define Interfaces.* The interface details between the test suite and the SUT must be defined. In the example below, the content and structure of the data flowing from the test manager element into the system under test as well as the data (i.e., motor speed) flowing from the system into the next test generator must be specified.
- *Step 5: Define Initial Test Data.* The initial test data must be specified. In the example below, a randomly selected initial speed of the motor constitutes this data.
- *Step 6: Define Test-Stopping Criteria.* A test-stopping criterion as well as the actual stopping mechanism must be defined in order to govern the stopping of the reactive testing process. In the example below, a successful test criterion could be that all tested variables (i.e., change time, settling time and maximum surge speed of the motor) have not been violated after certain (large) numbers of iterations of motor speed change requests.
- *Step 7: Perform Reactive Tests.* In this step, the actual reactive tests take place. In the example below, the test manager element sends repeated requests designed to change the speed of the motor; the real-time results (i.e., the motor speed dynamic measurements) are compared to the speed command, the results are stored within a database and a new speed data is generated by the next test generation element. This process continues until the stopping criterion is met, indicating either a successful or a failed end of test.
- *Step 8: Analyze Test Results.* In this step, the test results stored within the database are analyzed and a test pass/fail decision is made.

Reactive Testing Example Figure 5.33 depicts a system under test composed of a controller and a variable-speed electric induction motor, together with its test facility.[59] The controller inverts a three-phase input alternating current, first into direct current and then into a controlled voltage/frequency source using a digital converter.

[59]This example was inspired by Zander-Nowicka (2007).

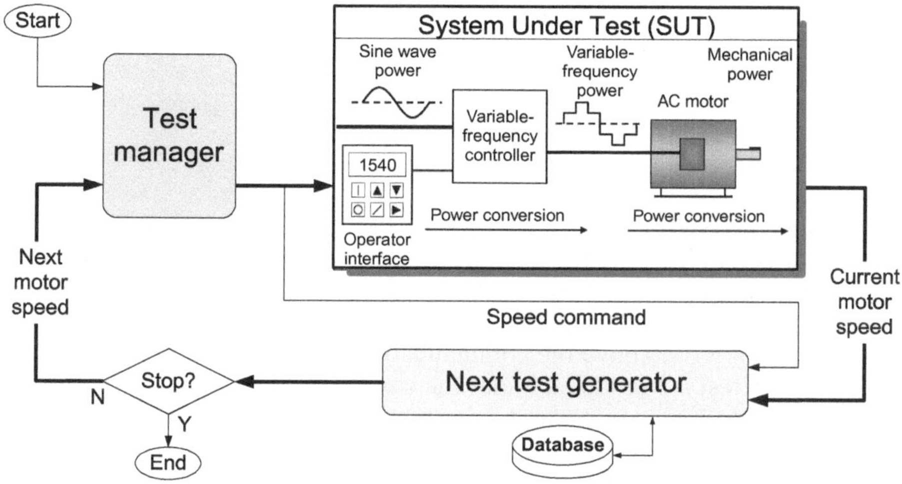

Figure 5.33 Variable-speed electric motor and reactive test facility.

The system allows adjusting the speed of the motor in the range of 0 to V_{max} Revolutions Per Minute (RPM), either manually or remotely by an external command.

The SUT in this example must meet three response characteristic requirements related to speed transitions from one value to another. More specifically, change time (t_C), settling time (t_S) and maximum surge speed (v_S) must be within specified limits (see Figure 5.34):

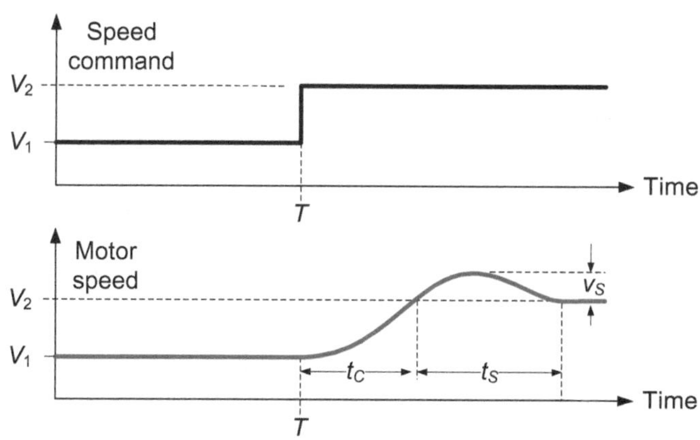

Figure 5.34 Variable-speed electric motor: Speed command and resulting motor speed.

$$t_C \leq K_1(|V_2 - V_1|) \quad \forall V_1, V_2$$
$$t_S \leq K_2(|V_2 - V_1|) \quad \forall V_1, V_2$$
$$v_S \leq K_3$$

Assuming $K_1 = 0.002$ s/RPM, $K_2 = 0.004$ s/RPM, $K_3 = 100$ RPM and the motor is commanded to increase its speed from 1000 to 1500 RPM, then the change time (t_C) must not exceed $0.002 \times (1500 - 1000) = 1$ s, the settling time (t_S) must not exceed $0.004 \times (1500 - 1000) = 2$ s and the maximum surge speed (v_S) must not exceed 100 RPM above the commanded 1500 RPM.

A reactive test is conducted under the control of the test manager element that commands the electrical motor in the SUT to transition from one speed value to another speed value. This information, together with data about the actual speed dynamics of the motor, is evaluated by the next test generator element and stored on a database for later analysis. Based on the evaluation result, a new test speed is computed using, for example, the genetic algorithm method (see search-based testing in this chapter). Here the fitness function of the genetic algorithm search increases upon finding motor speed commands leading to increased target test parameters (i.e., t_C, t_S and v_S). When the stopping criterion is met, then the reactive test process ends (either as a success or failure). Otherwise the cyclical process continues.

Further Literature

- Black (2007)
- Broy et al. (2005)
- Raheja and Allocco (2006)
- Zander-Nowicka (2007)

5.6.5 Temporal Testing

Purpose For many embedded systems, correct system functioning depends on temporal correctness as well as on logical correctness. Accordingly, the verification purpose of temporal behavior is to assess whether the Worst-Case Execution Time (WCET) does not exceed a system's specified time for performing an operation. Less prevalent, but still an aspect of temporal testing, is the verification that the Best-Case Execution Time (BCET) always meets the relevant minimum system timing interval. In other words, temporal testing evaluate whether relevant system operations are bounded within the BCET to WCET range.

Dynamic testing is the most important analytical method for verifying the temporal quality of embedded systems. During temporal testing we check if the implementation fulfills the specified requirements. Since a complete testing process (i.e., a set of tests with all possible input combinations) cannot be carried out in practical situations, the most appropriate test data must be selected according to some relevant criteria. Ultimately, the aim of temporal testing is to apply test inputs which will cause the system to violate performance timing requirements.

Rationale The motivation for temporal testing of real-time systems stems from the criticality of timing issues found in most embedded systems. Take, for example, an airbag in a passenger car. In order to protect passengers, an airbag must fully inflate within 40 ms from an initial impact. If the airbag will inflate in, say, 100 ms, then the system will be mostly ineffective in protecting the passengers.

Unfortunately, estimating temporal behavior is often unreliable due to errors introduced in computing execution times, estimating system loading and other unknown factors. In addition, specification complexity stemming from unforeseen effects of combinations of time and resource constraints as well as mistakes in scheduling analysis, make such estimation less useful.

The temporal testing of embedded systems is also complex due to requirements like timeliness, simultaneity and predictability, as well as the embodiment of digital and analog components that often characterize embedded systems. Also, technical characteristics like the strong connection with the system environment or the frequent use of parallelism, distribution and fault-tolerant mechanisms complicate the test. Nevertheless, temporal testing should be a mandatory part of the verification and validation process of certain embedded systems. It is a method that examines runtime behavior based on an execution in the application environment. Temporal testing is a way to consider dynamic aspects, which are especially important to rule out malfunctioning of embedded systems, for example, the synchronization of parallel processes or subsystems.

The temporal behavior of a real-time system is defective when the system is in a given state and the input data causes the system to violate specified timing constraints. In most cases, a temporal violation means that outputs are produced too late, relative to other components of the system or to the external environment of the system.

The task of the test engineer, therefore, is to find whether or not such system states and/or input combinations exist. In other words, the test engineer must generate a set of test cases that exercise system behaviors that are likely to reveal temporal defects. For example, in order to detect system timeliness defects, criteria must be defined for selecting the "right" test cases and appropriate time constraints must be extracted. In addition, in case of an event-triggered real-time system, the test engineer must consider factors like the nondeterministic execution order typically exhibited by such systems and the temporal impact exerted on the system by its environment.

Therefore, the contents of a temporal test case will typically include input data and expected result, event sequence, test process and system state and event sequence.

We will now discuss several methods for performing temporal testing of embedded, real-time systems. This includes (1) constrained random-based temporal testing, (2) stress-based temporal testing, (3) search-based temporal testing and (4) mutation-based temporal testing. Several researchers show and it is generally agreed that whereas the first two methods are easier and relatively inexpensive to implement, the last two methods are substantially more effective in identifying temporal failures in embedded or complex systems.[60]

Method 1: Constrained Random-Based Temporal Testing Random-based temporal testing is based on automatically creating test cases with randomized input, running the SUT with these test cases and measuring the temporal parameters exhibited by the system. On the one hand, this approach is quick, simple and straightforward to implement. It produces large amounts of easily created test cases using a pseudorandom number generator and it provides a ready mathematical basis for analysis.

On the other hand, random-based temporal testing might be a poor choice when dealing with complex systems or with complex adequacy criteria. The probability of selecting a defect revealing input by chance is, naturally, quite low. Therefore, one of the major issues in any random-based temporal testing approach is that it samples only a small fraction of all possible input space and a lot of important ranges of input could be missed completely as the input data distribution may not be distributed uniformly. Last but not least, not all random inputs may be applicable to the SUT. Certain input combinations are often illegal, could damage the system and thus should be avoided.

Since the combinatorial space of system inputs is so huge, we would like to restrict, in some way, the input space and not use a purely random method. For example, we can use principles of boundary value testing to divide each system's input into domains of interest. Thereafter, we can constrain the random input generator to produce input data representing different domains of interest rather than producing random inputs representing the same domain. Figure 5.35 captures a typical constrained random-based temporal testing procedure.

[60]Several computer scientists have experimented with determining WCET using strictly static software code analysis of real-time embedded systems. Some commercially available tools produce results ranging from 10 to 50% overestimation of WCET depending on the complexity of the programs involved and processor types (see e.g., AbsInt Angewandte Informatik, Germany, at http://www.absint.com).

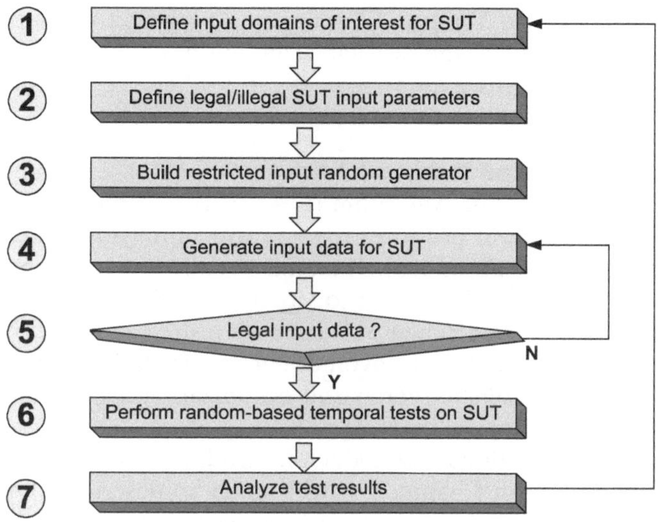

Figure 5.35 Typical constrained random-based temporal testing procedure.

The following depicts a typical constrained random-based temporal testing procedure:

- *Step 1: Define Input Domains of Interest for SUT.* First, the range of each input variable affecting the SUT is divided into domains of interests.
- *Step 2: Define Legal/Illegal SUT Input Parameters.* Next, the specific ranges and combinations of ranges of legal and illegal input parameters are defined.
- *Step 3: Build Restricted Input Random Generator.* The environment needed for executing the target SUT is created. The input random generator is built in such a way that input variables within individual test cases represent, to a large extent, different domains of interest.
- *Steps 4 and 5: Generate Input Data for SUT.* A predefined number of test cases are generated automatically. However, if a test case contains an illegal combination of input data, then it is scraped and another attempt is made to generate a test case with legal input values.
- *Step 6: Perform Random-Based Temporal Tests on SUT.* Next, the set of test cases is repeatedly executed to capture different behaviors of the potentially nondeterministic system under test.
- *Step 7: Analyze Test Results.* During test execution on the SUT, the various parameters of the testing are captured and then analyzed offline. The intent is to verify the behavior of the system under test. That is, to find whether all systems temporal behavior meets specifications.

Method 2: Stress-Based Temporal Testing Stress-based temporal testing is similar in many respects to the restricted random-based temporal testing. The difference lies in the efforts to apply test cases when the system is stressed, that is, many of its resources are utilized to the maximum.

Method 3: Search-Based Temporal Testing Search-based temporal testing using, for example, Genetic Algorithm (GA) searches automatically for temporal test inputs that will produce extreme execution times (i.e., either the longest or the shortest durations). The aim, of course is to discover whether such test input can cause the system to violate its temporal requirements.

A search method, often called evolutionary testing, can then be regarded as an optimization problem. Here, evolutionary computation forms the generic term for direct, probabilistic search and optimization algorithms is gleaned from the model of biological evolution.

Since this subject is discussed under the heading Search-Based Testing in this chapter as well as in Chapter 7, it will suffice to describe a typical search based temporal testing procedure using Genetic Algorithm (GA) (see Figure 5.36).

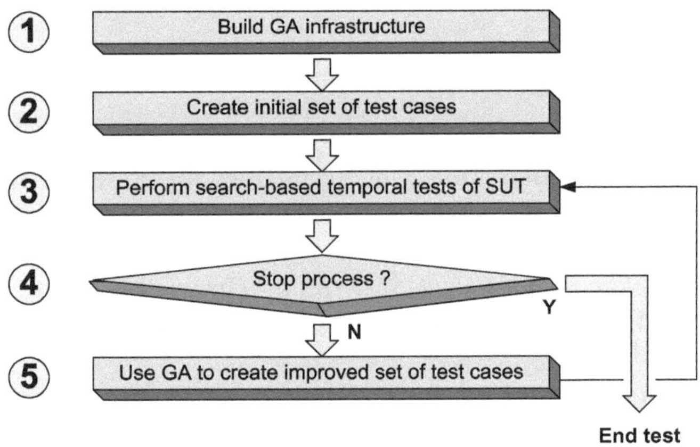

Figure 5.36 Typical search-based temporal testing procedure.

The following depicts a typical search-based temporal testing procedure:

- *Step 1: Build GA Infrastructure.* First, the infrastructure surrounding the SUT must be built. This should include the mechanical, electrical and communication environment to the system under test as well as the search engine which can generate new and improved test cases using, typically, GA means.
- *Step 2: Create Initial Set of Test Cases.* The initial set of test cases is usually generated at random. The test engineer must ensure that only legal input values shall be generated and presented to the SUT so it will function properly on the first evolutionary iteration.

- *Step 3: Perform Search-Based Temporal Tests of SUT.* The SUT is now subjected to the current test case and its temporal behavior is monitored.
- *Step 4: Evaluate Stopping Criteria.* The stopping criteria are evaluated against the test results, and the decision is made to either continue and go to the next stage or terminate the testing process. Such termination may be a result of actually achieving the predefined stopping criteria or a termination request issued by the test engineer.
- *Step 5: Use GA to Create Improved Set of Test Cases.* The search engine is activated in order to find an improved test case. The intent is, of course, to identify a set of input values that will violate the temporal constrains imposed on the system through its specifications. In case of a GA-based search engine, the entire sequence of selecting elitists, selecting parents for mating, combining their genes, generating mutants and replacing the old generation with the new one takes place. Thereafter the procedure continues at Step 3.

Method 4: Mutation-Based Temporal Testing Mutation-based temporal testing[61] utilizes temporal behavior of real-time application models within an appropriate system environment. This method is based on modeling the temporal behavior of the SUT and its environment. Naturally, each system model exhibits specific physical laws or causality constraints that, by design, limit certain task activation and other events from happening simultaneously.

The inputs to mutation-based temporal testing are a real-time system model (representing the real SUT) and a test criterion. A mutation-based testing criterion determines the level of thoroughness of testing and what kind of test cases should be produced, by specifying what mutation operators to use. Mutation operators change some property of the system model to mimic faults and deviations from assumptions that may lead to timeliness violations. Several mutation operators for testing of timeliness may be defined:

- *Execution Time Operators.* Execution time mutation operators increase the execution time of a task.
- *Lock Time Operators.* Lock time mutation operators increase the interval when a particular resource is locked. An increase in the time a resource is locked may increase the maximum blocking time for a higher priority task.
- *Unlock Time Operators.* Unlock time mutation operators delay the time required to unlock resources so it may become available to other system elements only after a certain delay.

[61]This section is based on Nilsson (2006).

A real-time model is invariably software based, so that a mutant generator tool can apply mutation operators onto these software modules creating a mutated copy of the original real-time model. Each mutated model is fed with inputs so that the resulting execution process can be analyzed. Different activation patterns are then modified using heuristics that guide the mutant to either require more time for execution or to exhibit abnormal temporal behavior (e.g., nonschedulability, missing specified time for performing an operation). If execution analysis actually reveals such abnormal temporal behavior in a mutated model, the model is identified, in the lingo of mutation, as "killed," and activation parameters that can kill mutated models are later used to create temporal test cases for the real target SUT.

Next, test execution is performed by executing the target system with the temporal test case and injecting the appropriate activation patterns. Since the target system often exhibits nondeterministic behavior patterns, multiple execution runs are required. The target system outputs are collected so that an analysis can be performed offline. Such analysis can reveal whether a temporal violation has occurred during a set of test runs

Figure 5.37 depicts a typical mutation-based temporal testing procedure:

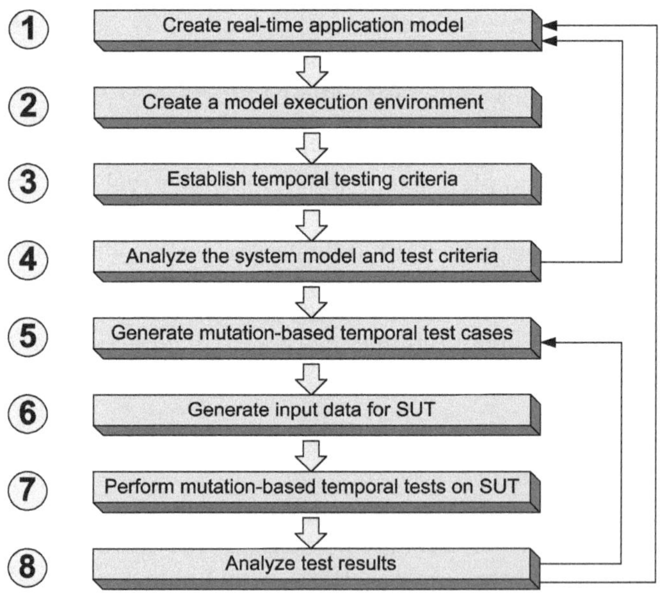

Figure 5.37 Typical mutation-based temporal testing procedure.

- *Step 1: Create Real-Time Application Model.* A system model is created by first building a real-time application model that mimics the temporal behavior of the target SUT. In addition, the corresponding triggering entities are modeled (by estimations or measurements).

- *Step 2: Create Model Execution Environment.* The environment needed for executing the real-time model is then created. This environment must fully correspond with the architectural properties and protocols that are present in the target SUT.
- *Step 3: Establish Temporal Testing Criteria.* Next, suitable mutation-based test criteria are selected based on the required levels of thoroughness and the allocated testing budget, available schedule and other resource constraints.
- *Step 4: Analyze System Model and Test Criteria.* Once the system model as well as the testing criteria is ready, it is possible to perform correctness and matching analysis and refine any of these elements (i.e., the real-time application model, the model execution environment and the temporal testing criteria).
- *Step 5: Generate Mutation-Based Temporal Test Cases.* As described below, a set of mutation-based test cases is generated automatically from the system model with the intent of fulfilling the defined test criteria.
- *Step 6: Generate Input Data for SUT.* Once a set of test cases is generated, corresponding sets of input data for individual tasks are acquired using compiler-based methods or temporal unit testing.
- *Step 7: Perform Mutation-Based Temporal Tests on SUT.* Next, the set of test cases is repeatedly executed to capture different behaviors of the potentially nondeterministic SUT.
- *Step 8: Analyze Test Results.* During test execution on the SUT, the various parameters of the testing are captured and then analyzed offline. The intent is to verify the behavior of the SUT (i.e., to find whether all system temporal behavior meet specifications) and, if needed, to further refine the system model or the set of test cases.

Further Literature

- Krstic and Cheng (1998)
- Nilsson (2006)
- Nilsson and Offutt (2007)
- Sthamer (1996)
- Wegener and Grochtmann (1998)
- Wegener and Frank (2001)

5.7 BLACK BOX—PHASE TESTING

Figure 5.38 depicts a typical set of testing activities distributed along a system's lifetime. Please note that sanity, exploratory and regression testing are performed throughout the system lifetime and are not specifically associated with any particular lifecycle.

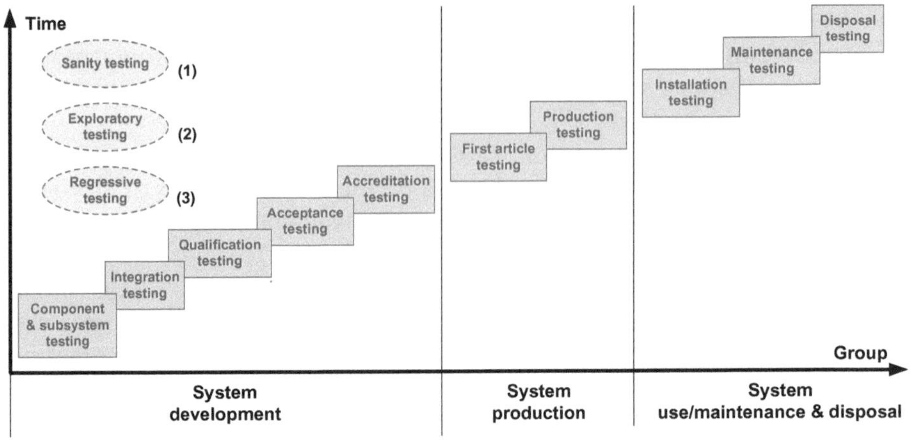

Figure 5.38 Testing activities distribution along a typical system lifetime.

5.7.1 Sanity Testing

Purpose Sanity testing is a rudimental testing process. The purpose is to evaluate quickly the general validity of a performance claim or the overall functionality of a system. In other words, to assure that a system or methodology works, in general, as expected.

Rationale Sanity testing, sometime called "smoke testing," is usually carried out prior to a more exhaustive round of testing at different levels of testing granularity (i.e., component, subsystem, system and system of systems levels). The rationale here is to perform cursory, fast and inexpensive testing, sufficient to show that the SUT is functioning reasonably well.

Method Sanity testing is similar in many respects to exploratory testing. It is an ad hoc and unscripted type of testing where the discovery or unexpected system behavior triggers further exploration and testing.

For example, a hardware engineer builds a new electronics board, connects it to its appropriate power source and checks first that the unit does not overheat or burn (thus, the term "smoke test"). Beyond this cursory look, hopefully, the board shows healthy signs of life. In another example, a software engineer may execute a new program and verify first that it does not crash the application or that the application enters into an endless loop. Then he or she verifies that the program responds appropriately to a few sets of input values.

Another example, applicable to the usage phase, relates to the purchasing of a new television set. The customer starts by performing a sanity test: He or she plugs the TV to the appropriate power source, as well as to an antenna or a cable TV and turns on the set, adjusts the volume control and tries out several TV channels. If nothing unpleasant happens, the TV passes the sanity test.

Further Literature

- Ahmed (2009)

5.7.2 Exploratory Testing

Purpose The purpose of exploratory testing is to verify system behavior by means of exploration and learning about the behavior of the system under test. The term "exploratory testing" is sometimes attributed to Dr. Cem Kaner, a professor of software engineering at Florida Institute of Technology.

Exploratory testing may be defined as "simultaneous learning and performing test design and test execution." As such, it is fundamentally different from scripted testing. Whereas scripted tests are conducted on the basis of a predefined manual or automated test procedures, exploratory tests are elaborated in a rudimentary manner and, typically, are not carried out precisely according to plan. In other words, exploratory testing does not entailed test plans, checklists, and the like. The testing strategy here involves systems functional exploration and uses past testing experience in order to make educated guesses about places and functionality that may be problematic.

Exploratory testing is located somewhere between purely scripted testing at one extreme and purely ad hoc testing on the other. It requires specific knowledge of test techniques and tools and is an individual exercise, difficult to pass on the knowledge gained and remarkably dependent on a test engineer's skills and knowledge of the SUT.

Rationale Exploratory testing, being a nonscripted testing approach, is attractive since little formal preparation is required prior to actual testing. Another major advantage stems from the fact that the test engineer is not bound to a specific course of action, dictated by a predesigned test procedure. This allows freedom to explore the system and concentrate on problem areas as they appear in a dynamic fashion. In summary, good exploratory testing involves investigating systems behavior vis-à-vis a mental model of the system present in the test engineer's mind.

On the other hand exploratory testing tends to be unstructured, even chaotic, and often test engineers do not document their testing process and observations. As a result, they may skip testing important portions of the system and be unable to recreate certain system defects by repeating specific sequence of test inputs. In addition, exploratory testing requires certain abilities and skills that often are beyond scripted testing. First and foremost, a test engineer must possess thorough knowledge of the system under test at hand. This requirement is due to the fundamental characteristics of unscripted testing where the expected behavior of the system under each test (i.e., the oracle), must be known to the test engineer as he or she proceeds with the testing process. Finally, test engineers must be able think critically, pose useful system questions and craft tests that systematically explore and analyze the system, as well as considering a multitude of risk issues relevant to the SUT.

Based on the above characteristics, exploratory testing is often employed in conjunction with other, more formal testing methods. Typically such a system is well known to the test engineers (e.g., a computer game designer acting as a test engineer, verifying a hardware or software test facility). Typically, there are limited or no specifications or requirements, and there is limited time to specify and design formal tests. In summary, the beauty of blending scripted and nonscripted testing methods stems from the fact that scripted tests are good at building confidence that the system has been thoroughly tested and meets its specifications. On the other hand, exploratory tests are good for discovering interesting and unexpected problems, since test engineers design tests in response to the reaction of the system, a process that is quite different from formally planning a test process.

Method As mentioned above, in scripted testing, tests are first designed and recorded. Then, at a later time, they are executed, possibly by persons other than the person who designed the original tests. In contrast, exploratory tests are designed and executed one right after the other by the same person based on a mental model of the SUT. This model includes what the system is and how it is supposed to behave.

Exploratory testing is usually a manual process conducted by professional skilled test engineers having the freedom, flexibility and enjoyment of the test process. The process is optimized to find failures by following individual hunches and continually adjusting plans, refocusing on the most promising risk areas while minimizing time spent on documentation.

Some test organizations endorse pair testing (two test engineers) as an effective strategy for encouraging discussion, promoting creativity and, in general, speeding up the testing process. Also, system tests jointly conducted by more than one person are often more effective, advancing more orderly progression of the test and tend to produce a more coherent documentation of the process.

In exploratory testing we stress the dynamic questioning of the SUT, such that each question answered properly increases our confidence in the system. Therefore, the testing process becomes a problem of choosing appropriate questions to get the best information we can. These questions are designed to (1) focus thinking on a problem by examining it from multiple angles and (2) seek to identify ways for finding the most appropriate solution.

As becomes clear, exploratory testing is not based on a hard and fast method but rather on the test engineers skills and experience coupled with heuristics. Readers should note that using exploratory testing is not applicable or correct for all situations. As mentioned, it is most applicable when a system must be tested in a short period of time and there are no clear and concise formal specifications; yet, intuitively, test engineers are familiar with the operational behavior of the SUT. By and large, test engineers should focus on system risks. They should ask themselves questions such as: What kind of risk can this system create for individuals, stakeholders, property and the environment? In other words, adopt a test strategy elevating risk concerns above other functional requirements. Nevertheless, test engineers should always be aware that heuristics are fallible guides for a testing process. One may use heuristics, but one should never fully rely on them.

One neglected aspect of exploratory testing is documentation. Test engineers should make it a habit of generating adequate documentation during the execution of the tests. This should include notes about what happened during testing, that is, what tests were conducted and what were the results, what was the status of the system and what resources (e.g., equipment, manpower) were utilized. Finally a list of open issues that must be dealt with in the preceding tests should be noted. Such minimal documentation is not too difficult to generate, and it may be used to assess the quality of the SUT as well as a basis for future planning.

Finally, a short note about the *ad hoc* versus *exploratory testing* controversy: The term ad hoc testing is sometimes used, in a rather derogatory way, to denote sloppy testing. That is, testing based on improvising or using intuition and experience rather than on planning the test process methodically. Some test engineers equate ad hoc testing with exploratory testing but, as we have seen above, exploratory testing has important and valuable features. Practitioners who have exploratory testing experience argue that even ad hoc testing is not quite the random, careless and unfocused approach to testing but rather an improvised testing that deals well with verifying a specific subject. Nevertheless, the controversy continues.

Further Literature

- Black (2002)
- Copeland (2004)
- Kaner et al. (2001)
- Shore and Warden (2007)

5.7.3 Regression Testing

Purpose Regression testing refers to a selective retesting process of a system that has been modified, to ensure that the new modifications have been properly introduced and that no other previously working functions of the system malfunction as a result of this modification. Typical modifications involve fixing system problems, adding new system features or changing and adapting the system to a new set of operational conditions

Regression testing may be performed at different lifecycle stages starting from unit level and continuing to the component, subsystem and system levels. The unit developers often do unit regression testing, whereas regression testing at higher level of system is assigned to the quality assurance personnel or test engineers. Customarily, these individuals perform their tasks using system or software change information, updated system or software requirements and design specifications as well as user manuals. In addition regression testers, especially at the subsystem and system levels use system regression testing procedures, pass/fail criteria and overall test strategy documentation.

Researchers suggest that the cost of regression testing often comprises over 20% of the total testing costs of a complex system development project. Therefore, system regression testing is often the target of cost reduction activities, in order to reduce system cost, as well as shorten delivery time. Regression testing is usually a major cost of system maintenance. In fact, during the maintenance phase, the cost of regression testing invariably far exceeds the cost of making the changes necessary to meet performance requirements. Several strategies have been suggested for improving the efficiency of regression testing, such as *regression test selection*, *test suite reduction*, and *test case prioritization*. The concern in using any of these strategies is that reducing the amount of testing will invariably increase the risk of leaving hidden faults in the system.

Rationale The goals of regression testing are to validate the updated aspects of the system as well as validate the whole system. The rationale of regression testing stems from two fundamental problems associated with modifying or maintaining systems.

Maintaining Correct System First, we must manage the risks associated with modifying the SUT. Typical problems in regression testing are: (1) a modification that did not solve the problem or did not provide the needed improvement, (2) a modification that had a side effect on the system, (3) error in rebuilding the system, so it does not behave in the expected way or (4) a faulty localization, that is, the modification was, in fact, performed on a different portion of the system (obviously not producing the expected result).

Regression testing of embedded and/or distributed systems presents particularly challenging problems. Test engineers contemplating regression testing must first classify the type of system modification in a systematic way, for example, requirements or specification changes, design changes, system implementation changes, various user manuals or other documentation changes and/or systems environment or technology changes. More importantly, system test engineers must clearly identify and map system parts which have been changed and determine how each particular change impacts the overall system. Ultimately, the major challenge in system regression testing is how to minimize the retesting efforts and achieve a sufficient confidence in the level of risk taking.

Maintaining Valid Test Suite Second, we must maintain a current and proper test suite. The reason for these dynamics stems from the fact that functional modifications in the SUT compel, in the majority of cases, a revision of test suite elements affected by this change. Therefore, the problem boils down to questions such as: Which test cases should be removed or replaced? Which test cases should be added? Since test case maintenance is inevitable, designers of test suites are advised to design and build them in a modular fashion and avoid unnecessary dependence. Also, generating concrete test cases from more generic test case specifications can significantly reduce system maintenance cost.

Method Regression testing involves reusing, and updating as necessary, test suites from previous versions of the system. A basic system regression procedure is described below:

- *Step 1: Analyze System Modifications.* During this step, the test engineer should analyze the exact nature and scope of system modifications. That is, what are the components or subsystems which have been changed and in what specific ways.
- *Step 2: Analyze System Modification Impacts.* During this step, the test engineer should analyze the specific system impact stemming from the above modifications. That is, in what ways (e.g., functional behavior, data flow, state transition control) does the system behave differently as a result of the above modifications.
- *Step 3: Define Regression Test Strategy and Criteria.* Next, the regression test strategy and test success criteria should be defined. A test strategy provides test engineers a guideline to perform system regression testing activities. It usually refers to a rational way to define regression testing scope, coverage and success criteria as well as order of the testing sequence. There are several frequently used regression test strategies, which will be discussed shortly.
- *Step 4: Update Test Case Suite for Current Regression Testing.* During this step, the test engineer should update the test case suite (i.e., the entire set of the test cases) in order to harmonize it with the updated system. This includes (1) selection of an appropriate subset of relevant test cases to be used during the current run of the regression test, (2) removal of obsolete test cases from the current test suite and (3) updating test cases that have been affected by the latest system modifications.
- *Step 5: Perform Regression Testing.* During this step, the test engineer should perform the regression testing at the appropriate system level in accordance with the selected regression test strategy and evaluate the results against the chosen test success criteria.
- *Step 6: Analyze and Report Regression Test Results.* During this step, the test engineer should analyze the overall test results and generate an appropriate report identifying the specific regression test strategy and criteria, the composition and content of the test suite used during the current test run and any remaining system problems.

Regression Test Strategies In general, all system regression test strategies involve a static analysis of the SUT, instrumentation of the system in order to collect relevant test data, execution of the system using the test suite and collecting system behavior information. The following describes established regression testing strategies:

- *Strategy 1: Test All.* Under this strategy, the entire test suite for a given system is executed each time a system has been changed. This strategy is rarely implemented since it constitutes an exceedingly expensive and time-consuming process.
- *Strategy 2: Prioritized and Rotating Regression Testing.* Under this strategy, the entire set of test cases is prioritized and executed on a piecemeal basis. That is, every time the system is modified, a different subset of the test suite is executed so, eventually, all the test cases are executed. Several ordering and priority schemes have been suggested, for example: (1) round robin scheme where priority is assigned to the least recently run test cases, (2) track record scheme where priority is assigned to test cases that have detected faults before and (3) structural scheme where priority is assigned to test cases covering components or subsystems that have not been recently executed
- *Strategy 3: Heuristic Minimization Regression Testing.* Under this strategy, the investment in time, money and other resources of regression testing are reduced by selecting a subset of the existing test suite cases based on information about the system, its modifications and their impact. This is done heuristically, by considering the economic impact of the test strategy against the risks of not fully testing the entire updated system. Often, the test cases are performed in a prioritized order so that those with the highest priority are executed earlier. If the resources allocated to regression testing are exhausted, then the less important subset of the test suite (i.e., test cases that are able to reveal less critical defects) are disregarded.
- *Strategy 4: Specification-Based Regression Testing.* Under this strategy, a selected subset of test cases is executed, primarily to verify the new functionality of the SUT. In addition, related functionality that may have been damaged during the system modifications is also tested.
- *Strategy 5: Control Flow and Data Flow Regression Testing.* Under this strategy, a selected subset of test cases is executed, primarily to verify the new control flows and data flows. Such selection necessitates a reasonable structural knowledge of the system, and in this respect it is akin to white-box testing. Of course, related control flows and data flows that may have been damaged during the system modifications are also tested.
- *Strategy 6: Risk-Based Regression Testing.* Under this strategy, a subset of test cases is selected, primarily on the basis of risks emanating from system defects that, possibly, have not been flushed out. The important

aspect here is that risk-based testing brings formal risk assessment methods (risk identification, risk analysis and risk prioritization) into the testing process. The overall aim of risk-based testing is to focus the test effort on the most critical areas of the system. This is done by analyzing the cost of performing each regression test, versus the risk of not performing it. The subject of modeling and optimizing the cost, time and risk of the overall VVT process is discussed at length in Chapters 6 and 7.

- *Strategy 7: Firewall-Based Regression Testing.* Under this strategy, a subset of test cases is selected, primarily on the basis of a firewall concept. A firewall area within a system refers to a changed element within a system, together with all other elements within the system that have possibly been impacted by the change. We define the test cases associated with the firewall as the test suite for a particular system modification. By executing only this test suite, we can reduce intelligently the system regression testing to a smaller scope. Different types of firewalls may be created based on various test models, for example, functional firewall, data flow firewall and state transition firewall.

  Figure 5.39 depicts an example of a particular system regression test where, in this case, subsystem B had been modified. The test engineer, analyzing this change, determines the impacted by this change and the relevant regression testing firewall zone.

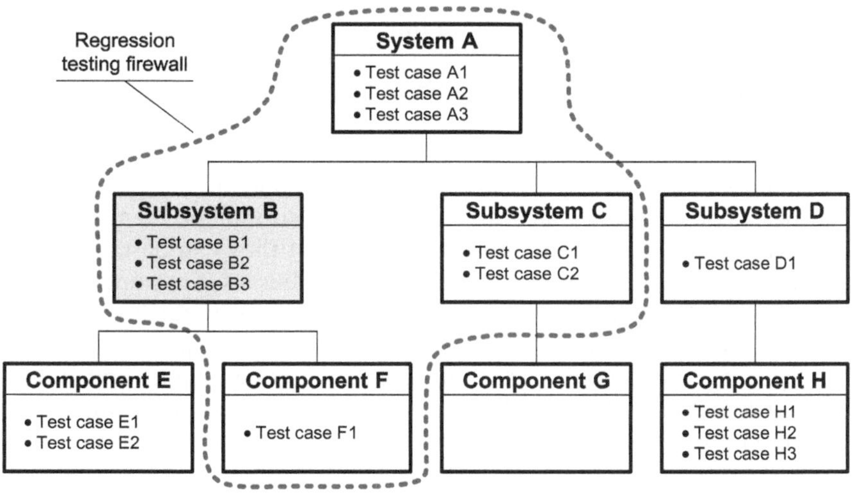

Figure 5.39 Firewall example of a system regression test.

Based on this analysis, he determines the particular test cases that constitute the test suite for this regression event (see Table 5.11).

TABLE 5.11 Analysis Results of Firewall System Regression Test Example

System Element	Modification Effect	Test Cases
System A	Impact	A1, A2, A3
Subsystem B	Change	B1, B2, B3
Subsystem C	Impact	C1, C2
Component F	Impact	F1

Further Literature

- Apiwattanapong and Harrold (2003)
- Beck (2002)
- Do and Rothermel (2006)
- Mao et al. (2007)
- Marick (1994)

5.7.4 Component and Subsystem Testing

Purpose A component may be described as "a hardware or software element used in the building of a system or subsystem." A subsystem may be characterized as "a major part of a system which itself has the characteristics of a system, usually consisting of one or more components."

Testing at the component level concentrates mainly on its functionality (black-box testing) and, to some extent, on its structure (white-box testing). In contrast, subsystem testing concentrates on verifying inter-subsystem functionality as well as intra-subsystem functionality.

Typically, subsystem testing will include not only functional correctness but also attributes such as reliability, robustness, performance, usability and security. By and large, subsystem tests are functionally grouped so that related functions can be tested by using similar setup procedures and stimuli.

Components and subsystems are tested by means of specially designed hardware or software harness. Such test environment is not part of the final version of the system; however, once built they can be used to perform regression testing whenever needed. Also, whereas components are often tested by individual developers in an informal way, testing of subsystem usually consists of a rigidly controlled sequence of steps designed to verify all critical functionalities. Actual testing of components and subsystems starts when these elements have been built; however, planning the process and creating the necessary supporting test infrastructure must be accomplished earlier.

Rationale The rationale for testing components and subsystems rather than testing the entire system at once stems from controllability and observability limitations inherited in large systems. Testing at the component or the subsystem level reduces the complexity of testing and simplifies the process of locating faults to specific parts of the system. This undertaking must be broken up and staged, so that it is carried out at different times and often by different groups of people. Testing any real-life system as a whole is likely to require a massive amount of resources and time and, probably, will miss a fair number of defects.

Test engineers must also consider and make decisions regarding what component or subsystem should be tested and how much. Since it is impractical to test everything and because we only have limited budgets and time to test a system, one needs to prioritize what gets tested. One strategy for testing is to test elements that have the greatest risk and the greatest cost of failure.

Method In order to perform component or subsystem testing, one must plan the test process, build a test requirement checklist, develop a set of test cases, develop the test infrastructure environment, perform the actual component and subsystems tests and, finally, evaluate, report and improve the tests. Here are typical steps for components and subsystem testing:

- *Step 1: Plan Test Process.* Testing should be undertaken after clearly defining objectives (and what the testing is to achieve). A well-defined understanding should be reached as to the features to be tested and the ones not to be tested. In addition, the general test approach as well as the test deliverables, test schedule and resource allocation should be addressed.

  Next, a test plan, used to organize and guide a systematic testing process, should be developed. For each stage of testing the plan should describe the methods needed to perform each test. In other words, a test plan is used to organize the activities of testing so that testing is synchronized with components and subsystem development and adequate resources, so that time, people, tools and the like are allocated to the testing process. Finally, a test plan should be utilized so other stakeholders can audit the testing process, critique it and determine whether or not it has been carried out properly.
- *Step 2: Develop Set of Test Cases.* Once the test plan is ready and approved, the actual set of test cases for a component or the subsystem is created. The intent, of course, is that the ensemble of test cases will fully exercise a component or the subsystem with the purpose of causing failures and detecting defects.

  Although experienced test engineers may claim that developing test cases is an art, there are some common heuristics to do so, for example, using relevant tests from an earlier development program, "restating" component or subsystem specifications into tests, utilizing failure reports to predict likely weakness and developing tests to uncover such pitfalls and considering potential functional risks and developing tests to uncover them. In summary, test case development involves combining relevant requirements into individual test cases, which will normally include:

a. Name or ID of the test
b. Objective of the test
c. Component or subsystem feature to be tested
d. Test environment infrastructure
e. Detailed test procedure
f. Input data needed to perform the test
g. Description of expected component or system output behavior and test result
h. Pass/fail criteria

- *Step 3: Develop Test Infrastructure Environment.* When developing a test infrastructure environment, we should consider first the mechanical, electrical and environmental harness that is needed to grasp, control and observe the component or SUT. More specifically, we must consider the controllability of the tested unit (i.e., inputs/outputs, operations, behavior) and observability of the tested unit (i.e., incoming and outgoing interfaces and traceability of interface events). The test infrastructure environment therefore should be composed of two types of elements:
  a. A collection of test stubs. These are partial implementations of mechanical, electrical and computerized components on which the tested element depends. Each stub consists of interfaces that are identical to the actual component but, usually, with simpler implementation. (Note that another fully implemented component or subsystem can also be used as a stub.)
  b. A collection of test drivers. These elements simulate the part of the system that controls (drives) the component or subsystem under test. It imparts the test inputs to the component under test and displays the test result.
- *Step 4: Perform Actual Component and Subsystem Tests.* At this step, the component or the subsystem is tested. The test infrastructure applies individual test cases, and the behavior of the component or subsystem is compared to the expected output that was defined in the specification.
- *Step 5: Evaluate, Report and Improve Tests.* As a final step, the test engineer is expected to check the test result and report any error discovered. The test engineer must be aware that the specification or the test case documentation or the test case itself is sometimes incomplete, ambiguous, contradictory or even wrong, whereas the component or subsystem under test may be fault free. Often, this is a further opportunity to find undertested or missing test requirements and perhaps an opportunity to write better test specifications and test cases.

Further Literature

- Fox (2005)
- Yang and EI-Haik (2008)

5.7.5 Integration Testing

Purpose Integration testing is the testing of systems elements (e.g., component, subsystem) to prove that they interact properly together. The intent is to expose faults in the interfaces and in the interaction between the elements. This process continues progressively until the entire system has been integrated, no element has adverse impact on another and the overall functionality of the system is achieved.

Integration testing is characterized as higher order testing compared with unit, component or subsystem testing in the sense that our aim is to detect defects primarily emanating from interactions between integrated elements. This type of testing presumes that a lower level testing has been successfully performed.

Rationale The harsh reality is that, more often than not, individual system elements that seem to work fine in isolation fail to work together properly due to functional, timing or other interface problems or just naturally interfere in the smooth operation of one another. Therefore, integration testing is needed in order to verify that the system under construction is working properly. That is, each component or subsystem performs its functionality correctly without disturbing other system components.

Very simple engineered systems may not require integration testing, but this is not the case for medium to large systems composed of hardware and software elements or systems having a large number of internal or external interfaces. In particular, integration testing is vital when the system being developed must meet real-time requirements or its architecture involves distributed computing elements or when it integrates Commercial Off-The-Shelf (COTS) with custom-developed hardware and software. Similarly, integration testing is often required when the system must meet high-quality requirements, for example, due to criticality, reliability and safety issues.

Method As mentioned, the primary goal of integration testing is to identify errors in a given configuration of a system. For that end, integration testing procedure follows these typical steps[62]:

- *Step 1: Prepare Integration Test Process.* This step includes defining the integration test objectives, determining the overall integration strategy, developing an integration test plan, design, create and verify the test case suite as well as the infrastructure for conducting the integration tests.
- *Step 2: Select Component to be Integrated.* This step includes selecting a component to be integrated with the current system. Ideally, this should be done based on the integration test plan, but often certain system

[62]Readers should distinguish between different integration process responsibilities. By and large system engineers undertake to connect system elements and see that they pass basic sanity tests (i.e., the system seems to be working). Test engineers undertake to verify whether the integrated system performs its functionality in accordance with its specifications.

components are not available as planned so certain ingenuity and flexibility is required to update the test plan and act accordingly.

- *Step 3: Integrate Selected Component with SUT.* This step includes connecting the selected component together with the rest of the system. This usually entails replacing a temporary driver or stub with an actual component or subsystem, as well as performing a rudimentary operational verification of the updated system.
- *Step 4: Perform Integration Tests.* This step includes executing the appropriate test suite. Ordinarily, this entails performing functional testing as well as structural and some performance testing.
- *Step 5: Update Relevant Testing Documentation.* This step includes updating the relevant testing documents, indicating the updated system configuration, information regarding the executed test cases and the results obtained.
- *Step 6: Perform Next Integration Test Cycle.* This step includes making a decision regarding the status of the integrated system and the available new components or subsystems for integration. Afterwards the next integration testing cycle is initiated (i.e., go to Step 2 above).

Customarily, the integration test plan is used as the vehicle to plan the technical and management implementation process. This document describes the testing pass/fail criteria, the specific testing activities and the funding, manpower and other resources necessary to undertake the integration test process. Specifically, a test plan should include the test approach as well as a detailed list of test cases for each stage of testing, the manner of recording and documenting the testing process at each stage and how results are to be analyzed. In addition, the test plan should include estimation of the required resources, infrastructure and tools.

System integration test strategy provides guidelines to test engineers on how to perform system testing activities in a rational way. Such strategy usually refers to a sequence of steps to integrate different components or subsystems. Different integration testing strategies are employed under different circumstances. We describe below the most prevalent ones: (1) top-down integration testing, (2) bottom-up integration testing, (3) sandwich integration testing and (4) big-bang integration testing.

Top-Down Integration Testing Top-down integration testing strategy is an incremental approach to the building of systems. Elements are integrated by downward moves through the system hierarchy, starting with the top element in the hierarchy. In most cases, this element is a simulated environment or a shell representing the system to which all other subsystems or components are connected. The simulated environment is used as a test driver, and test stubs are placed for the next lower level system components directly subordinate to it. These test stubs are temporary components or software programs used to emulate the behavior of the missing subsystems. Stubs are replaced, one at a time with subordinate subsystems in either depth-first (down the hierarchy) or breadth-first (across a hierarchical level) manner. The incremented system is tested and the process continues until the entire system is integrated and tested (see Figure 5.40).

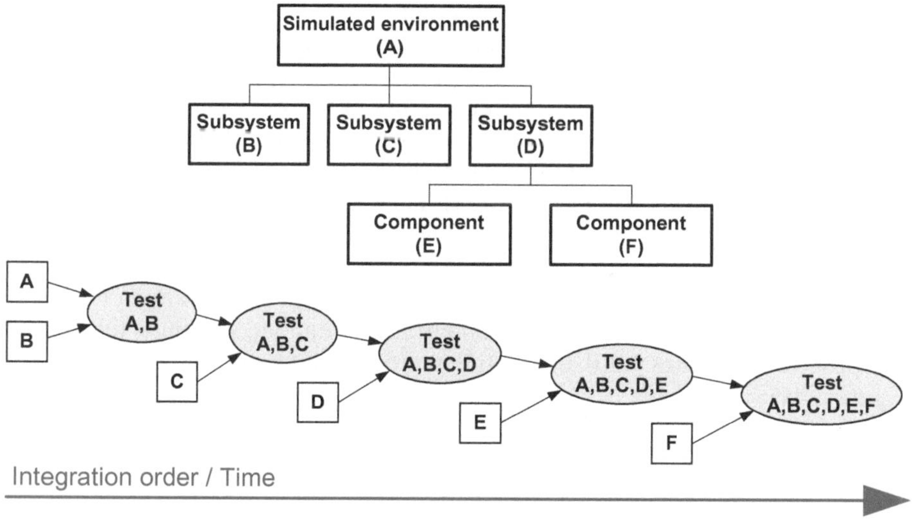

Figure 5.40 Top-down integration testing example.

The advantage of top-down integration testing strategy lies in the incremental approach that simplifies fault localization. In addition few or no system drivers are needed as the higher level components supply this control functionality. Another advantage of this strategy is that it is possible to obtain an early prototype depicting or simulating the general behavior of the final system. This can also help in early discovery of major design flaws.

The disadvantage of this strategy is the need for numerous temporary test stubs or mock objects. This requires investment in planning, designing, creating, testing, integrating and, sometimes also, maintaining these objects. Furthermore, testing upper level modules using test stubs is sometimes incomplete, as the stubs do not faithfully represent the actual component for which they temporary stand. Another disadvantage stems from low-level components, critical to system operation, which are integrated and tested late in the integration cycle. This, of course, presents an undesirable development risk.

A more sophisticated top-down approach to integration and testing of very complex systems is described earlier in Section 2.5.1 (Develop System Integration Laboratory). This approach, used in industry for a long time (i.e., over four decades in the aerospace industry), is called model-based integration testing. Several academic researchers (e.g., Braspenning, 2008) describe how executable models can replace unavailable component or subsystems, enabling early system integration and testing which expose system level problems earlier, resulting in lower costs and shorter delivery times.

Bottom-Up Integration Testing Bottom-up integration testing strategy is also an incremental approach to the building of systems. Modules are integrated by moving upwards through the system hierarchy, starting with the bottom (i.e., lowest level) components or subsystems.

Low-level components or subsystems are combined into a partial system that performs particular functions. Temporary test drivers are created in order to provide a hardware and software environment to the dynamically emerging partial systems. Test drivers, such as test stubs, are hardware and software elements that must be developed, tested and maintained. They simulate a test activation mechanism from a parent (higher level) object to a subordinate (lower level) object as well as a window for observing the behavior of the subordinate object. After this testing is completed, the test drivers are removed and the current cluster is combined or integrated with the next higher level of components or subsystems. The incrementally assembled system is tested and the process continues until the entire system is integrated and tested (see Figure 5.41).

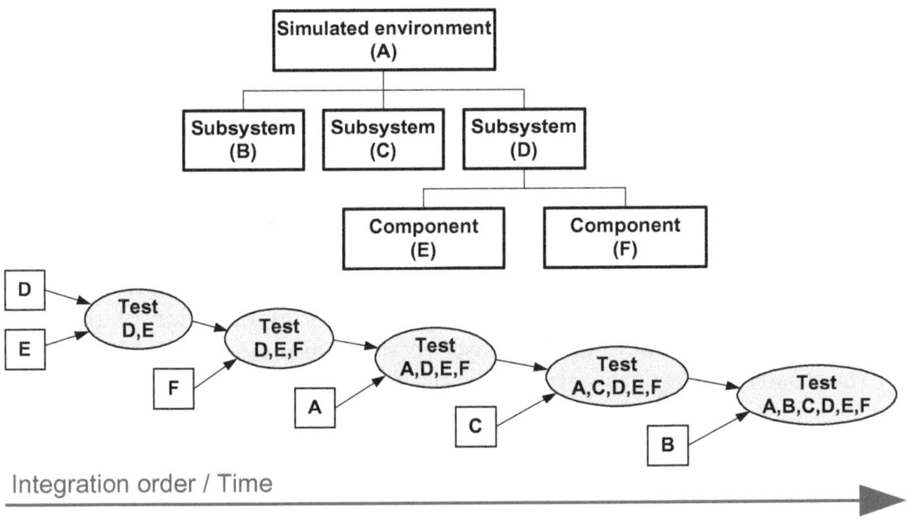

Figure 5.41 Bottom-up integration testing example.

The advantage of bottom-up integration testing strategy also lies in the incremental approach that simplifies fault localization. Another advantage of this approach is that integration and testing can be undertaken early, in fact, in parallel with the implementation of the system. Also testing of upper-level subsystems is easier and more thorough because lower components and subsystems are fully developed objects rather than stubs emulating them.

The disadvantage of this strategy is the need for numerous temporary test drivers, which require investment in planning, designing, creating, testing, integrating and, sometimes, also maintaining these support objects. Another disadvantage of bottom-up integration testing is the lack of a skeletal system until late in the integration test process. Demonstrating the system and identifying major requirements, design or implementation weaknesses are thus significantly delayed.

Sandwich Integration Testing Sandwich integration testing strategy is a combination of top-down and bottom-up strategies. Usually, two teams of engineers are working in parallel. One team uses the top-down approach to build the upper portion of the system and the other team uses the bottom-up approach to build the lower portion of the system. The process continues until there are two partial systems (upper and lower portions of the system) which then are combined, integrated and tested into the final system (see Figure 5.42). Similar to the previous strategies, this testing strategy is incremental, simplifying fault localization. The key advantage of sandwich integration testing, however, is that top and bottom portions of the system can be done in parallel, which could result in significantly shortening the integration and test period.

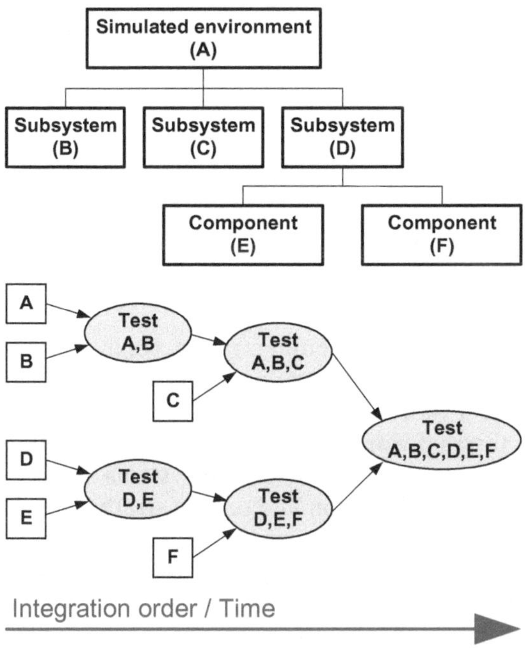

Figure 5.42 Sandwich integration testing example.

The disadvantage in this approach is that it requires more manpower resources and necessitates more coordination between the two teams. As integrating and testing the top portion requires stubs and integration and testing on the bottom portion requires drivers, the sandwich strategy inherits certain weaknesses from the previously discussed methods.

Big-Bang Integration Testing Big-bang integration testing strategy is an approach based on connecting and testing the entire set of components and subsystems into a single integrated system in one fell swoop (see Figure 5.43).

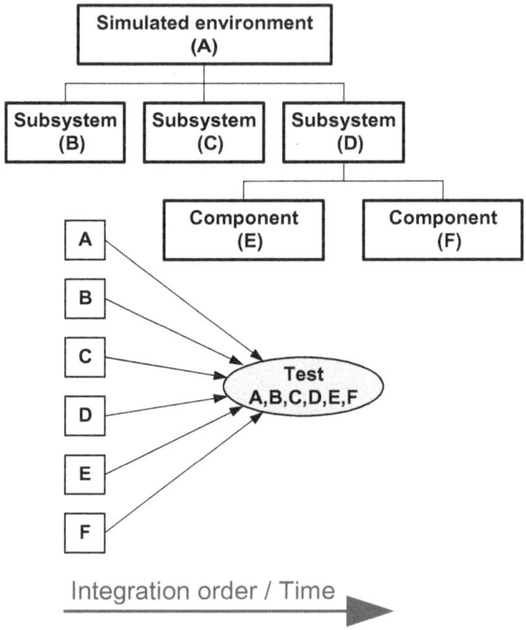

Figure 5.43 Big-bang integration testing example.

Obviously, this is a nonincremental strategy so, in an ideal situation, when the components and subsystems have been thoroughly tested, the integration and testing will take a short time as no test stubs or test drivers are required. Unfortunately, in the real world, this is not the case, and this strategy may be recommended only for very small or simple systems. The fact of the matter is that under the big-bang integration and testing strategy, it will be very difficult to localize system failures and figure out the exact mechanisms leading to individual faults. Another disadvantage of this approach is that, generally, integration testing can only begin when the vast majority of the components or subsystems are ready. Along the same line, users cannot see a skeletal system until very late in the integration and test cycle.

Further Literature

- Braspenning (2008)
- de Jong (2008)
- Duvall et al. (2007)
- Sommerville (2006)

5.7.6 Qualification Testing

Purpose The purpose of system qualification testing is to verify that a complete, integrated engineered system, composed of hardware and software elements, behaves in compliance with its specified requirements within its intended environment. The system must also be verified for proper interface with external system elements and the environment itself. System testing primarily focuses on verifying system functionality as well as certain "ancillary" test areas. Typically, these areas include many of the 16 following tests:

Performance testing	HSI testing	Temporal testing
Stress testing	Environmental testing	State machine testing
Security testing	EMI/EMC testing	Reliability testing
Interface testing	Destructive testing	Search-based testing
Usability testing	Random testing	Mutation testing
Recovery testing		

As systems become more complex and the number of possible test scenarios increase, system testing become a formidable challenge. This is so because testing consumes more time and money, and exhaustive testing is simply impossible, as testing time and resources are always limited. Therefore the challenge remains: How to assess system reliability and when to stop testing?

Rationale System qualification testing is required due to human inability to develop a system that operates perfectly on the first try. Therefore, the rationale of system qualification testing is to ensure that, after qualification testing and appropriate corrections, the system will adhere to the requirements specifications, as well as to assured system's utility and operating stability.

In practice, qualification testing provides only an overview of a system's behavior. It is generally impossible to test all aspects of the capabilities of the system under all the potential vagaries presented by its environment. Therefore, from the standpoint of instrumentation, manpower and other resources, it is not practical to test all the functional requirements at a system level. In fact, confidence in the system's behavior must also rely on extrapolating certain test results at subsystem or component levels.

Method Qualification testing is planned in accordance with the qualification/acceptance System Test Plan (SysTP) and is conducted in accordance with the qualification/acceptance System Test Description (SysTD). The test results are recorded in the qualification/acceptance System Test Report (SysTR).

The reality of system qualification testing is that not all problems with system performance will be found, no matter how thorough or systematic the testing process. This is so because of the enormous number of possible system behaviors and the inevitable limitations on testing resources (staff, time, tools, equipment, etc.). In addition, specifications are frequently incomplete, ambiguous and dynamic. Nevertheless, the testing process is vital and can be counted upon to work well when it is carefully planned and is performed by those who are independent of the developing team.

Usually, qualification testing begins only after the components, subsystems and integration tests have been completed and problems identified during those tests have been resolved. Similarly, qualification testing ends only when functional and nonfunctional tests satisfy system requirements or specifications and defects found are either eliminated or are redefined as acceptable system limitations.

Traditionally, system qualification testing employs functional as well as structural tests. Functional tests depend on a black-box approach to check that the system performs its functions as specified in the requirements. Testing the system, therefore, is based on exercising the system's inputs (using both valid and invalid inputs) and observing system outputs and performance. The SysTD must define the specific system output that should follow any input actions under a specific system state. Here, the testing coverage criteria as well as stopping criteria are based on system functional behavior. Additionally, system qualification testing should verify the error-handling capability of the system. Furthermore, the intersystem handling capability, that is, system interactions with external systems as well as the environment, must be verified as well.

Structural testing, using a white-box approach, is usually performed at the unit or component level. However, certain system level tests may be undertaken, which require a reasonable knowledge of the structure of the system under test. These white-box tests are linked to physical parts of the system under test. For example, the following tests could be considered as belonging to this category:

Performance testing	Recovery testing	Reliability testing
Stress testing	Destructive testing	Mutation testing
Interface testing	State machine testing	

Each test case within the test suite may be in either a control flow[63] testing category or a data flow[64] testing category.

In control flow testing, a test case verifies an individual control flow path segment that connects parts or nodes of the SUT. First, the functional specifications of the SUT are modeled as a directed graph. Here each element in the SUT is identified as a dedicated processing node. In addition, arcs between appropriate nodes represent the flow of control among the nodes. Next, the test engineer chooses input values that will cause the system to traverse each control path, such that each set of input control values per arc will define a test case for the system.

[63] Control flow refers to the order in which the individual commands, processes or operations by a system are executed. Control flow testing follows the same path.

[64] Data flow refers to the path that data takes through the system. Data flow testing follows the same path.

In data flow testing, a test case verifies an individual data flow path segment that connects parts or nodes of the SUT. The generation of test cases is similar, except that the arcs in the system model represent the flow of data among the nodes rather than the flow of control.

Test Automation In virtually every project, qualification testing is performed many times. Test cases, once created, are used again many times during regression testing. Testing is a repetitive process, tedious and error prone. Therefore, automating the system qualification test process, that is, running large number of test cases without human intervention, can result in fewer errors and can be accomplished in significantly shorter time.

However, the ROI question always comes up. Does the payoff from test automation justify the expense and effort involved? The test engineer must balance the advantages of test automation against its drawbacks: First, using test automation tools oblige test engineers to learn how to use them and get along with their peculiarities. Second, tests cases have a finite lifetime, and automated tests can lose relevancy when the SUT is modified. Updating test suites within automated testing is naturally more complex and therefore time consuming.

Further Literature

- Abramovici et al. (1994)
- Black (2002)
- Petschenik (2004)

5.7.7 Acceptance Testing

Purpose Acceptance testing is a formal testing process, often defined, performed and owned by the customer or the user of the system, with the aim of determining whether a system satisfies its acceptance criteria. Acceptance test is, therefore, a contractual mechanism enabling users to determine if the system built really meets their needs and expectations and whether to accept it. In other words, the purpose of acceptance testing is to confirm that the system meets the agreed-upon criteria, identify and resolve discrepancies, if there are any, and determine the readiness of the system for deployment at a user's site. These tests are often prespecified by the user and then carried out by either the user or a joint user and system developer team or by an Independent Verification and Validation (IV&V)[65] team. Of course, a user may refuse delivery of a system if acceptance test fails.

[65] IV&V is an audit process performed by an organization that is technically, managerially, and financially independent of the project and the developer to verify whether the system produced fulfills the specified requirements and to validate whether the project has met all of the stakeholders' requirements.

Acceptance testing is usually done when a system is specifically developed for a given user (e.g., corporation, government, military) by a contractor or a manufacturer and under specified contract. The unique aspect of acceptance testing is that, by and large, end users run the system in their environment using real-life, actual data to evaluate whether the system meets their criteria. In particular, acceptance testing focuses on input processing, use of the system in the user organization environment and whether or not the specifications meet the true needs of the user. The breadth of the testing is usually a subset of system qualification testing, and the test covers functionality as defined in formal documents, most often created by the user or approved by him, for example, Request For Proposal (RFP). Sometimes, auxiliary testing strategies are employed in addition, or in parallel, with a full-scale acceptance testing. This may include:

- *Benchmarking.* In this testing process, a predetermined set of test cases corresponding to typical system usage is performed in order to measure specific aspects of system behavior.
- *Pilot Testing.* In this testing strategy, users deploy the system as a small-scale experiment or in a controlled environment prior to a commitment on a full deployments'.
- *Parallel Testing.* In this testing strategy, the superseded as well as the new system are deployed in parallel and then the old system is gradually phased out.

Rationale The justification for acceptance testing is that it often uncovers requirement discrepancies as well as helps users to find out what they really want. As such, testing is owned and often carried out by the user. Acceptance testing offers an added opportunity to evaluate the system from users' perspectives rather than from system developers'.

Acceptance testing toward the end of the development cycle is also an important contractual milestone, often affecting payment schedule, transfer of the system into the user's possession and other management issues. In addition, formal system acceptance testing often occurs after each major corrective or upgrade of the system. Finally, as a lifecycle process, acceptance testing provides an opportunity to demonstrate to the user that the system is acceptable based on real-life data drawn from user sources. In addition, this process enables a formal mechanism for detection and correction of system problems.

Method System acceptance testing is an incremental process of approving or rejecting systems at the end of the development cycle or, sometimes, during maintenance. It is formally performed prior to the user accepting the ownership of that system using an acceptance test procedure. If test results of each test case consisting of the test suite meet systems specification, then the acceptance testing process is declared successful. Otherwise, the system may either be rejected or partially accepted on condition that it will be repaired upon an agreed date. Once the system has been accepted, the user will then sign off on the system as satisfying the contract.

Usually, the entry criteria for an acceptance testing are that the system qualification testing is completed and defects identified, are either fixed or documented and will be corrected within a specified timeframe. In addition, an acceptance plan has been prepared and adequate resources including test environment for the acceptance testing have been identified and are available. Likewise, exit criteria are based on completion of performing the acceptance test suite such that an acceptance decision has been reached and agreed to by the user.

The acceptance plan should include, among other items, a section on system acceptance criteria created jointly by the user and the system development team. It should define numeric values or ranges of values as well as pass or fail criteria related to each test case within the following requirement groups:

- *Group 1: Functionality Requirements.* These requirements relate to the technical or business aspect that the SUT must fulfill.
- *Group 2: Performance Requirements.* These requirements relate to the system operational aspects, such as time or resource constraints.
- *Group 3: Internal and External Interface Requirements.* These requirements relate to interfaces among components and subsystems as well as interfaces between the system and external systems and the environment.
- *Group 4: Overall System Quality Requirements.* These requirements specify quantitative system quality attributes such as reliability, testability, correctness and usability.

Inevitably, during the execution of the acceptance testing, some test criteria will not be completely satisfied. In such circumstances, the system developers and the user coordinate the resolution of any problem discovered until the user is satisfied. Thereafter, a system acceptance decision occurs after a formal verification proves that the delivered system meets all user requirements and its supporting documentation is adequate and consistent with the actual delivered system.

Roles in Acceptance Testing Depending on contractual matters, acceptance testing can become quite an involved type of system testing. Test engineers from a development organization who previously assumed a pivotal role may find themselves either completely noninvolved in testing or advisors to the user or active participants in the acceptance testing process. Acceptance testing is first and foremost the user's responsibility; therefore test engineers usually lead or are involved in the following activities:

- *Requirements and Acceptance Criteria.* The user is expected to be involved in developing the system requirements as well as the acceptance criteria throughout the system development process.
- *Acceptance Planning.* The user is expected to plan the system acceptance testing process. This includes among, other issues, how and by whom each acceptance activity will be performed and the testing schedule and associated resources, particularly, if gradual system capability and acceptance testing is contemplated.
- *Infrastructure Arrangement.* The user is expected to generate a test suite composed of individual test cases. In addition, the user is expected to arrange the hardware and software infrastructure and tool suite to conduct and manage the acceptance test process. Each test case must cover one or more system requirement together with appropriate test acceptance criteria.
- *System Testing.* The user is expected to assume primary responsibility for conducting a formal acceptance testing.
- *Decision Making.* The user should analyze actual acceptance test results, compare them with the expected ones and make a decision as to accepting or rejecting the system under test. More specifically, the user must determine whether the system is ready for production/deployment or specific additional work is needed before approval can be granted.

Further Literature

- Adzic (2009)
- Black (2002)
- Cimperman (2007)
- Windle (2007)

5.7.8 Certification and Accreditation Testing

Purpose Different scholars and organizations interpret the terms "certification" and "accreditation" in different ways. In systems engineering, certification[66] is the process of attesting that a certain product has passed performance or quality assurance tests or qualification requirements stipulated in regulations such as a building code or nationally accredited test standards or that it complies with a set of regulations governing quality or minimum performance requirements. Certification programs are often fostered or supervised by some certifying agency, such as government agencies and professional associations.

Accreditation[67] is the formal certification that a system is acceptable for use for a specific purpose. Accreditation is conferred by the organization best

[66]A slightly modified definition based on INCOSE (2007).
[67]This slightly modified definition is based on the Defense Acquisition University (DAU, 2005).

positioned to make the judgment that the system in question is acceptable. That organization may be an operational user, the program office or a contractor, depending upon the purposes intended.

In general, the purpose of certification and accreditation testing is to ensure that systems and services demonstrate that specified requirements are fulfilled and these characteristics are consistent from system to system and service to service.

In addition, certain systems and services must be certified and accredited as mandated by several U.S. federal laws, for example, the three leading types related to NIST, the National Information Assurance Partnership (NIAP) and the Defense Information Technology Security Certification and Accreditation Program (DITSCAP).

Rationale The rationale for performing certification and accreditation testing is that modern engineered systems are becoming more and more complex. Safety-critical software-rich systems, especially distributed, real-time embedded ones, permeate contemporary life and present enormous engineering challenge. Critical engineered system are used in various sectors: for example, transportation (e.g., aviation, automobile, trains, ships), energy (e.g., nuclear and conventional power stations and electrical distribution and control systems), manufacturing and chemical industries, communication, medical and the banking and financial sectors.

The rationale for performing certification and accreditation testing is that such activities should ensure, to the greatest possible degree, a secured, robust and available system that performs its intended mission with no essential flaws in design or implementation. Regrettably, the state of the art in certification and accreditation of engineered systems is that there are some support tools that, by and large, are not scalable to real-life systems. Furthermore, some of these support tools have no clear scientific or engineering foundation. The fact remains that virtually any real-life engineered system exhibits enormous behavioral permutations, so it is simply not possible to accredit it in absolute totality.

The current scientific and engineering challenge for factual certification and accreditation testing is to develop tools for assured applications which can guarantee robust, self-checking, self-healing, controllable systems, exhibiting comprehensive safety design and analysis. Such systems should provide evidence-based design assuring certifiably dependable behavior.

Method A typical certification and accreditation procedure necessary to assess systems and to provide reasonable assurance that the key concerns of such a system are carefully addressed is depicted in Figure 5.44. The key participants in the certification and accreditation process of engineered systems are relevant regulatory authorities or agencies, accredited test laboratories, system manufacturers or vendors and system users.

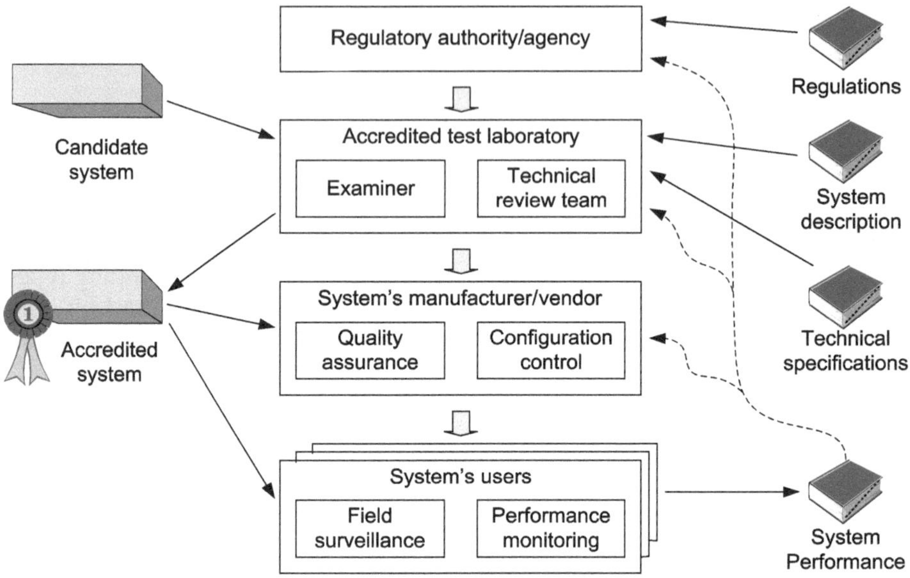

Figure 5.44 Typical system certification and accreditation process.

- *Regulatory Authority or Agency.* The primary role of a regulatory authority or agency is to ensure that the design, implementation and manufacturing of candidate systems submitted for certification and accreditation testing meet specified requirements in accordance with existing laws, regulations and technical standards.

  In addition, such bodies establish normative standards as well as refine and keep standards current. Also, they accredit testing laboratories provided that these laboratories show the required level of technical competence as well as the availability of qualified system assessors or evaluators. Finally, they monitor the performance of fielded accredited systems in order to follow up any noncompliant or defective systems and determine the appropriate response to each failure event.

  In order to achieve the above, regulatory authorities or agencies establish management structures and quality process control in order to initiate regulatory improvements, provide clarification or interpretations to the public and respond to petitions, appeals or complaints.

  Regulatory authorities or agencies frequently struggle with a multitude of fundamental certification and accreditation concerns, such as: What are the minimal criteria for systems to be considered acceptable? What, in fact, is the qualification of each accredited testing laboratory and staff? Will vendors deliver systems within manufacturing tolerances to those tested in the laboratory? Will users of certified and accredited systems, in reality, use the systems as intended?

- *Accredited Testing Laboratory.* The primary role of an accredited testing laboratory is to evaluate and test candidate systems based on existing regulations, individual system descriptions and technical specifications. The process is designed to eventually confer an appropriate certification or accreditation to that system.

  The accredited testing laboratory will receive candidate systems for testing and evaluation along with their system descriptions and technical specifications. The laboratory examiner for an engineered system will develop a test plan and then proceed to perform the actual certification or accreditation testing. The process and the results are often observed or witnessed by a technical review team as well as manufacturers or vendors. This is followed by the generation of a test report and system certification. Sometimes, certification will be issued together with certain qualifications regarding various aspects of system performance.
- *System Manufacturers or Vendors.* The primary role of the system manufacturer or vendor in the certification and accreditation process entails quality assurance, which verifies the quality of manufactured systems and processes. In addition, configuration control ensures the compatibility of all products and support documentation. Management within the manufacturing or vendor organization should ensure that all delivered certified or accredited systems meet and continue to meet the requirements over their useful life.
- *System Users.* The primary role of the users in the certification and accreditation process is to install, maintain and use these systems in the field in accordance with their intended use. In addition, users are expected to monitor the performance of their certified or accredited systems and regularly issue a system performance report regarding any failure or unexpected system behavior. This report is then made available to the relevant organizations (i.e., the regulatory authority or agency, accredited testing laboratory and the system manufacturer or vendor) for further evaluation and problem management.

Anecdote I: Mammography Systems Many modern medical devices are by definition safety critical. They can cause injury or death to patients or to medical staff. Defects associated with their design, manufacturing or operation can be grounds for multi-million-dollar litigation and court settlements.

A mammography system is a case in point where patient confidence is critical and facilities must meet the highest standards. Accreditation of these systems is considered a success story in the annals of system accreditations.

According to the U.S. National Cancer Institute and the American Cancer Society, mammography is a technique for early diagnosis of breast cancer. However, radiation from routine mammography, similar to that from a chest X-ray, poses cumulative risks of initiating and promoting breast cancer. Fortunately, annual clinical breast examination by a trained health professional together with monthly breast self-examination is safe, effective and low cost.

In 1987 the American College of Radiology (ACR) initiated a voluntary mammography accreditation program which later was transformed into specific legislation by the U.S. Congress called the Mammography Quality Standards Act (it was officially effective in 1994 and then extended in 2004). The U.S. Food and Drug Administration (FDA), which is the regulatory agency in these subjects, issued a final ruling on the matter which went into effect with new detailed requirements in 1999. Over the years, ACR members sent to the FDA over 30,000 pieces of information related to mammographic machine status as well as relevant proposals developed by radiologists, medical physicists and technologists, closing the certification and accreditation loop.

Anecdote II: Election Voting Equipment The effort to certify election voting equipment is not a success story. The effort started as an enactment of U.S. Congress Public Law 107-252, named "Help America Vote Act of 2002".[68] The importance of certification and accreditation of election voting equipment is self-evident. The consequence of having a person or a group manipulate the software embedded in such machines is of the highest national consequence.

In accordance with the above mandate, a collection of Voting System Standards (VSSs) were developed and administered by the National Association of State Election Directors (NASED) in 2002. However, this set of standards was not fully testable by independent testing laboratories. Consequently, the Voluntary Voting System Guidelines (VVSG) was developed by the Technical Guidelines Development Committee (TGDC) in conjunction with NIST in 2005.

As this book goes to print and $4 billion later, the promise of totally secured voting equipment has not been fulfilled. Some would say that the very nature of voting equipment certification is illusory and simply can not exist in the real world. More specifically, detractors of the certification and accreditation of voting equipment raise the following arguments:

- *Meeting Standards.* Based on the above-mentioned regulations, the maximum allowable error rate in the recording and processing of voting data is one in 10 million votes. With 4000 different ballot designs in the United States, the certification process is too lengthy and complex to meet these standards.
- *Computer System Modifications.* The reality of computer systems and especially of election voting equipment is that their software must be modified frequently (i.e., practically prior to each election). This is due to local, state and federal legislation changes affecting specific ballot requirements as well as fixes to eliminate newly discovered defects.

[68]The preamble to the law states its purpose: "To establish a program to provide funds to States to replace punch card voting systems, to establish the Election Assistance Commission (EAC) to assist in the administration of Federal elections and to otherwise provide assistance with the administration of certain Federal election laws and programs, to establish minimum election administration standards for States and units of local government with responsibility for the administration of Federal elections, and for other purposes".

Regrettably, system development and modification times are notoriously unpredictable and modified systems must be recertified in an accredited test laboratory in order to enable election officials to legally use these systems during an election. So, a looming specter for each election is that certified systems will not be ready for election time and voting will have to revert to hand-counted paper ballots.

- *Financial Considerations.* As it turns out, numerous municipalities and states cannot afford to update and recertify their voting equipment on a frequent cyclical basis.

In conclusion, we provide an illuminating summary of a testimony made by Dr. David Dill, Professor of Computer Science, Stanford University, before the Senate Committee on Rules and Administration, hearing on Voter Verification in the Federal Election Process, June 21, 2005[69]:

> In summary, paperless electronic voting is a technology that is fundamentally hostile to election transparency. No one can tell what is going on inside the machines, and there are no procedural changes that can remedy that flaw. Instead of seeking a technological quick fix to our election problems, we should return to paper ballot systems, and focus our energy on making our elections trustworthy by improving election practices. This can be done without reducing accessibility to voters with disabilities. Furthermore, it is the fiscally responsible choice.

Anecdote III: Columbia Space Shuttle Disaster Sometimes a certification and accreditation story can take a rather bizarre and tragic turn of events. On February 1, 2003, the space shuttle *Columbia* disintegrated during reentry into Earth's atmosphere and all seven crewmembers of flight STS-107 (Figure 5.45) perished.

(*a*) (*b*)

Figure 5.45 Space shuttle *Columbia* and (*b*) flight STS-107 crewmembers (left to right): Brown, Husband, Clark, Chawla, Anderson, McCool and Ramon (NASA pictures).

[69] Dr. David Dill's testimony is available at http://www.verifiedvoting.org/article.php?id=5789.

During the early phase of *Columbia's* launch, engineers observed a large piece of thermal insulation foam dropping from the external tank and hitting the leading edge of the left wing. Small foam debris dislodging from the external tank is a common sight during shuttle launche, however, here, a relatively large piece hit a critical wing area made of reinforced carbon–carbon composite material at a fairly high speed.

The incident was analyzed immediately after launch by NASA engineers using a certified simulator called Crater and, according to the Columbia Accident Investigation Board (CAIB F6.3-11, 2003) findings: "Crater initially predicted tile damage deeper than the actual tile depth. ... But engineers used their judgment to conclude that damage would not penetrate the dense layer of tile." In other words, a certified tool (granted, with its own significant limitations) indicates a probable catastrophic failure, but this finding is overruled on the basis of an engineering hunch.[70]

A physical test approximating the dynamic conditions of the actual event was conducted shortly after the accident. A similar piece of foam weighing approximately 0.5 kg was catapulted at a speed of approximately 230 m/sec onto a similar leading edge of a shuttle wing, creating a hole of about 20 cm. This demonstrated that the "foam debris is, in fact, the most probable cause creating the breach that led to the accident of the *Columbia* and the loss of the crew and vehicle."

NASA made numerous changes and safety improvements in the wake of the *Columbia* accident. In particular, the simulation and modeling accreditation policy has been revised in light of this fundamental problem: how to validate and accredit a modeling and simulation system in a cost-effective manner with little or no real-world data. Among other renovations, NASA accreditation of models and simulations use several accreditation layers depending on the specific purpose by the modeling and simulation users:

- *Informal Accreditation.* This type of accreditation does not include stringent mathematical formalism and the accreditation tools and approaches rely on human reasoning and subjectivity.
- *Static Accreditation.* This type of accreditation assesses the accuracy of the static model design or source code but does not require machine execution of the model. A static *walk-through* of the execution is sufficient to achieve accreditation.
- *Dynamic Accreditation.* This type of accreditation requires a model execution-and-accreditation process that evaluates the model based on its execution behavior.
- *Formal Accreditation.* This type of accreditation is based on a formal mathematical proof or correctness and requires significant effort. Formal accreditations are best suited to complex problems.

[70]This is not meant to denigrate engineering "gut feelings" which are important tools in any engineer's arsenal. However, such feelings cannot be trusted when contrary data are presented.

Further Literature

- Defense Acquisition University (2005)
- Hilderman et al. (2007)
- Hunter (2009)
- INCOSE-TP-2003-002-03.1 (DAU) (2007)
- Taylor (2006)
- Tobi (2006)

5.7.9 First Article Inspection (FAI)

Purpose The purpose of a First Article Inspection (FAI) is to compare an early manufactured physical sample or system with the specifications against which it was manufactured. The sample is usually one of the first parts or systems manufactured during the preproduction phase.

An FAI process consists of a complete, independent and documented physical and functional inspection and testing process undertaken to verify that the system under test meets all requirements and specifications. In addition, FAI consists of verifying that the methods employed during production are appropriate and fully conform to production specifications. In other words, the FAI process must assure that parts or systems can be manufactured continually and inexpensively, complying with the specification and with minimum disparity.

In the aerospace industry, this type of testing is done in accordance with aerospace standard AS9102A,[71] which establishes the requirements for performing and documenting a FAI. Primarily, the standard applies at the component level (i.e., parts, subassemblies and assemblies) but can also be effective in testing at higher levels, that is, subsystems and systems.

Rationale Within the context of this book, a FAI involves evaluating a supplier's initial, preproduction or sample system to ensure that the supplier can furnish a product conforming to all contract requirements as well as relevant quality standards and regulatory authority specifications.

The rationale for this activity stems from the methodical manner in which new systems are tested prior to transitioning to full-rate production. First article testing and approval ensure that suppliers can furnish products that conform to all contract requirements. Here, the term "approval" means the written notification from a customer to a supplier attesting to the fact that the test results of a given system have been successful so a full-scale production may proceed. The evaluation of the system under test will show if the supplier has correctly taken into consideration all the specifications. This is particularly important when the supplier has not previously furnished this type of system. Another relevant case is when the supplier did furnish such a system previously, but there have been subsequent changes in processes or specifications

[71] See SAE website, http://www.sae.org/technical/standards/AS9102A.

or that production has been discontinued for an extended period of time or an approved system developed a problem during its life.

Sometimes, a FAI is not justified. For example, such approval is not required in contracts for research and development effort only (i.e., no full-scale production is intended) or for products normally sold in the commercial market. Therefore, before requiring a first article testing and approval, the customer company should consider the nature of the contract as well as the impact of FAI on cost or time of delivery, the risk of foregoing such a test and the availability of other, less costly, methods of ensuring the desired quality.

Method A procedure for performing system FAI is presented below. It is based on Military Standard DI-NDTI-81307A and DI-QCIC-81110 as well as the Society of Automotive Engineers standard SAE-AS9102A, with specific orientation to systems rather than parts, subassemblies or assemblies.

Proposed Procedure: System First Article Inspection

Section 1: Scope and Objectives. This procedure defines the activities required to perform a FAI on new or modified customer systems. The objective of an FAI is to verify that all applicable specifications and requirements pertaining to a given system are satisfied by providing objective evidence of system conformance to specification. Another objective of an FAI is to provide evidence showing that the manufacturing process is capable of creating conforming systems. Approval to manufacture additional systems with the process being monitored will be given when the FAI indicates all contract requirements have been satisfied.

Section 2: Applicability. This subsection defines applicable systems for which an FAI is required. This usually includes components, subsystems and systems. In addition this subsection specifies the customers for whom the systems are produced.

Section 3: Related Documents. This subsection defines the relevant documents associated with the FAI process (e.g., system specifications and contractual requirements, relevant quality manuals and FAI reports or relevant government and regulatory authority documents and specifications).

Section 4: Process Flowchart for FAI. This subsection depicts a flowchart of the process for the FAI. For example, see Figure 5.46.

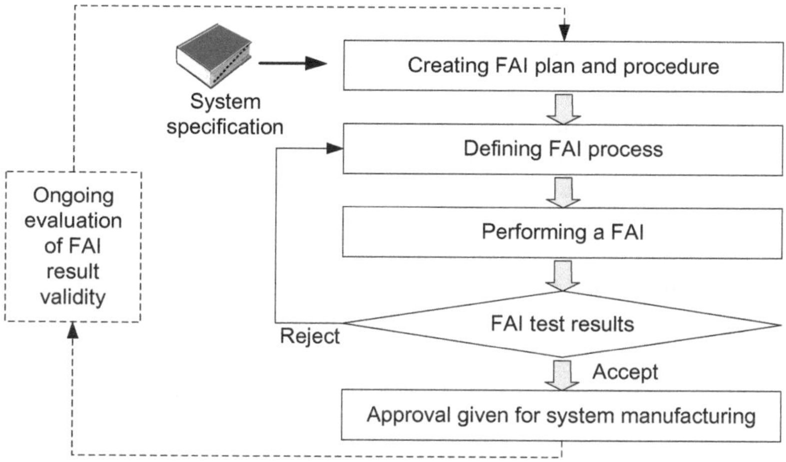

Figure 5.46 Process flowchart for FAI.

Section 5: Procedure. This subsection defines in detail the process flow for the FAI as follows:

- *Step 1: Create FAI Plan and Procedure.* During this step the FAI plan and procedure are defined. In addition, FAI is often performed in accordance with specified customer contract requirements.
- *Step 2: Define FAI Process.* During this step the FAI process is defined. This includes the process of inspecting new systems together with all relevant documentation. Typical documents to be inspected are system specifications, customer contract, customer and regulatory authority process specifications, system parts lists, bills of material, work orders, change orders and so on. Production personnel must also define the process that will be used to manufacture these systems. This may include customer models, documentation and customer change requests and machine tool data as well as standard machining practices.
- *Step 3: Perform FAI.* During this step the first article system is inspected to verify and document all applicable specifications and customer requirements that are produced and/or affected by the process. During this step, production personnel are not expected to continue manufacturing pending results of the FAI. However, production may continue at an appropriate rate under special circumstances, including considerations of system complexity and cost.
- *Step 4: FAI Test Results.* After completion of the FAI, a decision is made based on system conformance to specification. If the system does not conform to its specification and manufacturing process,

then the inspection is rejected and the system or the process must be modified to correct errors before a new first article cycle is initiated.

- *Step 5: Approval for System Manufacturing.* If the system does conform to its specifications under the specified manufacturing process, then the system is declared accepted and it is authorized for production.
- *Step 6: Ongoing Evaluation of FAI Results Validity.* Evaluation of the validity of FAI results takes place on an ongoing basis. A new first article is required under the following conditions: (1) major process modification has occurred, for example change in manufacturing machines and change in manufacturing process strategy, (2) revision in system specification that affects the manufacturing or current process and (3) major nonconformance identified during the manufacturing or use timeframe.

Section 6: Responsibilities. This subsection defines organizational responsibilities for the FAI process (e.g., customer, government and regulatory authority). This definition will cover the following:

- Responsibility for creating and maintaining FAI documentation.
- Responsibility for fabricating and supplying the first article as well as modifying the process as required.
- Responsibility for performing the first article system inspection and rendering an accept/reject decision.
- Responsibility for assigning a quality manager for the FAI process. Such a person should ensure that documents are maintained and regularly reviewed and updated.

Section 7: Record Retention. This subsection defines a record retention period of three years minimum for all documents. However, customers may stipulate longer retention times. Obsolete documents shall be removed and disposed of appropriately. By and large, all records should be available for customer or regulatory agency review.

Section 8: Document Control. This subsection defines the custodian of the documents related to the FAI process. Often, this is the quality manager which has document review and approval authority

Further Literature

- DI-NDTI-81307A (2006)
- DI-QCIC-81110 (1990)
- Myhrberg and Crabtree (2006)
- SAE (2008)

5.7.10 Production Testing

Purpose System manufacturing is a complex and precise process which invariably is not perfect. Each time a system or subsystem is fabricated, there is a finite probability that at least one of its components, wires or other elements will be faulty or exhibit abnormal behavior. Since users of a system expect each and every manufactured system to function properly, we need to have some method to identify all failing systems during the manufacturing phase. This process is called production testing and, given the complexity of today's systems, it is a critical part of manufacturing.

The purpose of conducting production tests is therefore to verify that the manufactured system meet its specifications. This testing is carried out with the basic assumption that the system was designed to specifications. Therefore only a limited subset of system functionality is tested during production. In addition, the test measures production line correctness and efficiency. This is achieved by gathering statistical information on sources of system faults as well as machine tool and assembly procedure failures.

Test engineers must define what tests are to be carried out during the production process. On the one hand, testing must cover as many potential problems as possible but, on the other hand, the intent is to test only manufacturing problems. So, the strategy is to test only to the extent necessary to ensure that the system has been manufactured correctly.

Within reason, production testing should be described in simple terms using nontechnical language where possible. Tests should be easy to perform by nontechnical people and should exercise a reasonable number of components and subsystems. In addition, it is not practical to test software in detail, but it should be possible to check the basic functionality of the software.

Rationale Experience shows that even if a system design is fully correct the manufacturing process invariably introduces some randomly distributed defects to the fabricated system. These defects come from many sources, in particular from defective components of the systems or poor workmanship or due to error introduced during the development period (e.g., incorrect requirement definition, faulty design, poor selection of components). Therefore, the rationale for performing production testing is to detect these anomalies prior to shipping systems to customers.

For example, a manufacturer of microchips produces Ultra Large Scale Integration (ULSI) having over one million transistors per microchip with 99.9% defects free. Each microchip therefore contains on average 0.1%

defective parts, that is, 1000 Defects Per Million (DPM). Imagine this product is not tested and in fact is released as is to customers with an average of 1000 defects per microchip—clearly, an alarming situation.

Testing should be carried out as an integral part of the manufacturing process to ensure quality. It is usually performed by the manufacturer or, sometimes, by an independent third party. The amount of production testing and the level of coverage should be inversely proportional to the stability of the system under test. That is, if the manufactured system is new, then usually more production testing will be needed and more defects are likely to be detected. Thus, higher test coverage is required in order to avoid releasing a system that is not properly debugged.

As discussed previously, the process of controlling inputs and observing outputs is called "sensitizing" a path to a particular fault. Sometimes, there is no combination of inputs that will sensitize a particular fault. Production testing also exhibits the problem that the test engineer may not be able to sensitize a path to a particular area he or she would like to test. This controllability and observability problem may be difficult to solve without the use of special instrumentation embedded within the system. The risk in accepting this solution is that the very existence of such probes may adversely affect the system under test.

Method Figure 5.47 depicts a manufacturing facility and a typical set of production test activities, which will be explained below:

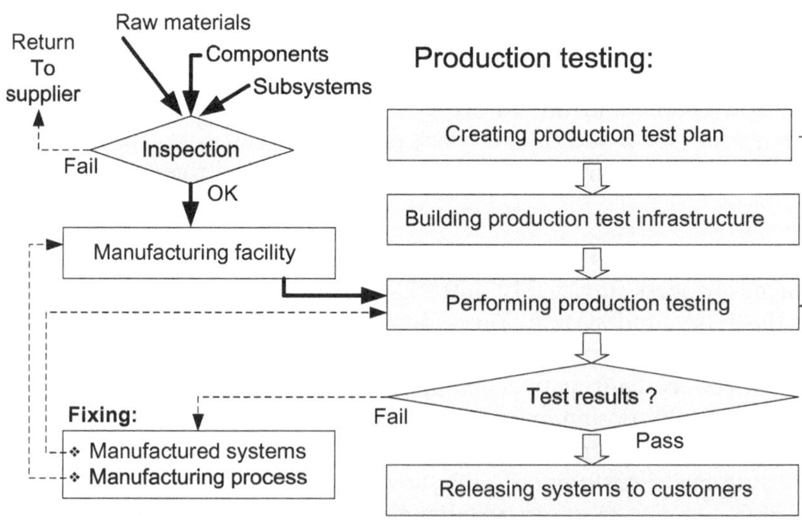

Figure 5.47 Manufacturing system and production test process.

Step 1: Create Production Test Plan. The VVT production testing starts by generating a production test plan. Planning the production test entails defining the scope, approach, resources and schedule of the testing activities as well as identifying the specific items to be tested, features to be tested, testing tasks to be performed, personnel responsible for each task and risks associated with this plan. In many ways, creating a production test plan is similar to the creation of a qualification/acceptance SysTP. This entails the following activities:

- *Test Items.* This includes identifying the test items as well as their version/revision level and supplying appropriate references and documentation to the test items.
- *Features to be Tested.* This includes identifying system features and combinations of system features to be tested as well as identifying the test design apparatus associated with each feature and each combination of features.
- *Features not to be Tested.* This includes identifying all features and significant combinations of features that will not be tested and the reasons for these decisions.
- *Approach.* This includes description of the overall approach to testing and the major activities, techniques and tools used for testing. It also includes estimation of the time required to do each activity and identification of any significant constraints on testing.
- *Items Pass/Fail Criteria.* This includes specifying criteria to determine whether a test item passed or failed during the testing process.
- *Test Deliverables.* This includes identifying the specific documents that must be generated and delivered to management, customers and other interested parties.
- *Testing Tasks.* This includes identifying the required pretesting activities and all intertask dependencies and special skills required.
- *Environmental Needs.* This includes specifying the necessary properties of the test environment. This covers physical characteristics of the facilities (e.g., hardware, software, communications and any other system or supplies needed for testing). All other special test tools, publications and so on should also be specified.
- *Responsibilities.* This includes identifying the groups responsible for test management, design, preparation, evaluation, witnessing and resolving problems. In addition, this includes identification of groups responsible for providing the test items and the environmental needs.
- *Staffing and Training Needs.* This includes specifying all test-staffing needs by skill level and identifying training options for providing the necessary skills.

- *Schedule.* This includes all relevant test milestones and timetables, that is, estimation of the time required to do each testing task and its schedule as well as association of needed resources per each timeframe.
- *Risks and Contingencies.* This includes identifying all the high-risk assumptions of the test plan and specifying appropriate contingency plans for each.

Production testing is time critical in the sense that during the production phase the VVT team must finish the testing process without disturbing the manufacturing process. This, in reality, means that only a subset of the design parameters may be tested.

Identifying this subset is one of the roles of the production test engineer. He or she can start by examining either each requirement in the System Requirements Specification (SysRS) or each test case in the qualification/acceptance SysTD document. Then, reducing test time can usually be achieved by identifying a minimum set of critical requirements/test cases which are fundamentally sensitive to system production issues. In addition, the test engineer can consider known system design weaknesses and create tests addressing these potential problems.

Step 2: Build Production Test Infrastructure. Building the test infrastructure entails the designing and purchasing or in-house building of tools and facilities to support the production test process. A key question to be addressed is whether to purchase COTS infrastructure and, especially, test equipment and test tools or develop and build the entire test infrastructure within the production organization. The advantage of producing in-house test equipment and tools is that one gets exactly what one wants. In addition, such strategy allows easy maintenance and repairs when it malfunctions. The disadvantages of this approach are often related to the time and cost involved in developing and building such equipment. Also, many production organizations do not have the resources to develop the test gear themselves. On the other hand, the advantage of purchasing COTS equipment and tools is that it is likely to perform tests in accordance with accepted standards, although it may be expensive and difficult to set up, use, maintain and repair.

Step 3: Performing Production Testing. Production testing can be divided into two groups of activities:

- *Incoming Inspections.* The purpose of incoming inspections is to ensure the quality of purchased raw materials, components and subsystems, that is, to avoid putting defective elements into the produced system. Incoming inspection is similar to production testing, but it is considered the responsibility of the receiving organization rather than the producing ones. The breadth of incoming inspection is generally tuned to the specific system application where the process investment strategy is that the cost of incoming inspections will not exceed the diagnosis and repair cost of the produced system.

- *Production Testing.* Production testing includes both the manufactured systems and the production line itself. This includes functional characteristics applied to selected portions of the produced system as well as some environmental characteristics such as effects of temperature and humidity on the behavior of the system. In addition, production line testing evaluates the manufacturing and fabrication machinery and process and collects, for example, the reliability and failure rate of the system under test as well as the production infrastructure.

Step 4: Determining Pass/Fail Status of Production Testing. Each component of the production testing—incoming inspections, manufactured systems and production line testing—may fail or succeed. If an incoming inspection fails, then the defective raw material or components must be rejected and returned to the supplier. If a manufactured system fails, then it must be corrected and the root cause of the failure (e.g., inappropriate raw materials, faulty components or subsystems, poor workmanship) must be ascertained and eliminated. If a production line testing fails, then the defects must be noted and fixed prior to releasing manufactured systems to the customers. Of course, data relating to all failures must be gathered and analyzed in order to correct any broad problems permeating the production system.

Step 5: Releasing Systems to Customers. Systems that successfully passed the production testing will be released to the customers.

Further Literature

- Bray and Stanley (1997)
- Breyfogle (1992)
- Papadakis (2006)

5.7.11 Installation Testing

Purpose In general, system installation is the process of relocating a system from its development environment, establishing it in a customer's or user's site and making it operational for the people who will use it. Usually, both the customer and the producer of the system are involved in the installation process where the customer gradually takes over responsibility as installation proceeds. However, in "turnkey" contracts, the system's producer assumes sole responsibility for system installation.

Common installation categories are new-system installation, system uninstallation and system reinstallation after repair or upgrade. The installation process itself can vary from simply updating a software package on an existing computer to constructing new facilities with complete infrastructure and installing newly produced systems. Regardless of the simplicity or complexity of the installation process, the system must be tested to ensure that the customer will be able to operate it successfully.

The purpose of installation testing therefore is to ensure that the installed system operates in accordance with all specified requirements within the system's operational environment. Often, the deployment of a newly installed system must be synchronized with an operating, live environment in order to minimize disruption to ongoing operations. In such cases, the deployment plans should include optional back-out procedures for rolling the target environment back on unsuccessful deployment.

Rationale In some cases, systems may be integrated at the place where they will actually be operated, but usually systems, especially computer-based ones, are designed, implemented and integrated at the developer's premises and then delivered to the customer for installation and deployment. The probability that such a system will automatically work harmoniously within the customer environment is slim. Therefore, the rationale for conducting installation testing is to verify that all necessary components of the system are indeed installed and to ensure that all installed features and options function properly.

Method Depending on the nature of the system being installed, the process may be simple (e.g., installing new application software in a personal computer) or complex, taking minutes, days, weeks or months to complete. The more complex installations must be carefully planned and carried out. The installation test engineer must appreciate the widespread problems associated with installation of new systems; in particular, physical problems (e.g., services at the installation facility are not ready or are unsuitable), compatibility problems (e.g., inappropriate interfaces between the installed system and other systems) and human problems (e.g., operators are not fully familiar with the system or resist its new operational features). This is especially relevant to installations of "few-of-a-kind" systems (e.g., system integration laboratory, plant assembly line). The following provides guidelines for the installation testing of typically complex systems:

- *Installation Planning.* The test engineer should verify that the manufacturer and the customer carry out installation planning. This will minimize the cost and time of the installation process as well as lessen the disruption to the customer's business. Such planning should take into account customer and manufacturer staff and equipment availability, customer intents for system introduction as well as phasing out any old systems. In addition, the test engineer should ensure that the training of relevant customer personnel who need to operate the new system is adequately planned.

- *Installation Preparations.* The test engineer should verify that the manufacturer together with the customer appraises the physical site where the system will be installed. The chosen site must be accessible to vehicles and people (and sometimes air and rail transport) and all needed utilities (water, electricity, communications, etc.) should be available to support the installed system and associated environment.
- *Data Harmonization.* The test engineer should verify that the manufacturer together with the customer harmonizes existing data elements in the old/original systems with data elements within the installed new system. Data harmonization is a dynamic process in which either the new system must be built to utilize original data or a new data structure must be redefined and then permeates throughout the organization. That is, when a new system is installed, it is often necessary to convert the original data used by the organization to the new data structure used by the new system.
- *Installation Process.* First, the manufactured system should be physically moved from the manufacturer's site to the customer's site and be installed there. Thereafter, the test engineer should verify that the manufacturer together with the customer carry out the installation of the hardware and software according to plan and in an appropriate manner.
- *Installation Testing.* Once the system is installed, the test engineer should start the testing sequence with, as close as possible, real data. He or she may encounter initial operational problems such as interface incompatibility between the installed system and other systems in the same environment, configuration mismatch requiring readjusting the installed system, data definition and format problems.
- *Training Operators.* In parallel with the above activities, the test engineer should verify that the manufacturer or an organization on his or her behalf carries out appropriate training of the personnel who will have to operate the new system.
- *Parallel Operations.* Often, it is necessary to run the new installed system in parallel with the existing old system during a transition period. The test engineer should verify that the parallel operation process is harmonized and processing results from the two systems are periodically compared. One way to achieve parallel operation is by dividing the system users such that some of them are serviced by the old system and some by the newly installed system. Sometimes, it may be necessary to maintain the old system as a backup for the newly deployed system.

Further Literature

- Lewis (2000)
- Pfleeger and Atlee (2009)
- Wasson (2005)

5.7.12 Maintenance Testing

Purpose The purpose of maintenance testing is to identify system or equipment problems, confirm that repair measures have been effective or establish that system adaptation to a new environment or system improvement has been successful.

To be more specific, preventive maintenance includes performing systematic inspection, detection, and correction of system failures before they actually occur. This is done in order to decrease the operational cost of the system and increase overall system maintainability and availability. Maintenance testing must validate that overall maintainability and availability have increased according to stated specifications. Corrective maintenance includes correcting problems that arise during the course of using the system. Maintenance testing must validate that the updated system does not exhibit the previously detected problems and no new defects have been introduced. Adaptive maintenance includes the hardware or software modification to the system necessitated due to changes in the physical environment of the system. Maintenance testing must validate that an updated system operates within the new environment according to stated specifications and all previously functional capabilities have been maintained. Finally, perfective[72] maintenance includes modifying and improving the system emanating from new and different user requirements. Maintenance testing must validate that the updated system fulfills new user requirements and that all previously functional capabilities have been maintained.

A comprehensive maintenance testing process can be illustrated via an in-service scheduled inspection of a passenger car. During such a process, certain components and perishables are replaced based on manufacturer recommendations as a preventive maintenance measure. Periodic vehicle inspections are usually mandated by regulatory authorities for testing safety features (e.g., brakes, lights) and environmental protection features (e.g., exhaust emissions) and faulty parts are replaced as a maintenance corrective measure. If, for example, the environment regulations are changed, then the vehicle may be fitted with a more effective catalytic converter, for example, as an adaptive maintenance measure. Finally, if the owner wants to upgrade his car and install, for example, an automatic parking system to perform parallel parking, then such a system may be incorporated into the vehicle as a perfective maintenance measure.

Rationale The justification for maintenance testing is that, by and large, systems naturally age and components deteriorate; however, equipment failure is not inevitable. An effective maintenance testing program identifies and recognizes factors leading to a system's decline. It provides measures for

[72]See definition later in this section.

reversing these effects and avoiding failures. A well-administered maintenance testing program can prevent accidents, save lives and minimize costly breakdowns or repairs. Users and the public at large are becoming progressively more dependent upon continued, uninterrupted system operation. The reliability and integrity of a system are based on an established program of maintenance and operational testing. The maintenance procedures and frequencies should follow those recommended by manufacturers, standards and relevant legislation.

Maintenance testing may be performed at different levels, for instance, at the system level (e.g., aircraft avionics) or at a subsystem level [e.g., Inertial Navigation System (INS)] or at a component level (e.g., INS power supply unit). In addition, maintenance testing tends to shift its emphasis over time. During the system's introductory stage, the emphasis is on user support. During the system's continuous-usage stage, the emphasis is on correcting faults. During the system's maturity stage, the emphasis is on enhancements to the system, and during the system's decline, the emphasis is on technology upgrades.

Method Generally, in system engineering, the term maintenance refers to any activity such as tests, measurements, replacements, adjustments and repairs intended to retain or restore a system to a specified state in which it can perform its required functions and, as a rule, maintenance testing uses system performance requirements as the basis for testing. Obviously, the method of testing is dependent on the category of maintenance testing:

- *Preventive Maintenance Testing.* Preventive maintenance is a strategy of replacing system elements on a scheduled basis, regardless of their condition, in order to reduce the risk of operational failure. The role of the test engineer is to verify whether the preventive maintenance is done in accordance with the relevant specifications and whether it does not introduce unexpected defects into the system. Preventive maintenance specifications should indicate when and which specific activities must be performed at any given time slot as well as the nature of allowable system and service interruption. Testing the system for proper functional behavior after a preventive maintenance is essentially a regression testing which must correspond to the type and nature of the preventive activities.
- *Corrective Maintenance Testing.* Corrective maintenance consists of activities aimed at restoring a failed system to operational status. This usually involves replacing or repairing the elements which are responsible for the failure of the system. The role of the test engineer includes inspection of the system under test on a regular basis in order to identify failure modes which lend themselves to condition monitoring. Such a process may allow reclassification of corrective activities into preventive mainte-

nance, in other words, enabling the corrective process to be performed prior to actual system failure and under a planned schedule in tandem with preventive maintenance. Once a problem has been identified in the operational system, the test engineer should also diagnose the system problem and identify the failed system element or otherwise the causes of the system failure. Once the system has been repaired, the test engineer must verify whether the repair action has brought the system into a full operating state and no other system functionality has been compromised. Finally, the test engineer must analyze and identify the reason for the failure so actions may be taken to eliminate or reduce the root cause and frequency of future similar failures.

- *Adaptive Maintenance Testing.* Adaptive maintenance is an activity of changing elements of a system in order to adapt it to changes in its physical, regulatory or operational environment. Generally, adaptive maintenance does not provide new system capabilities. The role of the test engineer is to verify whether the modified system meets the specifications of the new environment and no other system functionality has been compromised.
- *Perfective Maintenance Testing.* Perfective maintenance is an activity of changing elements of a system in order to enhance it in accordance with specific requirements stated by users and stakeholders. In general, enhancements may involve additional functional capabilities or improvements in the safety, reliability, efficiency or cost-effectiveness of the system. Similarly, the role of the test engineer is to verify whether the modified system meets the new specified requirements and no other system functionality has been compromised.

Maintenance Testing Data The test engineering group should maintain a record of all maintenance test actions and results. These data should be analyzed for trends and serve as the basis for decisions on appropriate testing frequency, need to replace or upgrade system elements, performance improvement opportunities and the like.

Commercially available database software tools should be used to support the maintenance testing process. Such tools help test engineers improve the availability of systems as well as reduce costs and system idle time during repairs. Specifically, such tools can assist in performing project maintenance planning and managing the execution of events, administer the organization's system assets (systems, components and tools inventories), interface with a relevant organization's departments, suppliers and customers as well as improve the organization' knowledge base. Examples of knowledge base management may include preserving a maintenance service history and calculating reliability data such as Mean Time Between Failure (MTBF) and Mean Time To Repair (MTTR) as well as generation and maintenance of repair documentation and best practices.

Further Literature

- Jarzabek (2007)
- Levitt (2002)
- Palmer (2005)
- Pigoski (1996)
- Polo et al. (2002)

5.7.13 Disposal Testing

Purpose Upon completion of its useful life, systems and their constituent elements are disposed of, one way or another. During this phase, systems should be dismantled, partially or completely recycled and shredded and finally any toxic materials found must be neutralized. The majority of engineered systems have no formal disposal requirements. However, most systems with hazardous materials or other safety issues have specific disposal requirements related to environmental protection, levels of materials recovery and methods of disposal.

The purpose of disposal testing is to participate in the system disposal process by (1) developing a VVT plan for the system disposal, (2) assessing the overall system disposal process plan, notwithstanding safety, environmental and economic issues as well as relevant statutory considerations, (3) assessing the optimality of the system disposal strategy in terms of environmental impact and the level of recycle-ability by means of simulation and (4) verifying that the ongoing system disposal process is performed according to the relevant system disposal plan as well as applicable environmental and health regulations and policies.

Rationale The justification for disposal testing is threefold. First, system disposal is often required by law in many industries and therefore must be verified by the relevant VVT team of the disposing organization. Second, proper disposal yields considerable economic savings by extracting from obsolete systems various substances, metals and other materials for recycling purposes. Finally, proper disposal of systems and protection of the environment as well as the existing ecosystems are a vital interests of society at large and therefore apply to each member of the VVT team.

Method Disposal methods are unique to individual industries as well as to specific legislations applicable to individual countries. Therefore, no specific method is described here. The reader is advised to go to Chapter 3 and also to the appropriate literature recommended below for further information.

Further Literature

- Blanchard and Fabrycky (2005)
- IEEE 1058-1998 (1998)
- Lippitt et al. (2000)
- NASA/SP-2007 6105 (2007)
- Ogata (2003)
- Spinner (1991)
- Richard (2002)
- SEF DoD (2001)
- Zahavi and Barlam (2000)
- Zienkiewicz and Morgan (2006)

5.8 REFERENCES

Abramovici, M., Breuer, A. M., and Friedman, D. A., *Digital Systems Testing & Testable Design*, rev. ed., Wiley-IEEE Press, New York, 1994.

Adzic, G., *Bridging the Communication Gap: Specification by Example and Agile Acceptance Testing*, Neuri Limited, 2009.

Ahmed, A., *Software Testing as a Service*, Auerbach Publications, 2009.

Apiwattanapong, T. O., and Harrold, M. J., Leveraging Field Data for Impact Analysis and Regression Testing, in *Proc. of the 9th European Software Engineering Conference*, 2003.

Basta, A., and Halton, W., *Computer Security and Penetration Testing*, Delmar Cengage Learning, 2007.

Beck, K., *Test Driven Development: By Example*, Addison-Wesley Professional, Reading, MA, 2002.

Beizer, B., *Software Testing Techniques*, 2nd ed., International Thomson Computer Press, 1990.

Belapurkar, A., Chakrabarti, A., Ponnapallis, H., Varadarajan, N., Padmanabhuni, S., and Sundarrajan, S., *Distributed Systems Security: Issues, Processes and Solutions*, Wiley, Hoboken, NJ, 2009.

Benso, A., and Prinetto, P. (Eds.), *Fault Injection Techniques and Tools for Embedded Systems Reliability Evaluation*, Springer, 2003.

Bin, E., Ziv, A., and Shmuel, U. (Eds.), *Hardware and Software, Verification and Testing: Second International Haifa Verification Conference*, HVC 2006, Haifa, Israel, October 23–26, 2006, Springer, 2007.

Black, R., *Managing the Testing Process: Practical Tools and Techniques for Managing Hardware and Software Testing*, 2nd ed., Wiley, Hoboken, NJ, 2002.

Black, R., *Pragmatic Software Testing: Becoming an Effective and Efficient Test Professional*, Wiley, Hoboken, NJ, 2007.

Blanchard, S. B., and Fabrycky, W. J., *Systems Engineering and Analysis*, 4th ed., Prentice Hall, upper Saddle River, NJ, 2005.

Braspenning, N., *Model-Based Integration and Testing: Bridging the Gap between Academic Theory and Industrial Practice*, VDM Verlag, 2008.

Bray, E. D., and Stanley, K. R., *Nondestructive Evaluation: A Tool in Design, Manufacturing, and Service*, CRC Press, Boca Raton, FL, 1997.

Breyfogle, W. F., *Statistical Methods for Testing, Development, and Manufacturing*, Wiley-Interscience, New York, 1992.

Broy, M., Jonsson, B., Katoen, J.-P., Leucker, M., and Pretschner, A. (Eds.), *Model-Based Testing of Reactive Systems: Advanced Lectures*, Springer, 2005.

Burnstein, I., *Practical Software Testing: A Process-Oriented Approach*, Springer, 2003.

Chan, A. H. (Ed.), *Accelerated Stress Testing Handbook: Guide for Achieving Quality Products*, Wiley-IEEE Press, Hoboken, NJ, 2001.

Charlton, G. S., and O'Brien, G. T. (Eds.), *Handbook of Human Factors Testing and Evaluation*, 2nd ed., CRC Press, Boca Raton, FL, 2001.

Cimperman, R., *UAT Defined: A Guide to Practical User Acceptance Testing*, Addison-Wesley Professional (Kindle edition), Reading, MA, 2007.

Copeland, L., *A Practitioner's Guide to Software Test Design*, Artech House, 2004.

David, R., *Random Testing of Digital Circuits: Theory and Application*, CRC Press, Boca Raton, FL, 1998.

Defense Acquisition University (DAU), *Systems Engineering Fundamentals*, DAU, 2005.

de Jong, I., *Integration and Test Strategies for Complex Manufacturing Machines: Integration and Testing Combined in a Single Planning and Optimization Framework*, VDM Verlag, 2008.

DI-NDTI-81307A, Data Item Description, First Article Qualification Test Plan and Procedures, Nov. 2006.

DI-QCIC-81110, Data Item Description, Inspection and Test Plan, Dec. 1990.

Do, H., and Rothermel, G., An Empirical Study of Regression Testing Techniques Incorporating Context and Lifetime Factors and Improved Cost-Benefit Models, paper presented at the International Symposium on Foundations of Software Engineering, Portland, OR, November 2006.

DoD 5200.28-STD, Trusted Computer System Evaluation Criteria, U.S. Department Of Defense, December 1985.

Dumas, S. J., and Redish, C. J., *A Practical Guide to Usability Testing*, Intellect, 1999.

Dustin, E., Rashka, J., and Paul, J., *Automated Software Testing: Introduction, Management and Performance*, Addison-Wesley, Reading, MA, 1999.

Duvall, P., Matyas, S., and Glover, A., *Continuous Integration: Improving Software Quality and Reducing Risk*, Addison-Wesley Professional, Reading, MA, 2007.

Fenton, E. N., and Pfleeger, L. S., *Software Metrics: A Rigorous and Practical Approach*, 2nd ed., Course Technology, 1998.

FIPS PUB 199, Federal Information Processing Standards Publication, Standards for Security Categorization of Federal Information and Information Systems, U.S. Computer Security Division, Information Technology Laboratory, National Institute of Standards and Technology, February 2004.

Fox, B., *Test and Evaluation Trends and Costs for Aircraft and Guided Weapons*, RAND Corporation, 2005.

Guastello, J. S., *Human Factors Engineering and Ergonomics: A Systems Approach*, Lawrence Erlbaum Associates, 2006.

Harel, D., and Naamad, A., The STATEMATE Semantics of Statecharts, *ACM Trans. Software Eng. Methodol.*, 5(4), 293–333, October 1996.

Hiermaier, J. S., *Structures Under Crash and Impact: Continuum Mechanics, Discretization and Experimental Characterization*, Springer, 2007.

Hilderman, V., Baghai, T., and Buckwalter, L. (Ed.), *Avionics Certification: A Complete Guide to DO-178 (Software), DO-254 (Hardware)*, Avionics Communications, 2007.

Hunter, D. R. *Standards, Conformity Assessment, and Accreditation for Engineers*, CRC Press, Boca Raton, FL, 2009.

IEEE 1058-1998, Standard for Software Project Management Plans, sponsored by IEEE Computer Society, New York, December 22, 1998.

INCOSE-TP-2003-002-03.1, C. Haskins (Ed.), Systems Engineering Handbook—A Guide for System Lifecycles Processes and Activities, Version 3.1, International Council on Systems Engineering, August 2007.

Jain, K. R., *The Art of Computer Systems Performance Analysis: Techniques for Experimental Design, Measurement, Simulation, and Modeling*, Wiley, New York, 1991.

Jarzabek, S., *Effective Software Maintenance and Evolution: A Reuse-Based Approach*, Auerbach Publications, 2007.

Kabisatpathy, P., Barua, A., and Sinha, S., *Fault Diagnosis of Analog Integrated Circuits*, Springer, 2005.

Kan, H. S., *Metrics and Models in Software Quality Engineering*, 2nd ed., Addison-Wesley Professional, Reading, MA, 2002.

Kaner, C., Bach, J., and Pettichord, B., *Lessons Learned in Software Testing*, Wiley, Hoboken, NJ, 2001.

Karr, C., and Freeman, M. L. (Eds.), *Industrial Applications of Genetic Algorithms*, CRC Press, Boca Raton, FL, 1998.

Kececioglu, B. D., and Sun, F.-B., *Environmental Stress Screening: Its Quantification, Optimization and Management*, DEStech Publications, 2003.

Krstic, A., and Cheng, K.-T., *Delay Fault Testing for VLSI Circuits*, Springer, 1998.

Lammermann, F., Baresel, A., and Wegener, J., Evaluating Evolutionary Testability with Software Measurements, in *Proceedings of the Genetic and Evolutionary Computation Conference, Part 2—GECCO 2004*, Seattle, WA, June 26–30, 2004.

Lammermann, F., Baresel, A., and Wegener, J., Evaluating Evolutionary Testability for Structure-Oriented Testing with Software Measurements, *Appl. Soft Comput.*, 8(2), 1018–1028, March 2008.

Lavagno, L., Martin, G., and Scheffer, L., *Electronic Design Automation for Integrated Circuits Handbook*, CRC, Press, Boca Raton, FL, 2006.

Lavi, Z. J., and Kudish, J., *Systems Modeling and Requirements Specification Using ECSAM: An Analysis Method for Embedded and Computer-Based Systems*, Dorset House Publishing, 2004.

Levitt, J., *Complete Guide to Predictive and Preventive Maintenance*, Industrial Press, 2002.

Lewis, W., *Software Testing and Continuous Quality Improvement*, Auerbach, 2000.

Lippitt, J., Webb, P., and Martin, W., *Hazardous Waste Handbook*, 3rd ed., Butterworth-Heinemann, 2000.

Mao, C., Lu, Y., and Zhang, J., Regression Testing for Component-Based Software via Built-In Test Design, in *Proc. of the ACM Symposium on Applied Computing* (SAC 2007), Seoul, Korea, March 11–15, 2007.

Mardiguian, M., *EMI Troubleshooting Techniques*, McGraw-Hill Professional, New York, 1999.

Marick, B., *The Craft of Software Testing: Subsystems Testing Including Object-Based and Object-Oriented Testing*, Prentice-Hall PTR, Englewood Cliffs, NJ, 1994.

Miettinen, K., Neittaanmäki, P., Mäkelä, M. M., and Périaux, J. (Eds.), *Evolutionary Algorithms in Engineering and Computer Science: Recent Advances in Genetic Algorithms, Evolution Strategies, Evolutionary Programming, Genetic Programming and Industrial Applications*, Wiley, New York, 1999.

MIL-HDBK-781A, Reliability Test Methods, Plans and Environments for Engineering, Development, Qualification and Production, U.S. Department of Defense, 1996.

MIL-HDBK-2164A, *Handbook Environment Stress Screening Process for Electronic Equipment*, Revision A, U.S. Department of Defense, June 19, 1996.

MIL-STD-461E, Interface Standard Requirements for the Control of Electromagnetic Interference Characteristics of Subsystems and Equipment, Revision E, U.S. Department of Defense, August 20, 1999.

Molyneaux, I., *The Art of Application Performance Testing: Help for Programmers and Quality Assurance*, O'Reilly Media, 2009.

Montrose, I. M., and Nakauchi, M. E., *Testing for EMC Compliance: Approaches and Techniques*, Wiley-IEEE Press, Hoboken, NJ, 2004.

Myers, J. G., Sandler, C., Badgett, T., and Thomas, M. T. (Eds.), *The Art of Software Testing*, 2nd ed. Wiley, Hoboken, NJ, 2004.

Myhrberg, V. E., and Crabtree, H. D., *A Practical Field Guide for AS9100*, ASQ Quality Press, 2006.

Musumeci, D. G. P., and Loukides, M. (Ed.), *System Performance Tuning*, 2nd ed., O'Reilly Media, 2002.

NASA/SP-2007 6105, *NASA Systems Engineering Handbook*, Revision 1, National Aeronautics and Space Administration, NASA Headquarters, Washington, DC, December 2007.

Nelson, B. W., *Accelerated Testing: Statistical Models, Test Plans, and Data Analysis*, Wiley-Interscience, Hoboken, NJ, 2004.

Nilsson, R., A Mutation-Based Framework for Automated Testing of Timeliness, Ph.D. Thesis, No. 10030, Department of Computer and Information Science, Linkoping University, Sweden, October 18, 2006.

Nilsson, R., and Offutt, J., Automated Testing of Timeliness: A Case Study, paper presented at the Second International Workshop on Automation of Software Test, (AST '07), Minneapolis, USA, May 20–26, 2007.

NIST 800-53, Recommended Security Controls for Federal Information Systems and Organizations, National Institute of Standards and Technology, February 2009.

Nordhoff, S. L., Freeman, D. M., and Siegmund, P. G., *Human Subject Crash Testing: Innovations and Advances*, SAE International, 2007.

O'Connor, T. D. P., *Practical Reliability Engineering*, 4 ed., Wiley, Hoboken, NJ, 2002.

Ogata, K., *System Dynamics*, 4th ed., Prentice-Hall, Upper Saddle River, NJ, 2003.

Palmer, R., *Maintenance Planning and Scheduling Handbook*, 2nd ed., McGraw-Hill Professional, New York, 2005.

Papadakis, P. E., *Financial Justification of Nondestructive Testing: Cost of Quality in Manufacturing*, Press, Boca Raton, FL, 2006.

Paul, R. C., *Introduction to Electromagnetic Compatibility*, 2nd ed., Wiley-Interscience, Hoboken, NJ, 2006.

Petschenik, N., *System Testing with an Attitude*, Dorset House, 2004.

Pfleeger, L. S., and Atlee, M. J., *Software Engineering: Theory and Practice*, 4th ed., Prentice-Hall, Upper Saddle River, NJ, 2009.

Pigoski, M. T., *Practical Software Maintenance: Best Practices for Managing Your Software Investment*, Wiley, New York, 1996.

Polo, M., Piattini, M., and Ruiz, F., *Advances in Software Maintenance Management: Technologies and Solutions*, IGI Global, 2002.

Porter, A., *Accelerated Testing and Validation*, Newnes, 2004.

Raheja, G. D., and Allocco, M., *Assurance Technologies Principles and Practices: A Product, Process, and System Safety Perspective*, 2nd ed., Wiley-Interscience, Hoboken, NJ, 2006.

Reorda, S. M., Peng, Z., and Violante, M. (Eds.), *System-Level Test and Validation of Hardware/Software Systems*, Springer, 2005.

Richard, C. P., *The Economics of Waste*, RFF Press, 2002.

Rubin, J., and Chisnell, D., *Handbook of Usability Testing: How to Plan, Design, and Conduct Effective Tests*, 2nd ed., Wiley, Hoboken, NJ, 2008.

SEF DOD, Systems Engineering Fundamentals, Department of Defense, supplementary text prepared by the Defense Acquisition University Press, Fort Belvoir, VA, 2001.

Shneiderman, B., and Plaisant, C., *Designing the User Interface: Strategies for Effective Human-Computer Interaction*, 4th ed., Addison-Wesley, Reading, MA, 2004.

Shneiderman, B., Plaisant, C., Cohen, M., and Jacobs, S., *Designing the User Interface: Strategies for Effective Human-Computer Interaction*, 5th ed., Addison-Wesley, Reading, MA, 2009.

Shore, J., and Warden, S., *The Art of Agile Development*, O'Reilly Media, 2007.

Society of Automotive Engineers (SAE), *Advances in Modeling and Testing of Materials and Vehicle Structures for Crash Safety Applications*, SAE, 2005.

Society of Automotive Engineers (SAE), *2008 SAE Handbook*, SAE, 2008.

Solomon, G. M., and Chapple, M., *Information Security Illuminated*, Jones and Bartlett, 2005.

Sommerville, I., *Software Engineering: Update*, 8th ed., Addison-Wesley, Reading, MA, 2006.

Spinner, P. M., *Elements of Project Management: Plan, Schedule, and Control*, Prentice-Hall, Englewood Cliffs, NJ, 1991.

Stamatis, H. D., *Six Sigma and Beyond: Design for Six Sigma*, Vol. VI, CRC Press, Boca Raton, FL, 2002.

Sthamer, H. H., The Automatic Generation of Software Test Data Using Genetic Algorithms, Ph.D. Thesis, Department of Electronics and Information Technology, University of Glamorgan, Wales, UK, 1996.

Taylor, L., *FISMA Certification & Accreditation Handbook*, Syngress, 2006.

Tobi, N., The EAC Voting Equipment Certification Ponzi Scheme, paper presented at We Count Conference, Cleveland, OH, 2006, available: http://www.bbvforums.org/forums/messages/73/PONZI-45278.ppt.

Tullis, T., and Albert, W., *Measuring the User Experience: Collecting, Analyzing, and Presenting Usability Metrics*, Morgan Kaufmann, 2008.

Voas, M. J., and McGraw, G., *Software Fault Injection: Inoculating Programs Against Errors*, Wiley, New York, 1998.

von Mayrhauser, A., Scheetz, M., Dahlman, E., and Howe, E. A., Planner Based Error Recovery Testing, Software Reliability Engineering, in *ISSRE 2000 Proceedings, 11th International Symposium on Software Reliability Engineering*, San Jose, CA October 8–11, 2000, pp. 186–195.

Wasserman, G., *Reliability Verification, Testing, and Analysis in Engineering Design*, CRC Press, Boca Raton, FL, 2002.

Wasson, S. C., *System Analysis, Design, and Development: Concepts, Principles, and Practices*, Wiley-Interscience, Hoboken, NJ, 2005.

Wegener, J., Baresel, A., and Sthamer, H., Evolutionary Test Environment for Automatic Structural Testing, *Inform. Software Technol.*, 43, 841–854, 2001.

Wegener, J., and Bühler, O., Evaluation of Different Fitness Functions for the Evolutionary Testing of an Autonomous Parking System, in *Proceedings of the Genetic and Evolutionary Computation Conference, Part 2—GECCO 2004*, Seattle, WA, June 26–30, 2004.

Wegener, J., and Frank, M., A Comparison of Static Analysis and Evolutionary Testing for the Verification of Timing Constraints, *Real-Time Syst.*, 21(3), 241–268, 2001.

Wegener, J., and Grochtmann, M., Verifying Timing Constraints of Real-Time Systems by Means of Evolutionary Testing, *Real-Time Systems*, 15(3), 275–298, 1998.

Windle, J., *User Acceptance Testing—A Practical Approach*, James Windle, 2007.

Wise, A. J., Hopkin, D. V., and Stager, P. (Eds.), *Verification and Validation of Complex Systems: Human Factors Issues*, Springer, 1993.

Yang, K., and El-Haik, B., *Design for Six Sigma: A Roadmap for Product Development*, 2nd ed. McGraw-Hill Professional, 2008.

Yarmolik, N. V., and Demidenko, N. S., *Generation and Application of Pseudorandom Sequences for Random Testing*, Wiley, New York, 1988.

Zahavi, E., and Barlam, D., *Nonlinear Problems in Machine Design*, CRC Press, Boca Raton, FL, 2000.

Zander-Nowicka, J., Reactive Testing and Test Control of Hybrid Embedded Software, paper presented at 5th Workshop on System Testing and Validation STV07, in conjunction with ICSSEA 2007, Paris, France, December 1, 2007.

Zienkiewicz, C. O., and Morgan, K., *Finite Elements and Approximation*, Dover Publications, 2006.

Part III

Modeling and Optimizing VVT Process

Chapter **6**

Modeling Quality Cost, Time and Risk

6.1 PURPOSE AND BASIC CONCEPTS

The purpose of this chapter is to examine difficult questions that concern anyone involved with verification, validation and testing, namely, what to test, when to test, how to test and how much to test and when to stop testing. In other words, how one should select a VVT strategy and how one should optimize it.

In this chapter, we address the above questions by presenting a VVT cost and risk model as well as a VVT time and risk model. Once such models are established, it is possible to optimize the VVT strategy to minimize quality cost, time and risk. Three VVT model variants are presented utilizing probabilistic, simulation (Monte Carlo–based) and fuzzy logic paradigms for estimating and optimizing system VVT cost, time and risk. Chapter 7 describes ways to acquire quality data and optimize the VVT strategy in the face of funding, time and other resource limitations and in accordance with different business objectives. Finally, Chapter 8 describes the methodology used to validate the quality model along with examples describing a system's quality improvements.[73]

[73] Chapters 6, 7, and 8 are partially based on published papers (see below). Permission to include updated portions of these papers has been granted by the original publishers and coauthors.

Hoppe, M., Engel, A. and Shachar, S., SysTest: Improving the Verification, Validation & Testing Process—Assessing Six Industrial Pilot Projects, *Syst. Eng. J.*, 10(4), 323–347, September 24, 2007, available online.

6.1.1 Historical Models for Cost of Quality

Generally, there is a direct correlation between VVT investment and system quality. Early in the 1950s, two quality luminaries proposed two different qualitative models, defining "quality cost" as the sum of VVT costs plus failure costs. Joseph Juran suggested that there is an optimal VVT strategy which will yield minimum quality cost, whereas Philip Crosby coined the slogan "Quality is free," advocating the notion that the more one invests in quality, the more savings are realized. Despite the beauty of Crosby's saying, many practicing engineers tend to agree with Juran. The main reason is that the cost of finding the "last defects" increases exponentially rather than linearly (see Figure 6.1).

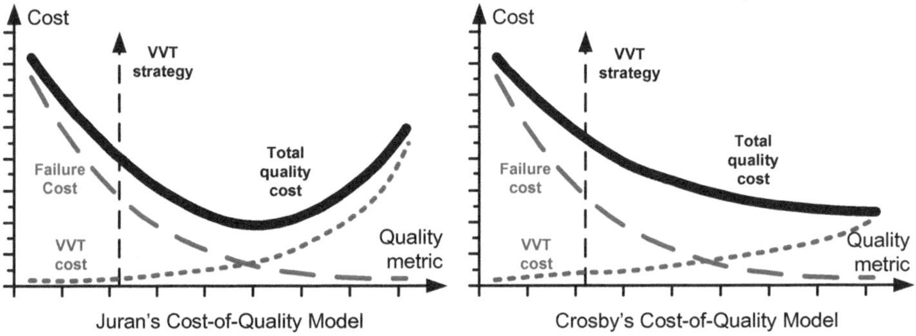

Figure 6.1 Qualitative cost-of-quality models.

The main weaknesses of both models is that they are qualitative and therefore do not help in designing practical VVT strategies. Furthermore, even if an optimal VVT strategy cost may be ascertained, large numbers of VVT strategies of equal cost are possible. This problem was addressed by designing a quantitative model to compute the cost of quality as a function of the VVT strategy and other relevant parameters (Engel and Barad, 2003). Thereafter, the practical usage of the model and the optimization of the VVT strategy to minimize project quality cost as well as project time were described (Engel and Shachar, 2006).

According to Clausing and Frey (2005), quality problems have two sources. The first source is mistakes or product defects which lead to the above-

---

Engel, A., and Last, M., Modeling Software Testing Costs and Risks Using Fuzzy Logic Paradigm, *J. Syst. Software*, 80(6), 817–835, June 2007, available online.

Barad, M., and Engel, A., Optimizing VVT Strategies—A decomposition approach, *J. Operation Res. Soc.*, 57(8), 965–974, August 2006.

Engel, A., and Shachar, S., Measuring and Optimizing Systems' Quality Costs and Project Duration, *Syst. Eng.*, 9(3), 259–280, June 22, 2006, available online.

Engel, A., and Barad M., A Methodology for Modeling VVT Risks and Costs, *Syst. Eng.*, 6(3), 135–151, May 8, 2003, available online.

described failure cost. The second source is a lack of robustness against the noises that occur in the field. In this chapter, we primarily deal with VVT strategy for avoiding product defects. However, ensuring robustness has to be an objective of VVT activity as well. Therefore, an optimal VVT strategy must also reduce quality problems due to a lack of system robustness.

6.1.2 Quantitative Models for Cost/Time of Quality

This chapter embraces recent scientific thinking which maintains that risk uncertainty emanates from inherent stochastic variability as well as from a fundamental lack of knowledge (Oberkampf et al., 2002). Accordingly, we develop methods for estimating system VVT cost, time and risk using both probabilistic and fuzzy paradigms (Klir and Yuan, 1995). The inherent stochastic variability is enhanced by adopting the Marczyk (1999) approach where a risk cost model contains a stochastic processor to represent the dynamic characteristics of the problem. The fundamental lack of knowledge is modeled by representing uncertain values using Triangular Fuzzy Numbers (TFNs) and linguistic computation advocated by numerous fuzzy logic authors (e.g., Kishk and Al-Hajj, 2000).

The estimation of actual, direct VVT cost is a rather trivial process. Assuming we are interested only in efforts of individuals involved in VVT, we simply sum up these cost components. The key problems to be addressed are (1) how to estimate additional costs which emanate from performing VVT and (2) how to estimate costs which emanate from not performing VVT. Accordingly, this chapter proposes a system quality cost/time model which hinges on four quantitative submodels associating a given VVT strategy with system behavior and failure consequences (see Figure 6.2):

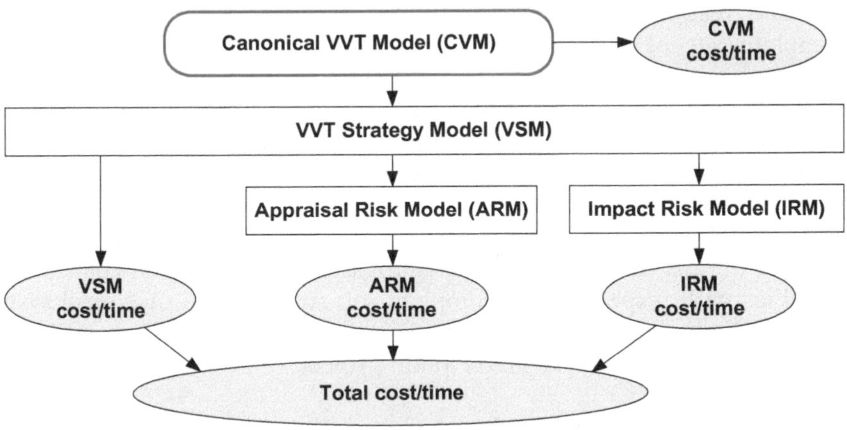

Figure 6.2 Overall quality cost/time and risk model.

1. *Canonical VVT Model (CVM).* An ideal, maximal set of VVT activities together with their performance cost and time.
2. *VVT Strategy Model (VSM).* A corresponding set of decision variables identifying the VVT performance levels.
3. *Appraisal Risk Model (ARM).* A set of VVT appraisal risks together with their appraisal costs, times, likelihood of occurrence and rates of decay.
4. *Impact Risk Model (IRM).* A set of VVT impact risks together with their impact costs, time, likelihood of occurrence and time of potential impact.

6.2 VVT COST AND RISK MODELING

6.2.1 Canonical VVT Cost Modeling

The CVM paradigm is introduced in order to capture the qualitative and quantitative implications of performing a system's VVT process. A CVM is a conceptual, almost philosophical entity with unique individual components for each industry, product and sometimes project. This model of the VVT process entails definition of a collection of VVT activities together with their costs and times where each activity is associated with a system's lifecycle phase.

The VVT activities are linked together defining a sequence of VVT tasks. In order to simplify the problem of modeling the system VVT process, we assume here a sequential process. From a feedback perspective, this model is simplistic, as it assumes a linear progression of the VVT process. Loops and feedbacks are left for future research. A CVM is defined as a complete set of activities and associated costs and time parameters designed to verify, validate and test a system throughout its lifecycle. It is worth noting that the CVM is an idealized concept. It is not likely to be carried out in practical applications in its all-inclusive form, since it would require excessive financial and time resources. Many industrial and governmental organizations perform about 15–25% of CVM and, in special circumstances (e.g., spaced manned missions), perhaps 25–50% of CVM is performed. The intent here is to create a yardstick for evaluating selected (partial) sets of activities with respect to the complete set. An example of typical system CVM activities is depicted in Figure 6.3. A list of CVM activities used in the Israel Aerospace Industries (IAI) pilot project of SysTest is provided in Appendix B.

The CVM exhibits the following characteristics:

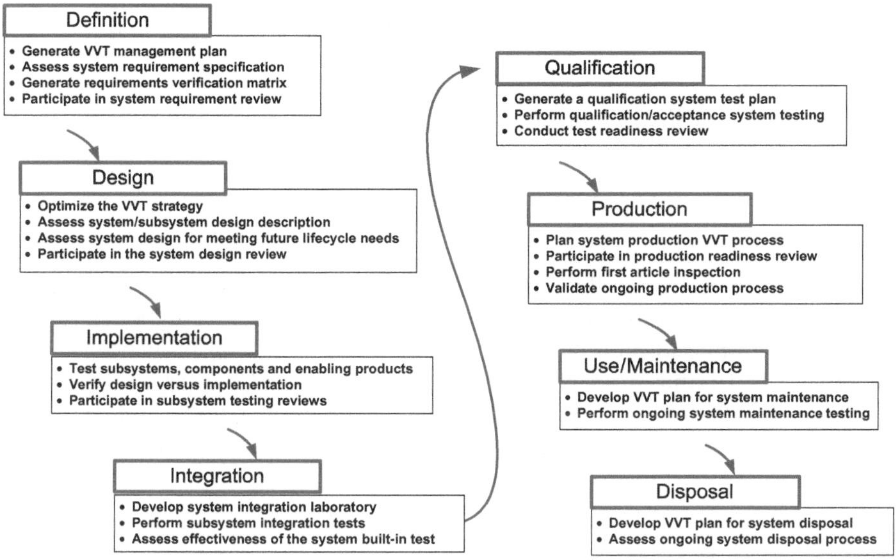

Figure 6.3 Example of typical system CVM activities.

1. VVT phases are chained serially, depicting the execution order of the overall VVT process. Similarly, the activities within each VVT phase are performed serially.
2. The VVT lifecycle phase {**L**} is a vector defined such that

$$L_i = i \qquad \{i = 1, 2, \ldots, z\} \tag{6.1}$$

   where, in the CVM model, $z = 8$.
3. Within the $\{L_i\}$ VVT lifecycle phase, there are $\{n_1, n_2, \ldots, n_i, \ldots, n_z\}$ sets of VVT activities. Each activity is designated as

$$V_{i,j}\{j(i) = 1, 2, \ldots, n_i, i = 1, 2, \ldots, z\} \tag{6.2}$$

4. The cost of performing activity $\{V_{i,j}\}$ is

$$\mathrm{Cv}_{i,j}\{j(i) = 1, 2, \ldots, n_i, i = 1, 2, \ldots, z\} \tag{6.3}$$

5. The terms "executing the CVM" and "performing the VVT process in accordance with (IAW) the CVM" entail serially performing all the VVT activities described above in the defined order. The ensemble of all CVM activities is designated V_{CVM}:

$$V_{\mathrm{CVM}} = \bigcup_{i=1}^{z} \bigcup_{j=1}^{n_i} V_{i,j} \tag{6.4}$$

6. The total CVM cost ($\mathrm{Cv}_{\mathrm{CVM}}$) incurred will be

$$C_{\mathrm{CVM}} = \sum_{i=1}^{z} \sum_{j=1}^{n_i} \mathrm{Cv}_{i,j} \tag{6.5}$$

The methodology for collecting cost data associated with performing each CVM activity is based on available data from the literature being supplemented by empirical data provided by an actual industrial pilot project conducted by IAI/Lahav under the SysTest project (see Chapter 7). Figure 6.4 describes the CVM paradigm. It depicts the lifecycle phases $\{L_i\}$, the VVT activities $\{V_{i,j}\}$, costs $\{Cv_{i,j}\}$ and time $\{Tv_{i,j}\}$ elements of the model.

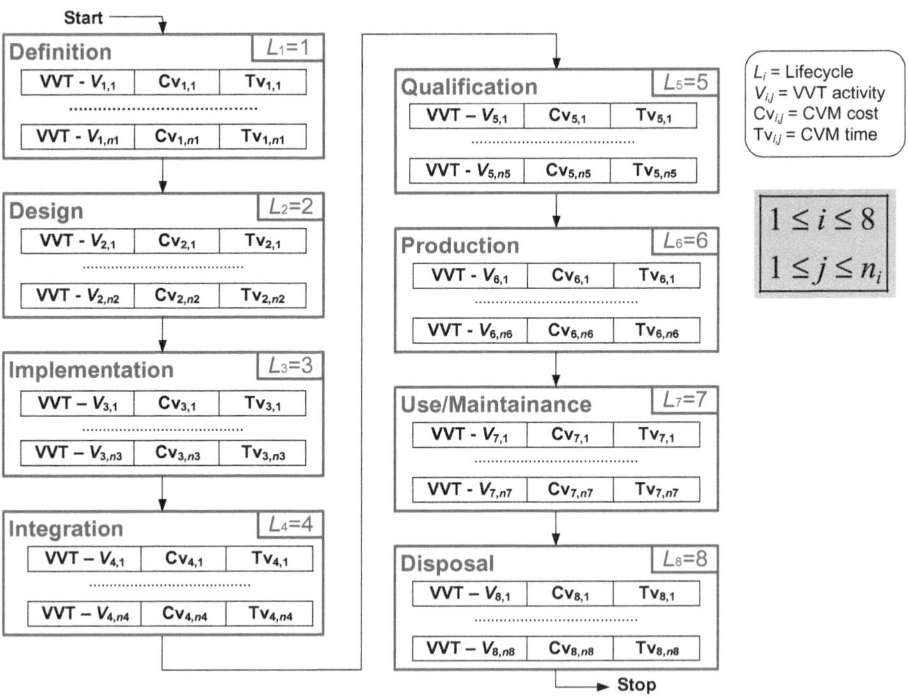

Figure 6.4 A Canonical VVT Model (CVM).

6.2.2 Modeling VVT Strategy as a Decision Problem

As mentioned, executing the all-inclusive CVM is not practical due to VVT funding limitations and Time-To-Market (TTM) considerations.

Therefore, industrial organizations elect to perform only a subset of the CVM, and within this subset, some VVT activities are only partially performed. We have called such policy a VVT strategy and have encapsulated this concept in a VVT Strategy Model (VSM). We define the cost of actually carrying out this VVT subset as the actual VVT cost.

The detailed design of an optimal VVT strategy for a given project will be discussed in Chapter 7; however, the VVT cost and risk dimension (approximately 40–60% of the overall engineering lifecycle cost) dictates a prudent consideration of this issue. The VVT strategy for rockets or spacecraft, which are produced in very small quantities and cannot be repaired after launch, is very different from VVT procedures implemented in a vehicle production line, which are produced in as much as tens of thousands of units per year. At the core, a VVT strategy should support organization business objectives. Examples of such objectives are as follows:

- To reduce product cost
- To reduce time to market
- To reduce internal and external failure costs
- To increase market share
- To increase quality of products
- To improve delivery time accuracy

Other considerations less directly linked to business objectives are meeting standards and statutory directives as well as following ethics and human values. In order to deal with realistic qualitative and quantitative modeling of the costs and risks associated with an incomplete set of VVT activities, some basic concepts are introduced.

VVT Horizon and Strategy

1. A *VVT horizon* is defined in this book as a specific sequence of system VVT phases for which the modeling results of VVT cost and risk data are of interest to the stakeholders of a given system.
2. A *VVT strategy* is defined in this book as a policy for a given system lifecycle under which a subset of CVM activities is fully performed another subset of activities is partially performed and the remaining activities are not performed at all.

Activity Performance Level

1. A *decision variable*, $X_{i,j}$, $0 \le X_{i,j} \le 1$, of a VVT strategy defines the performance level of any activity $V_{i,j}\{j(i) = 1, 2, \ldots, n_i, i = 1, 2, \ldots, z\}$. These represent the set of Activity Performance Levels (APLs) associated with the given strategy. It should be noted that $X_{i,j} = 1$ means that VVT activ-

ity $V_{i,j}$ is to be fully performed, whereas $X_{i,j} = 0$ means that VVT activity $V_{i,j}$ is not to be performed at all. A VSM is depicted in Figure 6.5.

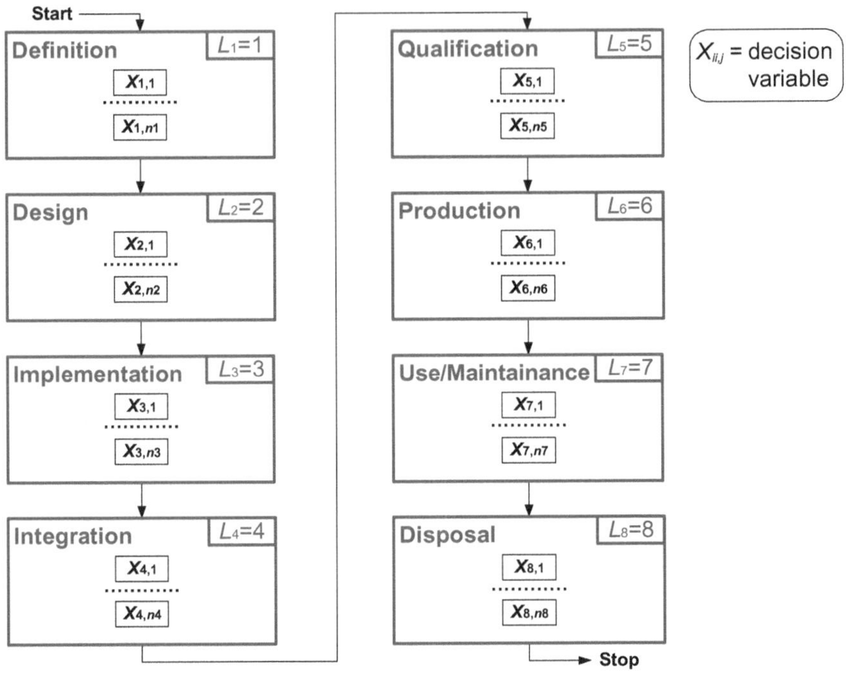

Figure 6.5 A VVT Stategy Model (VSM).

2. In principle, the decision variable can be modeled as a linguistic variable, that is, high, medium, low and so on (see Section 6.4).

Overall VVT Activity Set

1. Defining a VVT strategy within the CVM results in a realistic, qualitative and quantitative VVT activity and a cost model of a practical VVT process. As defined below, $\text{V\_item}_{i,j}$ is a symbolic representation of the individual VVT activity performed at respective APLs:

$$\text{V\_item}_{i,j} = V_{i,j}(X_{i,j}) \tag{6.6}$$

2. By extension, $V_{\text{VVT\_Strategy}}$ as defined below is a symbolic representation of the ensemble of VVT activities performed at their respective APLs:

$$V_{\text{VVT\_Strategy}} = \bigcup_{i=1}^{z} \bigcup_{j=1}^{n_i} \{\text{V\_item}_{i,j}\} \tag{6.7}$$

3. *Assumption.* The cost invested in an activity $V_{i,j}$ performed at level $X_{i,j}$, $0 \le X_{i,j} \le 1$, represents an $X_{i,j}$ fraction of the cost $\mathrm{Cv}_{i,j}$ for fully performing the activity. Accordingly, the individual VVT strategy cost $\mathrm{Cv\_item}_{i,j}$ incurred will be:

$$\mathrm{Cv\_item}_{i,j} = \mathrm{Cv}_{i,j} X_{i,j} \tag{6.8}$$

4. By extension, the total VVT strategy cost $Cv_{\mathrm{VVT\_Strategy}}$ incurred would be

$$\mathrm{Cv}_{\mathrm{VVT\_Strategy}} = \sum_{i=1}^{z} \sum_{j=1}^{n_i} \{\mathrm{Cv\_item}_{i,j}\} \tag{6.9}$$

$$\mathrm{Cv}_{\mathrm{VVT\_Strategy}} = \sum_{i=1}^{z} \sum_{j=1}^{n_i} \{\mathrm{Cv}_{i,j} X_{i,j}\} \tag{6.10}$$

6.2.3 Modeling Appraisal Risk Cost

System/VVT Interactions In many industrial and research organizations, systems are created by various system groups and then submitted for evaluation by one or more VVT teams. Let us define a specific system activity and its subsequent VVT activity as a system/VVT sequence. In an ideal system/VVT sequence the outcome of the VVT activity would suggest that the system product is correct.

In reality, some products exhibit defects due to an imperfection inherent in their development or manufacture or due to incorrect usage. In other words, there is a certain risk that a VVT activity will identify a defective product.[74] We have labeled the risk of detecting such deficiencies during the VVT process as an appraisal risk and we have encapsulated this phenomenon in the Appraisal Risk Model (ARM).

It is common practice in most industries to return a defective product for a corrective procedure and then to reevaluate it. As a rule, rectified products exhibit some decay in the probability of finding further defects. Executing a corrective procedure of a product found defective and reevaluating it is denoted here as a VVT iteration. We carry out VVT iterations until the evaluations do not reveal any more defects and call this course of action a corrective process. The length of a corrective process (i.e., the number of VVT iterations) depends on the risk that a VVT activity will detect any defects and on the decay factor of this risk. These parameters are typical of the system or field

[74]The reader should note that this model suggests that by merely testing a system one runs the risk of discovering faults which then must be fixed. If one does not test, this type of risk disappears altogether (of course, if one does not test, then the impact risk increases dramatically, but this is another matter).

in question as well as of the process maturity level of the organization. For example, software development in immature organizations exhibits high defect risk coupled with low rates of decay.

The level of a VVT performance reflects the relative effort of performing a given VVT activity and therefore it affects the appraisal risk. In one extreme case, if the VVT performance level is zero, then no VVT activity takes place and the appraisal risk becomes zero. Conversely, if the VVT performance level is 1, then the VVT activity is fully performed and the appraisal risk reaches its maximum value.

Appraisal Risk: Assumptions In analyzing the quality cost data associated with appraisal risks, we aim to make inferences about the mathematical model that we believe explains the physical circumstances appropriately. We then are making the following assumptions:

1. *Imperfection.* Some products and systems contain inherent flaws which can be detected during a VVT activity. As mentioned above, the probability ($p; 0 \le p \le 1$) of detecting such a flaw for a fully performed activity depends on the system or field in question and on the process maturity level of the organization.
2. *Decay Factor.* Following each rectification and reevaluation procedure, the probability of an error is reduced by a constant decay factor in the failure probability (Decay; $0 \le$ Decay ≤ 1). We have called the complementary factor L, $L = 1 -$ Decay ($L; 0 \le L \le 1$), the lingering factor. Accordingly, the probability of detecting deficiencies following each corrective procedure diminishes by a factor L.
3. *Independence.* The outcome (detecting or not detecting a defective product) of any system/VVT iteration is independent of the outcome of any previous VVT iteration.

Appraisal Risk Scenario We describe an appraisal risk scenario (see Figure 6.6) with the premise that a system product (SYS) is constructed.

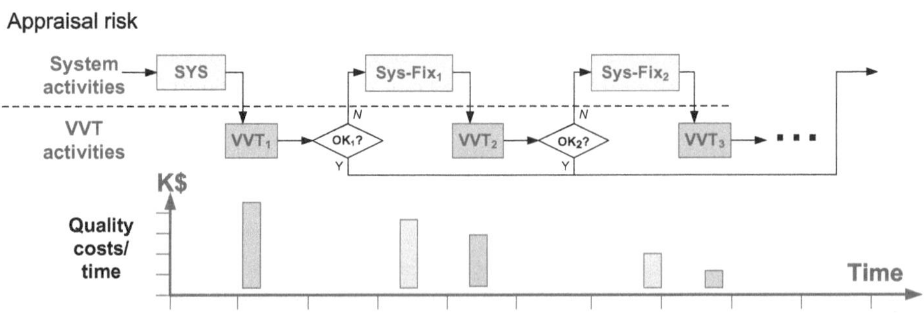

Figure 6.6 Appraisal risk scenario.

Next, a partial or full VVT evaluation ($X > 0$) is carried out (VVT_1). Depending on the probability of failure p, the product will be returned for rectification or fixing (thus becoming SYS-Fix_1) or will be identified as acceptable. A rectified product will be retested (via VVT_2) and then, depending on the above parameters subject to the lingering factor L, the product will be returned for rectification (thus becoming SYS-Fix_2), with the probability $p \times L$ that it will be identified as failure. The relevant parameters are as follows:

- p = probability of product or system failure
- L = lingering factor
- P_n = probability of n VVT iterations (i.e., the probability that the product is considered fault free for the first time on the nth VVT iteration), $n = 1, 2, \ldots$
- EN = expected number of system/VVT iterations
- Ca = cost of appraisal per iteration
- Ta = time of appraisal per iteration

Number of System/VVT Iterations

1. According to the above, we see that the number of VVT iterations is a stochastic variable whose distribution can be calculated for given parameters. Let us consider a System/VVT sequence of a fully performed VVT activity and calculate the probability associated with any number of VVT iterations and then calculate the expected number of VVT iterations EN (see Table 6.1).

TABLE 6.1 P_n as Function of n, p and L

N	P_n
1	$1 - p$
2	$p(1 - pL)$
3	$p^2L(1 - pL^2)$
4	$p^3L^3(1 - pL^3)$
5	$p^4L^6(1 - pL^4)$
$\vdots$	
N	$p^{n-1}L^{(n-1)(n-2)/2}(1 - pL^{n-1})$

2. Let us now calculate the expected number of system/VVT iterations EN (given $X = 1$):

$$\mathrm{EN} = \sum_{n=1}^{\infty} np_n \tag{6.11}$$

$$\text{EN} = \sum_{n=1}^{\infty} np^{n-1} L^{(n-1)(n-2)/2} \left(1 - pL^{n-1}\right) \tag{6.12}$$

3. *Note 1.* For $L = 1$ (no decay), the probability P_n that a product is considered fault free for the first time at the nth iteration follows the geometric distribution $p_n = p^{n-1}(1 - p)$ and its expected value EN is explicitly calculated as EN = $1/(1 - p)$.
4. *Note 2.* For $0 < L < 1$ there is no explicit result for EN. However, the calculation can be carried out by means of iterative approximation.

VVT Performance Level A fully performed VVT activity ($X = 1$) results in defect detection with probability p while not performing the VVT activity at all ($X = 0$) implies that a product is fault free. We have assumed that the performance level of a VVT activity (X; $0 \le X \le 1$) affects the probability of detecting faults linearly. That is, the probability of detecting a defect is multiplied by the level of the VVT performance ($pX; 0 \le p \le 1; 0 \le X \le 1$). Under these conditions, the expected number of VVT iterations given the performance level X becomes

$$\text{EN}|X = \sum_{n=1}^{\infty} n(pX)^{n-1} L^{(n-1)(n-2)/2} \left(1 - pXL^{n-1}\right) \tag{6.13}$$

Cost of Appraisal Risk

1. *Length of Corrective Process.* The first VVT operation in a system/VVT sequence represents normal VVT activity and is not considered a part of the appraisal risk cost. Therefore, we have subtracted one from the expected number of VVT iterations and have called the result the number of corrective iterations given the performance level X (NCI|X):

$$\text{NCI}|X = \text{EN}|X - 1 \tag{6.14}$$

2. *Note.* For $L = 0$ (100% decay), the expected number of system/VVT iterations is

$$E(NX) = 1 - pX + 2pX = 1 + pX \qquad \text{and} \qquad E(NX) - 1 = pX$$

3. *Cost Calculations.* We have assumed that the cost (Ca) of correcting and retesting a given system product depends only on the nature of this product and otherwise it is fixed. Therefore, the expected item appraisal risk cost of a system/VVT sequence carried out at performance level X, Ca_item|X, is a function of the appraisal cost per iteration (Ca) multiplied by the number of corrective iterations performed given the performance level X (NCI|X):

$$\text{Ca\_item}|X = \text{Ca NCI}|X \tag{6.15}$$

4. Substituting for EN in equation (6.14) yields

$$\text{Ca\_item}|X = \text{Ca}(\text{EN}|X - 1) \tag{6.16}$$

Example: Comparing Two Organizations Let us consider four cases of system/VVT sequences. An immature organization performing a noncritical VVT activity is compared to a very mature organization performing a very demanding, life-critical software application. Each of the two organizations contemplates two strategies for performing the VVT activity (see Table 6.2).

TABLE 6.2 Appraisal Risk Example Input Parameters

Parameters	Organization 1		Organization 2	
Organization maturity	Immature (CMMI[75] level 1)		Very mature (CMMI level 5)	
Application	Noncritical application		"Clean room" (most demanding, life-critical software application)	
Failure probability p	0.90		0.15	
Rate of decay D	0.20		0.95	
VVT performance X	Strategy A 0.5	Strategy B 1.0	Strategy A 0.5	Strategy B 1.0

Table 6.3 shows the number of system/VVT iterations of the two organizations as a function of the two contemplated VVT strategies:

TABLE 6.3 Number of System/VVT Iterations

Parameter	Organization 1		Organization 2	
	Strategy A	Strategy B	Strategy A	Strategy B
NCI|X	0.66	2.17	0.075	0.15

1. *First Inference.* Immature organization contemplating the above two strategies, will increase the number of corrective iterations by a factor of 2.17/0.66 = 3.29.
2. *Second Inference.* Very mature organization contemplating the above two strategies, will increase the number of corrective iterations by a factor 0.15/0.075 = 2.

[75]The CMMI is a model for appraising and improving the performance of development organizations. It stands for Capability Maturity Model Integration. It is published and developed by the Software Engineering Institute in Pittsburgh, PA.

3. *Third Inference.* If each of these organizations would fully perform the above VVT activity, the immature organization will conduct on average 2.17/0.15 = 14.47 times more system/VVT iterations than the very mature organization.

Appraisal Risk Model Figure 6.7 shows the lifecycle of Appraisal Risk Model (ARM) with phases L_i, $i = 1, 2, \ldots, z$. Each phase i in the ARM contains $j(i)$, $j(i) = 1, 2, \ldots, n_i$, appraisal risk tuples.

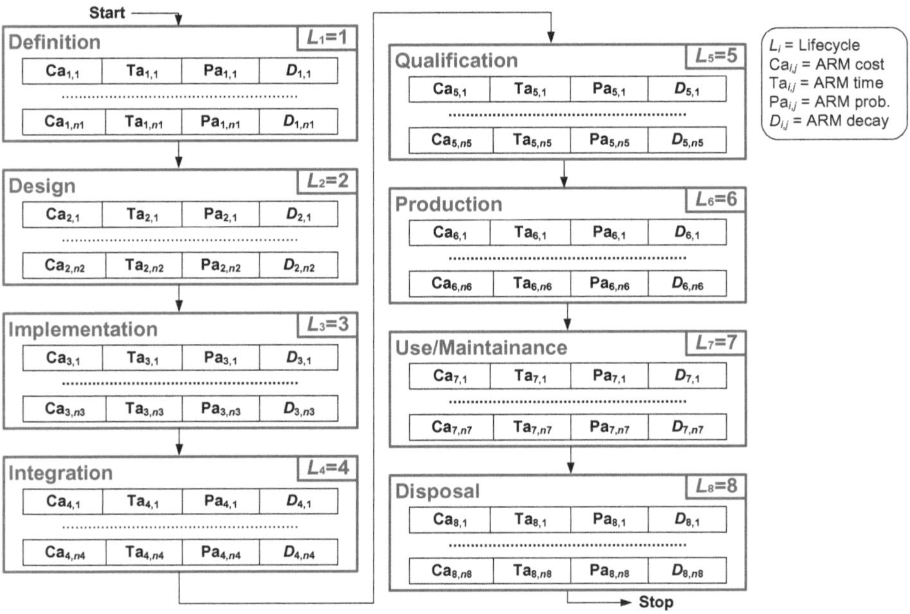

Figure 6.7 An Appraisal Risk Model (ARM).

Each ARM tuple is associated with a corresponding VVT activity ($V_{i,j}$) defined in the Canonical VVT Model (CVM). The meanings of the ARM elements are given in Table 6.4.

TABLE 6.4 Definition of Appraisal Risk Model (ARM) Tuples Elements

ARM Tuple Elements	Meaning
$Ca_{i,j}$	Cost of performing a system/VVT appraisal iteration following activity $V_{i,j}$ (appraisal cost per iteration)
$Ta_{i,j}$	Time required to carry out a system/VVT appraisal iteration following activity $V_{i,j}$ (appraisal time per iteration)
$Pa_{i,j}$	Probability of detecting a product or a system appraisal failure through VVT activity $V_{i,j}$, $0 \leq Pa_{i,j} \leq 1$ (appraisal risk probability)
$D_{i,j}$	Decay factor of appraisal risk probability $Pa_{i,j}$; $0 \leq D_{i,j} \leq 1$ (appraisal risk decay)

Lifecycle Appraisal Risk Let us determine the system lifecycle appraisal risk associated with a given VVT strategy ($X_{i,j}$; $0 \leq X_{i,j} \leq 1$).

1. *ARM Intermediary Values.* Utilizing the above equations for the ARM, we get the following intermediary values for each ARM item. The expected number of system/VVT iterations per item is given as

$$\mathrm{EN}_{i,j} = \sum_{n=1}^{\infty} n\left(\mathrm{Pa}_{i,j} X_{i,j}\right)^{n-1} L_{i,j}^{(n-1)(n-2)/2}\left(1 - \mathrm{Pa}_{i,j} X_{i,j} L_{i,j}^{n-1}\right) \tag{6.17}$$

where

$$L_{i,j} = 1 - D_{i,j} \tag{6.18}$$

2. The expected length of the corrective process per item is given as

$$\mathrm{NCI}_{i,j} = \mathrm{EN}_{i,j} - 1 \tag{6.19}$$

3. Expected appraisal cost per item is determined as

$$\mathrm{Ca\_item}_{i,j} = \mathrm{Ca}_{i,j} \mathrm{NCI}_{i,j} \tag{6.20}$$

4. The appraisal risk cost of a VVT strategy is defined as

$$\mathrm{Ca}_{\mathrm{VVT\_Strategy}} = \sum_{i=1}^{z} \sum_{j=1}^{n_i} \mathrm{Ca\_item}_{i,j} \tag{6.21}$$

$$\mathrm{Ca}_{\mathrm{VVT\_Strategy}} = \sum_{i=1}^{z} \sum_{j=1}^{n_i} \mathrm{Ca}_{i,j}\left(\mathrm{EN}_{i,j} - 1\right) \tag{6.22}$$

6.2.4 Modeling Impact Risk Cost

Selecting a VVT strategy entails designating the performance level of each VVT activity during the VVT process, including the partially performed activities, as well as those that are not performed at all. A basic assumption of this methodology is that any partially performed VVT activity or any VVT activity not performed at all constitutes a risk. These risks have uncertain effects on the system and, of course, they constitute undesirable expenditure that can be regarded as outcomes of implementing a selected VVT strategy. They are discernible only subsequent to the risk insertion (during the same lifecycle phase or at later lifecycle phases). We have labeled the risks emanating from the partial performance of the CVM as impact risks and we have encapsulated this situation in the Impact Risk Model (IRM).

VVT Impact Risk Scenario We describe an impact risk scenario (see Figure 6.8) with the premise that a system product is constructed (SYS); however, this system product is intentionally either not tested or only partially tested (X < 1). As a result, risk scenario impact may actually occur, causing unplanned expenditure or time delay. Impact may occur during the current phase or during any subsequent phase. The model assumes that multiple impacts may occur due to a single source (e.g., a VVT activity that was partially performed or not performed at all). A VVT Impact Risk Model (IRM) is defined as a model for ascertaining impact risk, given probability of impact and cost of impact. The IRM concept is comprised of risk identification attributes and risk variables:

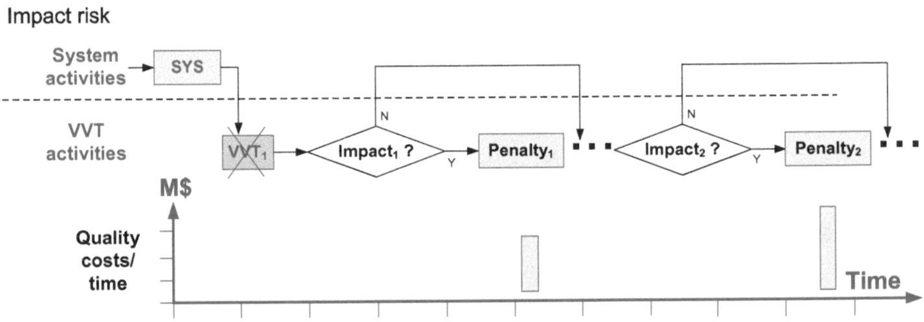

Figure 6.8 Impact risk scenario.

- *Risk Identification Attributes.* The risk identification attributes encompass (1) risk source, (2) risk qualitative description and (3) risk destination. A risk source $S_{i,j}$ is a specific activity $V_{i,j}$ which was not fully performed and may thus generate undesirable impact. For each such activity, the specific risks that can impact the system can be identified and qualitatively described. The risk destination is the system lifecycle phase within which a risk generated by a specific activity incompletely performed may impact the system. A risk destination may be reached from several risk sources and also several times from the same risk source.
- *Risk Variables.* There are two risk variables for any given risk: (1) probability that the risk will in fact impact the system and (2) impact severity.

Impact Risk Concept The impact risk exhibits the following characteristics:

1. When a particular VVT activity $V_{i,j}$ is not fully performed, $X_{i,j} < 1$, it gives rise to one or more risks and consequently is defined as a risk source $S_{i,j}$ so that

$$S_{i,j}\{j(i) = 1, 2, \ldots, n_i, i = 1, 2, \ldots, z\} \tag{6.23}$$

2. Therefore, any given VVT strategy other than performing the entire CVM gives rise to a collection of VVT risk sources associated with it.
3. Any risk source $S_{i,j}\{j(i) = 1, 2, \ldots, n_i, i = 1, 2, \ldots, z\}$ can give rise to one or more impact risks which may impact the system in the same or any subsequent lifecycle l. Several risk impacts emanating from a single risk source and occurring at the same phase l are identified as $k = 1, 2, \ldots$. It should be noted that the impact cannot occur in a lifecycle phase earlier than i, $l \geq i$. Accordingly,

$$\mathrm{Ri}_{i,j}^{(l,k)}\{j(i)=1,2,\ldots,n_i, i=1,2,\ldots,z, l\geq i, k=1,2,\ldots\} \tag{6.24}$$

4. Each impact risk $R_{i,j}^{(l,k)}$ can cause an impact cost $Ci_{i,j}^{(l,k)}$ in phase l, $l \geq i$, with probability $P_{i,j}^{(l,k)}$:

$$\mathrm{Ci}_{i,j}^{(l,k)}\{j(i)=1,2,\ldots,n_i, i=1,2,\ldots,z, l\geq i, k=1,2,\ldots\} \tag{6.25}$$

$$\mathrm{Pi}_{i,j}^{(l,k)}\{j(i)=1,2,\ldots,n_i, i=1,2,\ldots,z, l\geq i, k=1,2,\ldots\} \tag{6.26}$$

Impact Risk Model Figure 6.9 shows a VVT IRM and Table 6.5 gives definitions of the IRM tuple elements:

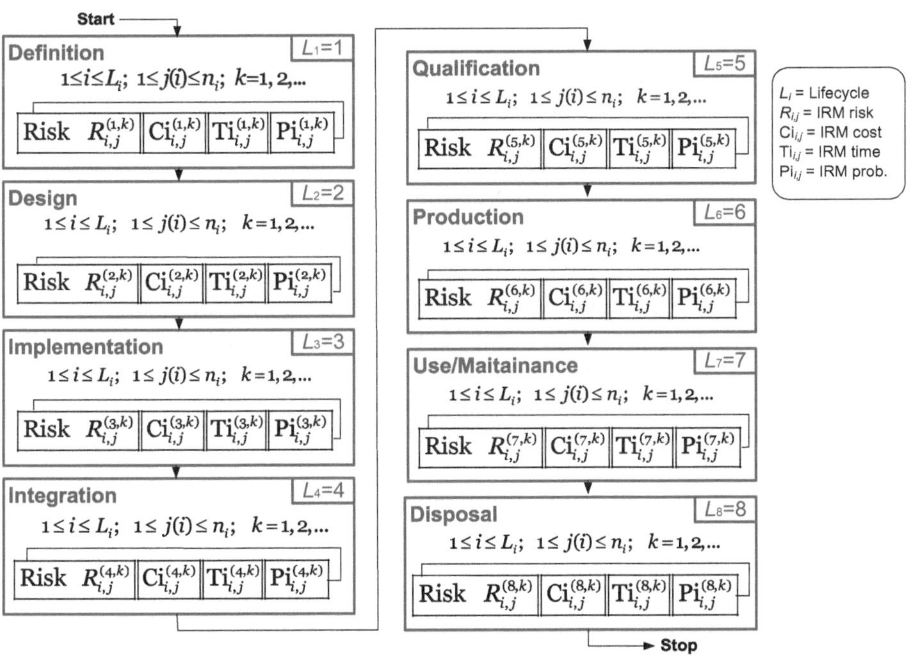

Figure 6.9 A VVT Impact Risk Model (IRM).

TABLE 6.5 VVT Impact Risk Model (IRM) Tuple Elements

IRM Tuple Elements	Definition
$R_{i,j}^{(l,k)}$	Name/description of specified impact risk k, $k = 1, 2, \ldots,$ occurring in phase l, $l \geq i$, when VVT activity $V_{i,j}$ is partially performed ($X_{i,j} < 1$)
$\mathrm{Ci}_{i,j}^{(l,k)}$	Cost associated with impact risk $R_{i,j}^{(l,k)}$ (impact cost)
$\mathrm{Ti}_{i,j}^{(l,k)}$	Time required to carry out a corrective action due to impact risk $R_{i,j}^{(l,k)}$ (impact time)
$\mathrm{Pi}_{i,j}^{(l,k)}$	Probability of impact risk $R_{i,j}^{(l,k)}$ occurrence, $0 \leq \mathrm{Pi}_{i,j}^{(l,k)} \leq 1$ (impact risk probability)

1. We have assumed a negative linear model in order to describe the functional relationships between the impact cost and the APLs. An item impact risk cost associated with a given VVT strategy ($X_{i,j}$; $0 \leq X_{i,j} \leq 1$) is given as

$$\mathrm{Ci\_item}_{i,j} = \sum_{l=i}^{z} \sum_{k=1,2\ldots} \left\{ \mathrm{Ci}_{i,j}^{(l,k)} \mathrm{Pi}_{i,j}^{(l,k)} (1 - X_{i,j}) \right\} \tag{6.27}$$

2. The system's lifecycle impact risk cost associated with a given VVT strategy ($X_{i,j}$; $0 \leq X_{i,j} \leq 1$) is given as

$$\mathrm{Ci}_{\mathrm{VVT\_Strategy}} = \sum_{i=1}^{z} \sum_{j=1}^{n_i} \mathrm{Ci\_item}_{i,j} \tag{6.28}$$

$$\mathrm{Ci}_{\mathrm{VVT\_Strategy}} = \sum_{i=1}^{z} \sum_{j=1}^{n_i} \sum_{l=i}^{z} \sum_{k=1,2\ldots} \left\{ \mathrm{Ci}_{i,j}^{(l,k)} \mathrm{Pi}_{i,j}^{(l,k)} (1 - X_{i,j}) \right\} \tag{6.29}$$

VVT Horizon and Strategy Example Figure 6.10 shows an example of the combined effect of appraisal and impact risk emanating from a specific VVT strategy within a defined horizon. In this example the VVT strategy is to perform a part or none of each VVT activity during the first four system lifecycle phases and to fully perform each VVT activity during the fifth system lifecycle phase. In this example, the test engineer defines the VVT horizon from phase Definition until phase Qualification, thereby ignoring the rest of the lifecycle. This strategy and horizon selection creates risks to the system, as depicted in the figure. We can see that, based on this VVT strategy, there is an appraisal risk affecting each system phase within the VVT horizon. In addition, each of the first four phases creates impact risks affecting the current phase as well as all the following phases within the VVT horizon.

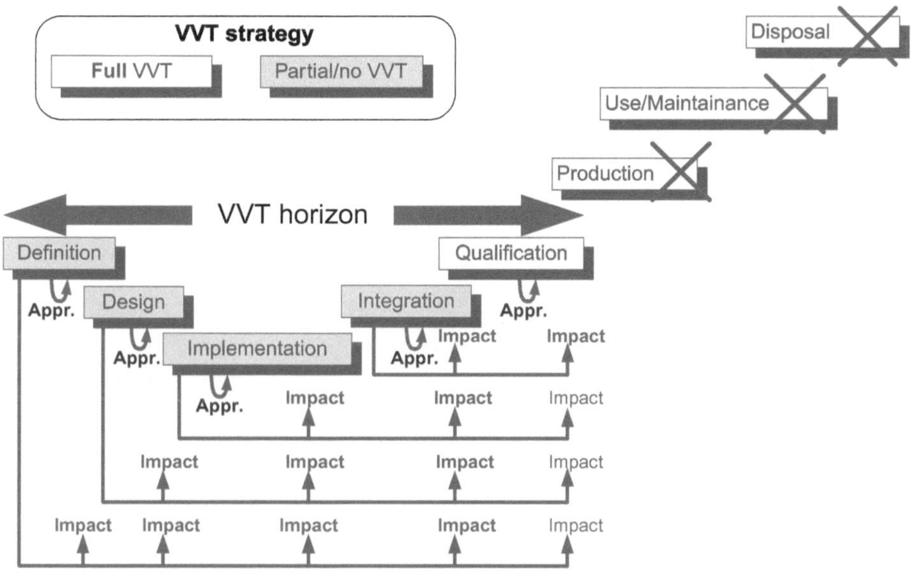

Figure 6.10 VVT horizon and strategy example.

Nonlinear APL Effects We have assumed so far that the effect of the APLs on the computation of the total VVT cost is linear. However, in general, the effects of the APL are not quite linear, in particular, the effect of the VVT strategy on the impact risk. The reason for this is that, in reality, performing testing at low levels will detect many glaring and simple errors with minimal effort, thus decreasing the impact risk. Conversely, testing at high levels requires substantial efforts for detecting obscure faults. Such faults tend to remain in the system, increasing the impact risks. This nonlinear situation is characterized by a function having an inflection point[76] about midway, which can be modeled by a piecewise linear function, as shown in Figure 6.11, and having the following features:

1. *Low APL.* If a VVT activity is performed at a level $X_{i,j}$ below the inflection point (but above zero), then a partial impact cost $\mathrm{Ci}_{i,j}^{l,k}(1-X_{i,j})$ will potentially occur. However, under a nonlinear assumption the impact cost $\mathrm{Ci}_{i,j}^{l,k}(1-Y_{i,j})$ will be lower than under a linear assumption $(Y_{i,j} > X_{i,j})$.
2. *High APL.* If a VVT activity is performed at a level $X_{i,j}$ above the inflection point (but below 1), then a partial impact cost will potentially occur. However, under a nonlinear assumption the impact cost will be higher than under a linear assumption $(Y_{i,j} < X_{i,j})$:

[76]An *inflection point* is a point on a curve at which the sign of the curvature (i.e., the concavity) changes.

X	Linear X	Nonlinear Y
0.0	0.000	0.000
0.1	0.100	0.200
0.2	0.200	0.350
0.3	0.300	0.425
0.4	0.400	0.475
0.5	0.500	0.500
0.6	0.600	0.525
0.7	0.700	0.577
0.8	0.800	0.650
0.9	0.900	0.800
1.0	1.000	1.000

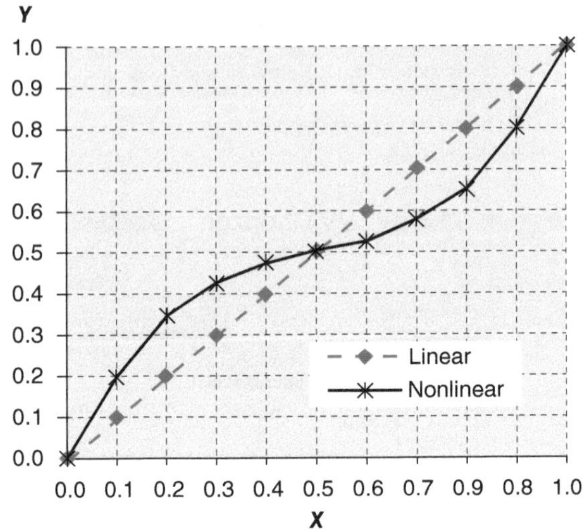

Figure 6.11 Linear and nonlinear VVT impact effects.

3. In a nonlinear impact computation, we have

$$Y_{i,j} = f(X_{i,j}) \qquad 0 \le Y_{i,j} \le 1 \tag{6.30}$$

4. A nonlinear item impact risk cost associated with a given VVT strategy ($X_{i,j}$; $0 \le X_{i,j} \le 1$) is

$$\text{Ci\_item}_{i,j} = \sum_{l=i}^{z} \sum_{k=1,2\ldots} \left\{ \text{Ci}_{i,j}^{(l,k)} \text{Pi}_{i,j}^{(l,k)} (1 - Y_{i,j}) \right\} \tag{6.31}$$

5. The system lifecycle nonlinear impact risk cost associated with a given VVT strategy ($X_{i,j}$; $0 \le X_{i,j} \le 1$) is

$$\text{Ci}_{\text{VVT\_Strategy}} = \sum_{i=1}^{z} \sum_{j=1}^{n_i} \text{Ci\_item}_{i,j} \tag{6.32}$$

$$\text{Ci}_{\text{VVT\_Strategy}} = \sum_{i=1}^{z} \sum_{j=1}^{n_i} \sum_{l=i}^{z} \sum_{k=1,2\ldots} \left\{ \text{Ci}_{i,j}^{(l,k)} \text{Pi}_{i,j}^{(l,k)} (1 - Y_{i,j}) \right\} \tag{6.33}$$

6.2.5 Modeling Total Quality Cost

Within the overall context of quality cost modeling, the total cost is the sum of (1) the VVT actual cost, (2) the VVT appraisal risk cost and (3) the VVT impact risk cost:

$$\text{C}_{\text{VVT\_Overall}} = \text{Cv}_{\text{VVT\_Strategy}} + \text{Ca}_{\text{VVT\_Strategy}} + \text{Ci}_{\text{VVT\_Strategy}} \tag{6.34}$$

1. Where the total VVT strategy cost is

$$\mathrm{Cv}_{\mathrm{VVT\_Strategy}} = \sum_{i=1}^{z}\sum_{j=1}^{n_i}\{\mathrm{Cv}_{i,j}X_{i,j}\} \tag{6.35}$$

2. The total VVT appraisal risk cost is

$$\mathrm{Ca}_{\mathrm{VVT\_Strategy}} = \sum_{i=1}^{z}\sum_{j=1}^{n_i}\{\mathrm{Ca}_{i,j}(\mathrm{EN}_{i,j}-1)\} \tag{6.36}$$

3. And the expected number of system/VVT iterations per item is

$$\mathrm{EN}_{i,j} = \sum_{n=1}^{\infty} n(\mathrm{Pa}_{i,j}X_{i,j})^{n-1} L_{i,j}^{(n-1)(n-2)/2}(1-\mathrm{Pa}_{i,j}X_{i,j}L_{i,j}^{n-1}) \tag{6.37}$$

4. The total VVT impact cost is

$$\mathrm{Ci}_{\mathrm{VVT\_Strategy}} = \sum_{i=1}^{z}\sum_{j=1}^{n_i}\sum_{l=i}^{z}\sum_{k=1,2\ldots}\{\mathrm{Ci}_{i,j}^{(l,k)}\mathrm{Pi}_{i,j}^{(l,k)}(1-X_{i,j})\} \tag{6.38}$$

5. Similarly, the total nonlinear impact risk cost is

$$\mathrm{Ci}_{\mathrm{VVT\_Strategy}} = \sum_{i=1}^{z}\sum_{j=1}^{n_i}\sum_{l=i}^{z}\sum_{k=1,2\ldots}\{\mathrm{Ci}_{i,j}^{(l,k)}\mathrm{Pi}_{i,j}^{(l,k)}(1-Y_{i,j})\} \tag{6.39}$$

6.2.6 VVT Cost and Risk Example

We will use the VVT cost estimation methodology and models described above in order to illustrate a simple practical derivation of VVT cost data. This example is derived from part of a pilot project (described in detail in Chapters 7 and 8) carried out by the Lahav division of the IAI during the SysTest project, which was designed to upgrade an avionic suite for a transport helicopter.

System and Quality Data The following system and quality data were collected during the SysTest project and harmonized with previously collected data:

1. A set of VVT activities performed during the development segment of the pilot project program (phases 1–5: Definition, Design, Implementation, Integration and Qualification). It identifies the identification number and description of each VVT activity $\{V_{i,j}\}$ as well as the VVT activity cost $\{\mathrm{Cv}_{i,j}\}$ and the VVT activity time $\{\mathrm{Tv}_{i,j}\}$ associated with fully performing each activity.

2. The VVT strategy adopted for the pilot project development. Specifically, it identifies the set of VVT activity performance level $\{X_{i,j}\}$ for each VVT activity.
3. A set of VVT appraisal risks associated with each VVT activity. Specifically, it provides information about the VVT appraisal cost $\{\text{Ca}_{i,j}\}$, VVT appraisal time $\{\text{Ta}_{i,j}\}$ as well as the probability of appraisal risk occurrence $\{\text{Pa}_{i,j}\}$ and the decay rate in the probability of the appraisal risk $\{D_{i,j}\}$.
4. A set of VVT impacts associated with each VVT activity. Specifically, it provides information about the impact risk scenario $\{R_{i,j}^{(l,k)}\}$, the probability that the impact will occur $\{\text{Pi}_{i,j}^{(l,k)}\}$, the phase in which the impact is likely to occur $\{l\}$ as well as the impact cost $\{\text{Ci}_{i,j}^{(l,k)}\}$ and the impact time $\{\text{Ti}_{i,j}^{(l,k)}\}$ if the impact does, in fact, occur.

Estimating VVT Cost and Risk To simplify this example, the VVT horizon selected covers only the first system lifecycle phase (Definition). The actual VVT strategy used during the pilot project was adopted for this example and the calculations of the VVT impacts used a nonlinear piecewise conversion function.

Direct Probabilistic Computations Figure 6.12 shows the result of direct probabilistic computations of the VVT cost and risk (in CU).[77]

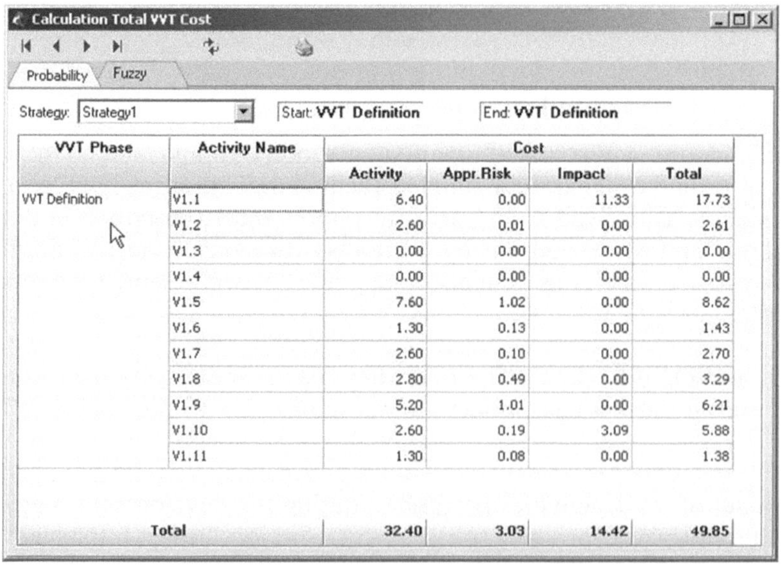

VVT Phase	Activity Name	Cost			
		Activity	Appr.Risk	Impact	Total
VVT Definition	V1.1	6.40	0.00	11.33	17.73
	V1.2	2.60	0.01	0.00	2.61
	V1.3	0.00	0.00	0.00	0.00
	V1.4	0.00	0.00	0.00	0.00
	V1.5	7.60	1.02	0.00	8.62
	V1.6	1.30	0.13	0.00	1.43
	V1.7	2.60	0.10	0.00	2.70
	V1.8	2.80	0.49	0.00	3.29
	V1.9	5.20	1.01	0.00	6.21
	V1.10	2.60	0.19	3.09	5.88
	V1.11	1.30	0.08	0.00	1.38
Total		32.40	3.03	14.42	49.85

Figure 6.12 Definition phase: quality costs using probabilistic computations.

[77]Actual costs are confidential and therefore identified as CU, or Cost Units.

The VVT strategy cost is 32.40 CU, the expected VVT appraisal risk is 3.03 CU and the impact cost is 14.42 CU for a total of 49.85 CU. The high impact cost can be attributed to the low overall VVT activity performance level selected for this phase of the pilot project.

Monte Carlo Simulation Monte Carlo simulations are extensively used for modeling complex system behavior in engineering, finance, network, bioinformatics, radiation therapy, project planning, physics and meteorology, to name a few (Jackel, 2002). In Monte Carlo simulation, a random number generator is used to simulate occurrence (or nonoccurrence) of given probabilistic events (here appraisal and impact risks) according to given probability distributions. A large number of hypothetical scenarios of risk impacts may thus be generated. These are then used to construct hypothetical overall risk costs and the distribution of these costs. For a given VVT strategy and VVT horizon the Monte Carlo simulation process performs the following operations:

- Calculates the deterministic VVT costs of the performed activities.
- Simulates the appraisal risk costs stemming from imperfection in the system development process.
- Simulates the impact risk cost stemming from the partially performed activities.
- Repeats the above process a large number of times.
- Eventually, sums up the overall VVT and risk impact costs and builds a cost distribution representation (histogram) of the results.

Figure 6.13 shows a single run of the Monte Carlo stochastic simulator.

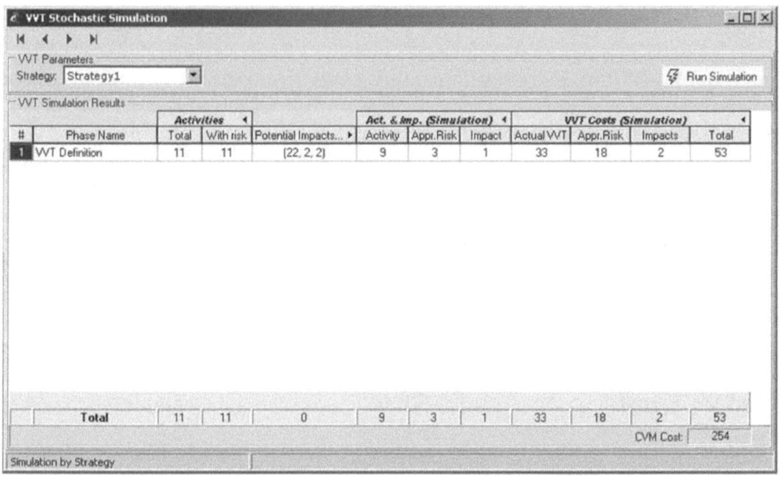

Figure 6.13 Single stochastic quality cost simulation: Definition phase.

As can be observed, by comparing the results obtained by direct probabilistic computation with the single run, the individual runs may differ from one another substantially. Chapter 8 presents a considerably richer single run of the Monte Carlo stochastic simulator.

Figure 6.14 shows a multiple run (3000 iterations) of the Monte Carlo stochastic simulator. The left vertical line represents the VVT strategy cost (33.0 CU), the right vertical line represents CVM cost (254.0 CU) and the total cost distribution is depicted graphically in between these two lines. The simulated total cost results (50.0 CU) are comparable to the results obtained in the probabilistic computations (49.85 CU). One advantage stemming from a simulated approach is the visualization of the cost distribution and the added information.

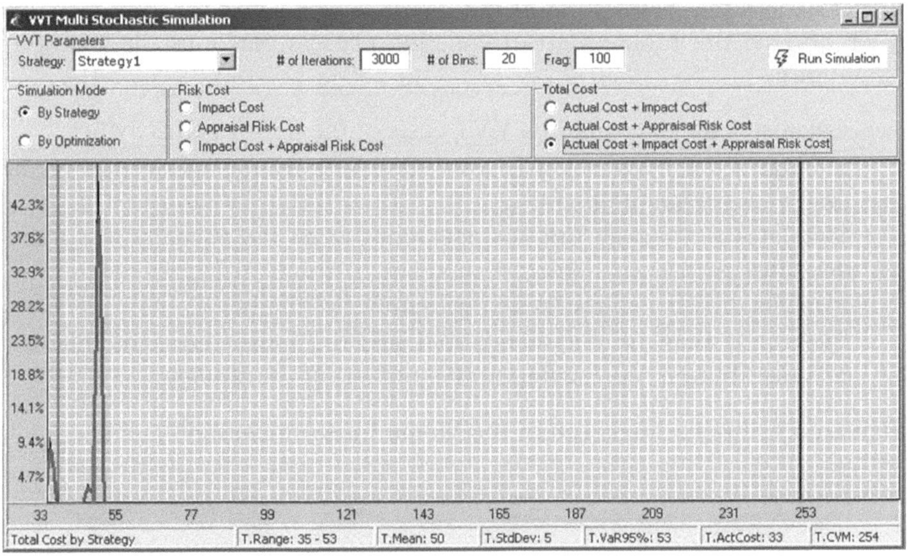

Figure 6.14 Multiple stochastic quality costs simulations: Definition phase.

Table 6.6 summarizes the key VVT cost data of this example.

TABLE 6.6 Overall Definition Phase Quality Cost Results

VVT Cost Data	Probabilistic Calculation (CU)	Monte Carlo Simulation (CU)	Percent
VVT activity cost	32	33	66
Mean appraisal risk cost	3	2	4
Mean impact risk cost	14	15	30
Mean total VVT cost	50	50	100
CVM cost	254	254	508
Standard deviation	—	5	10
VaR (95%)	—	53	106

When considering only the Definition phase, it is clear that the selected VVT strategy, leading to a mean total VVT cost of 50 CU, is significantly preferable to executing the entire CVM (costing 254 CU).

6.3 VVT TIME AND RISK MODELING

At the individual level, VVT time modeling mirrors VVT cost modeling. Therefore, we do not repeat the entire previously provided derivation. This section will however describe the unique features of VVT modeling related to the time domain. Specifically, it extends the VVT methodology elaborated before in order to perform estimation of the time required to carry out system VVT processes.

6.3.1 System/VVT Network

The basic issue of planning and estimating the duration of projects is well established and in common use. For example, the Critical Path Method (CPM) and the Program Evaluation and Review Technique (PERT) were developed in the late 1950s by DuPont and Remington Rand in order to manage plant maintenance and by Lockheed for the Polaris Missile program (Grant, 1983). We will cite a few relevant project-scheduling definitions and then expand our discussion to cover the system/VVT lifecycle domain.

Project Scheduling Definitions Among various definitions, we have chosen the following relevant ones:

1. A project is an interrelated set of activities that has definite starting and ending points and results in unique products or services.
2. Project duration is the elapsed time from the project start date through the project finish date.
3. A project network diagram is a schematic display of the logical precedence relationships of project activities. It is drawn from left to right to reflect project chronology.
4. An Activity-On-Node (AON) network is a network diagram whose nodes represent activities and arrows which represent the precedence relationships of activities in a project. (The complementary notation, which we are not using in this book, is Activity-On-Arc, or AOA).
5. A logical relationship is a dependency between two project activities. In this book, we have used the Finish-To-Start (FTS) relationship, namely the "from" activity must finish before the "to" activity can start.
6. PERT is an event-oriented network analysis technique used to estimate project duration. Often, PERT is used where events or activities incorporate components of stochastic duration.

7. A path is a connected sequence of activities in the network diagram which leads from the project start node to the project end node.
8. A critical path is the longest path in a project network diagram, which determines the total length of the project.
9. The Critical Path Method (CPM) is a technique used to estimate project duration by analyzing which sequence (or sequences) of activities are in the critical path. Originally, the CPM was designed for events or activities having deterministic duration.

Characterizing System/VVT Time Estimate Here, we are interested in lifecycle time estimates and, in particular, in the duration of each phase of the lifecycle. This means that we must deal with estimating the duration of each phase of the system's lifecycle starting with the Definition phase and ending either at the Disposal phase or at any phase defined by the VVT horizon. Although our primary interest is with VVT processes, estimation of time is necessarily intertwined with system activities. Therefore, we must consider the timing of the combined system and VVT schedules. Another concern is the logical interrelation network and the duration of each system and VVT activity. In addition, we must take into account the level of VVT performance ($X_{i,j}$) associated with each VVT activity $V_{i,j}$. Finally, we must factor in the stochastic phenomena of VVT appraisal risk time $\mathrm{Ta}_{i,j}$ and the VVT impact risk time $\mathrm{Ti}_{i,j}$.

Figure 6.15 shows an example of a system/VVT activity network diagram. As mentioned before, in an ideal world each system lifecycle phase follows the previous one. However, in order to compute a realistic overall project duration, we must deal with a nonideal situation.

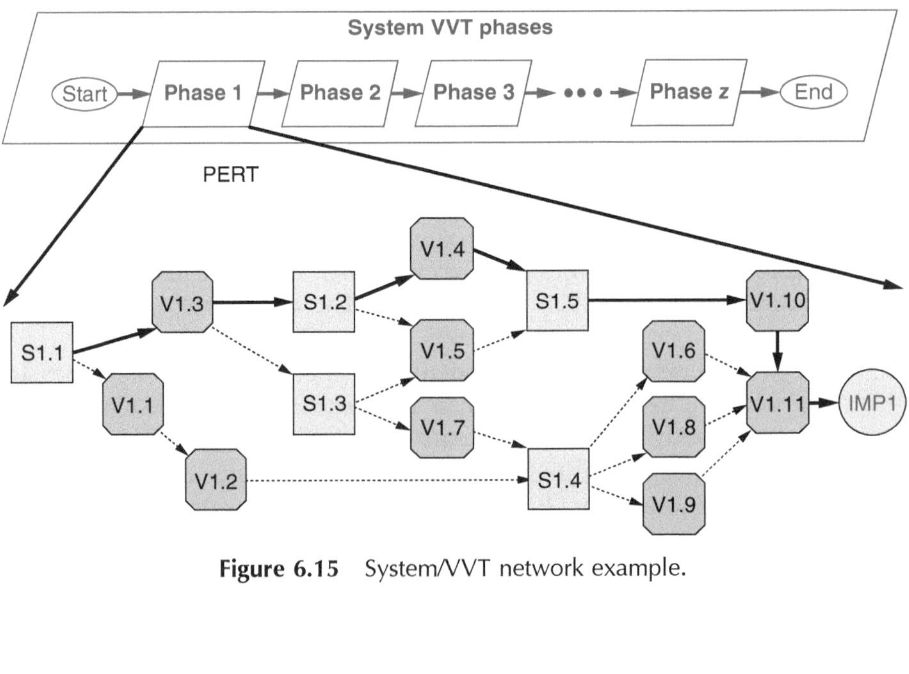

Figure 6.15 System/VVT network example.

Within each phase, system and VVT activities are connected in a complex network. Sometimes, VVT activities are carried out in order to support other VVT activities (e.g., VVT planning, VVT infrastructure building). On other occasions, VVT activities are performed after the completion of corresponding system activities. Similarly, further system activities are performed after VVT activities have been completed.

The system/VVT network example represents a typical set of system/VVT activities commonly adhered to by many industries during the Definition phase. System activities $\{S_{1.1}, S_{1.2}, \ldots, S_{1.5}\}$ are intertwined with VVT activities $\{V_{1.1}, V_{1.2}, \ldots, V_{1.11}\}$. Each VVT activity is performed at a selected level ($X_{i,j}$, $0 \leq X_{1,j} \leq 1$) and therefore gives rise to both VVT appraisal risks and VVT impact risks.

In principle, VVT impact risks affecting a given phase may occur at any time within that phase. However, we make the simplifying assumption that these VVT time impacts occur serially, taking place at the end of the phase and constituting a single event (IMP1) per phase.

VVT System Model We introduce a System Model (SM). Our interest in this model is primarily with the description of each system activity as well as the time required to perform each system activity. To a lesser degree, we are interested in the cost incurred in performing each system activity (see Figure 6.16 and Table 6.7).

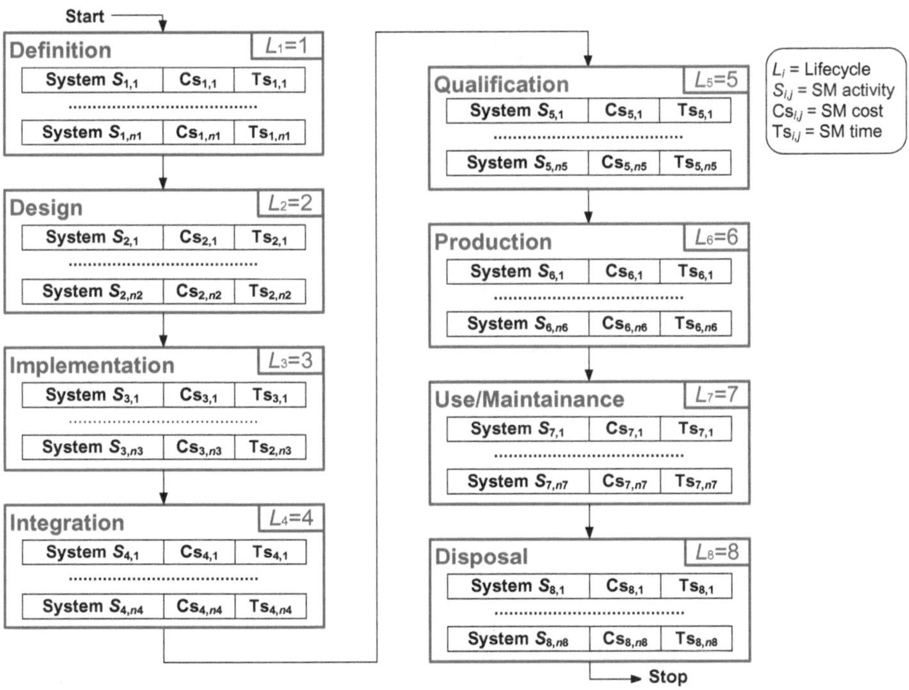

Figure 6.16 A VVT System Model (SM).

TABLE 6.7 A System Model (SM) Tuple Structure

Variable	Definition
$S_{i,j}$	Name/description of a specified system activity carried out in phase i related to system activity $j(i)$; $j(i) = 1, 2, \ldots, n_i$; $i = 1, 2, \ldots, z$
$\mathrm{Cs}_{i,j}$	Cost of performing system activity $S_{i,j}$
$\mathrm{Ts}_{i,j}$	Time required to carry out system activity $S_{i,j}$

6.3.2 Modeling Time of System/VVT Lifecycle

Time Estimation Variables System VVT time estimation must take the following into account:

1. System activity time $\mathrm{Ts}_{i,j}$, $j(i) = 1, 2, \ldots, n_i$, $i = 1, 2, \ldots, z$, required to perform system activities $S_{i,j}$ where i represents the system lifecycle phase and j represents the system activity index within that phase.
2. VVT activity time $\mathrm{Tv}_{i,j}$, $j(i) = 1, 2, \ldots, n_i$, $i = 1, 2, \ldots, z$, required to fully perform VVT activity $V_{i,j}$.
3. Level of VVT performance $X_{i,j}$, $0 \le X_{i,j} \le 1$, $j(i) = 1, 2, \ldots, n_i$, $i = 1, 2, \ldots, z$.
4. Appraisal time $\mathrm{Ta}_{i,j}$ required to carry out a system/VVT appraisal iteration following activity $V_{i,j}$, where $j(i) = 1, 2, \ldots, n_i$, $i = 1, 2, \ldots, z$.
5. VVT appraisal risk probability $\mathrm{Pa}_{i,j}$, $0 \le \mathrm{Pa}_{i,j} \le 1$, where $j(i) = 1, 2, \ldots, n_i$, $i = 1, 2, \ldots, z$.
6. Decay in the VVT appraisal risk probability $D_{i,j}$ so that $0 \le D_{i,j} \le 1$, $j(i) = 1, 2, \ldots, n_i$ and $i = 1, 2, \ldots, z$.
7. Impact time $\mathrm{Ti}_{i,j}^{(l,k)}$ required to carry out a corrective action due to impact risk $R_{i,j}^{(l,k)}$ where $j(i) = 1, 2, \ldots, n_i$, $l \ge i$, $i = 1, 2, \ldots, z$ *and* $k = 1, 2, \ldots$.
8. Probability $\mathrm{Pi}_{i,j}^{(l,k)}$, $0 \le \mathrm{Pi}_{i,j}^{(l,k)} \le 1$, of impact risk $R_{i,j}^{(l,k)}$.

Time Modeling Assumptions The following assumptions have been made:

1. *Lifecycle Phases.* Each system lifecycle phase follows the previous one so that a given lifecycle starts strictly after the end of the previous lifecycle and ends just prior to the beginning of the next lifecycle.
2. *System Activities.* System activities ($S_{i,j}$) are linked via a specified network arrangement and their time durations ($\mathrm{Ts}_{i,j}$) are assumed known and deterministic. In general, these activities generate products requiring VVT evaluation.
3. *VVT Activities.* VVT activities ($V_{i,j}$) are interspersed with system activities in a network arrangement. Their time durations ($\mathrm{Tv}_{i,j}$) are deterministic and either linearly or nonlinearly dependent upon the level of VVT performance ($X_{i,j}$; $0 \le X_{i,j} \le 1$).

4. *VVT Appraisal Risk.* An appraisal risk may require a VVT appraisal corrective process. We have called the time required to perform this corrective process as correction time (Ta). We have assumed that the correction time needed to perform a corrective and retesting process of a given system product is fixed and depends on the nature of the specific product. We have modeled it as occurring immediately after the relevant VVT activity. Therefore, corrective time appears to extend the duration of the relevant VVT activity.
5. *VVT Impact Risk.* Impact risks could emanate from partial VVT performance conducted during either previous or current lifecycle phases. They could cause unplanned delays and unexpected remedial processes extending the system schedule in a serial manner. We have called these delays and the time required to perform any remedial process the VVT impact risk time. For modeling simplicity, we have serially clustered all the materialized VVT impact times at the end of the relevant lifecycle phase. Therefore, impact time extends the duration of the phase in which it occurs.

Procedure for Time Modeling of Specific Phase The following steps have been used in order to compute the time of a specific system lifecycle phase l, $l = i°$:

1. Identify each system activity time $\text{Ts}_{i°,j}$ associated with each system activity $S_{i°,j}$ in phase $i°$.
2. Select the performance level $X_{i°,j}$ associated with each VVT activity $V_{i°,j}$ in phase $i°$.
3. Compute each VVT activity time item of each VVT activity $V_{i°,j}$ in phase $i°$ based on the level of VVT performance $X_{i°,j}$ and the time $\text{Tv}_{i°,j}$ associated with each VVT activity:

$$\text{Tv\_item}_{i°,j}\,|\,X = X_{i°,j}Tv_{i°,j} \qquad j(i°) = 1, 2, \ldots, n_{i°} \tag{6.40}$$

4. Compute each appraisal risk time item associated with each VVT activity in phase $i°$ based on the appraisal risk time per iteration $\text{Ta}_{i°,j}$ multiplied by the expected number of corrective iterations given performance level $X(\text{NCI}|X)$:

$$\text{Ta\_item}_{i°,j}\,|\,X = \text{Ta}_{i°,j}\text{NCI}_{i°,j}\,|\,X \qquad j(i°) = 1, 2, \ldots, n_i \tag{6.41}$$

where, according to Equation (6.14),

$$\text{NCI}_{i,j} = \text{EN}_{i,j} - 1 \tag{6.42}$$

and, according to Equation (6.13),

$$\text{EN}_{i,j} = \sum_{n=1}^{\infty} n\left(\text{Pa}_{i,j}X_{i,j}\right)^{n-1} L_{i,j}^{(n-1)(n-2)/2}\left(1 - \text{Pa}_{i,j}X_{i,j}L_{i,j}^{n-1}\right) \tag{6.43}$$

5. Compute each combined VVT impact risk time item affecting lifecycle phase i° by summing the relevant impact risk time that emanates from the individual VVT activity $V_{i^\circ j}$:

$$\text{Ti\_item}_{i^\circ,j}|X = \sum_{k=1,2,\ldots} \text{Ti}_{i^\circ,j}^{(i^\circ,k)} \text{Pi}_{i^\circ,j}^{(i^\circ,k)} (1 - X_{i^\circ,j}) \tag{6.44}$$

6. Similarly, we have used the piecewise linear function described within this chapter to compute nonlinear impact risk effects:

$$\text{Ti\_item}_{i^\circ,j}|Y = \sum_{k=1,2,\ldots} \text{Ti}_{i^\circ,j}^{(i^\circ,k)} \text{Pi}_{i^\circ,j}^{(i^\circ,k)} (1 - Y_{i^\circ,j}) \tag{6.45}$$

7. The computed data are searched in order to find the longest path using the Critical Path Method (CPM). Each time component associated with all critical path is summed as follows:

$$\text{Ts\_phase}_{i^\circ} = \sum_{j=\text{Critical\_path}} \{\text{Ts}_{i^\circ,j}\} \tag{6.46}$$

$$\text{Tv\_phase}_{i^\circ} = \sum_{j=\text{Critical\_path}} \{\text{Tv\_item}_{i^\circ,j}\} \tag{6.47}$$

$$\text{Ta\_phase}_{i^\circ} = \sum_{j=\text{Critical\_path}} \{\text{Ta\_item}_{i^\circ,j}\} \tag{6.48}$$

Aggregating Impact Times As stated earlier, impact cost aggregation is achieved by simply adding up all the impact costs which occurred during a given project phase. The above approach, however, is not simple when dealing with impact times. The reason is that individual impacts occur randomly (sometimes serially, sometimes in parallel with other VVT activities) during a given lifecycle phase. The duration of each impact time (the time required to fix a problem) is stochastic. In addition, the availability of staff to deal with the problem is also stochastic. Therefore, the question we are dealing with is: What will be the overall effect of a group of impact times on the total duration of a given lifecycle phase? One can envision several models to represent this problem:

1. *Maximal Aggregated Impact.* Adding all the individual impact times serially gives

$$\text{Ti\_phase}_{i^\circ} = \sum_{j=\text{Critical\_path}} \text{Ti\_item}_{i^\circ,j} \tag{6.49}$$

This approach produces a nonrealistic high overall aggregated impact time. In reality, many impact times are dealt with in parallel and therefore a realistic aggregated impact time will be shorter.

2. *Minimal Aggregated Impact.* Considering the largest impact time only gives

$$\text{Ti\_phase}_{i^\circ} = \max\{\text{Ti\_item}_{i^\circ,j}\} \qquad j = \text{Critical\_path} \tag{6.50}$$

This approach produces a nonrealistic low aggregated impact time. In reality, several small impact times will affect the overall aggregated impact time, and therefore a realistic aggregated impact time will be larger.

3. *Aggregating Impact Times Using Measurement Error Propagation Analogy.* Expert VVT engineers suggest modeling the above problem analogously to modeling measurement error propagation. Impact times are similar to measurement errors in the following properties: (1) they occur stochastically and (2) they are independent of one another. We can envision measuring N components, say resistors, where each measurement is subject to a known potential error ε_i. We ask the question: What will be the overall error if we measure all the resistors connected serially? Based on Taylor (1996), we can compute the total propagated error ε:

$$\varepsilon = \sqrt{\varepsilon_1^2 + \varepsilon_2^2 + \varepsilon_3^2 + \cdots + \varepsilon_N^2} = \sqrt{\sum_{i=1}^{N} \varepsilon_i^2} \tag{6.51}$$

We model the aggregated impact time by applying the above approach and assuming it occurs at the end of each lifecycle phase:

$$\text{Ti\_phase}_{i^\circ} = \sqrt{\sum_{j=\text{Critical\_path}} \left\{(\text{Ti\_item}_{i^\circ,j})^2\right\}} \tag{6.52}$$

This equation provides the most realistic result, which is between the above maximal and minimal aggregated impacts.

Earlier Phases Impact Times We must now consider that impact times emanating from earlier phases affect the system during the current phase i°:

$$\text{Ti\_early\_phase}_{i^0} = \sum_{i=1}^{i^0-1} \text{Ti\_phase}_i\text{\_phase}_{i^0} \tag{6.53}$$

Therefore

$$\text{Ti\_total\_phase}_{i^\circ} = \text{Ti\_phase}_{i^\circ} + \text{Ti\_early\_phase}_{i^\circ} \tag{6.54}$$

Total Phase Time The above four time components are summed to arrive at the overall phase i° time:

$$T_{i^\circ} = \sum\{\text{Ts\_phase}_{i^\circ} + \text{Tv\_phase}_{i^\circ} + \text{Ta\_phase}_{i^\circ} + \text{Ti\_total\_phase}_{i^\circ}\} \tag{6.55}$$

Procedure for Realistic Lifecycle Time Modeling The ideal system lifecycle describes each lifecycle phase as strictly following the previous phase. In practice, this is rarely the case. Often a new phase starts before the previous phase has finished. We have introduced a Premature Next Phase Start (PNPS) factor ($0 \leq \text{PNPS}_i \leq 1$) which identifies the incomplete portion of phase $\{i\}$ in which phase $\{i + 1\}$ starts. A set of PNPS factors exercised in the pilot project (see Chapters 7 and 8) is depicted in Table 6.8. Here, the Definition phase starts at the beginning of the system lifecycle, the Design phase starts when the Definition phase is 30% incomplete and so on. Based on the above, the overall time required for carrying out a system and VVT activity for the entire lifecycle is

$$T_{\text{VVT\_Strategy}} = T_1 + \sum_{i=2}^{z} T_i - (T_{i-1}\text{PNPS}_i) \qquad (6.56)$$

TABLE 6.8 Set of PNPS Used in Pilot Project

Lifecycle Phase Number		
Number	Name	PNPS Factor
1	Definition	0.0
2	Design	0.3
3	Implementation	0.4
4	Integration	0.1
5	Qualification	0.2

A comprehensive example of time and risk using the IAI/Lahav pilot project is provided in Chapter 8.

6.3.3 Time and Risk Example

We use the VVT time estimation methodology and models described above in order to calculate the overall time needed to carry out the first phase (Definition) of the IAI/Lahav pilot project conducted during SysTest. In other words, we used relevant system and quality parameters and set the VVT horizon to cover only the first system lifecycle phase. Finally, the calculations of the VVT impact risks use the nonlinear piecewise conversion function described in Section 6.2.6. Please note that this example is also a part of a full pilot project (described in detail in Chapter 8).

Figure 6.17 shows the system activities, the VVT activities and the critical path ($S_{1,1}$, $V_{1,3}$, $S_{1,2}$, $V_{1,5}$, $S_{1,5}$, $V_{1,10}$, $V_{1,11}$, IMP1) associated with the selected VVT strategy. The overall duration of this phase is 63.5 days.

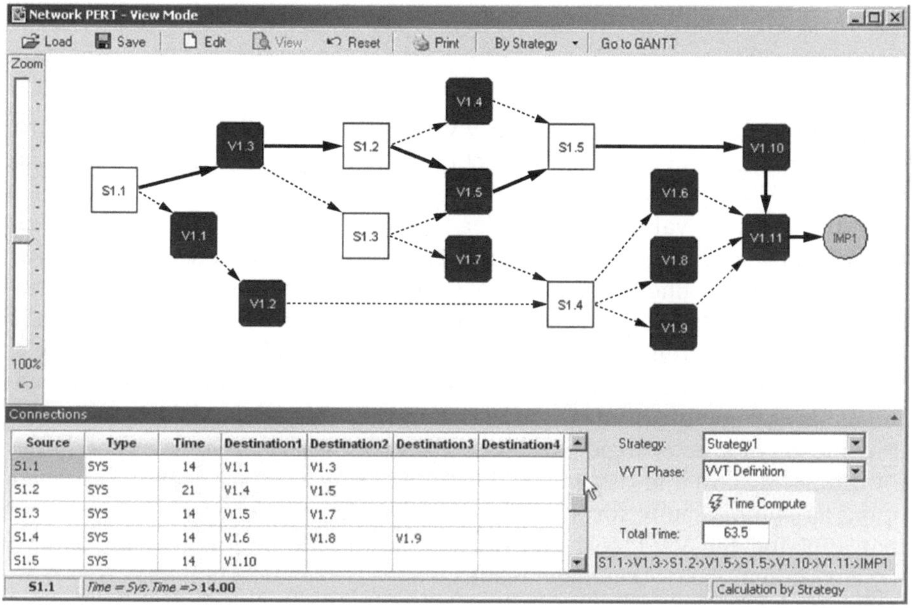

Figure 6.17 Overall phase 1 (Definition) PERT and critical path.

Figure 6.18 shows the overall phase 1 (Definition) GANTT. As can be seen, the critical path can be identified by a darker track and the length of the critical path is 63.5 days.

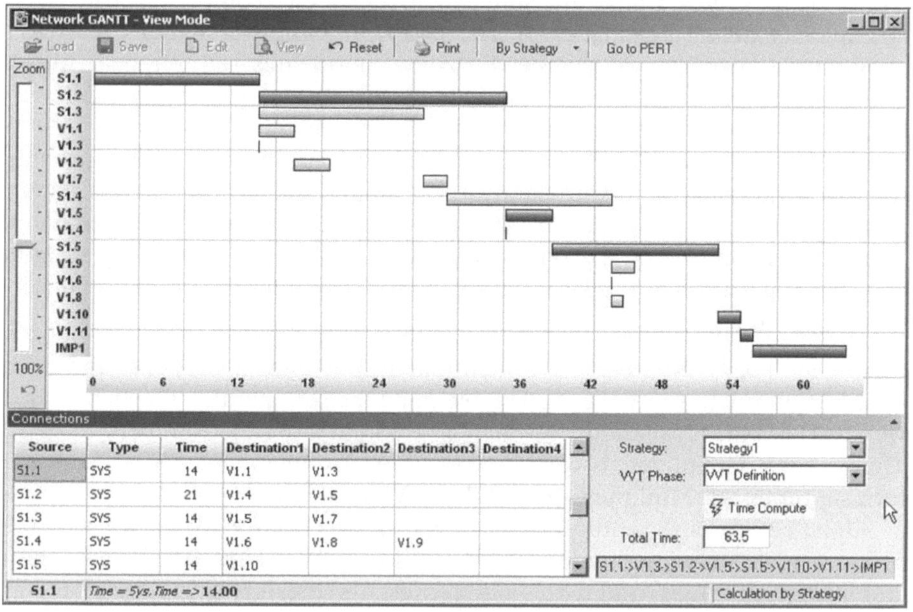

Figure 6.18 Overall phase 1 (Definition) GANTT and critical activities.

6.4 FUZZY VVT COST MODELING

6.4.1 Introduction

A key problem associated with the VVT cost, time and risk estimation procedures presented in the previous sections is that input data such as costs, risk levels and VVT performance levels are inexact by nature. As a result, it is difficult to obtain valid data in order to populate the various VVT models. However, using a fuzzy logic paradigm can reduce this predicament greatly.

Specifically, parameters such as VVT costs can be better quantified in terms of minimum, most-likely and maximum values. Likewise, parameters such as levels of risk occurrence can be better encapsulated in linguistic terms such as "high" or "low" rather than in exact values of probability. This chapter extends the methodology for estimating the cost and risk of system VVT by modeling it by means of a fuzzy logic paradigm. The proposed fuzzy logic methodology for estimating cost and risk will be illustrated using a similar example of developing an avionics suite for a transport helicopter. This chapter demonstrates that applying the VVT cost and risk estimation methodology using a fuzzy logic paradigm yields beneficial results.

6.4.2 General Fuzzy Logic Modeling

Fuzzy theory was introduced by Lotfi Zadeh (1965) of the University of California at Berkeley in the 1960s. Fuzzy logic is a superset of conventional (Boolean) logic which has been extended to handle the concept of partial truth—truth values which are somewhere between "completely true" and "completely false" (Kosko, 1996). The fuzzy principle states that everything in nature is a matter of degree. Fuzziness permeates our physical world. The importance of fuzzy logic derives from the fact that most modes of human reasoning and especially commonsense reasoning are approximate in nature. The arithmetic and calculus of fuzzy sets and fuzzy numbers provide us with a method for manipulating these imprecise representations. Fuzzy numbers and their associated arithmetic and calculus are the subject of many publications and several textbooks (e.g., Klir and Yuan, 1995). Many authors describe application of fuzzy set theory to model systems incorporating uncertain input data.

Triangular Fuzzy Numbers As mentioned above, most raw cost and time information related to VVT is available in an inexact manner. For this model, we shall assume availability of data in Triangular Fuzzy Numbers (TFNs). That is, each data element (e.g., the cost of performing a given VVT activity) forms a tuple: A; {Minimum, Most Likely, Maximum}.

Alpha-cut (α-cut) represents the level of uncertainty where α-cut = 0.0 corresponds to maximum uncertainty (i.e., somewhere between the minimum and maximum values) and α-cut = 1.0 represents certainty (i.e., a *crisp*[78] quan-

[78]The word "crisp" in fuzzy logic is used to describe exact, sharp and concise values without ambiguities. That is, a number typical to Aristotelian logic.

tity equal to the most likely value). Defuzzyfication is performed using the Center Of Gravity (COG) method (see Figure 6.19).

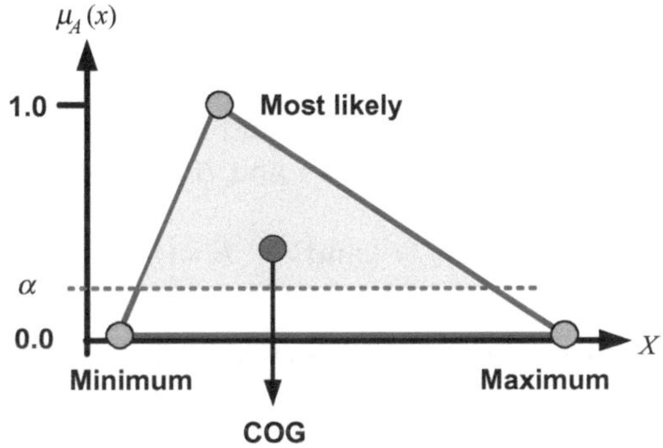

Figure 6.19 Fuzzification/defuzzyfication of approximate values.

Modeling Intrinsic Fuzzy Levels In order to model various types of fuzzy levels (e.g., VVT performance levels as well as appraisal and impact risk levels), we define intrinsic VVT fuzzy levels using natural language adjectives and adverbs and then translate from natural language words to fuzzy values using membership functions. The selected adjectives are *low* (L), *medium* (M) and *high* (H). In addition we add the adverb (hedge word) *very* (V). This set and the translation from natural language words to fuzzy values is depicted in:

- Figure 6.20a—adjective words only (L, M, H)
- Figure 6.20b—adjective combined with other adjective words and adjective combined with hedge word (VL, L, ML, M, MH, H, VH).

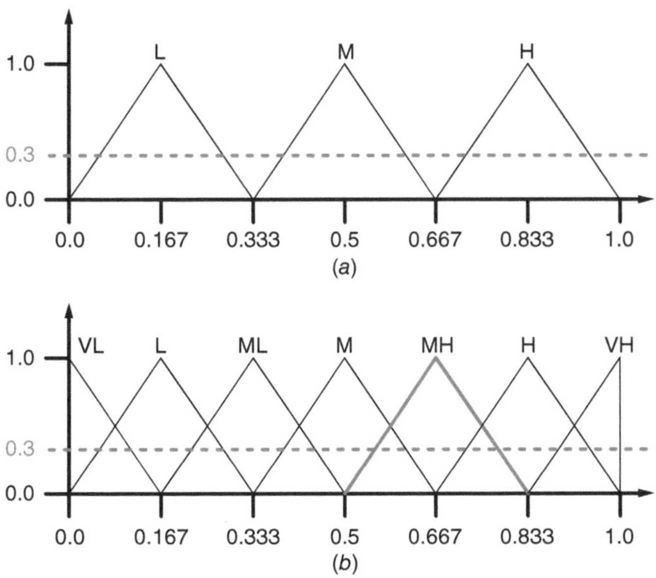

Figure 6.20 (*a*) Selected fuzzy value assignments with (*b*) modifying terms.

It was found that human experts find it very convenient to deal with the division of the intrinsic fuzzy level spectrum into seven segments along the line described above.

Fuzzy Calculus Arithmetic operations on the fuzzy numbers A and B are reduced to operations on the intervals A and B as defined by α-cuts. Let us consider two triangular fuzzy numbers A and B given by the tuples:

$$A = (a_1, a_2, a_3) \qquad \text{and} \qquad B = (b_1, b_2, b_3) \tag{6.57}$$

Then, there are the following definitions (Klir and Yuan, 1995):

1. *Fuzzy addition:*

$$A + B = (a_1 + b_1, a_2 + b_2, a_3 + b_3) \tag{6.58}$$

2. *Fuzzy subtraction:*

$$A - B = (a_1 - b_1, a_2 - b_2, a_3 - b_3) \tag{6.59}$$

3. *Fuzzy multiplication:*

$$AB = \{\min(a_1b_1, a_1b_3, a_3b_1, a_3b_3), a_2b_2, \max(a_1b_1, a_1b_3, a_3b_1, a_3b_3)\} \tag{6.60}$$

4. *Fuzzy division:*

$$\frac{A}{B} = \left\{\min\left(\frac{a_1}{b_1}, \frac{a_1}{b_3}, \frac{a_3}{b_1}, \frac{a_3}{b_3}\right), \frac{a_2}{b_2}, \max\left(\frac{a_1}{b_1}, \frac{a_1}{b_3}, \frac{a_3}{b_1}, \frac{a_3}{b_3}\right)\right\} \tag{6.61}$$

5. *Defuzzyfication:* based on the Center-Of-Gravity (COG) method,

$$\text{COG} = \frac{\int_{a1}^{a3} [A(x)x]dx}{\int_{a1}^{a3} [A(x)]dx} \tag{6.62}$$

6.4.3 Fuzzy Modeling of the VVT Process

The fuzzy VVT methodology extension for estimating VVT cost and risk may also be described through the corresponding four concepts: (1) fuzzy canonical VVT model, (2) fuzzy VVT strategy, (3) fuzzy appraisal risk model and (4) fuzzy impact risk model.

Fuzzy Canonical VVT Model (F-CVM) Figure 6.21 describes the F-CVM proposed. It depicts the lifecycle phases $\{L_i\}$, the VVT activities $\{V_{i,j}\}$, fuzzy

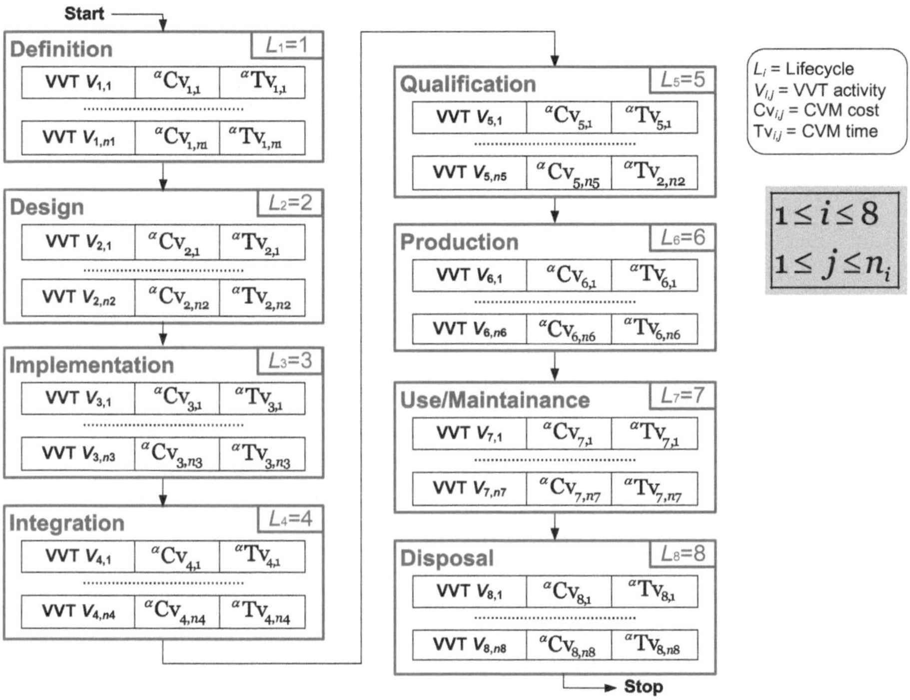

Figure 6.21 Fuzzy Canonical VVT Model (F-CVM).

costs $\{^{\alpha}Cv_{i,j}\}$ and fuzzy times $\{^{\alpha}Tv_{i,j}\}$. In the fuzzy paradigm, the cost and time of each VVT activity are provided in a fuzzified manner as a TFN, that is, minimum, most likely and maximum for a specified α-cut. Time issues are a straightforward extension and will not be described further.

Note: The reader should note that the F-CVM is an ideal concept. It represents a set of VVT activities which may be performed only if an unlimited amount of money and time is available. Many industrial and governmental organizations perform about 15–25% of CVM and, in special circumstances (e.g., spaced manned missions), perhaps 25–50% of CVM is performed.

The defuzzyfied "CVM cost," namely, the cost of performing the entire CVM procedure for a specified α-cut, using the COG method is

$$C_{\mathrm{CVM}} = \sum_{i=1}^{Z} \sum_{j=1}^{n_i} \mathrm{COG}\{^{\alpha}\mathrm{Cv}_{i,j}\} \tag{6.63}$$

where i, $1 \le i \le Z$, represents the entire systems lifecycle set of phases and j, $1 \le j \le n_i$, represents the different VVT activities within each phase.

Fuzzy VVT Strategy Model (F-VSM) As mentioned above, executing the all-inclusive CVM is not practical due to VVT funding limitations and time considerations. In order to deal with a realistic quantitative modeling of the costs and risks associated with an incomplete set of VVT activities, we have defined the concept of VVT strategy as a representation of the partial performance of the CVM.

The VVT strategy is, in fact, a set of decision variables which identify how much each VVT activity is to be performed—from none at all to partial to fully performed. In a crisp world, these decision variables ($X_{i,j}$; $0.0 \leq X_{i,j} \leq 1.0$) are precisely known. In a fuzzy world, we have defined a linguistic variable: ${}^{\alpha}\mathrm{Fx}_{i,j} \equiv$ Fuzzy VVT Performance Level. We have assigned each of the ${}^{\alpha}\mathrm{Fx}_{i,j}$ to one of the linguistic terms: {VL, L, ML, M, MH, H, VH}.

The VVT strategy, defines the performance level of any VVT activity $V_{i,j}\{j(i) = 1, 2, \ldots, n_i, i = 1, 2, \ldots, z\}$ within the system lifecycle (see Figure 6.22). In other words, ${}^{\alpha}\mathrm{Fx}_{i,j}$ = VL means that the VVT activity $V_{i,j}$ is performed at a very low level whereas ${}^{\alpha}\mathrm{Fx}_{i,j}$ = VH means that the VVT activity $V_{i,j}$ is performed at a very high level.

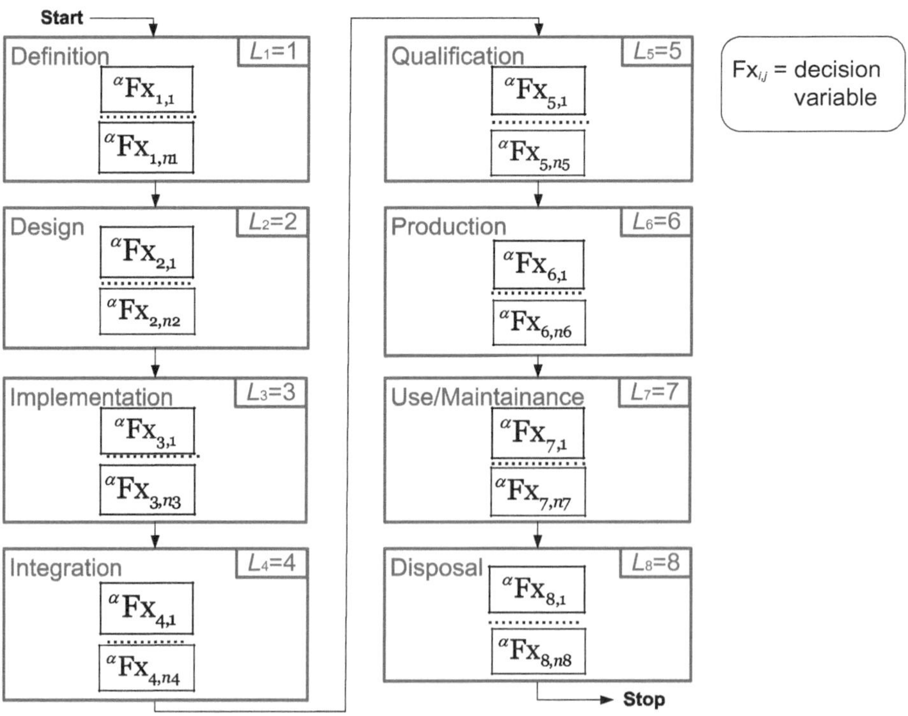

Figure 6.22 Fuzzy VVT Strategy Model (F-VSM).

A single fuzzy VVT activity cost is computed by multiplying the full cost of performing the VVT activity by the level of activity performance at the designated α-cut:

$$^{\alpha}\mathrm{Cv}_{\mathrm{vvt\_i,j}} = {^{\alpha}\mathrm{Cv}_{i,j}}\, {^{\alpha}\mathrm{Fx}_{i,j}} \tag{6.64}$$

This cost model makes two assumptions (1) each VVT activity $V_{i,j}$ is completely independent from any other VVT activity and (2) VVT performance level $^{\alpha}\mathrm{Fx}_{i,j}$ corresponds to a linear cost of the VVT activity.

Figure 6.23 shows an example of computing a single cost of partially performed VVT activity using software developed during the SysTest project. In this case, α-cut = 0.2 and VVT performance level equals medium–low (ML). A complete VVT activity cost is Minimum = 100 CU, Most likely = 200 CU and Maximum = 500 CU. The actual cost of the VVT activity in TFN terms is {20.0 CU, 66.6 CU, 233.5 CU} and in crisp terms (using the COG method for defuzzification) is 98.0 CU.

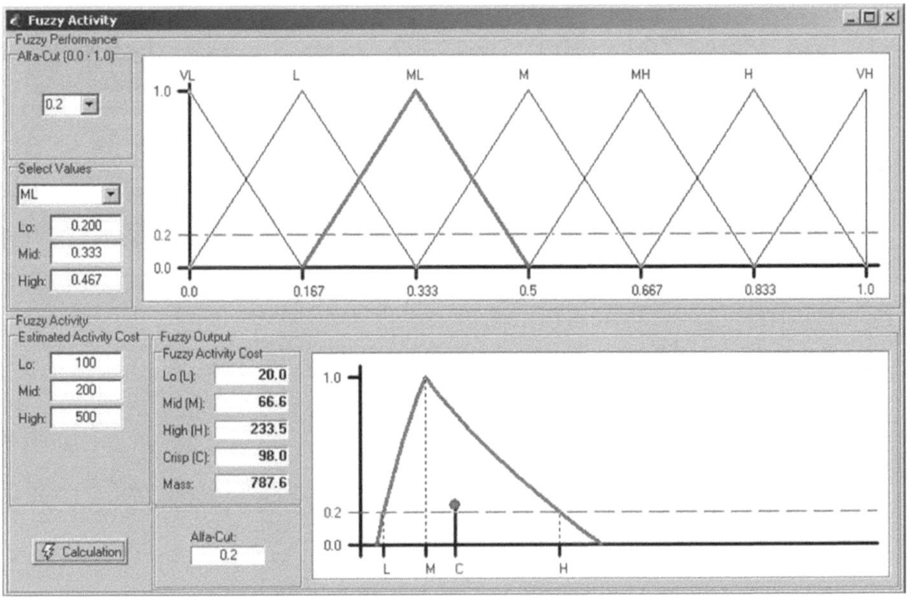

Figure 6.23 Example 1: computing VVT cost.

Fuzzy Appraisal Risk Model (F-ARM) As mentioned before, some products exhibit defects due to inherent imperfection in their development, their manufacturing or incorrect usage. In other words, there is a certain probability that a VVT activity will identify a defective product. We label the risk of detecting such deficiencies during the VVT process as an appraisal risk. This appraisal risk involves the probability that the system defect must be corrected at a certain cost and then retested. Depending on the type of system and the field in question, the probability of finding further problems tends to decrease after corrective iteration.

In a crisp world, the level of a VVT performance reflects the relative efforts of performing a given VVT activity, and therefore this VVT effort affects the appraisal risk. In one extreme case, if the VVT performance level is 0.0, then no VVT activity takes place and the appraisal risk becomes zero. Conversely, if the VVT performance level is 1.0, then the VVT activity is fully performed and the appraisal risk reaches its maximum value.

In a fuzzy paradigm, we have defined $^{\alpha}Ca_{i,j}$ as the full appraisal risk cost in TFN values. We also defined a linguistic variable, $^{\alpha}Fa_{i,j} \equiv$ Fuzzy Appraisal Risk Level, which may accept any value from the linguistic terms {VL, L, ML, M, MH, H, VH}. This appraisal risk level represents the likelihood of finding defects and its decay in a fuzzy logic paradigm. Figure 6.24 depicts a VVT Fuzzy Appraisal Risk Model (F-ARM).

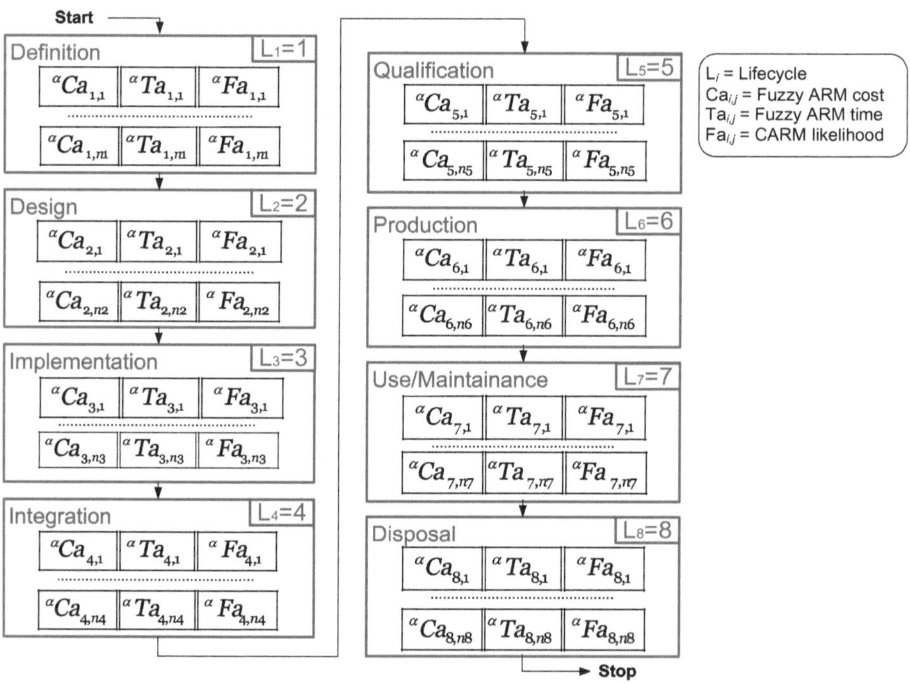

Figure 6.24 Fuzzy VVT Appraisal Risk Model (F-ARM).

A single fuzzy VVT appraisal cost is computed by multiplying the full appraisal risk cost by the level of appraisal risk and then by the VVT performance level at the designated α-cut:

$$^{\alpha}Ca_{\text{Appraisal\_i,j}} = {}^{\alpha}Ca_{i,j}\,{}^{\alpha}Fa_{i,j}\,{}^{\alpha}Fx_{i,j} \tag{6.65}$$

Figure 6.25 depicts an example of the appraisal risk associated with the first activity which was performed at level ML. The estimated appraisal risk cost is in the range of 50–350 CU, with a most likely value of 300 CU. The presumed level of the appraisal risk is high (H), yielding a fuzzy appraisal cost in the range 7.0–157.9 CU, with a most likely fuzzy value of 83.2 CU and a crisp (COG) value of 80.0 CU.

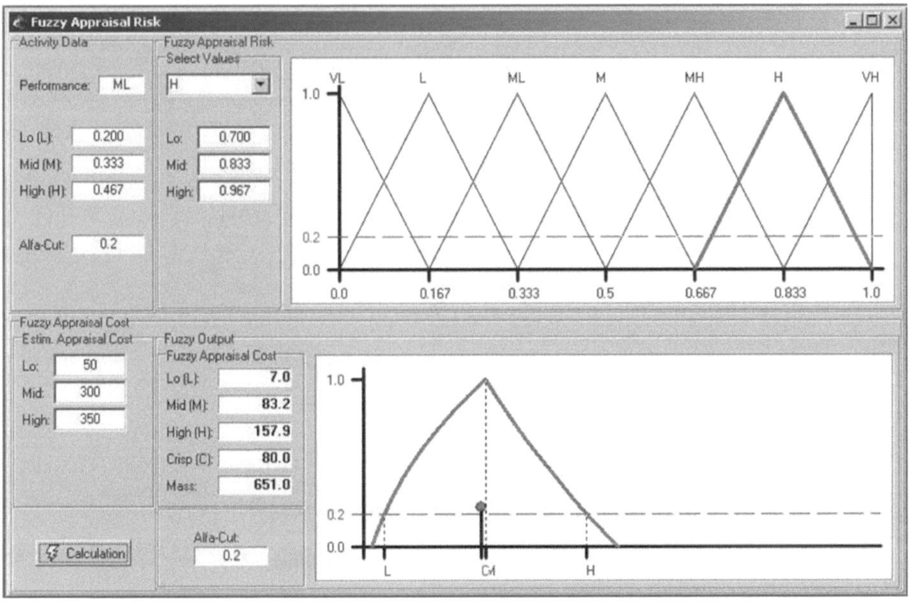

Figure 6.25 Example 2: computing a VVT appraisal risk cost.

Fuzzy Impact Risk Model (F-IRM) Selecting a VVT strategy entails designating the performance level of each VVT activity during the VVT process, including the partially performed activities and those that are not performed at all. The VVT strategy must be selected in accordance with the combined business objective and vision of the system's stakeholders.

A basic assumption of this book is that any partially performed VVT activity or any VVT activity not performed at all constitutes a risk. These risks have stochastic effects on the system and, of course, they constitute undesirable expenditures that can be regarded as the result of implementing a selected VVT strategy. They are discernible only subsequent to the risk insertion (during the same lifecycle phase or at later lifecycle phases). Another concern to bear in mind is that a single risk may generate multiple impacts, affecting the system at more than one lifecycle phase.

Figure 6.26 depicts a VVT Fuzzy Impact Risk Model (F-IRM). The risk identification attribute encompasses a risk qualitative description ($R_{i,j}^{(l,k)}$). A risk source is a specific VVT activity $V_{i,j}$ which was not fully performed and may thus generate undesirable impact. The risk destination l is the system lifecycle phase within which a risk may occur and the index k, $k = 1, 2, \ldots$, represents the impact risk number.

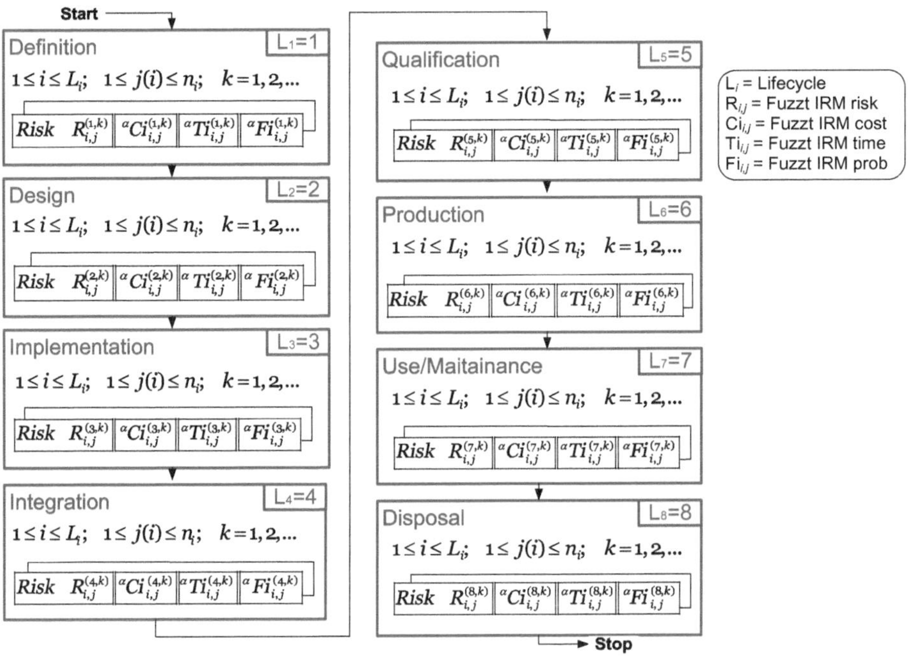

Figure 6.26 Fuzzy VVT Impact Risk Model (F-IRM).

In addition, the F-IRM encompasses three risk variables for any given risk: (1) impact severity in cost terms (${}^{\alpha}\mathrm{Ci}_{i,j}^{(l,k)}$), (2) time impact (${}^{\alpha}\mathrm{Ti}_{i,j}^{(l,k)}$), which we will ignore here, and (3) the degree or fuzzy level of impact risk severity (${}^{\alpha}\mathrm{Fi}_{i,j}^{(l,k)}$). VVT impact increases as the level of the VVT performance decreases. Therefore, we define a reciprocal level of VVT activity performance as $\overline{{}^{\alpha}\mathrm{Fx}_{i,j}}$ so that: $\overline{{}^{\alpha}\mathrm{Fx}_{i,j}} = 1 - {}^{\alpha}\mathrm{Fx}_{i,j}$. As previously mentioned, we have defined ${}^{\alpha}\mathrm{Ci}_{i,j}^{(L_i,k)}$ as the full impact risk cost in TFN values. In addition, we have defined a linguistic variable as ${}^{\alpha}\mathrm{Fi}_{i,j}^{(l,k)} \equiv$ Fuzzy Impact Risk Level, which accepts any of the linguistic terms {VL, L, ML, M, MH, H, VH}.

The total impact cost due to the partial performance of a single VVT activity is computed by summing the multiplication of the full impact risk costs by the fuzzy level of impact risks and then multiplying the result by the reciprocal of the fuzzy VVT performance level. At the designated α-cut this yields.

$$ {}^{\alpha}\mathrm{Ci}_{\mathrm{Impact\_i,j}} = \left\{ \sum_{l=1}^{\mathrm{Li}} \sum_{k=1,2,\ldots} {}^{\alpha}\mathrm{Ci}_{i,j}^{(l,k)}\, {}^{\alpha}\mathrm{Fi}_{i,j}^{(l,k)} \right\} \overline{{}^{\alpha}\mathrm{Fx}_{i,j}} \tag{6.66} $$

Figure 6.27 shows an example of the impact risk associated with an activity which was performed at level ML. The estimated impact risk cost is in the range of 1000–5000 CU, with a most likely value of 2000 CU. The presumed level of impact risk is medium (M), yielding a fuzzy impact cost in the range 292.8–1689.6 CU, with the most likely fuzzy value of 667.0 CU and the crisp (COG) value of 866.0 CU.

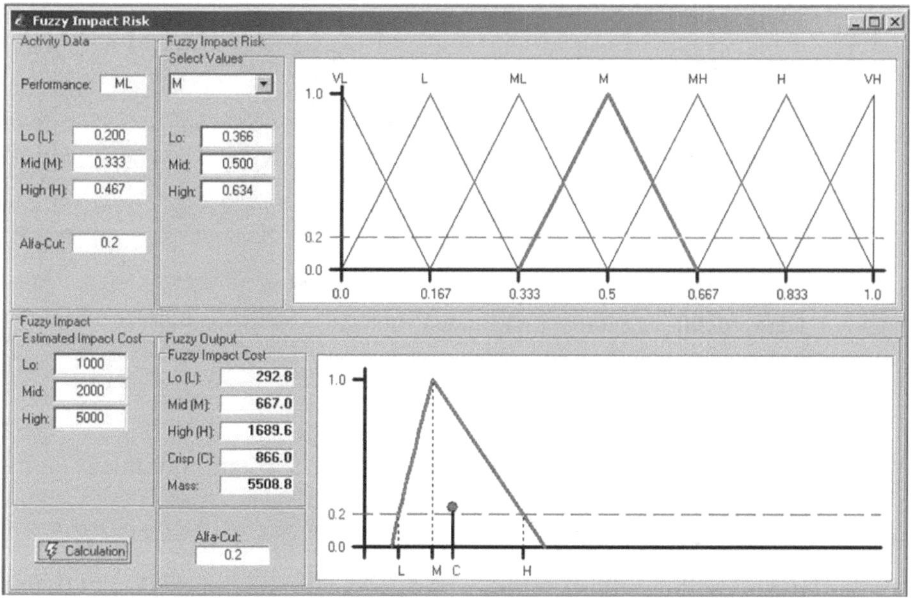

Figure 6.27 Example 3: computing impact risk cost.

Nonlinear Fuzzy APL Effects Similar to the point made in our discussion of the probabilistic paradigm in the section on modeling impact risk cost, the effects of the APLs are not quite linear. In particular, the computation of the effect of the VVT strategy on the impact risk cost exemplifies this situation. This nonlinearity has the following characteristics. In general, the less testing that is performed, the more VVT impact risk is expected. However, in the neighborhood of $Fx_{i,j} \approx$ VL, performing tests at low levels decreases the VVT impact risk to a disproportionably high extent since most of the glaring errors are detected with minimal effort.

Conversely, the more testing that is performed, the less the expected VVT impact. At the same time, in the neighborhood of $Fx_{i,j} \approx$ VH, performing testing at high levels decreases the VVT impact risk to a disproportionably low extent since the most intricate faults are difficult to detect and require a great deal of effort.

1. For example, this nonlinearity can be modeled by shifting the fuzzy impact APL values, as depicted in Figure 6.28. We have defined another set of nonlinear activity performance levels as Fxnl so that

$$\text{Fxnl} = \{\text{VL}_{\text{nl}}, \text{L}_{\text{nl}}, \text{ML}_{\text{nl}}, \text{M}_{\text{nl}}, \text{MH}_{\text{nl}}, \text{H}_{\text{nl}}, \text{VH}_{\text{nl}}\} \tag{6.67}$$

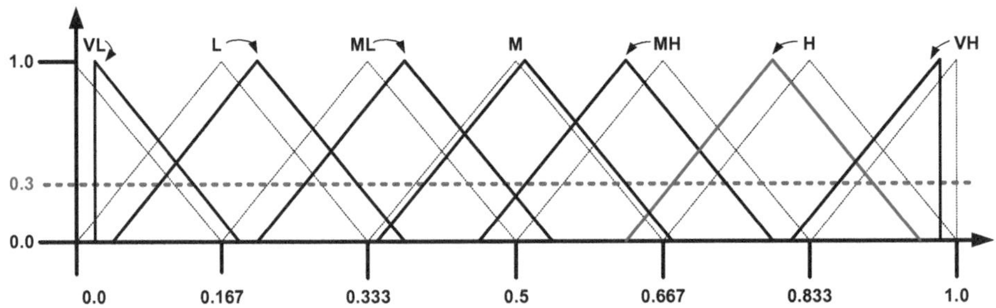

Figure 6.28 Nonlinear fuzzy impact activity performance levels.

2. We have also defined a corresponding set of constant, nonlinear shifts (NLS) as

$$\text{NLS} = \{\text{NLS}_{\text{VL}}, \text{NLS}_{\text{L}}, \text{NLS}_{\text{ML}}, \text{NLS}_{\text{M}}, \text{NLS}_{\text{MH}}, \text{NLS}_{\text{H}}, \text{NLS}_{\text{VH}},\} \tag{6.68}$$

3. Finally, the nonlinear activity performance levels (Lxnl) are generated by adding constant nonlinear shifts to the APL at the desired α-cut levels:

$${}^{\alpha}\text{VL}_{\text{nl}} = {}^{\alpha}\text{VL} + \text{NLS}_{\text{VL}}, \ldots, {}^{\alpha}\text{VH}_{\text{nl}} = {}^{\alpha}\text{VH} + \text{NLS}_{\text{VH}} \tag{6.69}$$

4. The total nonlinear impact cost due to partial performance of a given VVT activity is computed in a manner similar to the one depicted earlier for the probabilistic paradigm but using the fuzzy nonlinear VVT performance level:

$${}^{\alpha}\text{Ci}_{\text{Impact\_i,j}} = \left\{ \sum_{l=1}^{\text{Li}} \sum_{k=1,2,\ldots} {}^{\alpha}\text{Ci}_{i,j}^{(l,k)} \, {}^{\alpha}\text{Fi}_{i,j}^{(l,k)} \right\} \overline{{}^{\alpha}\text{Fxnl}_{i,j}} \tag{6.70}$$

Total Fuzzy VVT Strategy Cost In order to compute the overall VVT strategy cost, we must convert (1) the VVT activity cost, (2) the appraisal risk cost and (3) the impact risk cost into crisp values and then sum them over the entire system lifecycle phases (i; $i = 1, 2, \ldots, z$) and activities [j; $j(i) = 1, 2, \ldots, n_i$]:

1. The overall crisp VVT strategy cost is computed using a COG function which is summed over the entire lifecycle set of single VVT activity costs, yielding

$$\mathrm{Cv}_{\mathrm{VVT\_Strategy}} = \sum_{i=1}^{z} \sum_{j=1}^{n_i} \mathrm{COG}\{{}^{\alpha}\mathrm{Cv}_{\mathrm{VVT\_i,j}}\} \tag{6.71}$$

2. The overall crisp VVT appraisal risk cost is computed using the COG. The function is summed over the entire lifecycle set of single VVT appraisal risk costs, yielding

$$\mathrm{Ca}_{\mathrm{VVT\_Strategy}} = \sum_{i=1}^{z} \sum_{j=1}^{n_i} \mathrm{COG}\{{}^{\alpha}\mathrm{Ca}_{\mathrm{Appraisal\_i,j}}\} \tag{6.72}$$

3. The overall crisp VVT impact risk cost is computed using the COG function, which is summed over the entire lifecycle of single VVT impact risk costs, yielding

$$\mathrm{Ci}_{\mathrm{VVT\_Strategy}} = \sum_{i=1}^{z} \sum_{j=1}^{n_i} \mathrm{COG}\{{}^{\alpha}\mathrm{Ci}_{\mathrm{Impact\_i,j}}\} \tag{6.73}$$

4. The total VVT cost of executing the desired VVT strategy is the sum of the costs of the VVT activities, the VVT appraisal risk costs and the VVT impact risk costs:

$$C_{\mathrm{VVT\_Strategy}} = \mathrm{Cv}_{\mathrm{VVT\_Strategy}} + \mathrm{Ca}_{\mathrm{VVT\_Strategy}} + \mathrm{Ci}_{\mathrm{VVT\_Strategy}} \tag{6.74}$$

6.4.4 Fuzzy VVT Cost and Risk Estimation Example

Fuzzy VVT Raw Data A set of fuzzy VVT data was collected during the IAI/Lahav pilot project of SysTest. The data are utilized for the following examples:

1. A set of VVT activities performed during the development segment of the pilot project program (phases 1–5, Definition, Design, Implementation, Integration and Qualification). It depicts the identification number and description of each VVT activity $\{V_{i,j}\}$ as well as the fuzzy VVT activity cost $\{{}^{\alpha}\mathrm{Cv}_{i,j}\}$ and the fuzzy VVT activity time $\{{}^{\alpha}\mathrm{Tv}_{i,j}\}$ in TFN form associated with fully performing each activity.

2. A set of fuzzy VVT strategies adopted for the pilot project development. Specifically, it identifies the set of fuzzy VVT activity performance levels $\{^{\alpha}Fx_{i,j}\}$ for each VVT activity.
3. A set of VVT appraisal risks associated with each VVT activity. Specifically, it provides information about the VVT appraisal cost $\{^{\alpha}Ca_{i,j}\}$, VVT appraisal time $\{^{\alpha}Ta_{i,j}\}$ as well as level of appraisal risk occurrence $\{^{\alpha}Fa_{i,j}\}$.
4. A set of VVT impacts associated with each VVT activity. Specifically, it provides information about the impact risk scenario $\{R_{i,j}^{(l,k)}\}$, the likelihood that the impact will occur $\{^{\alpha}Fi_{i,j}^{(l,k)}\}$, the phase in which the impact is likely to occur $\{l\}$ as well as the impact cost $\{^{\alpha}Ci_{i,j}^{(l,k)}\}$ and the impact time $\{^{\alpha}Ti_{i,j}^{(l,k)}\}$ if the impact in fact occurs.

Fuzzy VVT Horizon and Strategy The actual VVT strategy used during the pilot project was used for this example. However, in order to simplify this example, the VVT horizon selected covers only the first system lifecycle phase (Definition). Figure 6.29 presents this VVT strategy in fuzzy vocabulary. It depicts the identification of each activity $V_{1,j}$ and the selected fuzzy VVT performance level value ($^{\alpha}Fx_{1,j}$; $^{\alpha}Fx_{1,j}$ = {VL, L, ML, M, MH, H, VH}) where α-Cut = 0.3 is arbitrarily selected for this example.

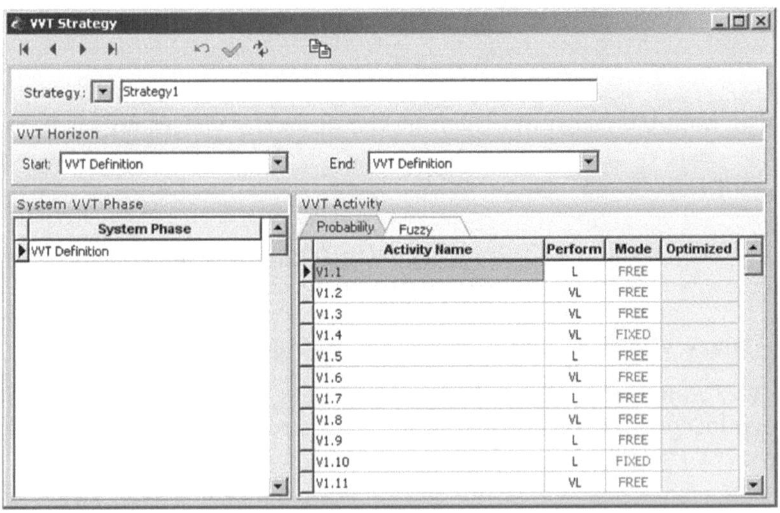

Figure 6.29 VVT fuzzy strategy: Definition phase.

Nonlinear VVT Impact Function The calculations of the VVT impact used realistic, nonlinear, shifted, fuzzy impact APL values, as depicted in Table 6.9.

TABLE 6.9 Nonlinear Shifted Fuzzy Impact APL Values

	Linear Impact				Nonlinear Impact		
APL	Min	ML	Max	Shift	Minimum	ML	Maximum
VL	0.000	0.000	0.167	0.04	0.040	0.040	0.207
L	0.000	0.167	0.333	0.08	0.080	0.247	0.413
ML	0.167	0.333	0.500	0.08	0.247	0.413	0.580
M	0.333	0.500	0.667	0.00	0.333	0.500	0.667
MH	0.500	0.667	0.833	−0.08	0.420	0.587	0.753
H	0.667	0.833	1.000	−0.08	0.587	0.753	0.920
VH	0.833	1.000	1.000	−0.04	0.793	0.960	0.960

Fuzzy VVT Cost Estimates Table 6.10 depicts the results of a VVT cost and risk fuzzy simulation for computing the VVT cost and risk of the selected VVT strategy and horizon. The computation is based on a nonlinear model using α-cut = 0.3.

TABLE 6.10 Fuzzy Cost: Definition Phase

VVT Cost Data	Fuzzy Calculations (CU)
VVT activity cost	28
Appraisal risk cost	3
Impact risk cost	16
Total VVT cost	47
CVM cost	241

α-Cut Sensitivity Analysis It is interesting to explore the sensitivity of the obtained results when the α-cut is allowed to fluctuate. Table 6.11 and Figure 6.30 show the estimated crisp cost results for different α-cut values.

TABLE 6.11 α-Cuts Versus VVT Costs

	Crisp α-Cut Costs (CU)				
α-Cut	Actual Cost	Appraisal Cost	Impact Cost	Total Cost	CVM Cost
0.0	33	4	16	53	242
0.1	31	4	16	51	242
0.2	30	3	16	49	242
0.3	28	3	16	47	241
0.4	27	3	16	45	241
0.5	25	3	15	43	241
0.6	24	3	15	42	241
0.7	23	2	15	40	241
0.8	21	2	15	39	241
0.9	20	2	15	38	241
1.0	19	2	15	36	241
Average	26	3	15	44	241
SD	4.7	0.8	0.5	5.6	0.5

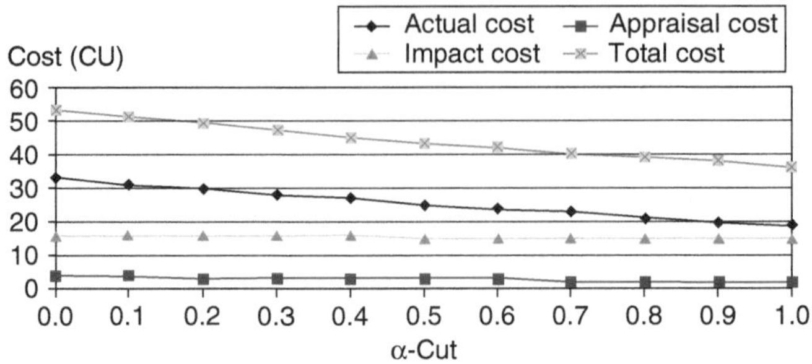

Figure 6.30 α-Cuts Versus VVT costs.

As can be seen, different α-cut values have limited effect on the VVT appraisal, impact and CVM costs. However, the actual VVT cost is significantly affected [standard deviation (SD) 4.7]. This phenomenon adds significantly to the total VVT cost.

As mentioned earlier, obtaining exact VVT cost and risk values in an industrial setting is generally not feasible. At best, engineers will be able to provide relevant real-life VVT data using TFNs. In such circumstances modeling system VVT costs and risks using a fuzzy logic paradigm is very effective.

6.4.5 Fuzzy Logic Versus Probabilistic Modeling

This section discusses a comparison between crisp and fuzzy VVT cost and risk modeling. This discussion covers (1) VVT overall cost comparison, (2) VVT model comparison and (3) VVT model resources and auxiliary information comparison.

VVT Overall Cost Comparison Table 6.12 and Figure 6.31 show a cost comparison between crisp and fuzzy modeling. Crisp VVT cost estimation uses both stochastic (Monte Carlo) simulation and probabilistic computation.

TABLE 6.12 VVT Costs: Stochastic, Probabilistic and Fuzzy Modeling (α-Cut = 0.3)

VVT Cost Data	Stochastic Simulations (CU)	Probabilistic Calculations (CU)	Fuzzy Calculations (CU)	Percent Difference: Monte Carlo Versus Fuzzy
VVT activity cost	32	33	28	12.5
Appraisal risk cost	3	3	3	0.0
Impact risk cost	14	14	16	−14.3
Total VVT cost	49	50	47	4.1
CVM cost	254	254	241	5.1

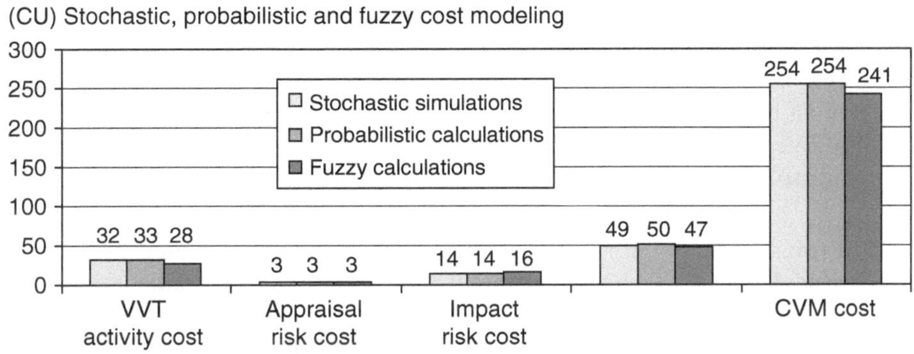

Figure 6.31 VVT cost comparison: stochastic, probabilistic versus fuzzy modeling.

All three estimation techniques use data obtained during the Definition phase of the above case study. It is apparent that the estimated total VVT costs using different methods yield results within about the 5% range.

VVT Model Comparison The following section provides VVT model comparisons:

1. *VVT Lifecycle Phases/Horizon.* Both crisp modeling and fuzzy VVT modeling employ the same system lifecycle definition. The definition consists of eight lifecycle phases where each phase is associated with a specialized set of VVT activities. Similarly, both modeling approaches utilize the same system lifecycle horizon.
2. *VVT Activity/Cost.* Both crisp modeling and fuzzy VVT modeling employ the same concept of VVT strategy, which customarily leads to partial performance of an idealized set of VVT activities (CVM). The following are distinct characteristics of each approach:

Probabilistic model	The VVT cost of a given VVT activity is calculated using (1) the concept of decision variables ($X_{i,j}$; $0.0 \le X_{i,j} \le 1.0$) where each $X_{i,j}$ defines the performance level of any VVT activity and (2) only a single VVT cost data.
Fuzzy model	The VVT cost of a given VVT activity is calculated using (1) the concept of a level of VVT activity performance (${}^{\alpha}Fx_{i,j}$; ${}^{\alpha}Fx_{i,j}$ = {VL, L, ML, M, MH, H, VH}), which defines the fuzzy performance level of any VVT activity and (2) a TFN VVT cost data tuple.

3. *VVT Appraisal Risk/Cost.* Both crisp modeling and fuzzy VVT modeling employ the concept of VVT appraisal risk. Appraisal risk represents the well-known fact that the VVT process often detects failures that must be corrected. The model ascertains the expected appraisal cost for a given VVT strategy. The following are distinct characteristics of each approach:

Probabilistic model	The VVT appraisal cost is calculated using both probabilistic computation and stochastic simulation. In both cases we use (1) VVT appraisal risk probability ($Pa_{i,j}$; $0.0 \le Pa_{i,j} \le 1.0$) and (2) only single VVT appraisal cost data.
Fuzzy model	The VVT appraisal cost is calculated using the fuzzy concept of (1) level of appraisal risk (${}^{\alpha}Fa_{i,j}$; ${}^{\alpha}Fa_{i,j}$ = {VL, L, ML, M, MH, H, VH}) and (2) a TFN VVT appraisal cost data tuple.

4. *VVT Impact Risk/Cost.* Both crisp modeling and fuzzy VVT modeling employ the concept of VVT impact risk. Impact risk represents the likelihood that a product, not adequately tested, will exhibit some malfunction in the future. The model ascertains the expected impact risk cost for a given VVT strategy. The following are distinct characteristics of each approach:

Probabilistic model	The individual VVT impact cost is calculated using the concept of (1) impact risk probabilities ($Pi_{i,j}^{(l,k)}$; $0.0 \le Pi_{i,j}^{(l,k)} \le 1.0$) associated with a given, partially performed VVT activity, (2) the reciprocal of the decision variable ($1 - X_{i,j}$) such that $X_{i,j}$; $0.0 \le X_{i,j} \le 1.0$ and (3) only a single VVT impact cost data.
Fuzzy model	The VVT impact cost is calculated using (1) the fuzzy concept of level of impact risk (${}^{\alpha}Fi_{i,j}^{(l,k)}$; ${}^{\alpha}Fi_{i,j}^{(l,k)}$ = {VL, L, ML, M, MH, H, VH}), (2) the reciprocal VVT activity performance ($1 - Fx_{i,j}$), where ${}^{\alpha}Fx_{i,j}$ = {VL, L, ML, M, MH, H, VH}, and (3) a single TFN VVT impact cost data tuple.

VVT Model Resources and Auxiliary Information Comparison Here we compare the required resources and the availability of dynamic cost information using crisp and fuzzy logic modeling.

Probabilistic Computation Probabilistic modeling requires minimal computation resources. However, no dynamic range may be obtained.

Stochastic Simulation Useful results can be obtained after several dozen iterations, but the stochastic VVT simulator reaches a robust and stable output after some 3000 iterations. Therefore, computer resources or, more specifically, the duration required for obtaining robust stochastic VVT cost and risk values are two orders of magnitudes larger than carrying out a similar estimation using a fuzzy paradigm (Maskey, 2001). Often, medium- to large-size engineering projects entail several hundred and even thousands of VVT activities. In such circumstances, the computer resources required to execute the VVT cost and risk model become significant.

Applying the stochastic simulator provides typical statistical auxiliary dynamic information related to the cost and risk data. Figure 6.32 shows this information plotted as a histogram with auxiliary dynamic statistical data (i.e., VVT cost range, average, cost distribution, VVT cost standard deviation and value at risk). In addition, relevant information can be superimposed (e.g., actual VVT cost, CVM cost).

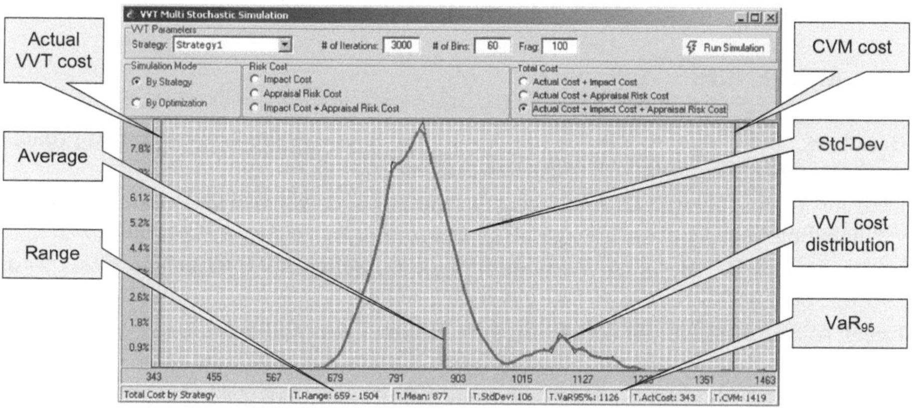

Figure 6.32 Dynamic stochastic data.

Fuzzy simulation Fuzzy computation requires a single iteration. This is comparable to performing crisp probabilistic cost computation (though somewhat more demanding computationally). Applying the fuzzy simulator provides auxiliary dynamic data information, namely (1) minimum value, (2) most likely value, (3) maximum value, (4) crisp value and (5) CVM value.

Table 6.13 and Figure 6.33 depict this auxiliary information for the above data.

TABLE 6.13 Dynamic Fuzzy Cost Data (α-Cut = 0.3)

VVT Cost Data	Minimum (CU)	Most Likely (CU)	Maximum (CU)	Crisp (CU)
VVT cost	5	19	65	28
Appraisal cost	3	12	48	3
Impact cost	8	15	25	16
Total cost	15	46	139	47
CVM cost	188	241	331	241

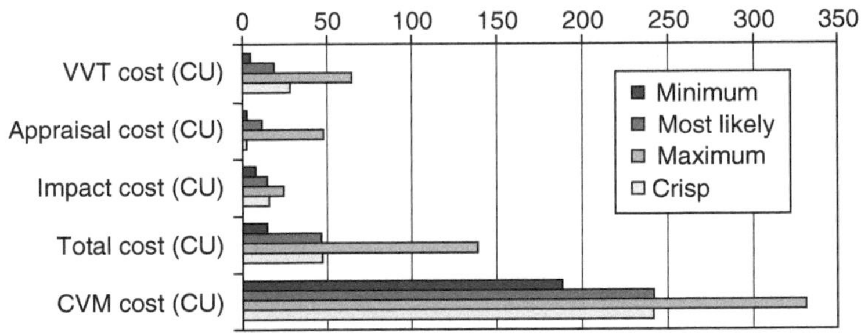

Figure 6.33 Dynamic fuzzy data.

6.5 REFERENCES

Barad, M., and Engel, A., Optimizing VVT Strategies—A Decomposition Approach, *J. Operation Res. Soc.*, 57(8), 965–974, August 2006.

Clausing, D., and Frey, D. D., Improving System Reliability by Failure-Mode Avoidance Including Four Concept Design Strategies, *Syst. Eng.*, 8(3), 245–261, Fall 2005.

Engel, A., and Barad, M., A Methodology for Modeling VVT Risks and Costs, *Syst. Eng.*, 6(3), 135–151, May 8, 2003, available online.

Engel, A., and Last, M., Modeling Software Testing Costs and Risks Using Fuzzy Logic Paradigm, *J. Syst. Software*, 80(6), 817–835, October 25, 2006, available online.

Engel, A., and Shachar, S., Measuring and Optimizing Systems' Quality Costs and Project Duration, *Syst. Eng.*, 9(3), 259–280, June 22, 2006, available online.

Grant, P. D., *Pert and CPM: Network Methods for Project Planning, Scheduling and Control*, Small-Scale Master Builder, 1983.

Hoppe, M., Engel, A., and Shachar, S., SysTest: Improving the Verification, Validation & Testing Process—Assessing Six Industrial Pilot Projects, *Syst. Eng. J.*, 10(4), 323–347, September 24, 2007, available online.

Jackel, P., *Monte Carlo Methods in Finance*, Hoboken, NJ, Wiley, 2002.

Kishk, M., and Al-Hajj, A., Fuzzy Modeling of Lifecycle Costs of Alternatives with Different Lives, paper presented at the Sixteenth Annual Conference of the Association of Researchers in Construction Management (ARCOMM 2000), Glasgow Caledonian University, September 6–8, 2000.

Klir, J. G., and Yuan, B., *Fuzzy Sets and Fuzzy Logic: Theory and Applications*, Prentice-Hall, Englewood Cliffs, NJ, 1995.

Kosko, B., *Fuzzy Engineering*, Prentice-Hall, Englewood Cliffs, NJ, 1996.

Marczyk, J., Principles of Simulation-Based Computer-Aided Engineering, FIM Publications, Dep. Legal B-38994-99, Barcelona, Spain, 1999.

Maskey, S., Uncertainty Analysis in Flood Forecasting and Warning System Using Expert Judgment and Fuzzy Set Theory, in *Proceedings of European Safety and Reliability International Conference, ESREL-2001*, Turin, Italy, 2001, September 16–20, pp. 1787–1794.

Oberkampf, L. W., Trucano, G. T., and Hirsch, C., Verification, Validation and Predictive Capability in Computational Engineering and Physics, paper presented at the Foundation for V&V in the 21st Century Workshop, John Hopkins University, Laural, MD, October 22–23, 2002.

Taylor, R. J., *An Introduction to Error Analysis: The Study of Uncertainties in Physical Measurements*, University Science Books, 1996.

Zadeh, L. A., Fuzzy Sets, *Inform. Control*, 8, 338–353, 1965.

Chapter 7

Obtaining Quality Data and Optimizing VVT Strategy

The purpose of this chapter is to discuss system quality cost information available in the literature and to show a practical and straightforward way to obtain such quantitative data. This process is illustrated utilizing a real-life pilot project conducted by the Israel Aerospace Industries (IAI)/Lahav division during the SysTest project.

Next, the chapter discusses the VVT-Tool, which was developed at IAI to support the modeling of quality cost and optimization of the VVT strategy discussed in this book. Finally, this chapter provides a collection of optimization techniques which may be used during the development of engineered systems, facilitating a reduction of 10–15% in engineering manpower or schedule.

7.1 SYSTEMS' QUALITY COSTS IN THE LITERATURE

System VVT encompass the ensemble of activities used in order to ensure the quality of engineered systems. The main objective of the VVT process is to identify system problems (e.g., in requirement definition, system design, system implementation, production processes, maintenance practices, system disposal).

Most of the system research (as opposed to software research) related to quality cost does not provide actual and specific data. Kumar et al. (1998) discuss the subject of collecting, measuring and reporting quality cost data in various countries. They also compare between different quality cost components based on empirical findings. Hwang and Aspinwall (1996) review different quality cost models, comparing both the strengths and the weaknesses of each model, and highlight their application in different areas. Several researchers describe models and applications in specific industries such as telecommunications (Hwang and Aspinwall, 1999), construction (Aoieong et al., 2002) and automotive (De Ruyter et al., 2002). Other researchers expand the traditional definition of quality cost to so-called invisible and, in particular, intangible costs (Chiadamrong, 2003).

Some system researchers provide real hard quality cost data. For example, Burns (1976) measured the quality costs in a machine tool company, and these costs were the equivalent of 5% of the sales turnover. The distribution of the total 100% quality costs was prevention 3.3%, appraisal 40.3% and failure 56.4%. In a study of a steel foundry, Moyers and Gilmore (1979) reported the quality costs at 38% of sales. The quality costs were allocated as prevention cost 6%, appraisal cost 14% and failure cost 80%. Wheelright and Hayes (1985) indicated that IBM's quality costs in the early 1980s were 30% of its manufacturing costs. Sorqvist (1998) surveyed 30 medium-to-large Swedish companies over a period of 3 years. His findings indicated average losses of 9–16% of a company's yearly financial turnover due to poor quality.

Specifically, the term *loss* refers to the cost of poor quality, which Juran (1952) defined as "the sum of all costs that would disappear if there were no quality problems." Simga-Mugan and Erel (2000) discuss quality cost categories and their distribution in the total quality cost. They describe a case study in the (Turkish) aeronautical defense industry showing that 13.16% of the total production cost is spent on quality. More specifically, detection cost is 51% of total quality cost whereas prevention and correction costs are 26 and 23% of total quality cost, respectively. Tang et al. (2004) concentrate on the cost of nonconformance (failure cost) in the construction industry. They describe a construction project of 38-story-high, twin residential block housing with 760 residential units. In this setting, the failure cost was 3.55% of the overall building cost. Huber (1999) shows that balancing testing cost and schedule with quality is a difficult task. In other words, one can optimize either cost alone or time alone. (In fact, we show in Chapter 8 that this is not a foregone conclusion.) However, quality problems discovered late in the system lifecycle necessitate expensive rework and may adversely affect the system's and the organization's reputation.

Giakatis et al. (2001) suggest that a distinction should be made between quality costs and quality losses. They define quality cost as "money spent on a successful prevention and appraisal activities" whereas quality loss is defined as "money spent on an unsuccessful prevention and unsuccessful appraisal activities plus costs of all failures" (development, manufacturing, maintenance, etc.). Therefore, instead of considering the total quality costs, it would be better for a company to try to reduce quality losses. We are under the impression that, in general, the Giakatis et al. notion has not been accepted by the quality community, possibly because the term "quality loss" was already "taken" by Genichi Taguchi, who asserted that the "Quality Loss

Function gives a financial value for customers' increasing dissatisfaction as the product performance goes below the desired target performance" (Taguchi et al., 1988).

Measuring the software VVT process has become a central issue for many software-intensive organizations and researchers (e.g., Ghiassi, 1994). Jones (1996) provides a massive amount of data gathered from over 6700 software projects. It includes raw data, such as the number of software faults discovered at a given lifecycle phase, as well as derived information, such as the cost of software rework. Observations suggest that the Definition phase of the software and system lifecycle is the most critical in terms of VVT. Real-life statistics from a number of U.S. military software-intensive projects show that 50% of the errors detected in the testing phase have been the result of mistakes made in the Definition phase. The approach proposed by Ittner (1996) is to optimize the process by performing regression tests on specific software components. The selection of candidates for regression testing should be based on the associated fault history. The paper provides specific values for cost of quality. Ittner collected relevant data from 49 manufacturing units belonging to 21 companies and reported that the cost of quality represents, on the average, 11.2% of sales. Of this amount, 56.7% is a result of internal and external failures.

Houston and Keats (1998) were interested in the cost of quality in the software industry in relationship to the organization's maturity level, as measured against the Test Maturity Model integration (TMMi). The TMMi is a detailed model for test process improvement and is complementary to the Capability Maturity Model Integrated (CMMI). Their findings were that organizations at maturity level 1 (the least capable) waste 55% of the software development budget on fixing errors. This figure is reduced to about 30% for organizations at maturity level 3 (the midpoint on the CMM scale). This is attributed to the fact that the vast majority of industrial organizations spend considerable funds to promote product quality using suboptimal VVT processes. Kim et al. (2000) deal with software regression testing. The key finding is that the cost of identifying, correcting and retesting these errors is approximately 40% of the overall rework cost of this phase. The approach proposed in the paper is to optimize the process by performing regression tests on specific and most critical software components.

Barry Boehm, the creator of the COnstructive COst MOdel (COCOMO) model for estimating software development cost, is a leading authority in the software industry. Boehm et al. (2000) utilized over 100 person-years of software cost estimation experience to calibrate and validate software cost data fitting both expert judgment and 161 specifically collected software project data points. In a distillation of this model, Boehm (2001) provided interesting findings on software quality costs: (1) Finding and fixing a software problem after delivery often costs 100 times more than finding and fixing it during the requirement and design phase. (2) Current software projects spend about 40–50% of their efforts on avoidable rework. (3) About 80% of avoidable rework comes from 20% of the defects. (4) About 80% of the defects come from 20% of the modules and about half of the modules are defect free. (5) About 90% of downtime comes from, at most, 10% of the defects. (6) It costs 50% more to develop high-dependency software products than to develop low-dependency software products. (7) A typical lifecycle cost distribution is 30% development and 70% maintenance.

A comprehensive set of system quality cost data has been gathered during the SysTest project (Engel and Shachar, 2006). SysTest was a research project partially funded by the European Commission (contract G1RD-CT-2002-00683) and conducted by a consortium of eight European companies and research institutes (see more information about the SysTest project in Appendix A). Table 7.1 and Figure 7.1 depict system quality cost data obtained from the following five industrial pilot projects conducted by the SysTest industrial partners: (1) IAI, (2) Centro Ricerche Fiat (CRF), (3) Daimler Chrysler AG (DC), (4) Hispano-Suiza (HS) and (5) Tetra Pak Carton Ambient (TPCA). The data are shown as a percentage of system development cost where mean quality cost is 58.9% with 5.5% standard deviation(SD).

TABLE 7.1 Quality Cost as Percentage of System Development Cost (%)

	IAI	CRF	DC	HS	TPCA	Mean	SD
VVT cost	41.0	45.8	38.3	38.0	50.6	42.7	4.8
Failure cost	20.3	19.0	11.3	18.0	12.3	16.2	3.7
Quality cost	61.3	64.8	49.6	56.0	62.9	58.9	5.5

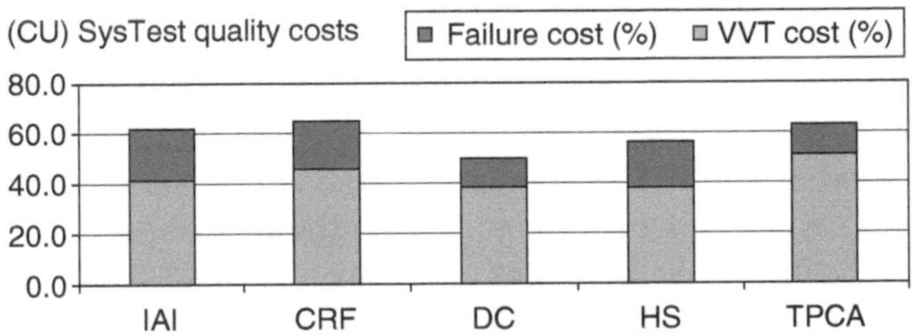

Figure 7.1 Quality cost as percentage of system development cost.

Researchers studying the Lean Aerospace Initiative (LAI, 2006) are interested in waste occurring during the product development stage. Womack and Jones (1996) extended the LAI work in order to improve the value delivery process in the manufacturing sector by eliminating waste. They classified all product-making activities into Value Adding (VA), to be continually perfected; Non–Value Adding (NVA), to be eliminated; and Required Non–Value Adding (RNVA), such as those required by contract or law. Several product development specifications are suggested by researchers in the field (e.g., Millard, 2001; Morgan and Henrion, 1992). However, no formal study is available on the portions of NVA and RNVA waste in aerospace programs (Oppenheim, 2004).

7.2 OBTAINING SYSTEM QUALITY DATA

7.2.1 Quality Data Acquisition

Usually, system quality data may be obtained from the following sources: (1) available databases of past projects, (2) measuring quality information during a specific project and (3) eliciting information from domain experts. Once raw data have been obtained, a method to aggregate and harmonize it must be selected in order to come up with unambiguous and clear quality data.

Delphi Process The purpose of eliciting data from experts is to bridge the gap between available database records and required information. Cooke (1991) provides an extensive survey and a critical examination of the literature on the use of expert opinion in scientific inquiry and policy making. The elicitation, representation and use of expert opinion have become increasingly important since advancing technology requires more and more complex decisions. Cooke considers how expert opinion is being used today, how an expert's uncertainty is represented, how people reason with uncertainty, how the quality and usefulness of the expert opinion can be assessed and how the views of several experts might be combined. Loveridge (2002) expands on Cooke's seminal work and covers topics such as the selection of people for expert committees, as this is much more critical than is generally appreciated. Vose (2008) describes modeling techniques based on various probability distributions (e.g., triangular, beta).

Eliciting data from experts can be a difficult and involved process. Keeney and von Winterfeld (1991) discuss the process of acquiring probabilities from experts in the complex nuclear power plant environment. Experts had to estimate failure probabilities associated with two critical valves in which a simultaneous failure could trigger a core meltdown catastrophe. The elicitation took several months to accomplish and the uncertainties were very large, often covering several orders of magnitude in the case of probability frequencies and 50–80% of the physically feasible range in the case of some uncertain quantities. In general, the above authors suggest a practical Delphi elicitation procedure comprised of the following steps:

1. Orientation, issue familiarization and training
2. Elicitation and collection of opinions
3. Aggregation and presentation of results
4. Group interaction, discussion and revision of findings (data scrubbing)
5. Conclusions

Data Elicitation in Triangular Distribution One variation of the Delphi process is to elicit quality data under a triangular distribution paradigm, that is, three values, minimum (a), most likely (m) and maximum (b), are supplied for each required quantity. Therefore, if 10 domain experts are asked one question (e.g., "What is the cost of VVT activities during the project definition phase as a percentage of the overall phase cost?"), then together they generate 30 values which must be aggregated and combined into a single response.

7.2.2 Quality Data Aggregation

Approach and Assumptions In order to aggregate the raw quality data, one may take the following approach:

1. Postulate that the investigated phenomenon is random, bounded within a certain minimum (a) and maximum (b) range.
2. Assume also that the random variable representing the phenomenon has a most likely value (m) within the range $a \geq m \geq b$.
3. Consider a collection of n experts as "measuring instruments" with built-in errors in their measurement abilities. The most likely (m) value represents the actual measurements of the instruments and the minimum (a) and maximum (b) values represent the respective lower and upper boundaries measured by the instruments.
4. Assume that each of the n instruments introduces an unbiased, random measuring error. Then aggregate the results using a numerical analysis, in this case, a Monte Carlo simulation (for a detailed description of Monte Carlo statistical methods, see Robert and Casella, 2005).

Based on the above, one can aggregate all the responses and present the results to the group in a second, "scrubbing" meeting. During that second meeting, each expert has a chance to review his or her original responses in light of the groups' aggregated data. Some debates may take place as to the exact meaning of certain questions. But, eventually, all experts came to a shared collective understanding of the issues. A few experts may change their original response along the line of the group. But, usually, the majority will not change their opinions, even when their data are quite different from that of the group. For example, see a distinct outlier in Figure 7.2. The data are then reaggregated and the participating experts are informed of the final results.

It should be noted that eliciting expert data within the system and quality domain is substantially simpler in comparison to the earlier reported work by Keeney and von Winterfeld (1991). First, experts supply actual cost and time estimates rather than probabilities (let alone conditional probabilities) of certain rare events. Second, these experts do these kind of estimations throughout their careers. Thus, they have ample opportunity to hone their estimating skill by observing historical project data over many years.

Aggregating Expert Data Each expert is assumed to have a probability $p_{k,i}$ of being correct, where $p_{k,i}$ is associated with response cluster k, representing individual expert i. For a total of n experts, the aggregated response cluster is represented by a generalized discrete distribution:

$$\{(p_{k,1}, f_{k,1}(x)), (p_{k,2}, f_{k,2}(x)), \ldots, (p_{k,n}, f_{k,n}(x))\} \tag{7.1}$$

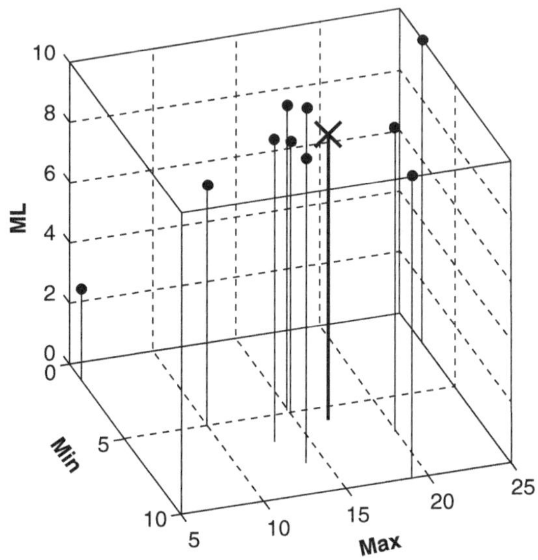

Figure 7.2 Response-cluster for ten experts.

This means that the aggregated probability density function of a value x in response cluster k for a total of n experts is

$$F_k(x) = \sum_{i=1}^{n} p_{k,i} f_{k,i}(x) \tag{7.2}$$

This equation satisfies the mathematical and behavioral approaches discussed by Clemen and Winkler (1999).

Note that the sum of several triangular distributions is not a triangular distribution, and this nonlinearity suggests that closed mathematical expressions for statistical moments of the aggregated distribution are impractical. Therefore, a credible data aggregation could be accomplished by means of a numerical analysis, for example by Monte Carlo simulation[79] (Vose, 2006). Also, note that each of the n experts dealing with all the response clusters was assumed to be correct equally likely; thus

$$p_{k,i} = \frac{1}{n} \quad \forall k, i \tag{7.3}$$

[79]There are several commercial tools that may support the aggregation and analysis of expert data, for example, @RISK (http://www.palisade-europe.com) and Crystalball (http://www.crystalball.com).

For example, Figure 7.3 depicts the aggregated probability distribution function of 10 experts' replies to the question whose response is depicted in Figure 7.2. After 10,000 Monte Carlo iterations per expert, the plot indicates a minimum value of 1.0, a maximum value of 25.0, a mean of 9.81 with standard deviation of 3.68.

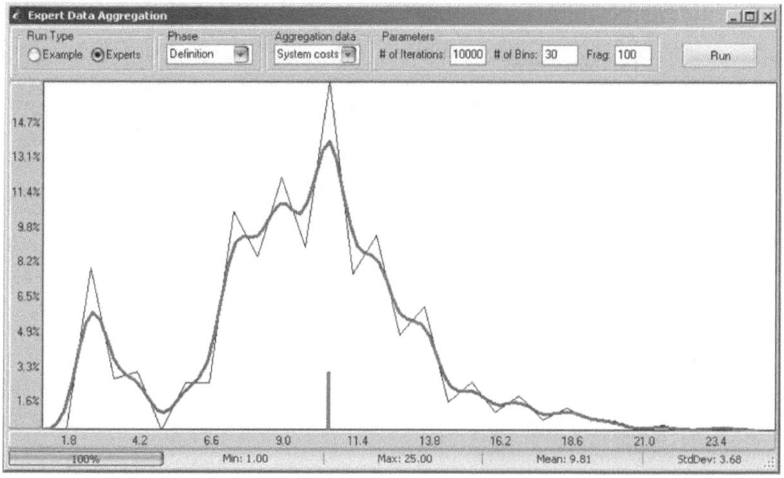

Figure 7.3 Aggregated plot for experts' replies depicted in Figure 7.2.

Final System and Quality Data Integration As mentioned above, there are often several sources of system and quality data. The final step therefore is to harmonize each source of data and merge it into a combined set. By harmonizing the data, we mean adjusting the data to reflect, as much as possible, similar projects. The most straightforward method for integration of similar data from different sources is to combine it, giving appropriate weight to each data source:

$$F(x) = \sum_{i=1}^{m} w_i f_i(x) \qquad \sum_{i}^{m} w_i = 1 \tag{7.4}$$

7.3 IAI/LAHAV QUALITY DATA—AN ILLUSTRATION

7.3.1 IAI/Lahav Pilot Project

We illustrate the process of obtaining system and quality data using an avionics upgrade pilot project that was conducted at the IAI/Lahav division in conjunction with the international research project SysTest (see Appendix A). The aim of the pilot project was to develop a new avionics system for a transport helicopter (see Figure 7.4) as well as building several helicopter prototypes for delivery. System cost and quality cost data were obtained from several previous IAI/Lahav avionics upgrade projects, quality cost and time measurements during the pilot project and cost and time estimations provided by domain experts (Engel and Shachar, 2006).

Figure 7.4 Pilot project transport helicopter.

The project replaced older designed avionics and a pilot control system with a modern system. The helicopter was also upgraded to support Search-And-Rescue (SAR) missions. In addition, IAI was to provide product support for several years. The following system characteristics, grouped into three categories, were identified:

Project Characteristics

- *Development Type.* typical IAI/Lahav avionics project.
- *Project Development Scale.* Full-scale development covering Definition, Design, Implementation, Integration and Qualification phases.
- *Project Lifecycle.* Includes several prototype productions and 15 years of avionics system support and spare parts.
- *Project Size.* Large (multiteam, multi-million-dollar development, production and maintenance efforts).
- *Project Complexity.* High (involves many and diverse organizational entities).
- *Historical Reference Data.* Available from similar projects.

System Characteristics

- *System Criticality.* Life- and mission-critical system.
- *System Complexity.* Contains complex system requirements and architecture.
- *System Precision.* Must meet high performance and precision requirements.

Project Risks

- *Schedule.* Very aggressive.
- *Developer/Customer/User communication.* Complex structure of partners, contractors and subcontractors.

7.3.2 Obtaining Raw System and Quality Data

We used historical IAI /Lahav project databases and data acquired during the pilot project itself as well as interviews with domain experts in order to obtain the required system and VVT data. The data were generated from the following sources:

- *Historical IAI/Lahav Projects.* Typical IAI /Lahav projects for upgrading aircraft avionics were selected as historical benchmarks. These data were adjusted to harmonize each past project with the IAI/Lahav pilot project.
- *Pilot Project Itself.* Relevant systems, VVT cost and time data as well as failure data were collected during the execution of the pilot project.
- *Domain Experts.* The typical Delphi procedure mentioned above was followed. All the experts were gathered for an initial meeting and given an explanation about the Delphi procedure as well as instructions about the nature and meaning of each question on the questionnaire. Then the experts were instructed to provide answers without interacting with each other.

  The *first question* was designed to ascertain the distribution of the overall engineering expenditures over the individual phases of a typical IAI/Lahav aircraft avionics upgrade development project:

  1. What is the typical IAI/Lahav engineering cost of each one of the project development phases as a percentage of the total development cost?

  The *next two questions* were designed to determine the quality engineering cost parameters, namely, the cost of performing VVT and the cost of failure:

  2. What is the typical IAI/Lahav engineering cost of VVT activities in each project phase as a percentage of the phase cost?
  3. What is the typical IAI/Lahav failure cost in each project phase as a percentage of the phase cost?

  The *last two questions* were designed to obtain the project time parameters, namely, the typical time of performing aircraft avionics upgrade and VVT activities per each development phase and the percentage of overlapping among lifecycle phases:

4. What is the typical IAI/Lahav time duration of system development and VVT activities in each project phase?
5. What is the typical IAI/Lahav percentage time overlap in each project phase ("early start" of the next phase before the current phase ends)?

7.3.3 Anchor System and Quality Data

The system and quality data obtained from the experts was aggregated using Monte Carlo techniques. Thereafter, we conducted a final system and quality data integration of the three sources mentioned above. As the difference between each data source was ±5%, it was decided to consider each source equally reliable and therefore these data were averaged. This information was defined as the "anchor data."

Figure 7.5 depicts the anchor system cost distribution covering the system development phases. One may note that about 10% of the total development effort was spent on the Definition phase. At the same time, each of the other phases consumed about twice or more of the overall effort. One may hypothesize that the high costs of system integration and qualification (totaling over 40% of the development cost) may be attributed to the low investment level in the system Definition phase. The figure also depicts that the standard deviation in experts' response ranges from 3.6 to 10.6%.

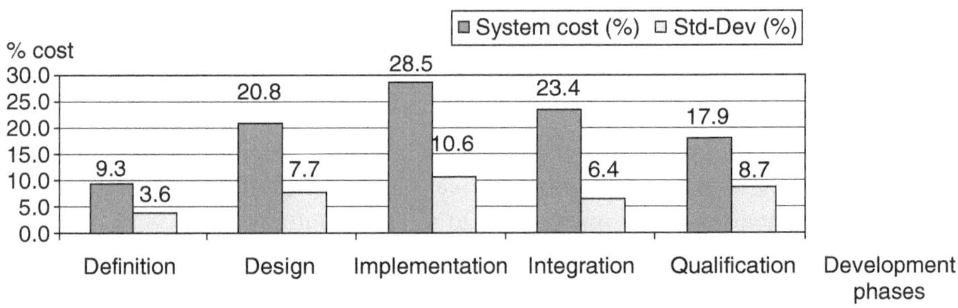

Figure 7.5 Anchor system development cost distribution.

Figure 7.6 depicts the anchor quality costs distributed over the development segment. As mentioned, quality cost is composed of VVT cost and failure cost. This information is presented as the percentage of overall development cost.

One may note that the overall engineering quality cost constitutes 61.3% of the engineering development cost. This cost is composed of approximately two-thirds VVT cost and one-third failure cost. Again, one may assume that the very low quality costs and especially VVT expenditure (just over 1% of overall system development cost) spent during the system Definition phase is directly responsible for the high quality costs incurred in the system Integration and Qualification phases (totaling over 40% of overall system development cost). The standard deviation of domain experts' responses was in the range of 0.8–5.7% for VVT costs and 0.5–2.3% for failure costs.

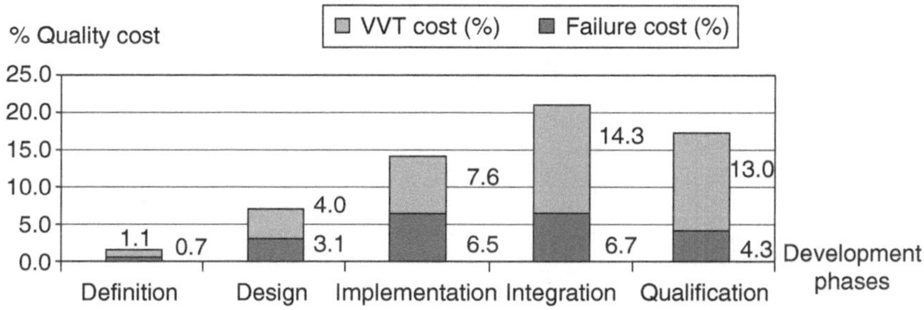

Figure 7.6 Anchor quality development cost distribution.

The duration of the development project was planned for one and a half years, or 550 days. The anchor duration and early start of each phase are shown in Table 7.2. The early start indicates how early a next phase starts relative to the end of the current phase. For example, the Design phase starts 30% before the end of the Definition phase.

TABLE 7.2 Anchor Development Duration and Phase Overlap

Phase	Duration (days)	Early Start (%)
Definition	110	0
Design	110	30
Implementation	210	40
Integration	140	10
Qualification	110	20

7.3.4 Generating the VVT Model Database

A group of three VVT model experts used a similar approach to develop the VVT model database. This included (1) a set of canonical VVT Model (CVM) activities and their associated cost and time parameters, (2) a set of risks in the Appraisal Risk Model (ARM) and their associated parameters and (3) a set of risk scenarios in the Impact Risk Model (IRM) and their associated parameters. The VVT model experts employed their "best judgment" to estimate data elements (i.e., each as a triangular tuple: minimum, most Likely and maximum) as applicable to the IAI/Lahav environment. In addition they used the Anticipated Failure Determination (AFD) method to generate system failure scenarios. Figure 7.7 depicts the relevant VVT model parameters gathered during the pilot project.

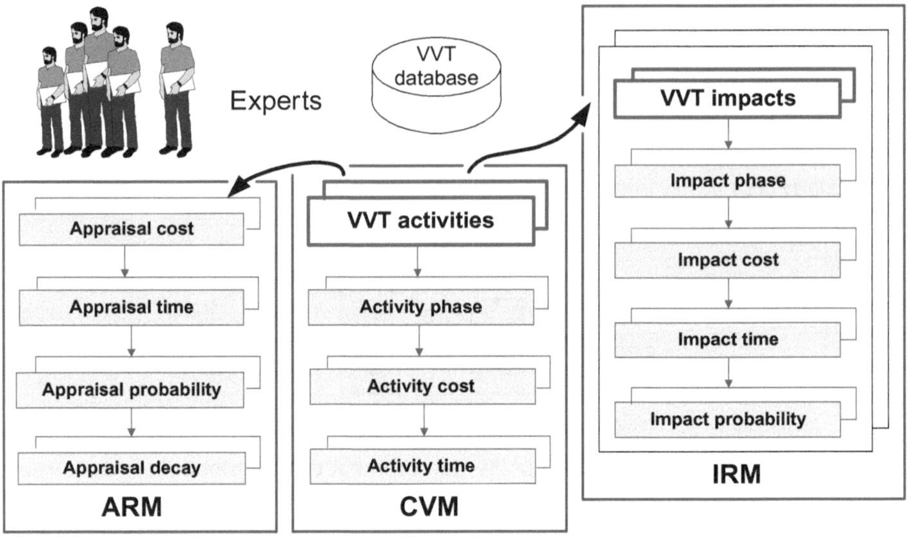

Figure 7.7 Relevant VVT system parameters.

7.4 THE VVT-TOOL

7.4.1 Background

The VVT-Tool is an experimental database utility created at the IAI in order to maintain system and quality data and implement the quality cost, time and risk models discussed in Chapter 6. Figure 7.8 depicts the VVT modeling tool input and output.

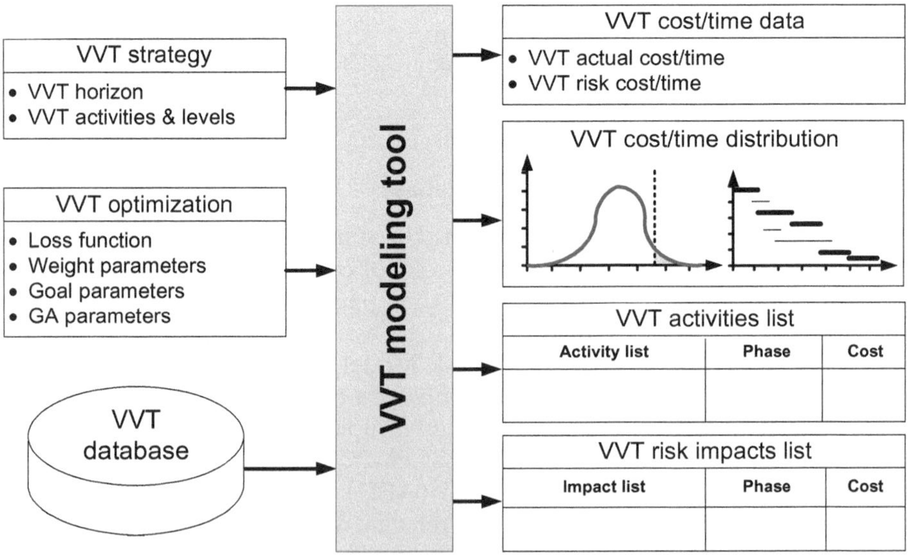

Figure 7.8 VVT modeling tool input and output.

The VVT-Tool is a software package which manipulates and visualizes VVT data (e.g., VVT activities, appraisal risks and impact risks along with their cost, time and other relevant parameters). As depicted in the figure, the VVT-Tool uses a commercial database (Sybase) which receives the VVT strategy and relevant VVT optimization parameters as inputs. The outputs from the VVT-Tool are the expected cost and time and their distribution as well as a list of VVT activities to be performed and a list of potential risks expected to impact the system during its lifetime.

The capabilities of the VVT-Tool are delineated in Table 7.3.

TABLE 7.3 VVT-Tool Capabilities

Group	Functionalities	
General VVT configuration	• System lifecycle/VVT configuration • Quality cost class • Risk probability model 882C • Risk cost model • System activity • VVT activity • Appraisal risk	• Risk impact • Variable constraints • Objective function and weights • Appraisal risk default • Fuzzy α-cut • Nonlinear impact • Timing scheduler
VVT strategy definition	• Probability configuration • Fuzzy configuration	• Probability configuration optimization • Fuzzy configuration after optimization
VVT cost simulation	• Single stochastic simulation • Multistochastic simulation	• Fuzzy simulation
VVT cost calculation	• VVT activities • Appraisal risk • VVT impact risk	• Total VVT cost • CVM cost • Lifecycle cost graph
VVT time calculation	• Network definition • Project schedule	• Lifecycle time graph
VVT optimization	• Probabilistic weighted optimization strategy • Goal optimization strategy • Fuzzy weighted optimization strategy	• Single-phase genetic algorithm • Multiphase genetic algorithm • Steady-state genetic algorithm
Report generation	• Configuration reports	• Strategy reports

7.4.2 Tool Availability

The VVT-Tool software is designed to estimate quality cost, time and risk of systems as well as to optimize the VVT strategy in order to attain certain project cost and time targets or reduce these targets according to prevailing business objectives. Provided various legal obstacles are resolved, it is hoped the tool could be made available to the public under the following provisions:

1. The VVT-Tool is noncommercial, research-based software. It has not been fully tested and therefore it may be made available to users under a Beta release policy. This means the software may be provided at no cost to users. However, no liability for the correct functioning of the software as well as maintaining the software is undertaken by the software developer or any other person or entity.
2. The VVT-Tool is bundled together with a user guide, sample data and a commercial Sybase database system. The commercial Sybase database may be run freely for certain duration (dependent on Sybase company policy). Thereafter, users who may be interested in using the VVT-Tool software on a continual basis must purchase a license to run the Sybase database.
3. The software is freely available to readers via an Internet site (**http://www.adisw.com**) in binary form under Beta release conditions. Under these conditions, a reader is free to install the VVT-Tool on his or her PC, execute it and modify the data in the database.

Disclaimer

The VVT-Tool software is provided on an "as-is" basis and any express or implied warranties, including but not limited to the implied warranties of merchantability and fitness for a particular purpose, are disclaimed. In no event shall Adi Mainly Software or any other person or entity associated with this software be liable for any direct, indirect, incidental, special, exemplary or consequential damages (including but not limited to procurement of substitute goods or services; loss of use, data, or profits; or business interruption) however caused and on any theory of liability, whether in contract, strict liability or tort (including negligence or otherwise), arising in any way out of the use of this software, even if advised of the possibility of such damage.

7.5 VVT COST, TIME AND RISK OPTIMIZATION

As mentioned before, it is virtually impossible to carry out a complete VVT process due to resource constraints, chiefly time and money. Nevertheless, once we can estimate the cost, time and risk of system VVT and model the process mathematically, we can provide VVT engineers and organizations in general with a tool to select a VVT strategy for specific business objectives.

Operations Research (OR) (i.e., optimization theory and practice) based on mathematical programming is a well-established field described in hundreds of books and thousands of articles. In their book, Avriel and Golany (1996) have bridged the gap between the theory of mathematical programming and the real-world practice of industrial engineering. This reference text presents issues in linear, integer, multiobjective, stochastic, network and dynamic programming.

7.5.1 Optimizing the VVT Process

Single-Objective Optimization Problems According to Avriel and Golany (1996), any Single Objective Optimization Problem (SOOP) is composed of three basic ingredients: (1) an objective function, which we want to minimize or maximize, (2) unknowns or variables, which affect the value of the objective function, and (3) constraints that permit the unknowns to take on certain values but exclude others.

The optimization problem then is: Find values of the variables that minimize or maximize the objective function while satisfying the constraints. For example, in choosing to optimize the VVT strategy, we seek to minimize the cost of carrying out the system's VVT while complying with regulatory agency directives (i.e., constraints). Unfortunately, real-life problems are more complex and most realistic optimization problems require the simultaneous optimization of more than one objective function.

Multiobjective Optimization Most problems in engineering, economics and so on have several (frequently conflicting) objectives to be satisfied. Many of these problems are frequently treated as single-objective optimization problems by transforming all but one objective into constraints. Consider the following: Osyczka (1985) defines the Multiobjective Optimization Problem (MOP) as the problem of finding: "A vector of decision variables which satisfies constraints and optimizes a vector function whose elements represent the objective functions. These functions form a mathematical description of performance criteria, which are usually in conflict with each other. Hence, the term Optimizing means finding such a solution which would give the values of all the objective functions acceptable to the decision maker."

Triantaphyllou (2000) deals with Multi-Criteria Decision Making (MCDM). He asserts that the key problem is how to appraise a set of alternative solutions based on independent criteria. Although this problem is often applicable, there are limited practical solution methods available. Essentially, MCDM is an attempt to solve a MOP. In general, a MOP can be formally defined as follows: Find a vector

$$X^* = [x_1^*, x_2^*, \ldots, x_n^*] \tag{7.5}$$

which satisfies the m inequality constraints

$$g_i(X) \geq 0 \qquad i = 1, 2, \ldots, m \tag{7.6}$$

and the p equality constraints

$$h_i(X) = 0 \qquad i = 1, 2, \ldots, p \tag{7.7}$$

and which optimizes the vector function

$$F(X) = [f_1(X), f_2(X), \ldots, f_k(X)] \tag{7.8}$$

The notion of optimum in MOP was defined by Vilfredo Pareto in 1896. It is called the definition of Pareto optimality: A state $A(X^*)$ (a set of object parameters) is said to be Pareto optimal if there is no other state $B(X)$ dominating the state $A(X^*)$ with respect to a set of objective functions. A state $A(X^*)$ dominates a state $B(X)$ if $A(X^*)$ is better than $B(X)$ in at least one objective function and not worse with respect to all other objective functions.

In reality, this concept usually leads to a set of solutions called the Pareto optimal set. The vector X^* corresponding to multiple solutions which are included in the Pareto optimal set are called nondominated. The plot of the objective function whose nondominated vectors in the Pareto optimal set is called a Pareto front. The Pareto optimal set is usually an infinite set. The decision maker therefore has to choose the desired solution from the set. This input usually consists of ranking or weighting the objectives so that a MOP can be converted into a SOOP. Two of the many MOP techniques (weighting objectives method and goal programming) are particularly suited to the problem at hand.

MOP Building Blocks VVT cost and risk optimization are composed of the following primary building blocks:

1. VVT strategy cost $Cv_{VVT\_Strategy}|X$ and time $Tv_{VVT\_Strategy}|X$, which are deterministic and dependent upon the corresponding VVT performance level (X).
2. VVT appraisal cost $Ca_{VVT\_Strategy}|X$ and time $Ta_{VVT\_Strategy}|X$, which are stochastic and dependent upon the corresponding VVT performance level (X).
3. VVT impact cost $Ci_{VVT\_Strategy}|X$ and time $Ti_{VVT\_Strategy}|X$, which are stochastic and dependent upon the corresponding VVT performance level X.
4. Total VVT cost $C_{VVT\_Strategy}|X$, which is composed of the sum of the VVT strategy cost, VVT appraisal cost and VVT impact cost.
5. Total VVT time $T_{VVT\_Strategy}|X$, which is composed of the sum of the VVT strategy time, VVT appraisal time and VVT impact time associated with the critical path.

Optimization Assumptions The following assumptions have been made:

1. The risk impact is independent of one another, that is, the selected performance level $X_{i,j}$, $0 \leq X_{i,j} \leq 1$, of any VVT activity $V_{i,j}$, $j(i) = 1,2,\ldots,n_i$, $i = 1,2,\ldots,z$, may only affect the impact associated with this particular VVT activity.
2. The impact probability $\mathrm{Pi}_{i,j}^{l,k}$, $j(i) = 1,2,\ldots,n_i$, $i = 1,2,\ldots,z$, $l \geq i$, $k = 1,2,\ldots,$ is constant, that is, it is independent of level $X_{i,j}$ at which VVT activity $V_{i,j}$ is performed.
3. There is a negative linear relationship between the VVT activity $V_{i,j}$ performance level $X_{i,j}$, $0 \leq X_{i,j} \leq 1$, $j(i) = 1,2,\ldots,n_i$, $i = 1,2,\ldots,z$, and its risk impact cost $\mathrm{Ci}_{i,j}$.
4. There is a positive linear relationship between the VVT activity $V_{i,j}$ performance level $X_{i,j}$, $0 \leq X_{i,j} \leq 1$, $j(i) = 1,2,\ldots,n_i$, $i = 1,2,\ldots,z$, and its actual cost $\mathrm{Cv}_{i,j}$.
5. There is a positive linear relationship between the VVT activity $V_{i,j}$ performance level $X_{i,j}$, $0 \leq X_{i,j} \leq 1$, $j(i) = 1,2,\ldots,n_i$, $i = 1,2,\ldots,z$, and the $V_{i,j}$ risk appraisal probability $P_{i,j}$.
6. Following each rectifying and reevaluation procedure, the probability of an error is reduced by a constant decay factor in the failure probability Decay, $0 \leq \text{Decay} \leq 1$. We denote the complementary factor L, $L = 1 - \text{Decay}$, $0 \leq L \leq 1$, the lingering factor.

VVT Optimization Objectives We are seeking to optimize the VVT strategy with respect to the following objectives: (1) VVT activity cost, (2) expected VVT appraisal cost, (3) VVT appraisal cost variance, (4) expected VVT impact cost, (5) VVT impact cost variance and (6) VVT risk cost Value-at-Risk (VaR). Corresponding time domain objectives can be defined when time optimization is performed. The features and optimization of each cost objective is explained below:

1. *VVT Activity Cost.* As regards the CVM, the VVT strategy cost ($\mathrm{Cv}_{\mathrm{VVT\_Strategy}}$) is deterministic and dependent solely on the VVT selected strategy. The objective is to minimize the cost of performing the VVT process, which is the VVT strategy cost:

$$\min\{\mathrm{Cv}_{\mathrm{VVT\_Strategy}} | X\} \tag{7.9}$$

2. *Expected VVT Appraisal Cost.* The VVT appraisal risk cost ($\mathrm{Ca}_{\mathrm{VVT\_Strategy}}$) is a stochastic component which is dependent on the number and cost of failed VVT activities. The objective is to minimize the expected VVT appraisal cost:

$$\min E\{\mathrm{Ca}_{\mathrm{VVT\_Strategy}} | X\} \tag{7.10}$$

3. *Variance of VVT Appraisal Cost.* The variance of the VVT appraisal cost ($V\{\text{Ca}_{\text{VVT\_Strategy}}\}$) represents the dispersion or the uncertainty of the risk measure. The objective is to minimize this uncertainty in order to reduce the possibility of facing extremely risky, high-appraisal-cost situations:

$$\min V\{\text{Ca}_{\text{VVT\_Staregy}} | X\} \tag{7.11}$$

4. *Expected VVT Impact Cost.* The VVT impact cost ($\text{Ci}_{\text{VVT\_Strategy}}$) is a stochastic component dependent upon the individual impact cost and the impact probability as well as the definition of the VVT strategy. The objective is to minimize the expected VVT impact cost:

$$\min E\{\text{Ci}_{\text{VVT\_Strategy}} | X\} \tag{7.12}$$

5. *Variance of VVT Impact Cost.* The variance of the VVT impact cost ($V\{\text{Ci}_{\text{VVT\_Strategy}}\}$) represents the dispersion or the uncertainty of the risk measure. The objective is to minimize this uncertainty in order to reduce the possibility of facing extremely risky, high-impact-cost situations:

$$\min V\{\text{Ci}_{\text{VVT\_Staregy}} | X\} \tag{7.13}$$

6. *VVT Risk Cost VaR.* A frequently used financial risk measure is Value-at-Risk (VaR) (Jorion, 2001). VaR is defined as the maximum loss over a target horizon within a given confidence interval. In systems engineering terminology, VaR measures the worst loss under normal system conditions over a specific time interval at a given confidence level. The objective is to minimize risk losses above 95%. We can say with 95% probability that the risk cost will not exceed a certain value. The objective is to minimize the VaR cost:

$$\min \text{VaR}_{95\%}\{\text{Cv}_{\text{VVT\_Strategy}} | X + \text{Ca}_{\text{VVT\_Strategy}} | X + \text{Ci}_{\text{VVT\_Strategy}} | X\} \tag{7.14}$$

Comparison of VVT Optimization Methods This section describes several optimization methods implemented in the VVT domain. Specifically, five illustrative optimization approaches are described having the characteristics in Table 7.4. Table 7.5 shows the advantages and disadvantages of all of the VVT optimization methods described in this chapter.

TABLE 7.4 Illustrative Optimizations and Characteristics

<table>
<tr><th rowspan="2"></th><th colspan="2">Optimization Number</th><th rowspan="2">VVT domain</th></tr>
<tr><th>Method</th><th>Category</th></tr>
<tr><td>1</td><td>Loss function optimization</td><td>Mathematical analysis</td><td rowspan="3">VVT cost</td></tr>
<tr><td>2</td><td>Weight optimization</td><td rowspan="4">Numerical analysis</td></tr>
<tr><td>3</td><td>Goal optimization</td></tr>
<tr><td>4</td><td rowspan="2">Genetic algorithm optimization</td><td>VVT time</td></tr>
<tr><td>5</td><td>VVT cost and time</td></tr>
</table>

TABLE 7.5 Advantages and Disadvantages of VVT Optimization Methods

	Advantages	Disadvantages
1	• Ensures global optimum value	• Assumes independence of VVT activities • Assumes maximal VVT appraisal risk decay $D_{i,j}$, $D_{i,j} = 1$ • Optimizes a single objective
2	• Optimizes multiple objectives • Concepts are easy to grasp	• Assumes VVT activity independence • Results can express local optimality • Difficult to control target results
3	• Optimizes multiple objectives • Easy to control the goal results	• Assumes VVT activity independence • Results can express local optimality
4	• Assumptions about VVT data are not required	• Results can express local optimality • Optimizes single objective
5	• Optimizes multiple objectives • Assumptions about VVT data are not required	• Results can express local optimality

7.5.2 Loss Function Optimization—VVT Cost

Barad and Engel (2006) suggest a novel analytical model which explores several mechanisms in order to optimize the VVT strategy. These are then integrated into a generalized optimization objective expressing Taguchi's expected loss function. This approach is detailed below.

Theoretical Discussion In the same way as the risk impact costs and probabilities are independent, so are the decision variables related to each activity. Thus, the decision process can be split up into

$$N = \sum_{i=1}^{Z} n_i$$

separate decisions, one for each activity. If the decision objective is minimum expected total cost, the decomposed problem becomes analogous to an inspection policy in a one-stage production process with fixed expected fraction defectives p, inspection cost per item C, and cost incurred by a defective unit reaching the customer M.

The fraction of defectives and the cost M are respectively equivalent to the risk probability and risk impact cost in our model (see Chapter 6). Barad (1986) has shown that for a linear cost structure such as the above and a fixed expected fraction of defectives, sampling (a mixed strategy) cannot be an optimal inspection policy. Under these conditions, even if inspection is not perfect, an optimal inspection policy can only be represented by a pure strategy, either "no inspection" or 100% inspection.

From an analogy perspective, a pure strategy is expressed in our model by $X_{i,j} = 0$ or $X_{i,j} = 1$ and a sampling strategy by $0 < X_{i,j} < 1$. Barad used the break-even economic concept to prove her arguments and to suggest an optimal sampling inspection policy for the expected fraction of defectives, which are not fixed but vary according to a given prior distribution.

Here, our first decision objective is minimum expected total cost. We apply the break-even concept to define an optimal strategy for any activity $V_{i,j}$, $j(i) = 1,2,\ldots,n_i$, $i = 1,2,\ldots,Z$, and show that the optimal values of $X_{i,j}$ consist solely of $X_{i,j} = 0$ or $X_{i,j} = 1$. Next, we consider an additional decision objective, that is, minimum variance of total cost. Eventually, we combine the two objectives through the well-known Taguchi loss function.

Our generalized objective will be an optimal strategy leading to the minimum expected loss function. It will be shown that by, adopting this strategy, optimal values for $X_{i,j}$, $0 \le X_{i,j} \le 1$, representing the optimal performance level of each activity can be directly determined by a simple formula (as a function of the parameters $\mathrm{Cv}_{i,j}$ $\mathrm{Ci}_{i,j}$ $\mathrm{Pi}_{i,j}$). To accomplish the above, we use the assumptions stated earlier.

First Optimization Objective: Minimum Expected Total Cost For the sake of simplicity, let us use the term $\mathrm{Ct}(X_{i,j})$ to denote total cost instead of $\mathrm{Ct}_{\mathrm{VVT\_Strategy}}|X$.

1. Let $\mathrm{Ct}(X_{i,j})$ be the total cost for performing VVT activity $V_{i,j}$ at level $X_{i,j}$, $0 \le X_{i,j} \le 1, j(i) = 1, 2, \ldots, n_i, i = 1, 2, \ldots, Z$. Thus $E[\mathrm{Ct}(X_{i,j})]$ is the expected value of $\mathrm{Ct}(X_{i,j})$.
2. For each VVT activity $V_{i,j}, j(i) = 1, 2, \ldots, n_i, i = 1, 2, \ldots, Z$, the objective is

$$\min_{0 \le X_{i,j} \le 1} \{E[\mathrm{Ct}(X_{i,j})]\} \tag{7.15}$$

3. According to the linearity assumptions stated above, the total VVT cost is composed of the cost of performing the VVT process, the VVT appraisal cost and the VVT impact cost:

$$\mathrm{Ct}(X_{i,j}) = \mathrm{Cv}_{i,j} X_{i,j} + \mathrm{Ca}_{i,j} N_{i,j} | X_{i,j} + (1 - X_{i,j}) \sum_{l=i}^{Z} \sum_{k=1,2,\ldots} \mathrm{Ci}_{i,j}^{l,k} \mathrm{Pi}_{i,j}^{l,k} \tag{7.16}$$

where $N_{i,j}|X_{i,j}$ is the number of corrective iterations related to VVT activity $V_{i,j}$ performed at level $X_{i,j}$. We can see that the stochastic variables in (7.16) are composed of appraisal costs $\mathrm{Ca}_{i,j}\ \mathrm{N}_{i,j}|X_{i,j}$ and impact costs $\mathrm{Ci}_{i,j}^{l,k}$, $l \geq i$, $i = 1, 2, \ldots, Z$, $k = 1, 2, \ldots$, which occur with probability $\mathrm{Pi}_{i,j}^{l,k}$. In practice, the lingering factor is very small. Therefore, assuming $L_{i,j} = 0$, we have the result

$$E[N_{i,j}|X_{i,j}] = \mathrm{Pa}_{i,j}X_{i,j}$$
$$V[N_{i,j}|X_{i,j}] = \mathrm{Pa}_{i,j}X_{i,j}(1-\mathrm{Pa}_{i,j}X_{i,j}) \tag{7.17}$$

4. Hence the expected value is

$$E[\mathrm{Ct}(X_{i,j})] = \mathrm{Cv}_{i,j}X_{i,j} + \mathrm{Ca}_{i,j}\mathrm{Pa}_{i,j}X_{i,j} + (1-X_{i,j})\sum_{l=i}^{Z}\sum_{k=1,2,..}\mathrm{Ci}_{i,j}^{l,k}\mathrm{Pi}_{i,j}^{l,k} \tag{7.18}$$

5. Rearranging gives

$$E[\mathrm{Ct}(X_{i,j})] = BX_{i,j} + \sum_{l=i}^{Z}\sum_{k=1,2,..}\mathrm{Ci}_{i,j}^{l,k}\mathrm{Pi}_{i,j}^{l,k} \tag{7.19}$$

where

$$B = \mathrm{Cv}_{i,j} + \mathrm{Ca}_{i,j}\mathrm{Pa}_{i,j} - \sum_{l=i}^{Z}\sum_{k=1,2,..}\mathrm{Ci}_{i,j}^{l,k}\mathrm{Pi}_{i,j}^{l,k} \tag{7.20}$$

6. The break-even condition, for which all alternatives are economically equal, is

$$B = 0 \tag{7.21}$$

7. It is easily seen that the optimal strategy for any activity is $V_{i,j}$, $j(i) = 1, 2, \ldots, n_i$, $i = 1, 2, \ldots, Z$, becomes:

$$\min\{E[\mathrm{Ct}(X_{i,j})]\} = \tag{7.22}$$

$$E[\mathrm{Ct}(X_{i,j}=1)] = \mathrm{Cv}_{i,j} + \mathrm{Ca}_{i,j}\mathrm{Pa}_{i,j} \quad \text{for } B<0 \tag{7.23}$$

$$E[\mathrm{Ct}(X_{i,j}=0)] = \sum_{l=i}^{Z}\sum_{k=1,2,..}\mathrm{Ci}_{i,j}^{l,k}\mathrm{Pi}_{i,j}^{l,k} \quad \text{for } B>0 \tag{7.24}$$

Second Optimization Objective: Minimum Variance of Total Cost

1. Let $V[\mathrm{Ct}(X_{i,j})]$ be the variance of $\mathrm{Ct}(X_{i,j})$.
2. For each activity $V_{i,j}$, $j(i) = 1, 2, \ldots, n_i$, $i = 1, 2, \ldots, Z$, the objective is now

$$\min_{0 \leq X_{i,j} \leq 1}\{V[\mathrm{Ct}(X_{i,j})]\} \tag{7.25}$$

3. Applying basic probability calculations on (7.16) we obtain

$$V[\mathrm{Ct}(X_{i,j})] = \mathrm{Ca}_{i,j}^2 \mathrm{Pa}_{i,j} X_{i,j}(1 - \mathrm{Pa}_{i,j} X_{i,j}) + (1 - X_{i,j})^2 D \tag{7.26}$$

4. Rearranging gives

$$V[\mathrm{Ct}(X_{i,j})] = \mathrm{Ca}_{i,j}^2 [\mathrm{Pa}_{i,j} X_{i,j} - \mathrm{Pa}_{i,j}^2 X_{i,j}^2] + (1 - X_{i,j})^2 D \tag{7.27}$$

where

$$D = \sum_{l=i}^{Z} \sum_{k=1,2,..} [\mathrm{Ci}_{i,j}^{l,k}]^2 [\mathrm{Pi}_{i,j}^{l,k}][1 - \mathrm{Pi}_{i,j}^{l,k}] \tag{7.28}$$

5. Further rearranging yields

$$V[\mathrm{Ct}(X_{i,j})] = EX_{i,j} - FX_{i,j}^2 + (1 - X_{i,j})^2 D \tag{7.29}$$

where

$$E = \mathrm{Ca}_{i,j}^2 \mathrm{Pa}_{i,j} \qquad \text{and} \qquad F = \mathrm{Ca}_{i,j}^2 \mathrm{Pa}_{i,j}^2 \tag{7.30}$$

6. Our objective is now to find optimal values for $X_{i,j}$, $j(i) = 1, 2, \ldots, n_i$, $i = 1, 2, \ldots, Z$, so that the variance of the total cost is minimized:

$$\min_{0 \le X_{i,j} \le 1} \{V[\mathrm{Ct}(X_{i,j})]\} \tag{7.31}$$

7. By calculating the derivative of Equation (7.29), equaling it to zero and checking the second derivative we obtain a simple equation for the optimal value of $X_{i,j}$:

$$X_{i,j\,(\mathrm{Optimum})} = \frac{D - E/2}{D - F} \tag{7.32}$$

Note: It is apparent from Equation (7.29) that the variance for the two pure strategies $X_{1,j} = 0$, $X_{1,j} = 1$, can be calculated as follows:

$$\begin{aligned} V[\mathrm{Ct}(X_{i,j} = 0)] &= D \\ V[\mathrm{Ct}(X_{i,j} = 1)] &= E - F \end{aligned} \tag{7.33}$$

Generalized Optimization Objective: Minimum Expected Loss Function We note that if (7.23) prevails ($B < 0$), $X_{i,j} = 1$ is the optimal decision for the first optimization objective as defined in (7.15), that is, the minimum expected total cost. Substituting $X = 1$ in Equation (7.27), we obtain $V[\mathrm{Ct}(X_{i,j})] = \mathrm{Ca}_{i,j}^2 \mathrm{Pa}_{i,j}[1 - \mathrm{Pa}_{i,j}]$. In most cases the value of this expression is very small and can be disregarded. Accordingly, for $X = 1$, the second objective, that is, the minimum variance of total cost as defined in (7.31), is also achieved. This is not the case for (7.24). Hence, let us now seek generalized optimal decision integrating objectives (7.15) and (7.25) when condition (7.24) prevails, that is, for $B > 0$:

1. To consider objectives (7.15) and (7.25) together, we will make use of the well-known Taguchi loss function $L(Y)$ according to which each deviation of a quality characteristic from a target value gives rise to customer dissatisfaction, which is proportional to the squared deviation from the target (e.g., Ross, 1996). Accordingly,

$$L(Y)=C(Y-T)^2 \tag{7.34}$$

where Y is the numerical value of the product quality characteristic, T is its numerical target value and C is a constant. Assuming that the quality characteristic Y is a random variable with expected value $E[Y]$ and variance $V[Y]$, the expected value of the loss function, $E[L(Y)]$, over the product population becomes

$$E[L(Y)]=CE\left[(Y-T)^2\right] \tag{7.35}$$

2. It is easily proved that

$$CE\left[(Y-T)^2\right]=C\left[V(Y)+(E(Y)-T)^2\right] \tag{7.36}$$

3. Taguchi used this expression as the main argument for showing that by reducing the variance of any quality characteristic, customer satisfaction is improved. In our case the quality characteristic Y is superseded by $\mathrm{Ct}(X_{i,j})$, the total cost for performing activity $V_{i,j}$ at level $X_{i,j}$, $0 \le X_{i,j} \le 1$, $j(i) = 1, 2, \ldots, n_i$, $i = 1, 2, \ldots, Z$, and the target value T is superseded by the minimum expected total cost. Since we are looking for an optimal solution under condition (7.24), that is, for $(B > 0)$, the target T is

$$\min\{E[\mathrm{Ct}(X_{i,j})]\}=E[\mathrm{Ct}(X_{i,j}=0)]=\sum_{l=i}^{Z}\sum_{k=1,2,..}\mathrm{Ci}_{i,j}^{l,k}\mathrm{Pi}_{i,j}^{l,k} \tag{7.37}$$

4. Assigning $C = 1$ and substituting $\mathrm{Ct}(X_{i,j})$ as defined in (7.16) for Y and the above $\min\{E[\mathrm{Ct}(X_{i,j})]\}$ for T in Equation (7.36), the expected loss function for $\mathrm{Ct}(X_{i,j})$ will be

$$E\{L[\mathrm{Ct}(X_{i,j})]\}=V[\mathrm{Ct}(X_{i,j})]+\left\{E[\mathrm{Ct}(X_{i,j})]-\sum_{l=i}^{Z}\sum_{k=1,2,..}\mathrm{Ci}_{i,j}^{l,k}\mathrm{Pi}_{i,j}^{l,k}\right\}^2 \tag{7.38}$$

5. From Equation (7.19) we obtain

$$E[\mathrm{Ct}(X_{i,j})]-\sum_{l=i}^{Z}\sum_{k=1,2,..}\mathrm{Ci}_{i,j}^{l,k}\mathrm{Pi}_{i,j}^{l,k}=X_{i,j}B \tag{7.39}$$

6. Substituting (7.27) and (7.38) in (7.39), we obtain

$$V[\mathrm{Ct}(X_{i,j})] = \mathrm{Ca}_{i,j}^2[\mathrm{Pa}_{i,j}X_{i,j} - \mathrm{Pa}_{i,j}^2X_{i,j}^2] + (1 - X_{i,j})^2 D \quad (7.40)$$

$$E\{L[\mathrm{Ct}(X_{i,j})]\} = \mathrm{Ca}_{i,j}^2[\mathrm{Pa}_{i,j}X_{i,j} - \mathrm{Pa}_{i,j}^2X_{i,j}^2] + (1 - X_{i,j})^2 D + [BX_{i,j}]^2 \quad (7.41)$$

or

$$E\{L[\mathrm{Ct}(X_{i,j})]\} = EX_{i,j} - FX_{i,j}^2 + (1 - X_{i,j})^2 D + (BX_{i,j})^2 \quad (7.42)$$

7. Our objective is now to find optimal values for $X_{i,j}$, $j(i) = 1, 2, \ldots, n_i$, $i = 1, 2, \ldots, Z$, so that the expected loss is minimal:

$$\min_{0 \le X_{i,j} \le 1} \{E(L[\mathrm{Ct}(X_{i,j})])\} \quad (7.43)$$

8. By calculating the derivative of Equation (7.42), equaling it to zero and checking the second derivative, we obtain a simple equation for the optimal value of $X_{i,j}$, namely,

$$X_{i,j\,(\mathrm{Optimum})} = \frac{D - E/2}{B^2 + D - F} \quad (7.44)$$

Numerical Example We will continue to use the VVT cost and risk data from the IAI/Lahav pilot project of SysTest. In order to simplify this example, (1) the VVT horizon will only continue to apply to the first system lifecycle phase (Definition) and (2) the VVT cost parameters will be the same ones used during the pilot project. Also we assume linear impact risk cost behavior.

Optimizing for Minimum Expected Total Cost We will now optimize the VVT strategy in order to obtain a minimum expected total cost using Equation (7.22). Table 7.6 presents the numerical results of the two pure strategies: $X_{1,j} = 0$, $X_{1,j} = 1$, which, according to Equation (7.22), are the only potential optimal strategies.

As can be observed, the optimal strategy for obtaining minimum expected total cost is not to perform any of the 11 VVT activities ($X_{1,j}^* = 0$, $j = 1, 2, \ldots, 11$). This optimal strategy yields a total expected cost of 22 CU. The main reason for this situation is that the VVT horizon only covers the Definition phase and most of the VVT impacts tend to occur during later phases. The cost variance of this strategy is 63.4 CU, and accordingly, its standard deviation is 7.96 CU.

TABLE 7.6 Optimizing for Minimum Expected Total Cost

VVT ID	VVT Cost	Expected Appraisal Cost	Subtotal	Expected Impact Cost	Cost Difference, B	Optimal Strategy, $X^*_{i,j}$	Minimum Expected Cost[CU], $E[C_t(X^*_{i,j})]$	Var
	Expected Cost for Given Strategy ($X_{i,j}$): $E[Ct(X_{i,j}=1)]$			$E[Ct(X_{i,j}=0)]$				
V1.1	32.0	0.0	32.0	17.0	15.0	0.0	17.0	62.2
V1.2	26.0	0.0	26.0	0.0	26.0	0.0	0.0	0.0
V1.3	20.0	6.0	26.0	0.0	26.0	0.0	0.0	0.0
V1.4	19.0	2.0	21.0	0.0	21.0	0.0	0.0	0.0
V1.5	38.0	5.0	43.0	0.0	43.0	0.0	0.0	0.0
V1.6	26.0	2.0	28.0	0.0	28.0	0.0	0.0	0.0
V1.7	13.0	0.0	13.0	0.0	13.0	0.0	0.0	0.0
V1.8	28.0	5.0	33.0	0.0	33.0	0.0	0.0	0.0
V1.9	26.0	5.0	31.0	0.0	31.0	0.0	0.0	0.0
V1.10	13.0	1.0	14.0	5.0	9.0	0.0	5.0	1.2
V1.11	13.0	1.0	14.0	0.0	14.0	0.0	0.0	0.0
Total	254.0	27.0	281.0	22.0	259.0		22.0	63.4

Optimizing for Minimum Variance of Expected Total Cost We will now optimize the VVT strategy in order to obtain a minimum variance of the expected total cost. We will use Equation (7.32) and the same raw VVT data in order to obtain optimal VVT performance levels ($X^*_{1,j}$) (see Table 7.7).

TABLE 7.7 Optimizing for Minimum Variance of Expected Total Cost

VVT ID	Optimal Strategy, $X^*_{i,j}$	Minimum Variance, $V[C_t(X^*_{i,j})]$	VVT Cost	Expected Appraisal Cost	Expected Impact Cost	Expected Total Cost
			$E[C_t(X^*_{i,j})]$			
V1.1	1.00	0.0	32.0	0.0	0.0	32.0
V1.2	1.00	0.1	26.0	0.1	0.0	26.1
V1.3	0.53	9.0	10.5	3.0	0.0	13.5
V1.4	0.60	2.2	11.4	1.5	0.0	12.9
V1.5	0.60	9.0	22.9	3.0	0.0	25.9
V1.6	0.60	2.2	15.7	1.5	0.0	17.2
V1.7	1.00	0.3	13.0	0.5	0.0	13.5
V1.8	0.53	6.3	14.7	2.5	0.0	17.2
V1.9	0.53	6.3	13.7	2.5	0.0	16.2
V1.10	1.00	0.0	13.0	1.0	0.0	14.0
V1.11	0.60	0.2	7.8	0.5	0.0	8.3
Total		35.6	180.7	16.1	0.0	196.8

As can be seen, the total minimum variance is 35.6 CU, and therefore the standard deviation is nearly 6 CU. The overall expected cost of this strategy is 196.8 CU. This result is better than completely performing the VVT process but far less attractive than that of the minimum expected cost strategy.

Optimizing for Minimum Loss Function We will now optimize the VVT strategy in order to obtain a minimum loss function. We will use Equation (7.44) as well as the same raw VVT data and get optimal VVT performance levels ($X^*_{1,j}$) that minimize the total expected loss of the VVT process (see Table 7.8).

TABLE 7.8 Optimizing for Minimum Loss Function

VVT ID	Optimal Strategy, $X^*_{i,j}$	Minimum Expected Loss, $E[L(C_t(X^*_{i,j}))]$	VVT Cost	Expected Appraisal Cost	Expected Impact Cost	Expected Total Cost	Variance, $V[C_t(X^*_{i,j})]$
			$E[C_t(X^*_{i,j})]$				
V1.1	0.23	48.1	7.3	0.0	13.5	20.8	37.2
V1.2	0.00	0.0	0.0	0.0	0.0	0.0	0.0
V1.3	0.00	0.0	0.0	0.0	0.0	0.0	0.0
V1.4	0.00	0.0	0.0	0.0	0.0	0.0	0.0
V1.5	0.00	0.0	0.0	0.0	0.0	0.0	0.0
V1.6	0.00	0.0	0.0	0.0	0.0	0.0	0.0
V1.7	0.00	0.0	0.0	0.0	0.0	0.0	0.0
V1.8	0.00	0.0	0.0	0.0	0.0	0.0	0.0
V1.9	0.00	0.0	0.0	0.0	0.0	0.0	0.0
V1.10	0.01	1.2	0.1	0.0	4.7	4.8	1.2
V1.11	0.00	0.0	0.0	0.0	0.0	0.0	0.0
Total		49.3	7.4	0.0	18.2	25.6	38.4

The optimal VVT strategy yields a minimal total expected loss of 49.3 CU. Additionally, this strategy produces an overall expected cost of 25.6 CU and a cost variance of 38.4 CU. We can see that the overall expected cost is quite close to 22 CU, the result of the minimum expected cost strategy, whereas its variance is much lower than 63.4 CU, the variance obtained by the latter strategy.

7.5.3 Weight Optimization—VVT Cost

Using a numerical analysis method based on weighting objectives is intuitive and is in common practice. In this method, the MOP is converted into a SOOP by using a weighted sum of the original multiple objectives. The equivalent minimization problem is then given by

$$\min f(x) = \sum_{i=1}^{N} w_i f_i(x) \qquad (7.45)$$

subject to

$$g_j(x) \le 0 \qquad j = 1, 2, \ldots, J \qquad (7.46)$$

$$h_k(x) = 0 \qquad k = 1, 2, \ldots, K \qquad (7.47)$$

where the W_i's are the weighting coefficients satisfying the conditions

$$0 < w_i \le 1 \quad \forall i \qquad \text{and} \qquad \sum_{i=1}^{N} w_i = 1 \qquad (7.48)$$

The weighting coefficients are chosen a priori. If the problem is convex, a complete set of Pareto solutions can be obtained by varying the weighting coefficients.

Practical Optimization Constraints The following practical optimization constraints were defined for this example:

1. *Full Set.* In order to reduce the computation time associated with the optimization process, each decision variable $X_{i,j}$, $0 \le X_{i,j} \le 1$, is constrained so that it assumes one value from the set $X_{i,j} = \{0.0, 0.1, 0.2, 0.3, 0.4, 0.5, 0.6, 0.7, 0.8, 0.9, 1.0\}$.
2. *Predetermined (Fixed) Values.* In general, decision variables are allowed to assume different values during the optimization process. However, under certain circumstances, we have to perform (or avoid performing) a certain VVT activity regardless of purely optimization considerations. In this example, VVT activity $V_{1,4}$ is never carried out due to organization policy and therefore the performance level must be "fixed" (in this case $X_{1,4} = 0$). Similarly, VVT activity $V_{1,10}$ must be performed at some nominal level due to nominal contractual obligations. Again, the performance level has to be fixed (in this case $X_{1,10} = 0.2$).

Weighting Method Optimization In explaining weighting method optimization, we continue to use the same VVT cost and risk data. Specifically, in order to simplify this example, (1) the VVT horizon will continue to cover only the first system lifecycle phase (Definition) and (2) the VVT cost parameters will be the same ones used during the pilot project.

Multiobjective Weight For this multiobjective weight example (utilizing numerical analysis technique), the objective functions f_i, $i = 1, 2, \ldots, 5$, shown in Table 7.9 are linearly normalized so that $0 \le f_i \le 1$, $\forall i$ and their sum equals

1 ($\sum_{i=1}^{5} f_i = 1$). Additionally, in this example, we are trying to minimize the actual VVT cost as well as the expected VVT impact cost. We are less interested in reducing (1) the expected appraisal risk cost, (2) the standard deviation of the VVT strategy cost and (3) the VaR.

TABLE 7.9 Multiobjective Weights

Objective	VVT Actual Cost (1)	Expected VVT Impact Cost (2)	Expected VVT Appraisal Cost (3)	Standard Deviation of VVT Cost (4)	VaR of VVT Cost (5)
Weight	0.4	0.3	0.1	0.1	0.1

Weight Optimization Figure 7.9 depicts the VVT multiobjective weighted optimization input layout. As can be seen, the decision variable is limited in terms of the set of 11 values it is able to accept ($X_{i,j}$ = {0, 0.1, 0.2, 0.3, 0.4, 0.5, 0.6, 0.7, 0.8, 0.9, 1}).

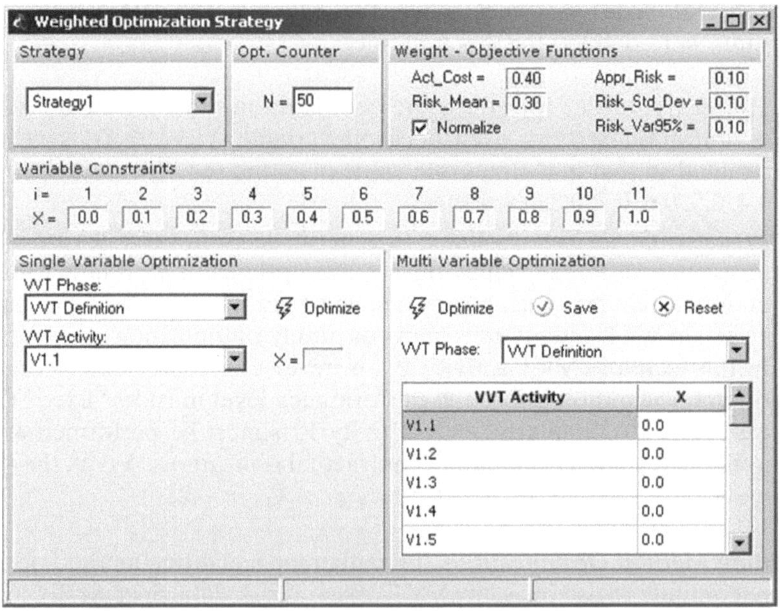

Figure 7.9 VVT multiobjective weighted optimizer input.

Figure 7.10 shows the original versus the optimized values of the decision variables. This optimization procedure has utilized nonlinear modeling of impact risk cost. This model reflects the real-life system behavior more accurately.

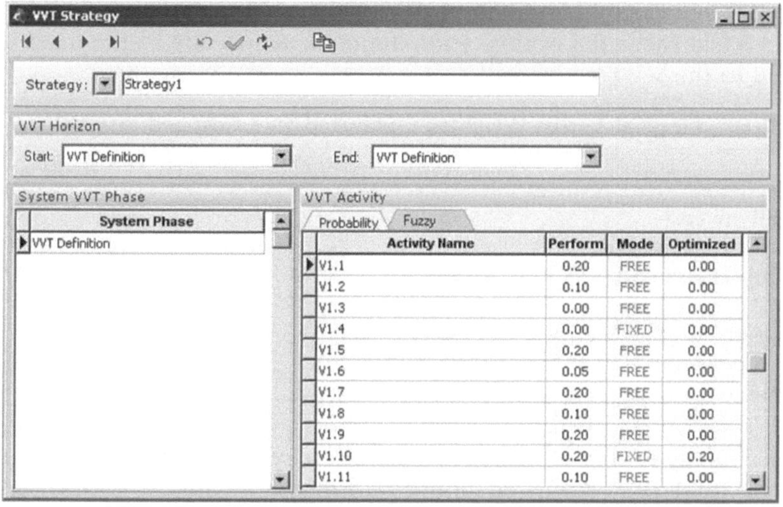

Figure 7.10 Original versus weighted optimized decision variables.

VVT Optimization Results Figure 7.11 shows the VVT cost distribution results derived from the stochastic simulator after 3000 iterations utilizing the above optimized VVT strategy. Table 7.10 shows the original (preoptimization VVT cost and risk values derived earlier) versus the optimized VVT cost and risk results using the weighted optimized VVT strategy. In this example, the optimized VVT strategy has reduced the expected total VVT cost by 54%.

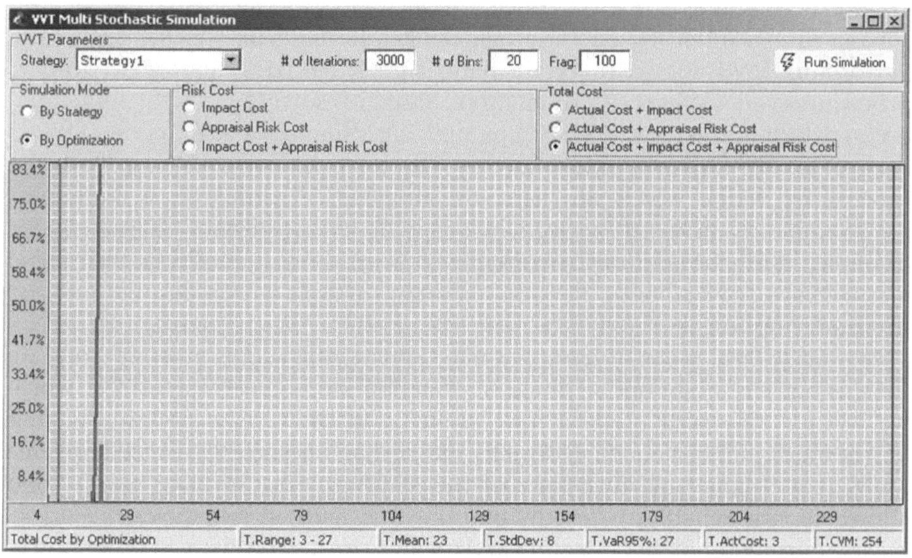

Figure 7.11 VVT cost based on weighted optimized VVT strategy.

This was achieved by identifying a VVT strategy that reduces performance level to 3 CU (here the optimization limits its action due to the fixed values assigned to the VVT activities $V_{1,4}$, $V_{1,10}$).

TABLE 7.10 Original Versus Weighted Optimized VVT Cost and Risk Values

VVT Cost	Original (CU)	Optimized (CU)	Reduction (%)
VVT activity cost	33	3	91
Mean appraisal risk cost	2	0	100
Mean impact risk cost	15	20	–33
Mean total VVT cost	50	23	54
CVM cost	254	254	0
Standard deviation	5	8	–60
VaR (95%)	53	27	49

This strategy is very close to the pure strategy of optimizing for minimum expected total cost discussed above, which yields almost the same total expected cost of 23 CU. This optimization approach becomes obvious when we consider that the optimization ignores VVT impact that could affect the system outside the VVT horizon. We can also observe the main weakness of a weighted optimization approach. Even though the optimization was supposed to reduce the VVT impact cost, it actually generated a VVT strategy that increased cost. The next section describes an optimization approach which is often more suitable in industrial settings.

7.5.4 Goal Optimization—VVT Cost

Goal Programming (GP) is a collection of numerical analysis techniques for solving MOPs when the decision maker likes to focus on specific goals (e.g., performing VVT within a certain budget goal). In particular, lexicographic procedures are a subset of GP methods based on solving multiple, sequential optimization problems rather than minimizing a single scalar function. It treats each objective function independently. The term *lexicographic* refers to the process by which words are listed in alphabetical order. In a dictionary, for example, each word is listed in order of its constituent characters, for example, in alphabetic (or lexicographic) order. Consequently, lexicographic methods are applicable to decision-making problems with a discrete number of priority options.

Within several types of lexicographic GP, one that seems most appropriate for this book is the bounded objective function method (Hwang, 1979), which we will call the goal method for simplicity. In practice, a system engineer has certain goals to meet. Normally, he or she should not exceed specific limits (e.g., budget, performance of the VVT process, level of quality and allowable risk, schedule). One by one the goal method minimizes the single, most important objective function and creates constraints with all other objective functions as follows:

Find: $x \in X$
To minimize: $F_s(x)$
Subject to: $\alpha_i \leq F_i(x) \leq \beta_i$; $i = 1, 2, \ldots, k$; $i \neq s$

The decision maker selects $F_s(x)$, which is called the primary objective function, as well as the values of α_i, β_i, which represent, respectively, the lower and upper limits of the objective function $F_s(x)$. Like the lexicographic methods, these added constraints reduce the region of the feasible solution (see Figure 7.12).

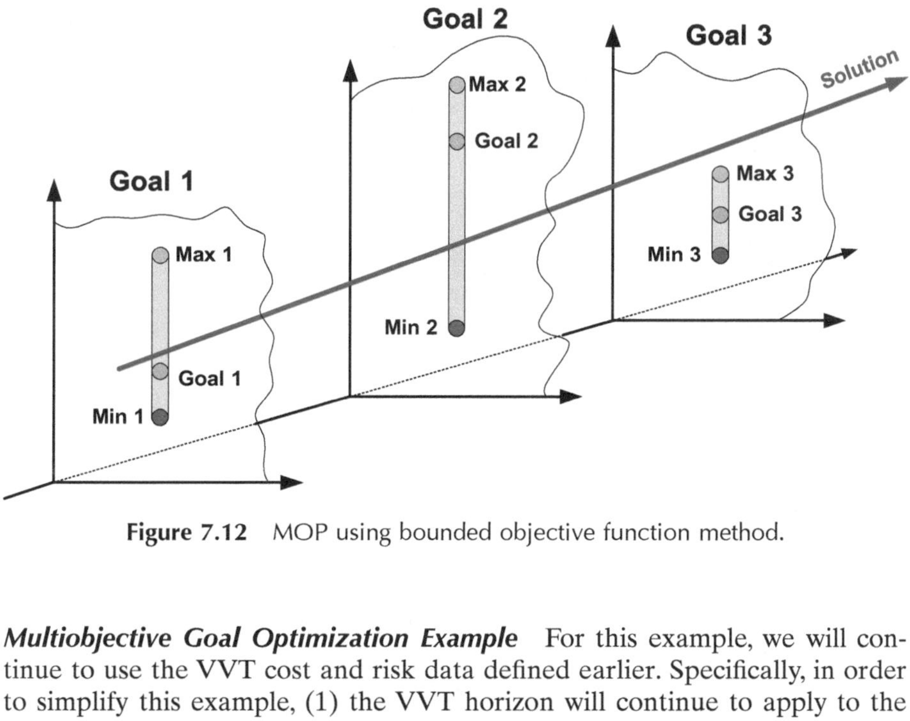

Figure 7.12 MOP using bounded objective function method.

Multiobjective Goal Optimization Example For this example, we will continue to use the VVT cost and risk data defined earlier. Specifically, in order to simplify this example, (1) the VVT horizon will continue to apply to the first system lifecycle phase (Definition) and (2) the VVT cost parameters will be the same ones which were used in the pilot project.

In practical, industrial situations we tend to seek no more than two or three risk goals. In this example we look for a VVT strategy leading to the goals depicted in Table 7.11. Usually, we try to minimize the total VVT expected cost. However, in this example, we assume severe limitations on the actual engineering VVT budget. Hence, we are looking for a VVT strategy which, first and foremost, ensures specific VVT budget allocation for VVT actual cost. As a secondary goal, we seek a VVT strategy which will set the VVT impact within a certain range. Lastly, at the lowest priority, we like to reduce the appraisal cost to a predefined level.

TABLE 7.11 Goal Optimization Parameters

Objectives	Priority	Desired Values		
		Goal	Minimum	Maximum
VVT activity cost	1	12	10	15
VVT impact cost	2	15	8	17
VVT appraisal cost	3	2	1	4

Here, again, VVT activity $V_{1,4}$ and VVT activity $V_{1,10}$ are fixed and the optimization procedure utilizes a nonlinear model.

The VVT multiobjective goal optimization layout is shown in Figure 7.13. As we see, a VVT strategy was found such that all three objectives have been achieved. If we will implement this VVT strategy, the VVT actual cost will be 15 CU, the expected VVT impact cost will be 17 CU and the expected VVT appraisal cost will be only 1 CU.

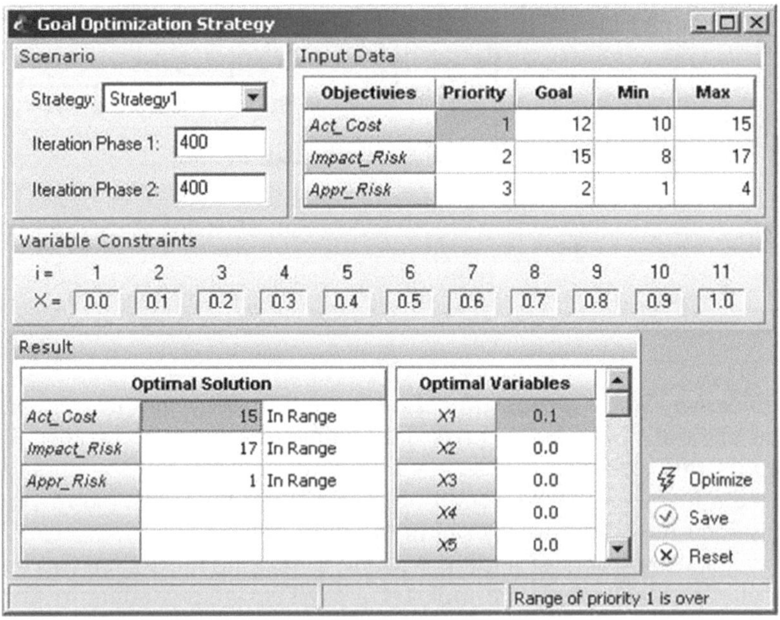

Figure 7.13 VVT multiobjective goal optimizer input.

Figure 7.14 shows the original versus the optimized VVT strategy (decision variables). Figure 7.15 shows the results obtained when executing the VVT stochastic simulation 3000 times using the goal optimized VVT strategy.

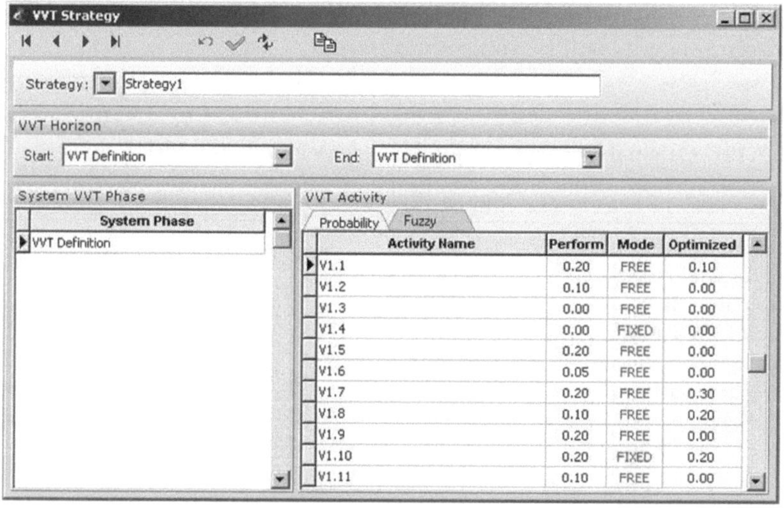

Figure 7.14 Original versus goal optimized decision variables.

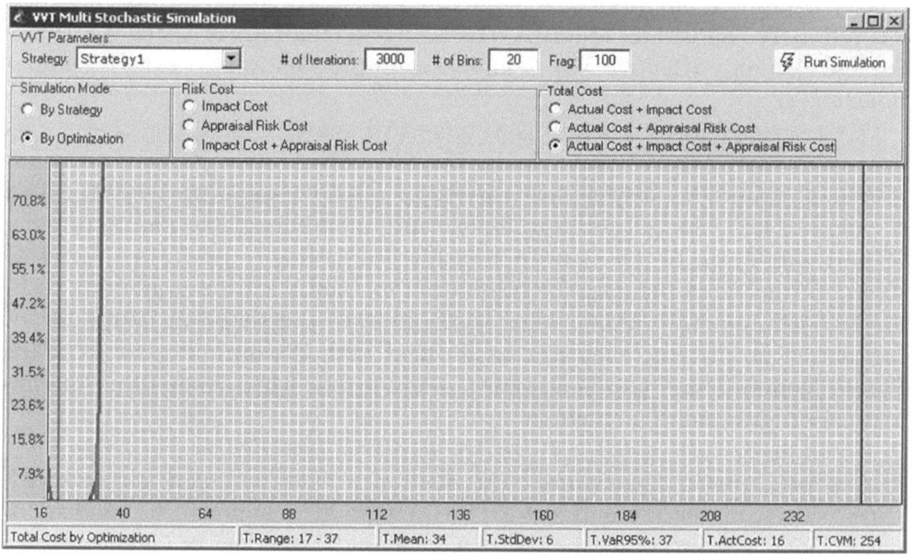

Figure 7.15 Goal optimized overall VVT cost.

Table 7.12 shows the original versus the goal optimized VVT cost and risk. Again, it is clear that when we consider only the Definition phase, the strategy can be "improved" significantly. As we will see in Chapter 8, optimizing the VVT strategy for the five phases of the product development stage, the optimization results will be radically different.

TABLE 7.12 Original Versus Goal Optimized VVT Cost and Risk Values

Total VVT Cost	Original (CU)	Optimized (CU)	Reduction (%)
VVT activity cost	33	16	52
Mean appraisal risk cost	2	1	50
Mean impact risk cost	15	17	–13
Mean total VVT cost	50	34	32
CVM cost	254	254	0
Standard deviation	5	6	–20
VaR (95%)	53	37	30

7.5.5 Genetic Algorithm Optimization—VVT Time

Evolutionary algorithms are powerful techniques used to find solutions to many multiple objectives, real-world search and optimization problems. Deb (2001) compiled the first complete and updated text on Multi-Objective Evolutionary Algorithms (MOEAs) covering all major areas. Using such techniques is a highly effective way of finding multiple solutions in a single simulation run.

A Genetic Algorithm (GA) is a numerical analysis technique that seems particularly suitable for solving multiobjective optimization problems because they deal simultaneously with a variety of possible solutions (the so-called population set). This allows us to find several members of the Pareto optimal set in a single run of the algorithm, instead of having to perform a series of separate runs, as is the case for traditional mathematical programming techniques. Additionally, GAs are less susceptible to the shape or continuity of the Pareto front (e.g., they can easily deal with discontinuous or concave Pareto fronts). These two issues are a real concern when using other mathematical programming techniques.

Theory of Genetic Algorithms This section, which has been adapted with permission from Obitko and Slavik (1999), introduces some of the fundamentals of GAs. The intent is to provide a basic understanding of GAs. A subset of evolutionary computing, GAs are classified by some researchers as belonging to artificial intelligence. Genetic algorithms are motivated by Darwin's theory of evolution.

Computer simulations of evolution started in 1954 with the work of Nils Aall Barricelli. Later, evolutionary computing was introduced in the 1960s (Rechenberg, 1994) and these ideas were then developed by other researchers. GAs were proposed in the early 1970s by John Holland and developed by him and his students and colleagues (Holland, 1992). Goldberg (1989) described numerous GAs and covers all of the important topics in the field, including crossover, mutation, classifier systems and fitness scaling. Koza (1992) used genetic algorithms to evolve programs in order to perform certain tasks.

Biological Background All living organisms consist of cells in which there are specific sets of chromosomes. Chromosomes serve as a pattern for the construction of the entire organism. A chromosome consists of genes that each encodes a particular trait or capability of the organism for, example, the ability to run fast.

During reproduction, a recombination (or crossover) occurs. Genes from parents combine to form a completely new chromosome. The newly created offspring may then be mutated. Mutation means that the elements of DNA are slightly changed. These changes are mainly caused by errors in inheriting genes from one's parents. The fitness of an organism is measured by the success of the organism to survive.

Search Space When we are solving a problem with multiple solutions, we usually look for a solution which will be better than all the others. The space of all feasible solutions, for example, the set of all solutions, is called the search space. Each element in the search space represents a single possible solution. Each possible solution can be assessed by its survival fitness.

By using GAs, we are looking for the best solution from a number of other possible solutions—represented by one point in the search space. Looking for a solution is equal to looking for some extreme value (minimum or maximum) in the search space. At times, the search space can be well defined, but usually we only know a few points in the search space. In the process of using GAs, finding solutions generates other possible solutions as evolution proceeds.

There are several possible methods for finding a suitable solution, but these methods do not necessarily provide the best solution. Some of these methods are hill climbing, simulated annealing and GAs. The solutions found by these methods are usually considered to be good or very good solutions, though it is sometimes impossible to reach an absolute optimum.

Genetic Algorithm Genetic algorithms use an evolutionary process beginning with a set of initial solutions (represented by chromosomes) called the population. This population is then used to form a new population on a cyclical basis. The strategy is based on the anticipation that the new population will be better than the old one. New solutions (offspring) are selected according to their suitability—the more suitable they are, the more chance they have to reproduce. This is repeated until some condition, for example, improvement of the best solution, is satisfied. As can be seen, the outline of the basic GA is very general. Many parameters and settings can be implemented differently in various problems. The basic GA is outlined in Table 7.13.

TABLE 7.13 Basic Genetic Algorithm

Step	Stage	Actions
1	Start	Generate random population of n chromosomes (suitable solutions for the problem).
2	Suitability	Evaluate the suitability $f(x)$ of each chromosome x in the population.
3	New population	Create a new population by repeating the following steps until the new population is complete.
4	Elitism	Select one or more of the most suitable parent chromosomes and transfer them to the new population.
5	Selection	Select two parent chromosomes from a population according to their suitability (the more suitable they are, the higher chance they have to be selected).
6	Crossover	With a crossover probability, a selected set of parents form new offspring. If no crossover was performed, the offspring will be the exact copy of parents.
7	Mutation	With a mutation probability, mutate new offspring at each locus (the position in the chromosome).
8	Replace	Place the new offspring in the new population and replace/remove part or all the old population. The current population is used to continue the algorithm.
9	Test	If the end condition is satisfied, stop, and return the best solution in the current population.
10	Loop	Proceed to step 2.

Encoding Chromosomes A chromosome should contain information about the solution it represents. Encoding chromosomes is the first step required in order to solve a problem with the GA and depends on the problem at hand. There are many ways to encode chromosomes and we will describe two common encoding approaches.

Binary Encoding Binary encoding is the most common, mainly because of its relative simplicity. In binary encoding, every chromosome is a string of different pieces, 0 or 1. The following is an example of chromosomes with binary encoding

Chromosome A	1 0 1 1 0 0 1 0 1 1 0 1 0
Chromosome B	1 1 1 1 1 1 1 0 0 0 0 1 1

Binary encoding gives many possible chromosomes even with a small number of alleles[80]. On the other hand, this encoding is often not natural for many problems and sometimes corrections have to be made after crossover and/or mutation.

[80]Wikipedia definition: "An allele (from the Greek αλληλος, *allelos*, meaning "other") is one of a series of different forms of a gene."

Value Encoding Direct value encoding is used in problems where some specific values such as real numbers are required. In value encoding, every chromosome is a sequence of some value. Values can be anything connected to the problem, such as (real) numbers, characters or any objects. The following is an example of chromosomes with value encoding:

Chromosome A	0.4	0.5	1.0	1.0	0.5	0.2	0.4	0.6
Chromosome B	0.0	0.4	0.9	0.5	1.0	1.0	0.5	0.2

Value encoding is a good choice for some special problems. For this encoding it is necessary to develop a problem-specific crossover and mutation. We used this type of encoding to solve the VVT optimization problem.

Determining Population Size Several GA researchers struggled with the question of determining the optimal population size. In a 2002 paper, Gotshall and Rylander seek to find the optimal population size needed in order to obtain rapid convergence with high accuracy. The paper concludes that:

1. For arbitrarily large population size the accuracy of the GA approaches but does not reach 100%.
2. Increase in population size causes the number of generations required for convergence to increase.
3. Optimal population for a given problem implementation is the point at which the benefit of low numbers of generations as regards convergence balances with the benefit of increased accuracy as the population size increases.

A complementary view suggests that optimal population size is related to the size of the chromosomes. It means that if one encodes a GA problem with 32 genes per chromosome, then the population will be larger than when one encodes a chromosome with 16 genes. A good approximation for a reasonable population size suggests that the size of the population will be approximately equal to the number of genes in a chromosome. We adopted this approach for this book.

Selecting the Best in a Population As mentioned above, chromosomes are selected from the population in order to contribute their genes to the next generation. Parents are selected according to their suitability. The better the chromosomes are, the more chance there is to be selected. There are numerous methods for selecting the best chromosomes for parenting, for example, roulette wheel selection, Boltzmann selection, tournament selection, rank selection and steady-state selection.

The selection process approach adopted for this book is roulette wheel selection. Here, each chromosome in the population will be placed in one groove of the roulette wheel. The size of the section on the roulette wheel is proportional to the value of the suitability function of every chromosome.

The higher the value is, the larger the selection is. A marble is thrown onto the roulette wheel and the chromosome is selected where it stops. Clearly, the chromosomes with the highest fitness value will be selected more often. This process may be described by the following algorithm:

- Calculate the sum (S) of all the chromosomes in the population.
- Create an imaginary roulette wheel having S slots, corresponding to the existing population.
- Generate a random number (R) from the interval (0, S).
- Select the chromosome associated with the R slot of the roulette wheel.

Genetic Algorithm Operators Elitism, crossover and mutation are three basic operators of the GA. The performance of GAs is very dependent on them. The type and implementation of the operators depend on the encoding and the specific problem at hand. There are many ways of performing elitism, crossover and mutation. We will briefly give some examples and suggestions on how to perform them.

Elitism When creating a new population by crossover and mutation, there exists a certain chance that we will lose the best chromosome. Therefore, the purpose of elitism is to ensure that the best chromosomes are always transmitted to the new generation. Elitism is the name of the method which copies a few of the best chromosomes to the new population and the rest of the population is constructed by way of crossover and mutation. Elitism invariably increases the performance of GAs, because it prevents a loss of the one or more best-found solution. As a rule, adopted also in this book, approximately 5–10% of the population will be treated as the elite of the current population and thus copied onto the new population.

Crossover Crossovers operate on selected genes from the parents' chromosomes and create new offspring. The simplest way to do this is to choose some crossover point at random and then to copy everything before this point from the first parent and then copy everything after the crossover point from the other parent. Crossover can be illustrated as follows:

Chromosome A	1	1	0	1	1	0	0	1	0	1	1
Chromosome B	1	0	0	1	1	1	1	0	1	0	0
Offspring A	1	1	0	1	1	1	1	0	1	0	0
Offspring B	1	0	0	1	1	0	0	1	0	1	1

There are other ways to make a crossover, for example, we can choose one, two or more crossover points. A crossover can be complicated and it depends mainly on the encoding of chromosomes.

Crossover probability defines how often the crossover is performed. If there is no crossover, then the offspring will be the exact copies of parents. If there is crossover, offspring are made from parts of both parents' chromosomes. If crossover probability is 100%, then all offspring are created by the crossover. If it is 0%, a whole new generation is made from exact copies of chromosomes from the old population. A crossover is made with the expectation that new chromosomes will contain good parts of old chromosomes and therefore the new chromosomes will be better. However, it is good to let some part of the old populations survive to the next generation. The crossover rate is normally about 80–95%.

Mutation Mutation is intended to prevent the algorithm from sinking into a local optimum. In other words, mutation introduces randomness and divergence into the population. Most often mutation will create unattractive solutions. But every now and then it will lead to a breakthrough solution. Mutation randomly changes an offspring which has resulted from a crossover operation. In the case of binary encoding, we can randomly switch a few chosen bits from 1 to 0 or from 0 to 1. Mutation can be then illustrated as follows:

Original offspring	1 1 0 1 1 1 1 0 0 0 0
Mutated offspring	1 1 0 0 1 1 1 0 1 0 0

In the case of value encoding, mutation can be performed by randomly modifying selected gene values, for example:

Original offspring	1.2	5.6	2.8	4.1	5.5
Mutated offspring	1.2	5.6	6.5	4.2	5.5

Mutation probability defines how often parts of a chromosome will be mutated. If the mutation probability is 100%, a whole chromosome is changed. If the probability is 0%, then mutation will not occur at all. Mutation therefore should not occur very often because then GAs will in fact change to a random search. In general, if the population is small or the fitness function is extremely uneven, the mutation rate has to be high (~10%). However, most GAs use a large population, mirroring the complexity of the problem, so a very small mutation rate (1% or less) is often used.

VVT Time Optimization Optimizing VVT duration is more complex than optimizing VVT cost, since we cannot separate the decision processes to optimize one VVT activity at a time. In other words, to compute time duration we must identify the critical path (and ignore other system or VVT activities). To calculate the critical path duration, we must take into account the entire VVT strategy associated with that critical path.

Optimizing the VVT time is not necessarily always the same as minimizing VVT duration. A more general approach, common in industry, is to seek a VVT strategy matching with a time goal (T_{Goal}) for each lifecycle phase. The reason is that often a certain amount of VVT time is planned as part of commonsense project management. Decreasing the VVT efforts to an absolute minimum tends to increase the likelihood of defective products. Accordingly, for a given lifecycle phase l, $l = i^0$, we seek to find a VVT strategy $\{X_{i^\circ,j}\}$, $0 \le X_{i^\circ,j} \le 1$, where $j(i^\circ) = 1,2, \ldots, n_{i^\circ}$, that will yield a critical path length closest to a time goal T_{Goal,i°. This can be described as follows:

$$\min\{\text{ABS}[T_{\text{Goal},i^\circ} - \tau(X_{i^\circ,\tilde{j};\,\tilde{j}\equiv\{\text{Critical\_Path}\}})]\} \tag{7.49}$$

Subject to

$$0 \le X_{i^\circ,\tilde{j}} \le 1 \quad \forall \tilde{j}$$
$$T_{\text{Goal},i^\circ} \ge 0$$
$$1 \le i^\circ \le z$$

(For this Book $z = 8$.) This section will describe the utilization of a GA in order to optimize the VVT time for a given system lifecycle phase.

Encoding VVT Strategy Every chromosome should contain a definition of the VVT strategy which we choose to optimize. Chromosomes consist of genes and each gene encodes a particular trait. We have translated this concept into our problem domain by associating each VVT performance level $X_{i,j}$ with a certain gene in the chromosome. Because the VVT strategy is composed of a set of decision variables $[X_{i,j}; 0 \le X_{i,j} \le 1; j(i) = 1, 2, \ldots, n_i, i = 1, 2, \ldots, z]$, it will be natural to use value encoding where each gene is represented by the value of $X_{i,j}$. For example, a given lifecycle phase 1 VVT strategy could be encoded by the following chromosome:

Gene ID	$X_{1,1}$	$X_{1,2}$	$X_{1,3}$	…	X_{1,n_1}
Gene value	0.4	0.1	0.7	…	1.0

VVT Population Size Selection of population size affects the rate of GA convergence in a significant way. Because the size of the VVT strategy (as well as the size of the chromosome) per lifecycle phase is in the 10–20 range, we found that a similar population size yielded a reasonable convergence rate for the optimization algorithm.

VVT Population Initialization On initiation of the optimization procedure, each gene within each chromosome in the current population is assigned a random value in the range $0 \le X_{i,j} \le 1$.

VVT Strategy Selection At this point, we have to ascertain the "fitness" or "value" of each chromosome within the current population. Once this information is available, we have to select individual parents in order to perform elitism, crossover and mutation operations. The following procedure is carried out on the current population for each lifecycle phase l, $l = i^0$:

1. *Critical Path.* Compute the duration of each system and VVT activity and determine the length of the critical path ($T_{i^\circ,j}$) for each chromosome/VVT strategy (j).

2. *Optimal Value.* Because this is a minimization problem, the optimal value of each chromosome is the reciprocal of the critical path time ($\text{Value}_{i^\circ,j} = 1/T_{i^\circ,j}$).
3. *Integer Value.* For simplicity of computation, multiply the optimal values by a sufficiently large number (say, 10,000) in order to shift the solution space into the integer world ($\text{Integer\_value}_{i^\circ,j} \Leftarrow 10{,}000 \times \text{Value}_{i^\circ,j}$).
4. *Sum Value.* Construct an imaginary roulette with a wheel size (S_{i°) equal to the sum of the integer values ($S_{i^\circ} = \sum_j \text{Integer\_value}_{i^\circ,j}$).
5. *Random.* Generate a random number (R_{i°) in the range (0, S_{i°).
6. *Selection.* Select the chromosome in the range of R_{i° as the next parent for the crossover process.

VVT Strategy Operators The following VVT strategy operators were used:

1. *Elitism.* As mentioned above, elitism ensures that one or more of the best VVT strategies within the current poptulation will be transmitted to the new population. The procedure to achieve this is as follows. The present population was ordered according to its optimal value ($\text{Value}_{i^\circ,j}$) and the upper 10% was copied from the current population to the future population.
2. *Crossover.* As mentioned above, the purpose of a crossover is to initiate natural selection processes. This translates into our VVT time optimization domain as follows: (1) Select VVT strategies with a random preference to their value as defined above and (2) randomly combine VVT performance levels of two existing parent VVT strategies in order to create two new child VVT strategies. The procedure to achieve this is as follows.
   - Two parent VVT strategies were extracted from the current population based on the selection method described above. Crossover operation was randomly performed with a probability of 0.9. If no crossover was performed, the first VVT strategy offspring was set equal to the first VVT strategy parent and vice versa.
   - If a crossover procedure was required, then a first VVT strategy child was constructed by copying everything before this crossover point from the first VVT strategy parent and then copying everything after the crossover point from the other VVT strategy parent. Similarly, a second VVT strategy child was created by copying everything before this point from the second parent and then copying everything after the crossover point from the first parent.
3. *Mutation.* As mentioned above, the purpose of mutation is to avoid locking the VVT strategy solution onto a local maximum, which may not be necessarily a global optimum. The procedure to achieve this is as follows. Mutation was conducted on the two VVT strategy children emanating from the crossover operation. For each VVT performance

level $X_{i,j}$ (gene) in each VVT strategy (chromosome), a new value was generated with probability of 0.01. Therefore in 1% of the new population the VVT performance level was recalculated using a random generator: $X_{i,j} \leftarrow$ Random (0,1).

Constrained Solution Sometimes, the GA must be constrained as various VVT activities must meet additional criteria. For example, performing certain VVT activities at certain levels are mandated by customer requirements or by regulatory agencies. Similarly, other VVT activities cannot be performed at an optimal level due to lack of equipment or expertise. In both cases, the optimization algorithm ignores these activities so they may be performed at predefined levels.

Single-Phase VVT Time Optimization Example We will use the VVT time and risk data defined above in a simple example where (1) the VVT horizon will only cover the first system lifecycle phase (Definition) and (2) the VVT time parameters will be the same ones that were used in the pilot project. The GA software within VVT-Tool was run seeking to find a VVT strategy which would bring the phase 1 duration to a goal of 50.0 days. Here, again, VVT activity $V_{1,4}$ and VVT activity $V_{1,10}$ were fixed (were not optimized) and the optimization procedure utilized nonlinear modeling. The GA determined an improved VVT strategy but was manually stopped after 167 genetic generations. (See Figures 7.16–7.19.)

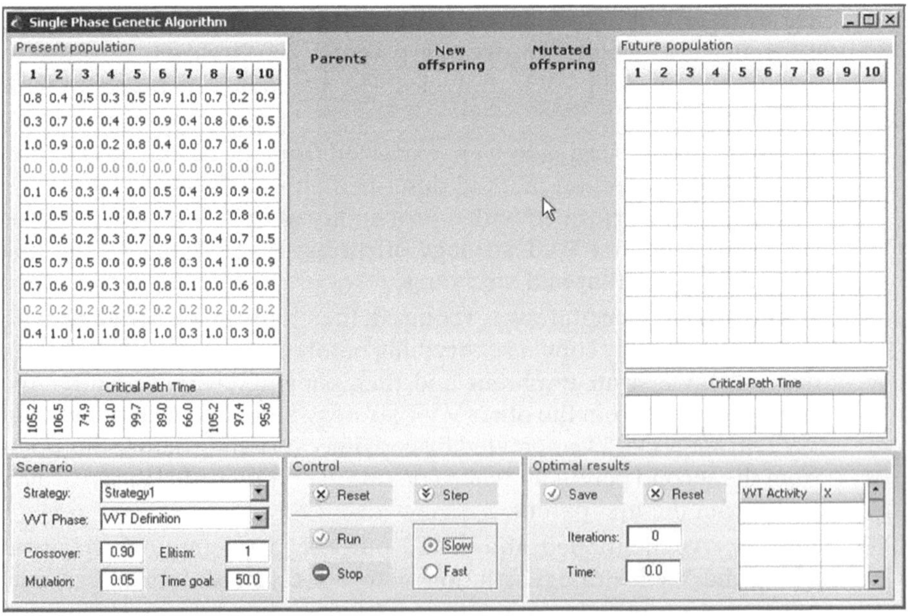

Figure 7.16 GA optimization at initialization.

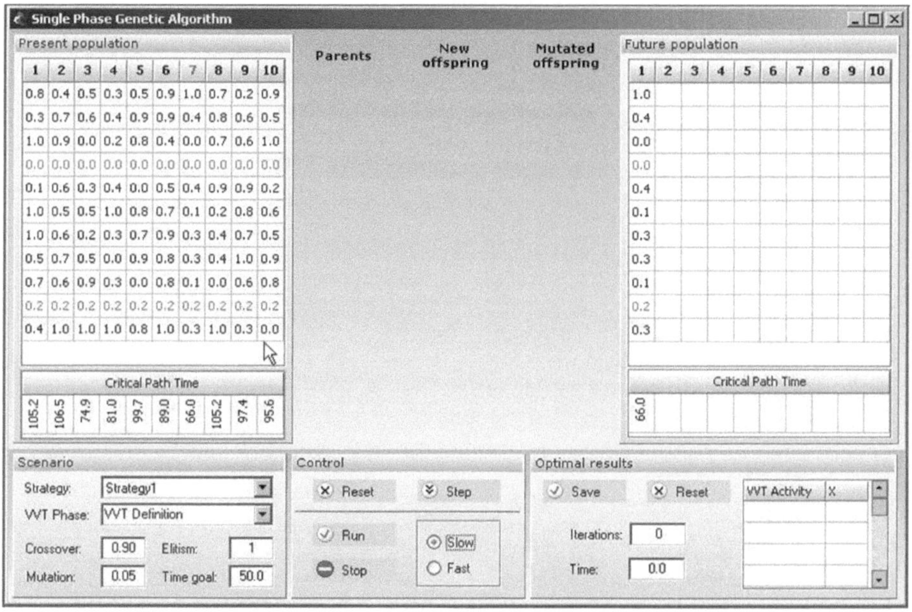

Figure 7.17 GA optimization after first elitism step.

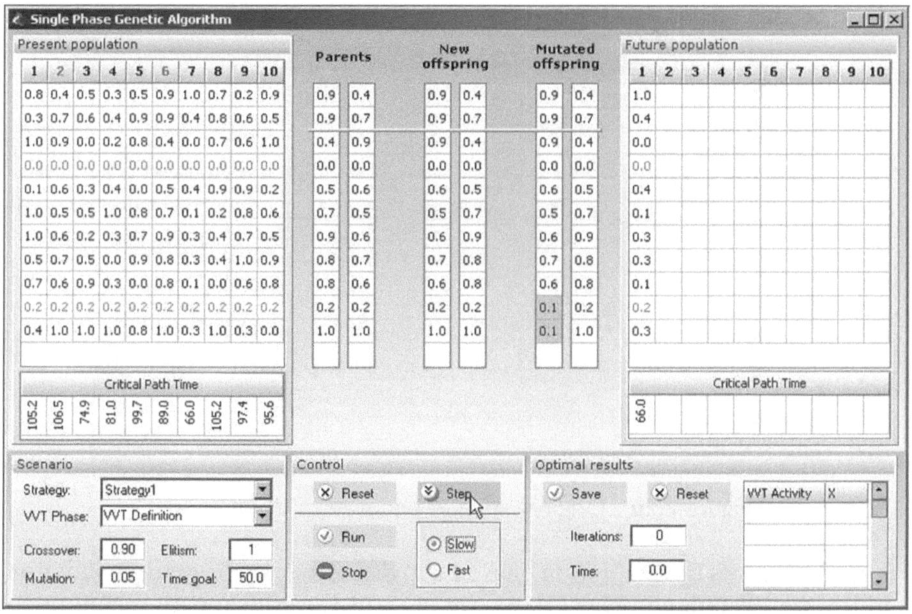

Figure 7.18 GA optimization during first crossover and mutation steps.

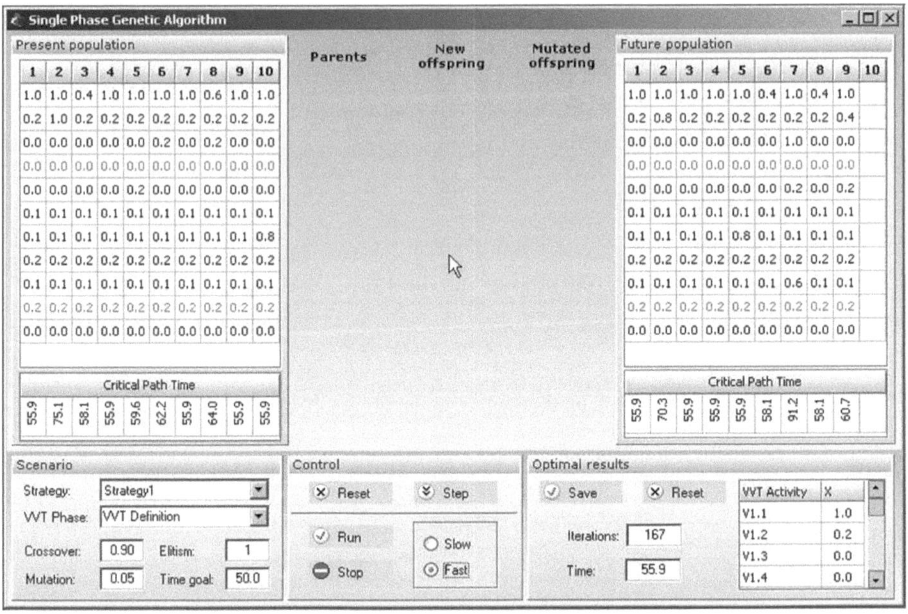

Figure 7.19 GA optimization after stabilization.

Figure 7.20 depicts the original and optimized VVT strategy. The original critical path has been changed under this optimized VVT strategy.

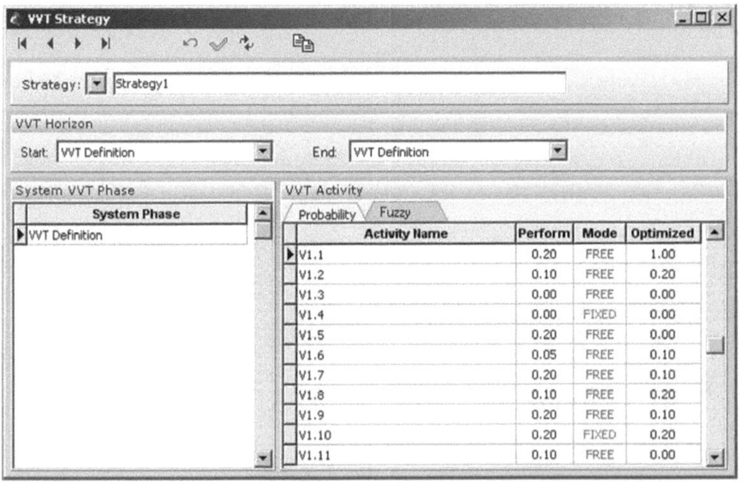

Figure 7.20 Original and GA optimized VVT strategy.

Figure 7.21 shows the Definition phase 1 activity network [Program Evaluation and Review Technique (PERT)]. As can be seen, the critical path is composed of the following ordered set of system and VVT activities: $\{S_{1,1}, V_{1,1}, V_{1,2}, S_{1,4}, V_{1,8}, V_{1,11}, \text{IMP1}\}$. The original phase 1 schedule was 63.5 days.

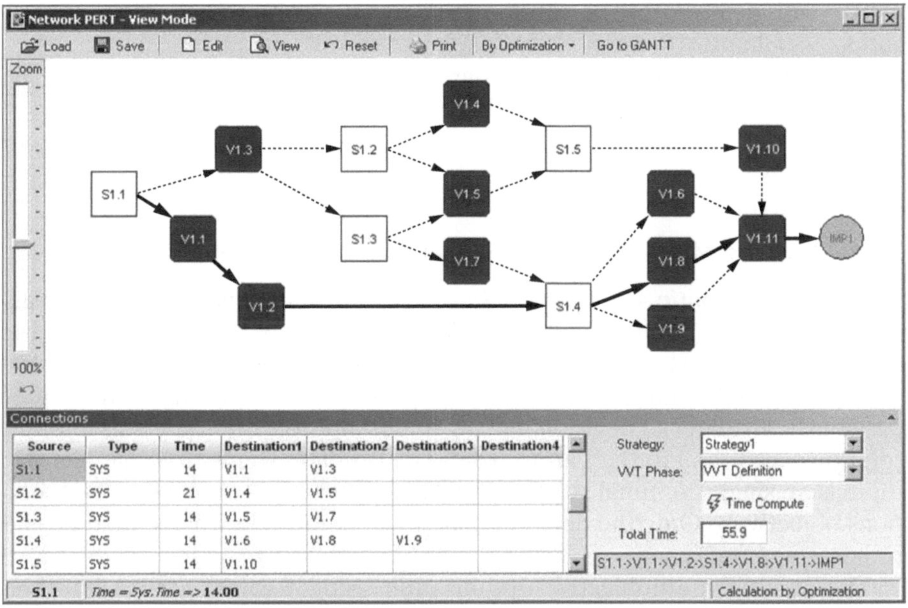

Figure 7.21 GA optimized VVT PERT.

Figure 7.22 shows this optimized VVT Gantt timetable, derived from the GA optimized VVT strategy, having a critical path duration of 55.9 days. In this case, there is a schedule optimization of 12%.

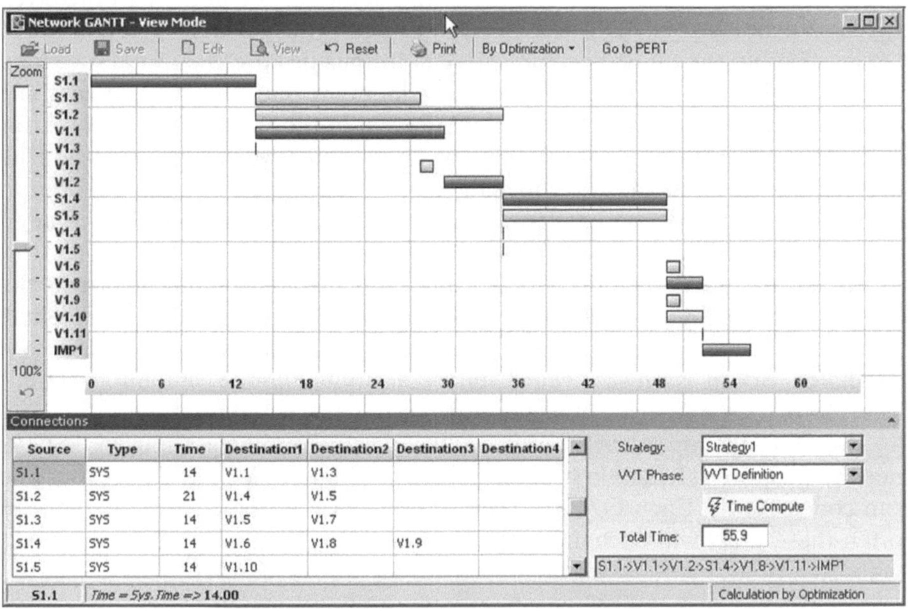

Figure 7.22 GA optimized VVT Gantt.

We will demonstrate a substantially more significant schedule optimization in Chapter 8 when the VVT horizon will encompass five development lifecycles (Definition, Design, Implementation, Integration and Qualification).

7.5.6 Genetic Multi-Domain Optimization—VVT Cost and Time

Steady-State Genetic Algorithms Since absolute system verification and validation are virtually unattainable, the VVT engineer is normally expected to optimize the VVT strategy with respect to two key parameters: (1) the amount of time available and (2) the amount of money available. As we have seen previously, virtually all decision making under multiple criteria yields multiple solutions according to the Pareto front. The problem stems from the fact that there is no single "optimal solution" and it is difficult to compare several solutions objectively. Only the decision maker can determine what should be "his or her optimum."

The area of multiobjective optimization using Evolutionary Algorithms (EAs) has been explored for some time. The first multiobjective GA implementation, called the vector evaluated genetic algorithm, was proposed by Schaffer (1985). Since then, many evolutionary algorithms for solving multiobjective optimization problems have been developed.

Chafekar et al. (2003) describe a Steady-State Genetic Algorithm (SSGA) applicable to realistic engineering design domains. This algorithm involves simultaneous optimization of multiple and sometimes conflicting objectives. In these problems, instead of a single optimum, there usually exists a set of trade-off solutions called nondominated solutions or Pareto optimal solutions. Therefore, if a solution exhibits improvement in one objective, then some other objective will suffer. No specific solution in the search space is superior to the other Pareto optimal solutions when all objectives are considered. The user is then responsible for choosing a particular solution from the Pareto optimal set. Some of the challenges faced in the application of GAs to engineering design domains are briefly discussed below.

The search space could be very complex with many limitations and the feasible (physically realizable) area of the search space could be very small. Determining the quality (fitness) of each point could involve the use of a simulator or an analysis, which takes a nonnegligible amount of time. Therefore, the number of objectives has to be minimized.

Two methods for solving limited multiobjective optimization using SSGAs may be proposed. In the first method several single-objective GAs should be run concurrently. Each GA will optimize one of the objectives and every so often these GAs will exchange information about their respective objectives with each other.

In the second method, a single GA will run multiple objectives in a sequence, switching at certain intervals between objectives. The main idea in the second method is to use a single GA that optimizes multiple objectives in a sequential order. Every objective is optimized for a certain number of iterations; then a switch will be made and the next objective will be optimized. The population is not changed when objectives are switched. This process continues until all the objectives are reached or the maximum number of iterations is completed. An illustration of an SSGA convergence toward two goals (quality cost and time) is shown in Figure 7.23. Here, the GA performs four iterations moving toward the cost goal. This is followed by the next four iterations moving towards the time goal and vice versa. Often, moving toward one goal tends to worsen the solution of the other goals, but after a certain number of genetic cycles an optimal multiobjective solution will emerge.

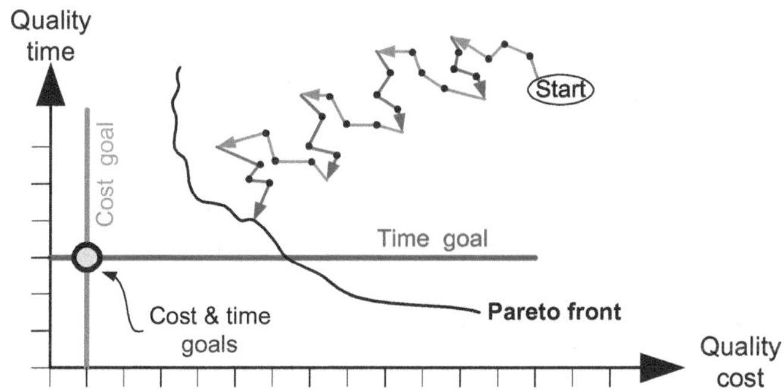

Figure 7.23 Illustration of SSGA convergence toward cost and time goals.

Optimizing VVT Cost and Time: Example We will use the VVT data presented in the previous sections in order to demonstrate a multigoal optimization. Here, VVT cost includes the expected total VVT cost within the Definition phase. The VVT time includes the total length of the critical path along the VVT activity network within the selected VVT horizon.

Figure 7.24 shows a multiobjective optimization using a SSGA method. It attempts to optimize the VVT cost toward its cost goal of 30 CU for 5 iterations and then attempts to optimize the VVT time toward its time goal of 55 days for the next 5 iterations and vice versa. Here, again, VVT activity $V_{1,4}$ and VVT activity $V_{1,10}$ are fixed and the optimization procedure utilizes nonlinear modeling.

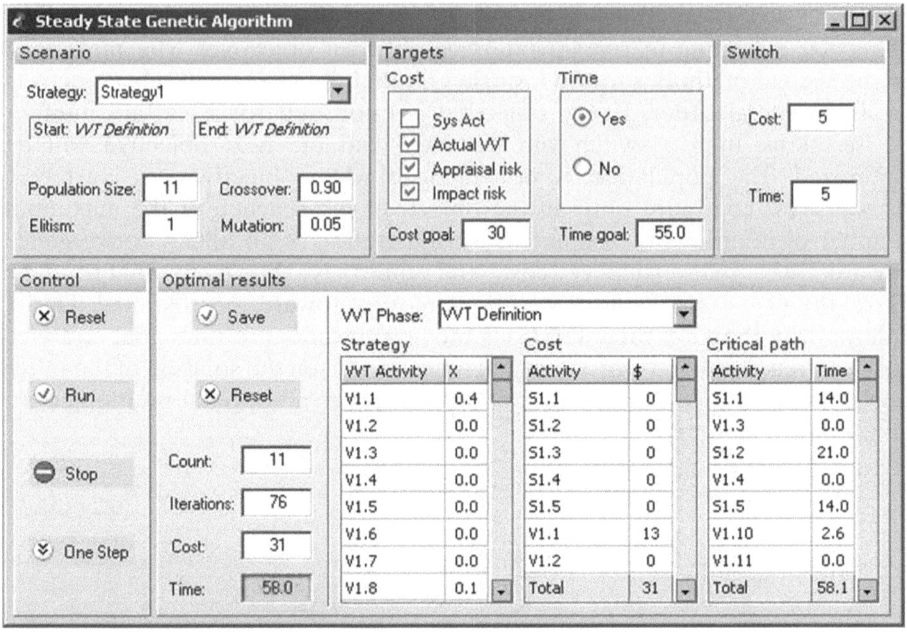

Figure 7.24 Multiobjective optimizer: SSGA.

The two goals are nearly reached after 76 iterations. Figure 7.25 shows the original and the optimized VVT strategies for the Definition phase.

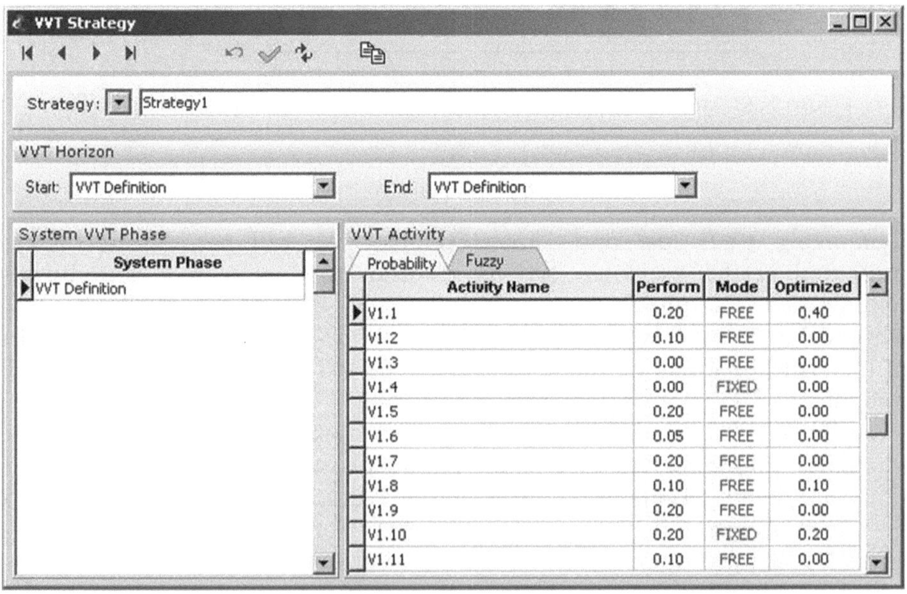

Figure 7.25 Original and SSGA optimized VVT strategy.

Figure 7.26 shows the result of the optimized cost simulation. As can be seen, after 3000 iterations of the stochastic simulator, the overall expected optimized cost is 31 CU.

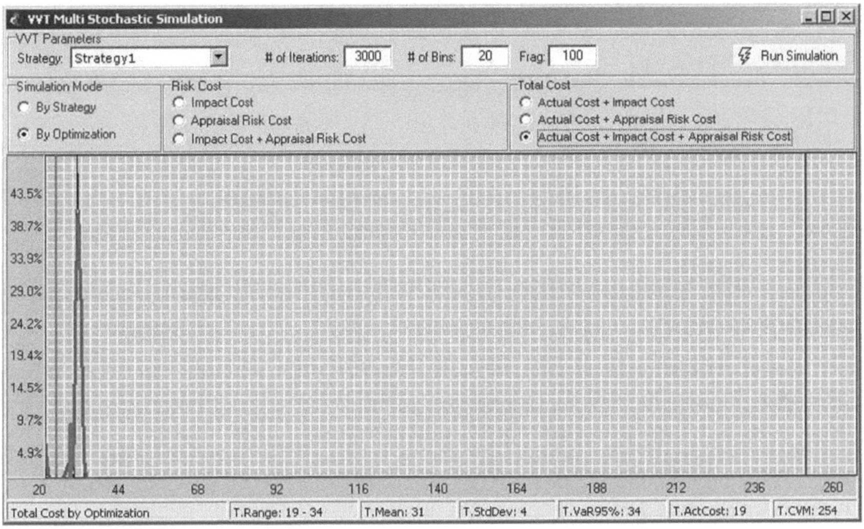

Figure 7.26 VVT cost simulation under optimized conditions.

Figure 7.27 shows the Definition phase time network (PERT). As can be seen, the critical path is composed of the following ordered set of system and VVT activities: $\{S_{1,1}, V_{1,3}, S_{1,2}, V_{1,4}, S_{1,5}, V_{1,10}, V_{1,11}, \mathrm{IMP}_1\}$.

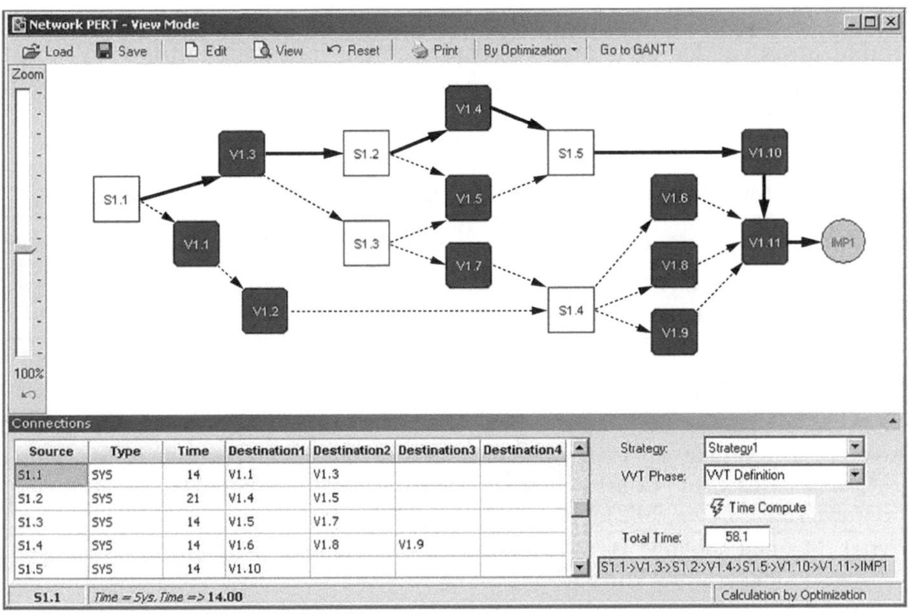

Figure 7.27 SSGA optimized schedule (PERT) chart.

Figure 7.28 shows the optimized Definition phase schedule (Gantt) chart. As can be seen, the overall expected optimized time is 58.1 days.

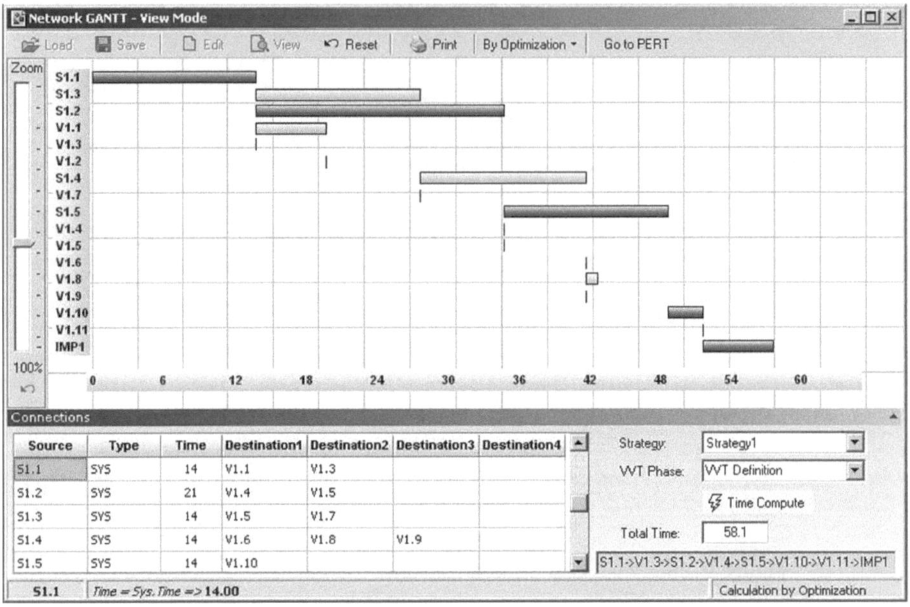

Figure 7.28 SSGA optimized schedule (Gantt) chart.

Table 7.14 shows the original and optimized cost and time. As we can see, the optimization yielded 38% and 9% improvement in the mean total VVT cost and phase duration. We will observe a substantial increase in optimization results in Chapter 8, which covers the entire system development stage.

Table 7.14 Multi-Objective Initial Values, Goals and Optimized Values

	Original	Optimized	Change	Percent Change
Mean total VVT cost (CU)	50	31	19.0	38
Total phase duration (days)	63.5	58.0	5.5	9

7.6 REFERENCES

Aoieong, T. R., Tang, L. S., and Ahmed, M. S., A Process Approach in Measuring Quality Costs of Construction Projects: Model Development, *Construction Management Econ.*, 20(2), 179–192, March 2002.

Avriel, M., and Golany, B., *Mathematical Programming for Industrial Engineers*, Marcel Dekker; New York, 1996.

Barad, M., Using Break-Even Quality Level for Selecting Acceptance Sampling Plans Given a Prior Distribution, *Int. J. Production Res.*, 24(1), 65–72, 1986.

Barad, M., and Engel, A., Optimizing VVT Strategies—A Decomposition Approach, *J. Operation Res. Soc.*, 57(8), 965–974, August 2006.

Boehm, W. B., Software Defects Reduction Top 10 List, *IEEE Comput.*, 34(1), 135–137, January 2001.

Boehm, W. B., Horowitz, E., Madachy, R., and Reifer, D., *Software Cost Estimation with COCOMO II*, Prentice-Hall, Englewood Cliffs, NJ, 2000.

Burns, C. R., Quality Costing Used as a Tool for the Reduction in the Machine Tool Industry, *Qual. Assurance*, 2, 25–32, 1976.

Chafekar, D., Xuan, J., and Rasheed, K., Constrained Multi-Objective Optimization Using Steady State Genetic Algorithms, paper presented at the Genetic and Evolutionary Computation Conference (GECCO-2003), Chicago IL, 12–16, July 2003.

Chiadamrong, N., The Development of an Economic Quality Cost Model, *Total Qual. Manag. Bus. Excellence*, 14(2), 999–1014, November 2003.

Clemen, R. T., and Winkler, R. L., Combining Probability Distributions from Experts in Risk Analysis, *Risk Anal.*, 19, 187–204, 1999.

Cooke, M. R., *Experts in Uncertainty: Opinion and Subjective Probability in Science*, Oxford University Press, Oxford, 1991.

Deb, K., *Multi-Objective Optimization Using Evolutionary Algorithms*, Wiley, Hoboken, NJ, 2001.

De Ruyter, S. A., Cardew-Hall, J. M., and Hodgson, D. P., Estimating Quality Costs in an Automotive Stamping Plant through the Use of Simulation, *Int. J. Prod. Res.*, 40(15), 3835–3848, October 2002.

Engel, A., and Shachar, S., Measuring and Optimizing Systems' Quality Costs and Project Duration, *Syst. Eng.*, 9(3), 259–280, 2006.

Ghiassi, M., Dual Programming Approach to Software Testing, *Software Qual. J.*, 45(3), 45–58, 1994.

Giakatis, G., Enkawa, T., and Washitani, K., Hidden Quality Costs and the Distinction between Quality Cost and Quality Loss, *Total Qual. Manag. Bus. Excellence*, 12(2), 179–190, March 2001.

Goldberg, E. D., *Genetic Algorithms in Search, Optimization and Machine Learning*, Addison-Wesley Professional, Reading, MA, 1989.

Gotshall, S., and Rylander, B., Optimal Population Size and the Genetic Algorithm, paper presented at the 2nd WSEAS International Conference on Soft Computing, Optimization, Simulation and Manufacturing Systems 2002 (SOSM 2002), Cancun, Mexico, 12–16, May 2002.

Holland, H. J., *Adaptation in Natural and Artificial Systems: An Introductory Analysis with Applications to Biology, Control, and Artificial Intelligence*, Bradford Book, 1992.

Houston, D., and Keats, J. B., Cost of Software Quality: A Means of Promoting Software Process Improvement, *Qual. Eng.*, 10(3), 563–574, 1998.

Huber, J., Efficiency and Effectiveness Measures to Help Guide the Business of Software Testing, paper presented at the Int. Conf. Software Management Appl. Software Measure, San Jose, CA, February 1999.

Hwang, H. G., and Aspinwall, M. E., Quality Cost Models and Their Application: A Review, *Total Qual. Manag. Bus. Excellence*, 7(3), 267–282, June 1996.

Hwang, H. G., and Aspinwall, M. E., The Development of a Quality Cost Model in a Telecommunications Company, *Total Qual. Manag. Bus. Excellence*, 10(7), 949–965, September 1999.

Hwang, L. C., Multiple Objective Decision Making, Methods and Applications: A State of the Art Survey, *Lecture Notes in Economics and Mathematical Systems*, No. 164, Springer-Verlag, Berlin, 1979.

Ittner, C. D., Exploratory Evidence on the Behavior of Quality Costs, *Oper. Res.*, 44(1), 114–130, 1996.

Jones, C., *Applied Software Measurement*, McGraw-Hill, New York, 1996.

Jorion, P., *Value-at-Risk*, McGraw Hill, New York, 2001.

Juran, J., *Quality Control Handbook*, McGraw-Hill, New York, 1952.

Keeney, L. R., and von Winterfeld, D., Eliciting Probabilities from Experts in Complex Technical Problems, *IEEE Trans. Eng. Manag.*, 38(3), 191–201, 1991.

Kim, M. J., Porter, A., and Rothermel, G., An Empirical Study of Regression Test Application Frequency, paper presented at the 22nd Int. Conf. Software Eng., ICSE 2000, June 4–11, 2000, Toronto, Canada, ACM, New York, 2000, pp. 126–135.

Koza R. J., *Genetic Programming: On the Programming of Computers by Means of Natural Selection (Complex Adaptive Systems)*, Bradford Book, 1992.

Kumar, K., Shah, R., and Fitzroy, T. P., A Review of Quality Cost Surveys, *Total Qual. Manag. Bus. Excellence*, 9(6), 479–486, August 1998.

Lean Aerospace Initiative (LAI), available: http://web.mit.edu/lean/, 2006.

Loveridge, D., Experts and Foresight: Review and Experience, Paper 02-09, PRES, University of Manchester, Manchester, UK, June 2002.

Millard, R. L., Value Stream Analysis and Mapping for Product Development, Master's thesis in Aeronautics and Astronautics, Massachusetts Institute of Technology, Cambridge, MA, June 2001.

Morgan M. G., and Henrion, M. (Eds.), *Uncertainty: A Guide to Dealing with Uncertainty in Quantitative Risk and Policy Analysis*, Cambridge University Press, Cambridge, MA, 1992.

Moyers, D. R., and Gilmore, L. H., Product Conformance In the Steel Foundry Jobbing Shop, *Qual. Prog.*, 12, 17–19, May 1979.

Obitko, M., and Slavik, P., Visualization of Genetic Algorithms in a Learning Environment, Spring Conference on Computer Graphics, SCCG'99, pp. 101–106, Bratislava Comenius University, 1999.

Oppenheim, W. B., Lean Product Development Flow, *Syst. Eng.* 7(4), 352–376, 2004.

Osyczka, A., Multicriterion Optimization for Engineering Design, in *Design Optimization*, J. S. Gero (Ed.), Academic, New York, 1985, pp. 193–227.

Rechenberg, I., *Evolutionsstrategie*, Frommann Holzboog, 1994.

Robert, P. C., and Casella, G., *Monte Carlo Statistical Methods*, 2nd ed., Springer, New York, 2005.

Ross, P. J., *Taguchi Techniques for Quality Engineering: Loss Function, Orthogonal Experiments, Parameter and Tolerance Design*, McGraw-Hill, New York, 1996.

Schaffer, D. J., Multi-objective optimization with vector evaluated genetic algorithms. In *Proceedings of the International Conference on Genetic Algorithms and Their Applications*, pp. 93–100, J. J. Grefenstette, Ed., Pittsburgh, PA, July 24–26, 1985.

Simga-Mugan, C., and Erel, E., Distribution of Quality Costs: Evidence from an Aeronautical Firm, *Total Qual. Manag. Bus. Excellence*, 11(2), 227–234, March 2000.

Sorqvist, L., On Poor Quality Costing, Ph.D. Thesis, Royal Institute of Technology, Department of Production Engineering, Stockholm, 1998.

Taguchi, G., Elsayed, A. E., and Hsiang, C. T., *Quality Engineering in Production Systems*, McGraw-Hill, New York, 1988.

Tang, L. S., Aoieong, T. R., and Ahmed, M. S., The Use of Process Cost Model (PCM) for Measuring Quality Costs of Construction Projects: Model Testing, *Constr. Manag. Econ.*, 22(3), 263–275, March 2004.

Triantaphyllou, E., *Multi-Criteria Decision Making Methods: A Comparative Study (Applied Optimization, Vol. 44)*, Kluwer Academic, 2000.

Wheelright, S. C., and Hayes, R. H., Competing through Manufacturing, *Harvard Bus. Rev.*, 99–109, January/February 1985.

Womack, P. J., and Jones, T. D., *Lean Thinking: Banish Waste and Create Wealth in Your Corporation*, Simon & Schuster, New York, 1996.

Vose, D., Correct Way of Incorporating Differences in Expert Opinion, Decisioneering, Inc., available: http://www.crystalball.com/support/risktips/risktip-4.html, 2006.

Vose, D., *Risk Analysis: A Quantitative Guide*, 3rd ed., Wiley, Hoboken, NJ, 2008.

Chapter 8

Methodology Validation and Examples

8.1 METHODOLOGY VALIDATION USING A PILOT PROJECT

In this chapter we describe the validation of the methodology for estimating VVT cost, time and risks. Methodology validation is illustrated by means of an industrial pilot project development conducted at the Lahav division of the Israel Aerospace Industries (IAI) during the timeframe 2003–2004. Figure 8.1 depicts this process. The IAI/Lahav typical avionics upgrade Anchor system and VVT data were created from (1) records of past projects, (2) data gathered during the pilot project itself and (3) estimates made by domain experts by means of a Delphi process.

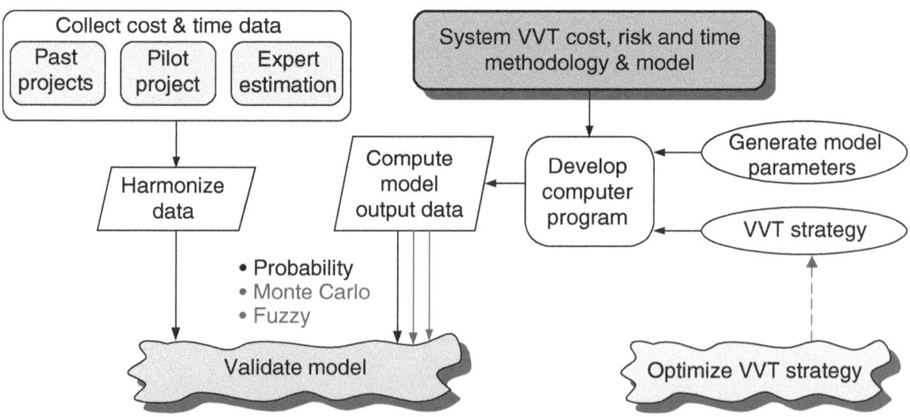

Figure 8.1 VVT methodology validation and VVT strategy optimization.

Verification, Validation, and Testing of Engineered Systems, Avner Engel

However, the VVT cost, time and risk model database was created by relevant VVT data obtained from current literature and domain experts. Additionally, the VVT strategy representing the one used during the pilot project was employed. The VVT estimation model embodied in the VVT-Tool was executed. Then, the computed VVT model output was compared with the anchor VVT data and was found to be quite similar. The above procedure was carried out using an initially probabilistic paradigm and then a fuzzy logic paradigm. This chapter describes the validation of the VVT cost and time models as well as the validation of the fuzzy VVT cost model.

8.1.1 VVT Cost Model Validation

Anchor VVT Cost Results The overall anchor cost representing the pilot project was 1227 CU. Table 8.1 and Figure 8.2 show this cost distribution in the project development phases.

TABLE 8.1 Project Development Distribution at IAI/Lahav

Cost Categories	Definition	Design	Implementation	Integration	Qualification	Total
System cost (%)	9.3	20.8	28.5	23.4	17.9	100.0
System cost (CU)	114.2	255.6	350.1	287.2	219.9	1227.0

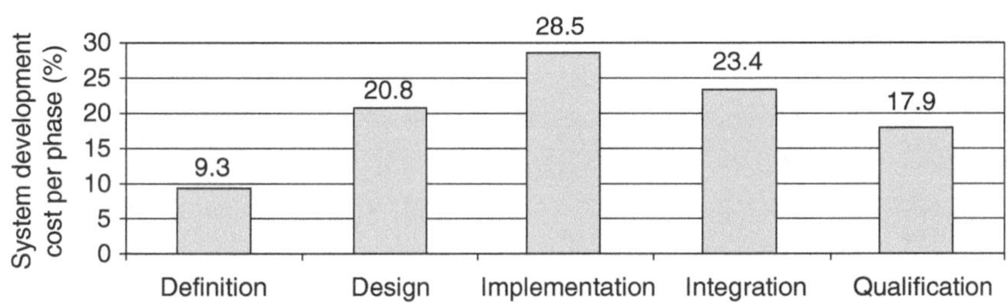

Figure 8.2 Project development cost distribution at IAI/Lahav.

Table 8.2 and Figure 8.3 show the quality cost distribution during the pilot project. It can be seen that over 60% of the development cost has been expended on quality cost. In addition, the cost of performing VVT was twice as high as the cost of reworking. One explanation for these dual findings is the relative limited quality investment during the project's Definition phase.

TABLE 8.2 Quality Cost Distribution by Phases at IAI/Lahav

	Definition	Design	Implementation	Integration	Qualification	Total
Failure cost [%]	0.7	3.1	6.5	6.7	4.3	21.3
VVT cost [%]	1.1	4.0	7.6	14.3	13.0	40.0
Quality cost [%]	1.8	7.1	14.1	21.0	17.3	61.3
Quality cost [CU]	21.8	86.9	173.5	257.4	212.6	752.2

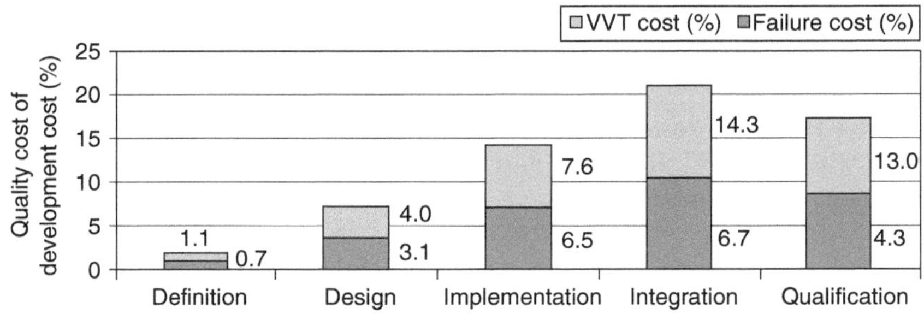

Figure 8.3 Quality cost distribution at IAI/Lahav.

Model Estimated Cost Data The VVT-Tool software package was used to simulate the nondeterministic nature of the problem using a Monte Carlo approach. A single random Monte Carlo simulation result, covering the five development phases of the pilot project, is shown in Figure 8.4. This VVT cost estimation represents a single stochastic datum.

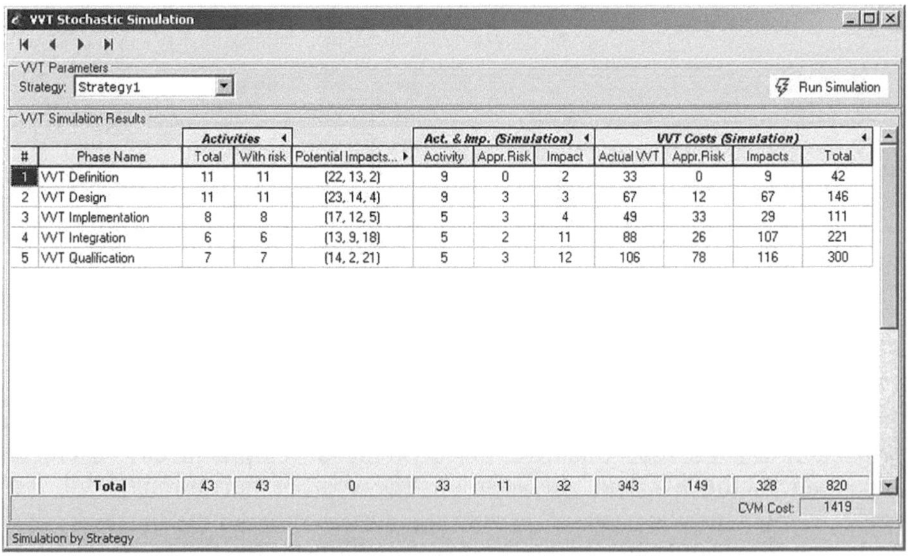

Figure 8.4 Single run of Monte Carlo cost simulator.

The figure identifies the number of VVT activities that were performed within each phase as well as the number of activities with risk [namely their corresponding Activity Performance Level (APL) $X_{i,j}$, $X_{i,j} < 1$], the number of performed VVT activities and their associated stochastic appraisal risks and the number of impacts during this single run. Finally, all the cost components incurred during each phase of this single stochastic run are depicted.

Figure 8.5 depicts the total quality cost density distribution (histogram) derived from modeling the pilot project development quality data. This result was obtained after performing 3000 Monte Carlo iterations within the VVT-Tool. The X axis represents the cost and the Y axis represents the percentage of iterations associated with each of 60 histogram bins. The vertical line on the left-hand side of the figure represents the fixed VVT cost of 343 CU (i.e., Cost Units) associated with the pilot project VVT strategy. Similarly, the vertical line on the right-hand side represents the Canonical VVT Model (CVM) cost of 1419 CU. Note the large variance of the expected total cost associated with the pilot project VVT strategy and, most ominously, the right-hand side "tail of catastrophic risk" representing costly outliers which may approach or even exceed (albeit very rarely) the CVM cost.

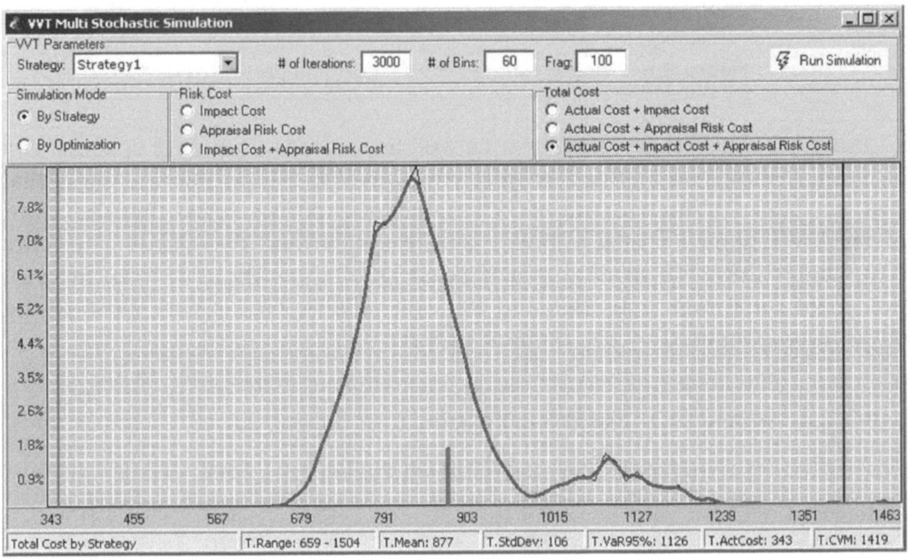

Figure 8.5 Cost distribution after 3000 iterations of Monte Carlo simulator.

The overall cost estimations obtained during the Monte Carlo simulation as well as the cost results obtained via direct probabilistic calculations are shown in Table 8.3 and Figure 8.6. As can be seen, the VVT activity cost (343 CU) is significantly less than the CVM cost (1419 CU). Likewise the total quality cost (877 CU) is also appreciably less than the CVM cost.

TABLE 8.3 Model Estimated VVT Cost Categories

	Simulation		Probability	
Cost Categories	CU	Percent of Total	CU	Percent of Total
VVT activity cost	343	39.1	343	39.1
Mean VVT appraisal cost	62	7.1	62	7.1
Mean VVT impact cost	472	53.8	472	53.8
Mean total VVT cost	877	100.0	877	100.0
CVM	1419	161.8	1419	161.8
Standard deviation (SD)	106	12.1		
VaR95%	1126	128.4		

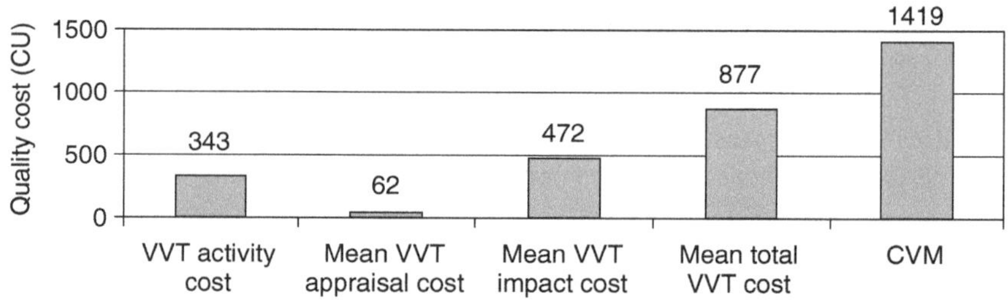

Figure 8.6 Model estimated VVT cost categories (CU).

Table 8.4 shows the system cost as well as the overall quality cost. As can be seen, the overall planned project cost was 700 CU. However, the actual cost was 1227 CU, an overrun of 75%. Domain experts suggest that that third of the VVT impact cost was borne by the system and two-thirds as quality cost. Under the VVT strategy practiced at IAI/Lahav, about 41% of the cost was spent on system activities and 59% on quality. This is very close to the 40/60 ratio estimated by the domain experts.

TABLE 8.4 System Level and Overall Quality Cost Levels

	Planned	Actual	Appraisal	Impact	Total	Percent of Total
System cost (CU)	400	350		157	507	41.3
Quality cost (CU)	300	343	62	315	720	58.7
Total (CU)	700	693	62	472	1227	100.0

Table 8.5 and Figure 8.7 show the VVT cost distribution as estimated by the VVT cost model. One can clearly observe the near-linear increase in the VVT activity phase cost. In addition, we can observe that very high VVT impact is incurred during the Integration and Qualification phases, when the system is put together and many untested functionalities and interfaces show

up as defects. We can speculate that the cost overruns of the project, as well as the high VVT impact cost, partially stem from the limited VVT activities (and hence cost) incurred during the Definition and Design phases.

TABLE 8.5 Estimated VVT Costs per Development Phases (CU)

Cost Categories	Definition	Design	Implementation	Integration	Qualification
VVT activity cost	33	67	49	88	106
Mean appraisal cost	2	12	11	12	25
Mean impact cost	15	34	40	202	181
Mean total cost	50	113	101	301	312
CVM cost	254	455	224	169	317
SD	5	21	6	17	57
VaR95%	53	153	114	327	479

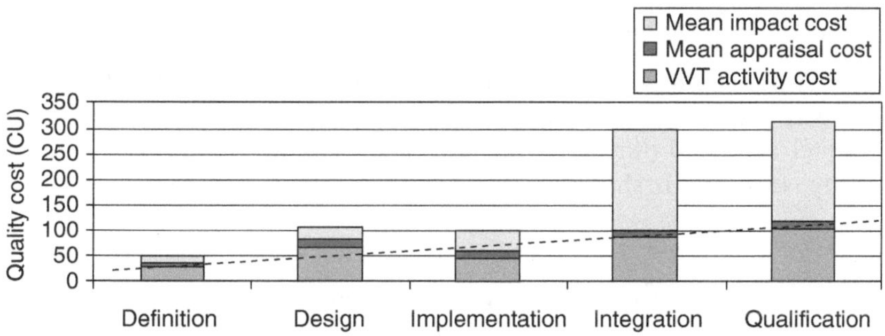

Figure 8.7 Estimated VVT costs per development phases.

Validating VVT Cost Model Table 8.6 and Figure 8.8 show comparisons between costs estimated by the VVT model and the anchor costs.

TABLE 8.6 Estimated VVT Model Cost Versus Anchor Cost (CU)

Cost Categories	Definition	Design	Implementation	Integration	Qualification	Total
VVT activity	33	67	49	88	106	343
VVT appraisal	2	12	11	12	25	62
2/3 VVT impact	10	23	27	135	121	315
Total model cost	45	102	87	235	252	720
Total anchor cost	22	87	173	257	213	752
Error (CU)	23	15	−87	−23	39	−32
Error (%)	106.3	17.0	−50.0	−8.8	18.4	−4.3

In some phases, the two values are more apart and in others they are much closer. However, the total cost difference estimated by the VVT model and measured by the experts is –4.3% and the correlation coefficient between the two sets is 0.9844.

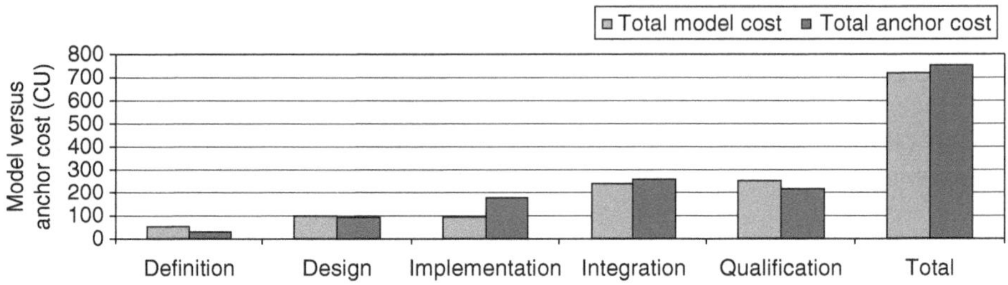

Figure 8.8 Estimated VVT model cost versus anchor cost (CU).

8.1.2 VVT Time Model Validation

Anchor System and VVT Time Results The duration of the development project was planned for one and a half years, or 554 days. This target schedule was in fact achieved during the pilot project. The anchor system/VVT times and early start data is shown in Table 8.7 and Figure 8.9.

TABLE 8.7 System and VVT Time Duration and Phase Overlap

Phase	Duration (days)	Early Start (%)
Definition	110	0
Design	110	30
Implementation	210	40
Integration	140	10
Qualification	110	20

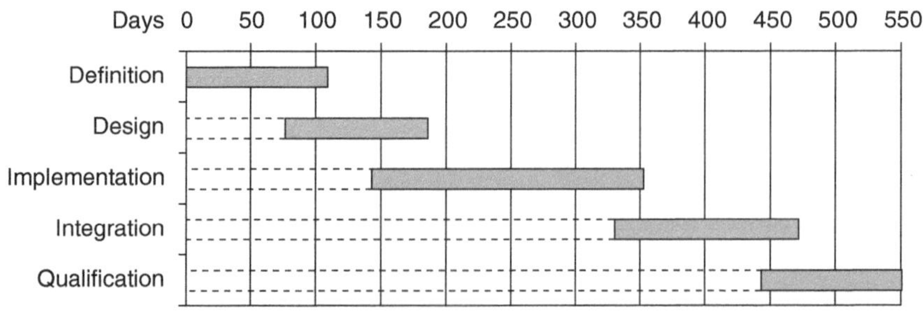

Figure 8.9 System and VVT time duration (days) and phase overlap.

Model Estimated Time Data Figures 8.10–8.19 illustrate the VVT-Tool model computation of the Program Evaluation and Review Technique (PERT) and Gantt charts for all phases of the pilot projects.

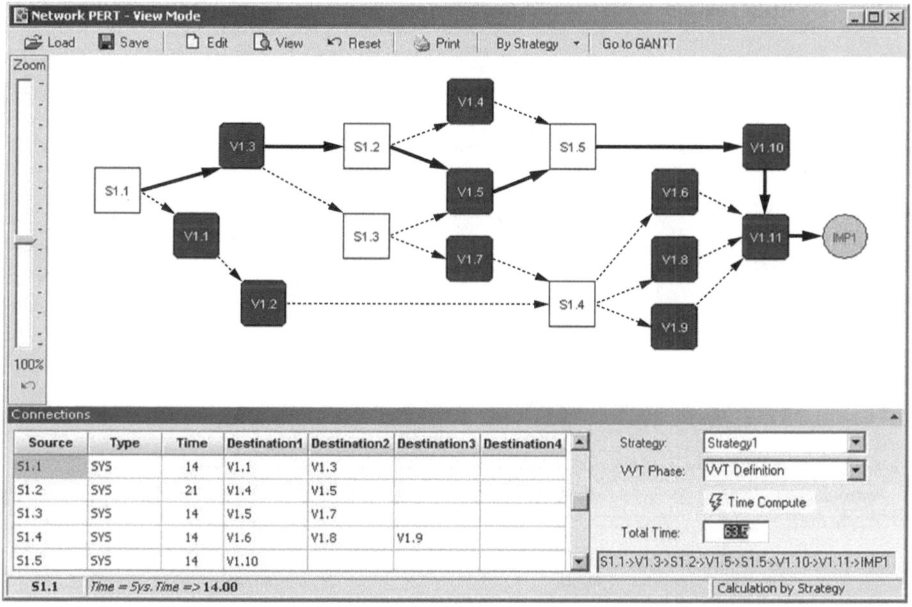

Figure 8.10 Estimated model: Definition phase PERT.

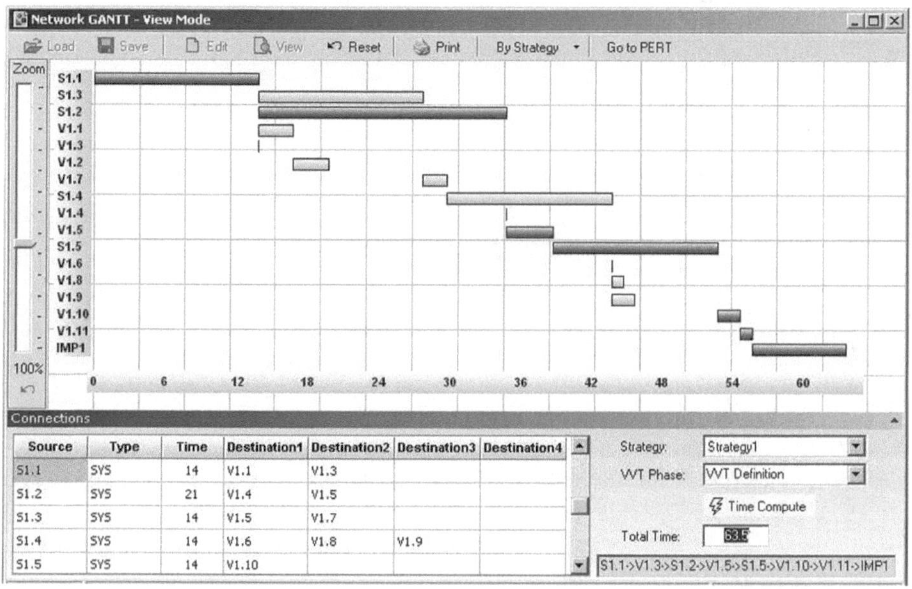

Figure 8.11 Estimated model: Definition phase Gantt.

Each PERT chart shows the network of the system activities as well as the VVT activities. It also identifies the critical path within the given development phase. The critical path is dependent upon the VVT strategy, as the duration of each VVT activity as well as the extra duration emanate from each appraisal risk and impact risk.

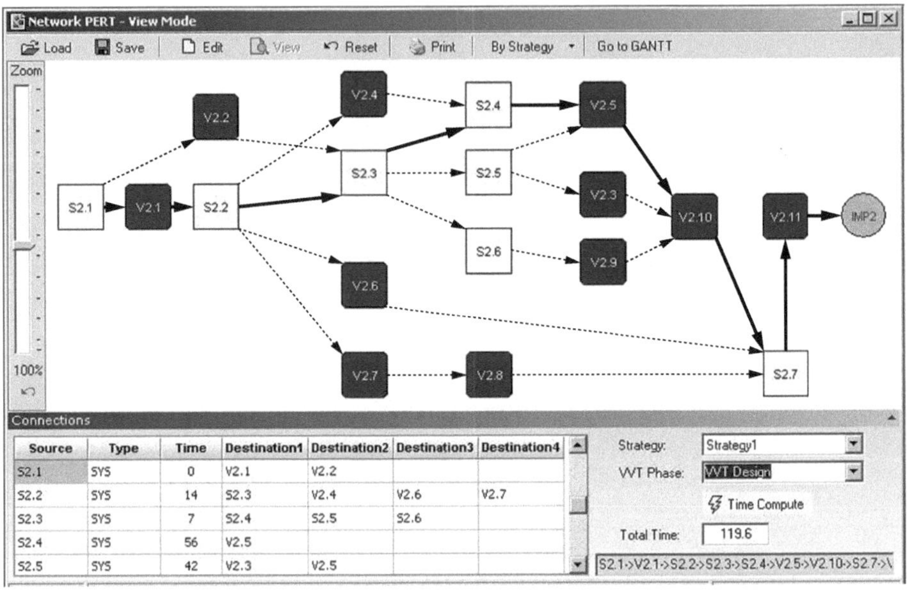

Figure 8.12 Estimated model: Design phase PERT.

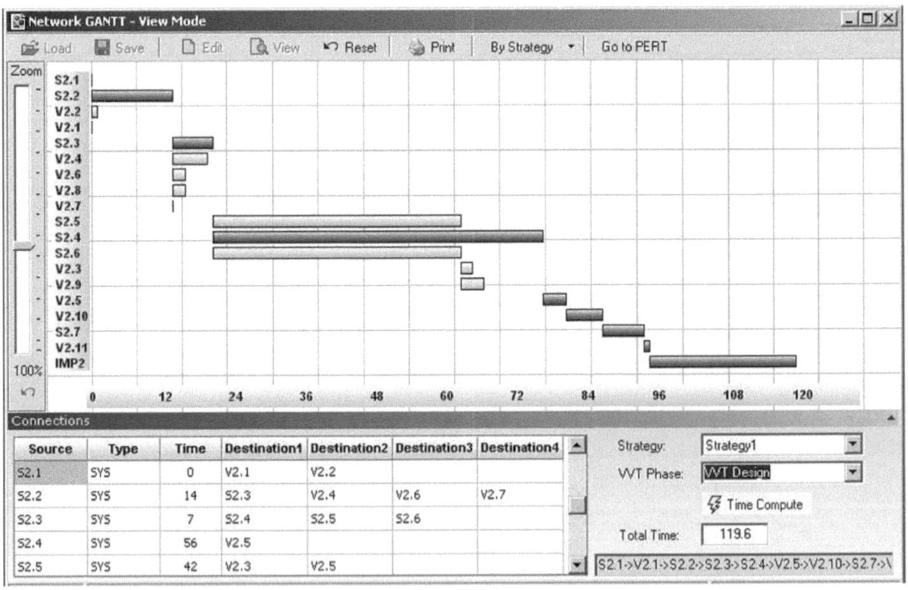

Figure 8.13 Estimated model: Design phase Gantt.

This is a function, among others, of the corresponding APL ($X_{i,j}$). Each Gantt chart shows the duration in days of each system activity as well as each VVT activity. Time calculations take into account the actual VVT time as well as the expected appraisal time and impact time. In addition, each chart identifies the critical path within the given development phase.

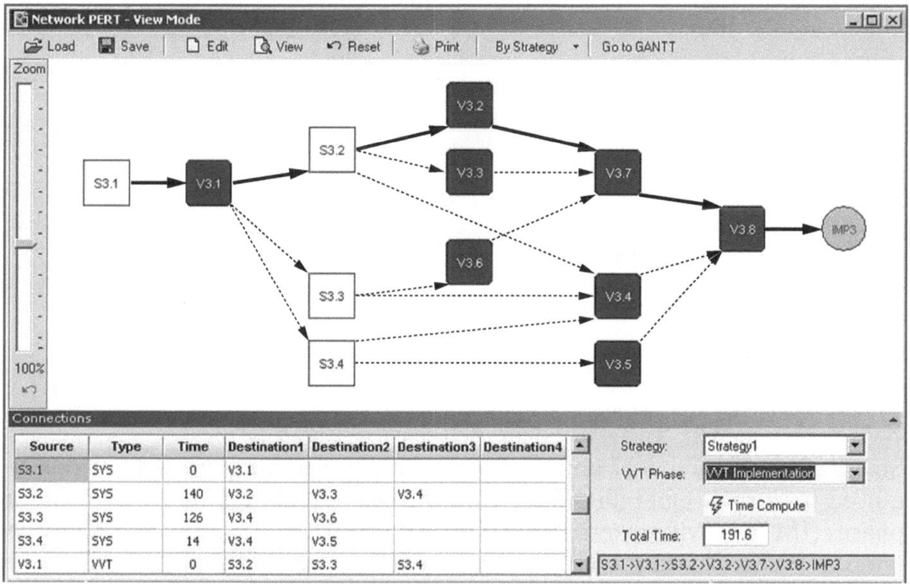

Figure 8.14 Estimated model: Implementation phase PERT.

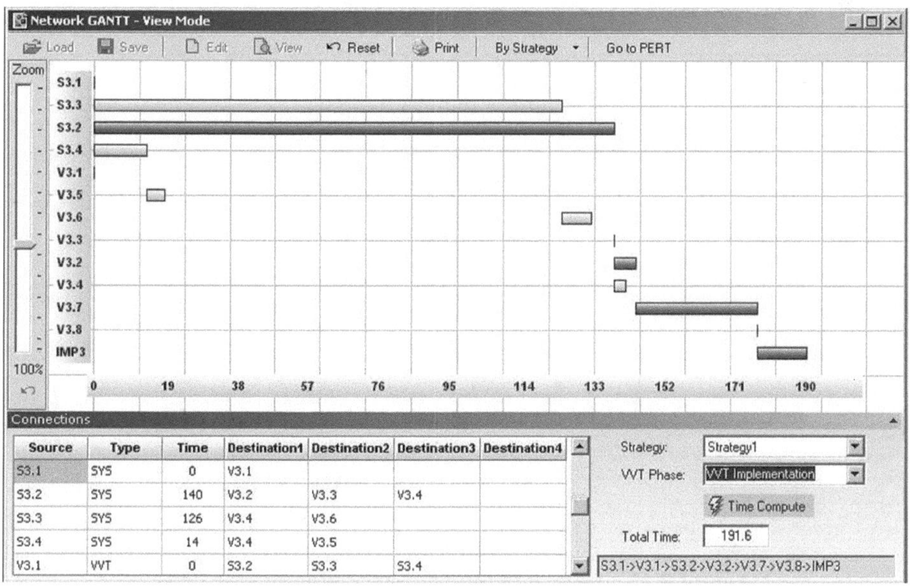

Figure 8.15 Estimated model: Implementation phase Gantt.

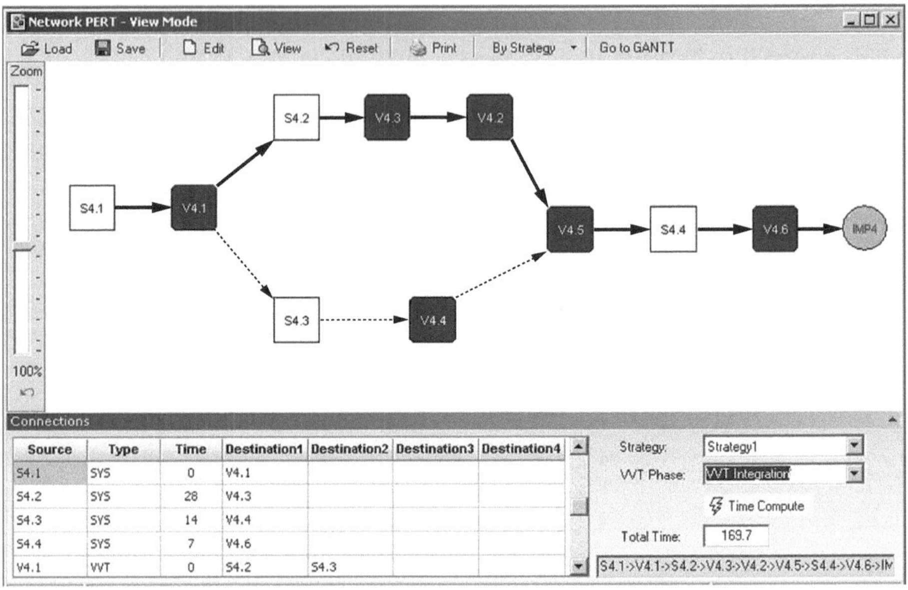

Source	Type	Time	Destination1	Destination2	Destination3	Destination4
S4.1	SYS	0	V4.1			
S4.2	SYS	28	V4.3			
S4.3	SYS	14	V4.4			
S4.4	SYS	7	V4.6			
V4.1	VVT	0	S4.2	S4.3		

Figure 8.16 Estimated model: Integration phase PERT.

Significantly, Figures 8.17 and 8.19 depict the exasperating impact periods required to correct emerging defects during the Integration and Qualification phases (IMP4 + IMP5 = 69.4 + 27.9 = 97.3 days). These defects are directly related to the inadequate VVT strategy carried out during the IAI/Lahav pilot project.

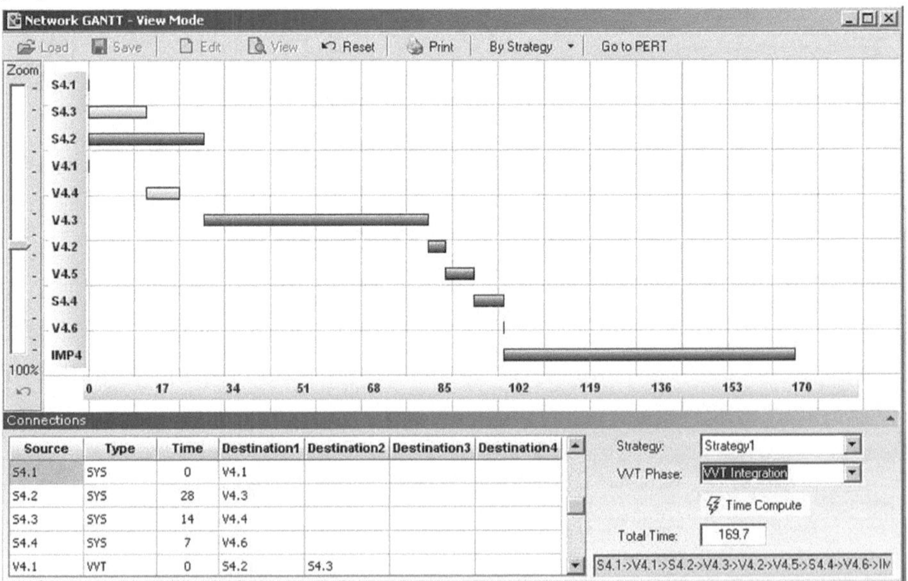

Source	Type	Time	Destination1	Destination2	Destination3	Destination4
S4.1	SYS	0	V4.1			
S4.2	SYS	28	V4.3			
S4.3	SYS	14	V4.4			
S4.4	SYS	7	V4.6			
V4.1	VVT	0	S4.2	S4.3		

Figure 8.17 Estimated model: Integration phase Gantt.

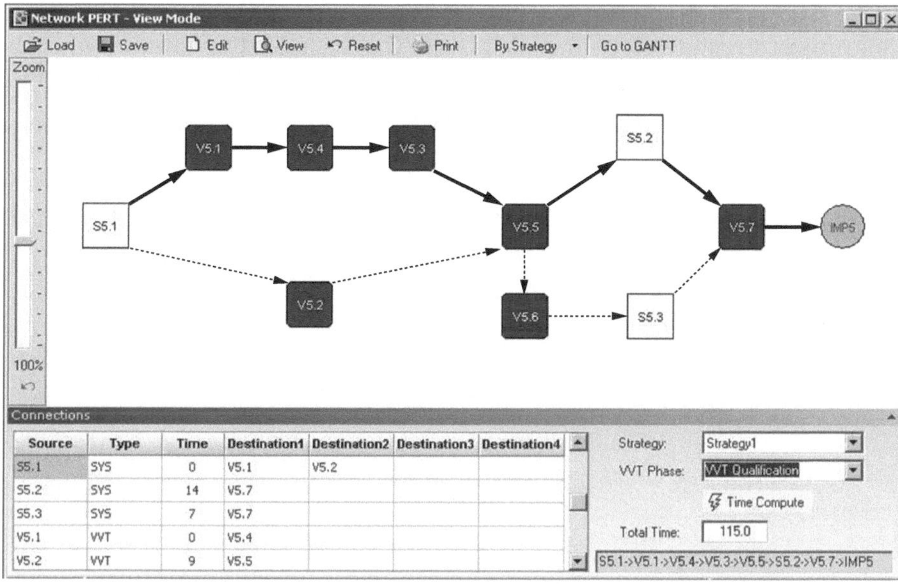

Figure 8.18 Estimated model: Qualification phase PERT.

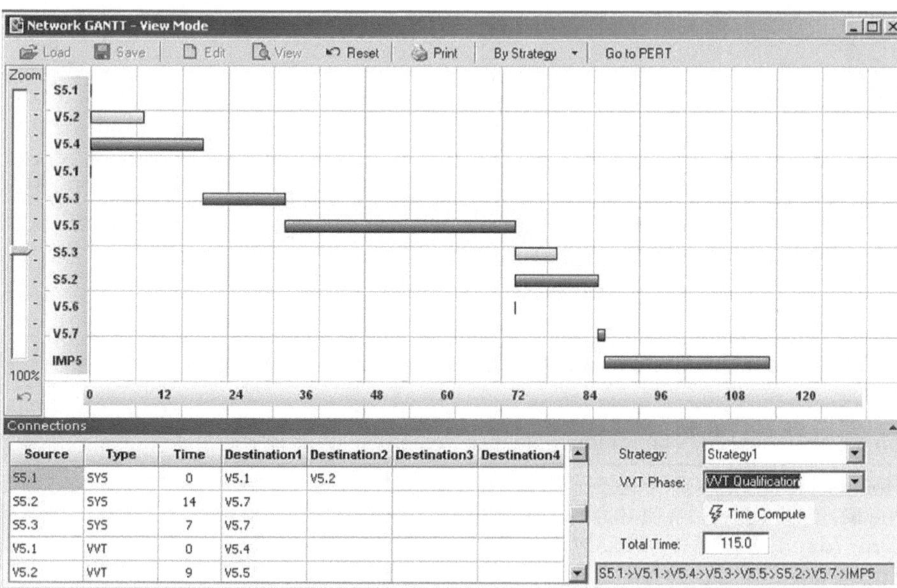

Figure 8.19 Estimated model: Qualification phase Gantt.

The last Gantt chart (Figure 8.20) depicts the duration in days of the entire pilot project as modeled by the VVT-Tool. This includes the duration of each development phase as well as the "early start" component provided by the domain experts and used by the VVT model.

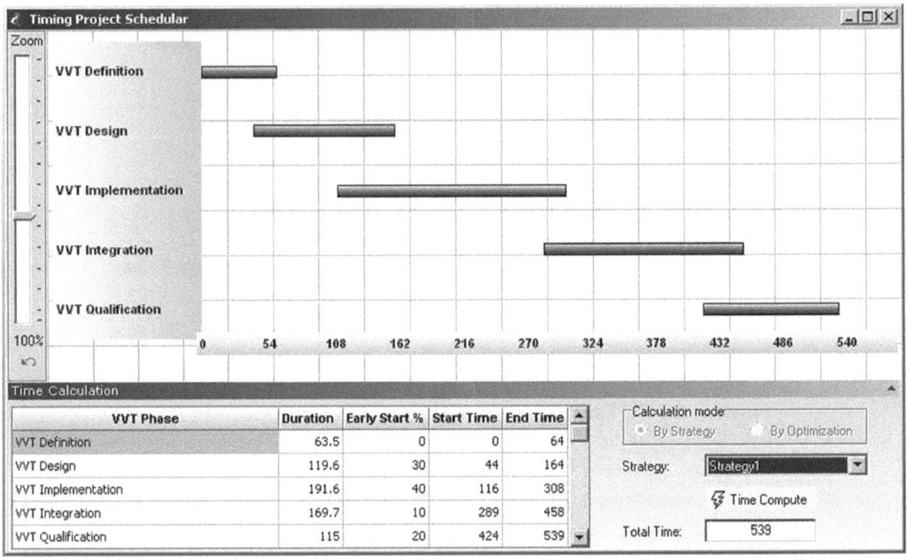

Figure 8.20 Estimated model: Overall system development Gantt.

Validating VVT Time Model Table 8.8 and Figure 8.21 show the phased correlation between times predicted by the VVT simulation model and anchor times. Here, too, in some phases, the two values are different and in others they are quite similar. However, the total time difference estimated by the VVT model and measured by the experts is –3.7% and the correlation coefficient between the two sets is 0.9885.

TABLE 8.8 Estimated VVT Model Time Versus Anchor Time (days)

Time Categories	Definition	Design	Implementation	Integration	Qualification	Total
Model time (days)	63.5	119.6	191.6	169.7	115.0	539.0
Anchor time (days)	110.0	110.0	210.0	140.0	110.0	554.0
Error (days)	–46.5	9.6	–18.4	29.7	5.0	–20.6
Error (%)	–42.3	8.7	–8.8	21.2	4.5	–3.7

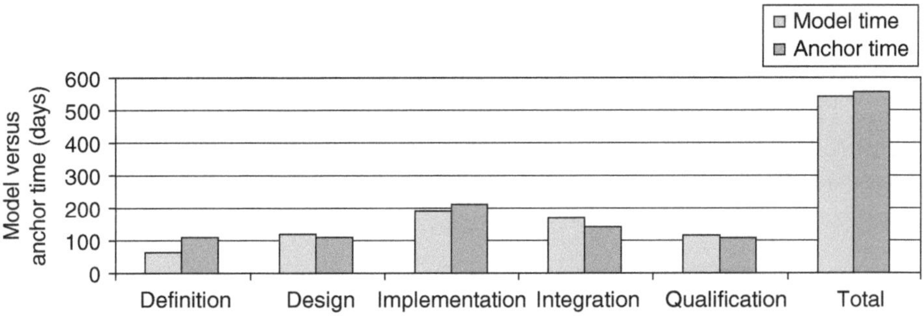

Figure 8.21 Estimated VVT model time versus anchor time (days).

8.1.3 Fuzzy VVT Cost Model Validation

Model Estimated Fuzzy Cost Data Figure 8.22 shows the estimated crisp cost results per pilot project development phase as computed by the fuzzy simulator with α-cut = 0.3. Note that the VVT activity cost is 340 CU, the VVT appraisal cost is 56 CU, the VVT impact cost is 479 CU, the total quality cost is 877 CU and the CVM cost is 1348 CU.

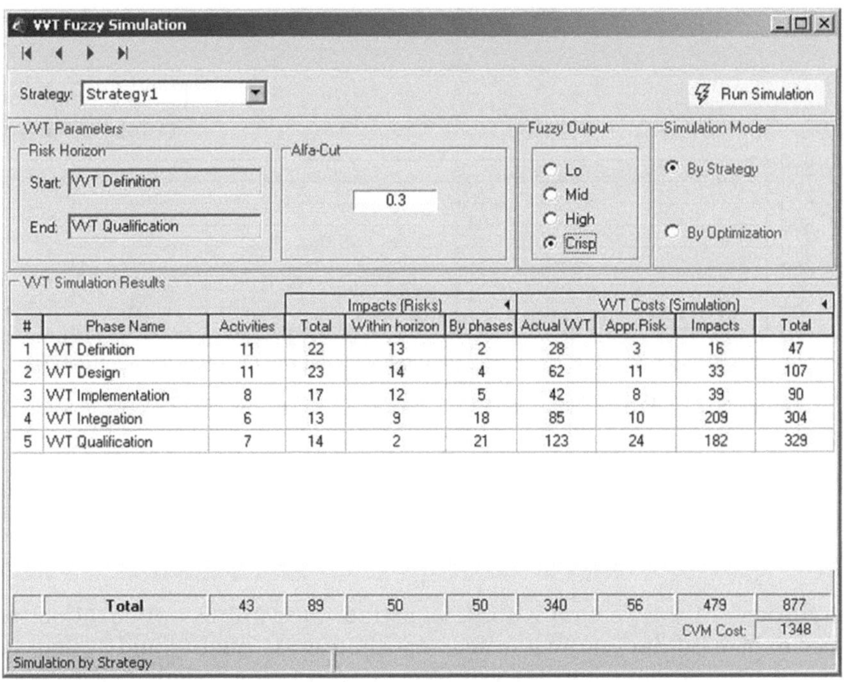

Figure 8.22 Fuzzy cost model estimation in crisp values (CU).

Validating Fuzzy VVT Cost Model Table 8.9 and Figure 8.23 show comparisons between costs computed by the VVT fuzzy model and the anchor costs. In some phases the two values are dissimilar and in others they are much closer. However, the total cost difference estimated by the VVT model and measured by the experts is –4.9% and the correlation coefficient between the two sets is 0.9791.

TABLE 8.9 Fuzzy Estimated VVT Model Cost Versus Anchor Cost (CU)

Cost Categories	Definition	Design	Implementation	Integration	Qualification	Total
VVT activity	28	62	42	85	123	340
VVT appraisal	3	11	8	10	24	56
2/3 VVT impact	11	22	26	139	121	319
Total fuzzy cost	42	95	76	234	268	715
Total anchor cost	22	87	173	257	213	752
Error (CU)	20	8	–97	–23	56	-37
Error (%)	91.0	9.3	–56.2	–8.9	26.2	–4.9

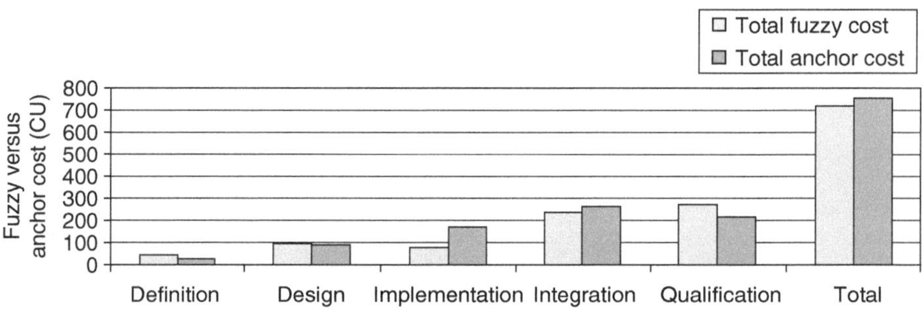

Figure 8.23 Fuzzy estimated VVT model cost versus anchor cost (CU).

8.2 OPTIMIZING THE VVT STRATEGY

The purpose of this section is to demonstrate the various optimization techniques explained earlier. This is accomplished using quality and system data obtained from the industrial pilot project development conducted under the SysTest project at the Lahav Division of the Israel Aerospace Industries (IAI) during the timeframe 2002–2005.

8.2.1 Analytical Optimization of Cost

Analytic cost optimization was based on the following: (1) the VVT horizon covers five system lifecycle phases (Definition through Qualification) and (2) the VVT cost parameters are the same ones used during the pilot project. We also (1) have assumed that linear impact risk cost behavior and (2) have allowed each individual VVT APL ($X_{i,j}$) freedom to acquire any optimal value (i.e., in this section we assume a "free" mode for all decision variables).

Optimizing for Minimum Expected Total Cost We shall now optimize the VVT strategy in order to obtain a minimum expected total cost. Table 8.10 presents the numerical results of the two pure strategies, $X_{i,j} = 0$, $X_{i,j} = 1$, which are the only potential optimal strategies.

TABLE 8.10 Optimizing for Minimum Expected Total Cost

	Expected Cost for Given Strategy $[X_{ij}]$							
VVT ID	VVT Cost $E[C_t(X_{i,j} = 1)]$	Expected Appraisal Cost $E[C_t(X_{i,j} = 1)]$	Subtotal $E[C_t(X_{i,j} = 1)]$	Expected Impact Cost, $E[C_t(X_{i,j} = 0)]$	Cost Difference, B	Optimal Strategy, $X^*_{i,j}$	Minimum expected Cost, (CU) $E[C_t(X^*_{i,j})]$	Variance
V1.1	32.0	0.0	32.0	42.0	–10.0	1.0	32.0	0.0
V1.2	26.0	0.0	26.0	53.0	–27.0	1.0	26.0	0.1
V1.3	20.0	6.0	26.0	17.0	9.0	0.0	17.0	621.1
V1.4	19.0	2.0	21.0	14.0	7.0	0.0	14.0	3,967.2
V1.5	38.0	5.0	43.0	19.0	24.0	0.0	19.0	122.5
V1.6	26.0	2.0	28.0	13.0	15.0	0.0	13.0	146.4
V1.7	13.0	0.0	13.0	0.0	13.0	0.0	0.0	0.0
V1.8	28.0	5.0	33.0	0.0	33.0	0.0	0.0	0.0
V1.9	26.0	5.0	31.0	17.0	14.0	0.0	17.0	62.2
V1.10	13.0	1.0	14.0	14.0	0.0	1.0	14.0	0.0
V1.11	13.0	1.0	14.0	8.0	6.0	0.0	8.0	14.1
V2.1	6.0	1.0	7.0	8.0	–1.0	1.0	7.0	0.2
V2.2	106.0	0.0	106.0	32.0	74.0	0.0	32.0	159.7
V2.3	38.0	6.0	44.0	20.0	24.0	0.0	20.0	20.9
V2.4	42.0	5.0	47.0	17.0	30.0	0.0	17.0	253.7
V2.5	26.0	2.0	28.0	26.0	2.0	0.0	26.0	13,092.2
V2.6	13.0	2.0	15.0	7.0	8.0	0.0	7.0	22.1
V2.7	58.0	9.0	67.0	0.0	67.0	0.0	0.0	0.0
V2.8	22.0	2.0	24.0	7.0	17.0	0.0	7.0	97.5
V2.9	93.0	18.0	111.0	34.0	77.0	0.0	34.0	118.5
V2.10	38.0	12.0	50.0	50.0	0.0	1.0	50.0	8.0
V2.11	13.0	12.0	25.0	17.0	8.0	0.0	17.0	597.9
V3.1	6.0	1.0	7.0	16.0	–9.0	1.0	7.0	0.2
V3.2	32.0	5.0	37.0	25.0	12.0	0.0	25.0	323.9
V3.3	13.0	2.0	15.0	21.0	–6.0	1.0	15.0	0.6
V3.4	77.0	1.0	78.0	10.0	68.0	0.0	10.0	4.8

TABLE 8.10 Continued

VVT ID	Expected Cost for Given Strategy $[X_{ij}]$: VVT Cost	Expected Appraisal Cost	Subtotal $E[C_t(X_{i,j}=1)]$	Expected Impact Cost, $E[C_t(X_{i,j}=0)]$	Cost Difference, B	Optimal Strategy, $X^*_{i,j}$	Minimum expected Cost, (CU) $E[C_t(X^*_{i,j})]$	Variance
V3.5	19.0	3.0	22.0	17.0	5.0	0.0	17.0	62.2
V3.6	13.0	2.0	15.0	19.0	−4.0	1.0	15.0	2.3
V3.7	51.0	12.0	63.0	14.0	49.0	0.0	14.0	97.5
V3.8	13.0	6.0	19.0	27.0	−8.0	1.0	19.0	1.7
V4.1	6.0	3.0	9.0	21.0	−12.0	1.0	9.0	5.5
V4.2	22.0	5.0	27.0	17.0	10.0	0.0	17.0	28.2
V4.3	64.0	8.0	72.0	15.0	57.0	0.0	15.0	36.2
V4.4	38.0	5.0	43.0	7.0	36.0	0.0	7.0	22.1
V4.5	26.0	0.0	26.0	13.0	13.0	0.0	13.0	44.2
V4.6	13.0	0.0	13.0	0.0	13.0	0.0	0.0	0.0
V5.1	10.0	1.0	11.0	7.0	4.0	0.0	7.0	22.1
V5.2	83.0	18.0	101.0	7.0	94.0	0.0	7.0	22.1
V5.3	38.0	3.0	41.0	0.0	41.0	0.0	0.0	0.0
V5.4	13.0	3.0	16.0	0.0	16.0	0.0	0.0	0.0
V5.5	70.0	16.0	86.0	0.0	86.0	0.0	0.0	0.0
V5.6	90.0	14.0	104.0	0.0	104.0	0.0	0.0	0.0
V5.7	13.0	0.0	13.0	0.0	13.0	0.0	0.0	0.0
Total	1,419.0	204.0	1,623.0	651.0	972.0		574.0	19,977.9

As can be observed, the optimal strategy for obtaining minimum expected total cost is to fully perform 10 VVT activities and to not perform any of the other VVT activities. This optimal strategy yields a minimal total expected cost of 574 CU (578 CU under the stochastic simulation). The cost variance of this strategy is 19,977.9 CU and, accordingly, its standard deviation is 141.3 CU. We can execute the stochastic simulator using this optimized VVT strategy yielding comparable results (see Table 8.11 and Figures 8.24 and 8.25).

TABLE 8.11 Stochastic Simulation Results for Minimum Expected Total Cost

Cost Categories	Minimum Expected Total Cost: Original (CU)	Optimized (CU)	Change (CU)	Change (%)
VVT activity cost	343	166	177	52
Mean appraisal cost	62	33	29	47
Mean impact cost	472	378	94	20
Mean total cost	877	578	299	34
CVM cost	1419	1419	0	0
SD	106	141	−35	−33
VaR95%	1126	962	164	15

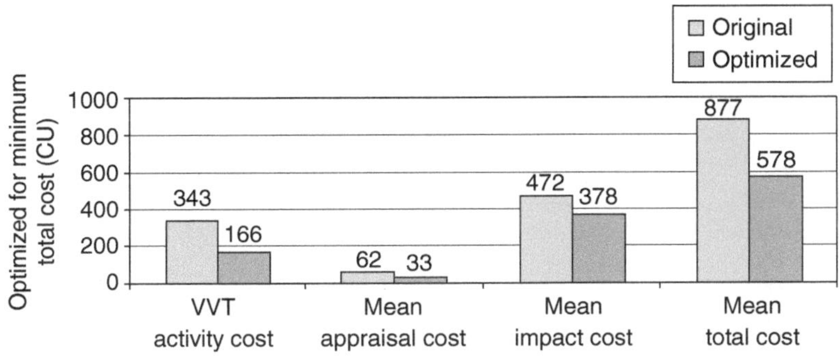

Figure 8.24 Comparisons under minimum expected total cost.

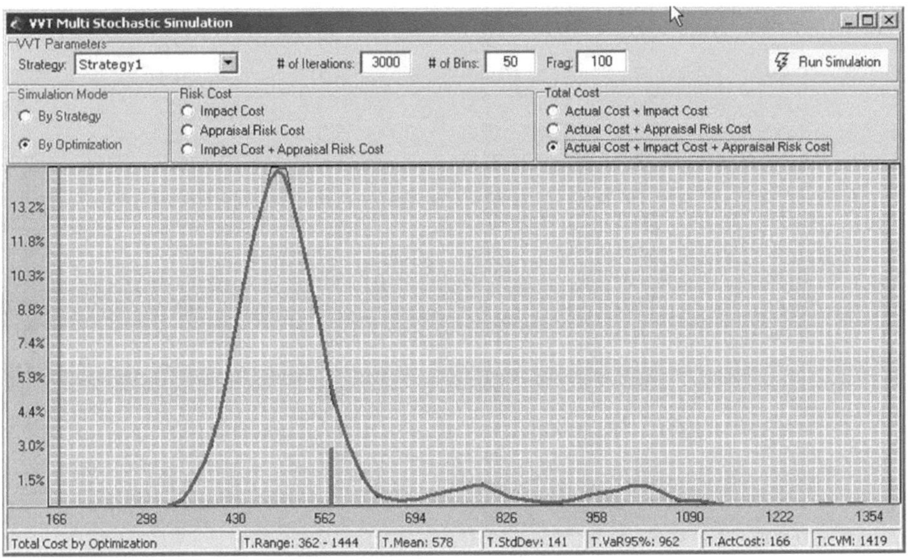

Figure 8.25 Stochastic simulation results: minimum expected total cost.

Optimizing for Minimum Variance of Expected Total Cost We now optimize the VVT strategy in order to obtain a minimum variance of the expected total cost. We will use the same raw VVT data in order to obtain optimal VVT performance levels ($X^*_{i,j}$) (see Table 8.12).

As can be seen, the minimal variance is 707.2 CU and therefore the standard deviation is nearly 26.6 CU. The overall expected cost of this strategy is 1196.1 CU. This result is better than fully performing the VVT process but far less attractive than that of the minimum expected cost strategy.

TABLE 8.12 Optimizing for Minimum Variance of Expected Total Cost

VVT ID	Optimal Strategy, $X^*_{i,j}$	Minimum Variance, $V[C_t(X^*_{i,j})]$	VVT Cost	Expected Appraisal Cost	Expected Impact Cost	Expected Total Cost
			$E[C_t(X^*_{i,j})]$			
V1.1	1.00	0.0	32.0	0.0	0.0	32.0
V1.2	1.00	0.1	26.0	0.1	0.0	26.1
V1.3	1.00	1.7	20.0	5.7	0.0	25.7
V1.4	1.00	1.3	19.0	2.5	0.0	21.5
V1.5	1.00	5.1	38.0	5.0	0.0	43.0
V1.6	1.00	1.3	26.0	2.5	0.0	28.5
V1.7	1.00	0.3	13.0	0.5	0.0	13.5
V1.8	0.53	6.3	14.7	2.5	0.0	17.2
V1.9	1.00	1.2	26.0	4.8	0.0	30.8
V1.10	1.00	0.0	13.0	1.0	0.0	14.0
V1.11	1.00	0.1	13.0	0.8	0.0	13.8
V2.1	1.00	0.2	6.0	0.7	0.0	6.7
V2.2	0.99	4.7	104.5	0.5	0.4	105.4
V2.3	0.00	20.9	0.0	0.0	20.0	20.0
V2.4	1.00	5.1	42.0	5.0	0.0	47.0
V2.5	1.00	2.3	26.0	1.5	0.0	27.5
V2.6	1.00	1.3	13.0	2.5	0.0	15.5
V2.7	0.53	20.3	30.5	4.5	0.0	35.0
V2.8	1.00	1.3	22.0	2.5	0.0	24.5
V2.9	0.26	132.0	23.8	4.6	25.3	53.7
V2.10	0.13	71.0	4.9	1.6	43.9	50.4
V2.11	1.00	8.0	13.0	12.4	0.0	25.4
V3.1	1.00	0.2	6.0	0.7	0.0	6.7
V3.2	1.00	5.1	32.0	5.0	0.0	37.0
V3.3	1.00	0.6	13.0	1.7	0.0	14.7
V3.4	0.67	2.8	51.7	0.4	3.1	55.2
V3.5	1.00	2.3	19.0	3.3	0.0	22.3
V3.6	1.00	2.3	13.0	1.5	0.0	14.5
V3.7	0.00	97.5	0.0	0.0	14.1	14.1
V3.8	1.00	1.7	13.0	5.7	0.0	18.7
V4.1	1.00	5.5	6.0	3.4	0.0	9.4
V4.2	1.00	5.1	22.0	5.0	0.0	27.0
V4.3	0.16	37.1	10.3	1.3	12.6	24.2
V4.4	0.00	22.1	0.0	0.0	6.7	6.7
V4.5	0.98	1.7	25.5	0.3	0.2	26.0
V4.6	1.00	0.3	13.0	0.5	0.0	13.5
V5.1	1.00	0.9	10.0	1.3	0.0	11.3
V5.2	0.49	95.6	40.8	8.9	3.4	53.1
V5.3	1.00	9.0	38.0	3.0	0.0	41.0
V5.4	0.60	4.0	7.8	2.0	0.0	9.8
V5.5	0.53	72.3	36.8	8.5	0.0	45.3
V5.6	0.53	56.3	47.4	7.5	0.0	54.9
V5.7	1.00	0.3	13.0	0.5	0.0	13.5
Total		707.2	944.7	121.7	129.7	1196.1

We can now use this optimized strategy and run the stochastic simulation which yields comparable results (see Table 8.13 and Figures 8.26 and 8.27).

TABLE 8.13 Stochastic Simulation: Minimum Variance of Expected Total Cost

	Minimum Variance of Expected Total Cost			
Cost Categories	Original (CU)	Optimized (CU)	Change (CU)	Change (%)
VVT activity cost	343	950	−607	−177
Mean appraisal cost	62	150	−88	−142
Mean impact cost	472	117	355	75
Mean total cost	877	1217	−340	−39
CVM cost	1419	1419	0	0
SD	106	15	91	86
VaR95%	1126	1239	−113	−10

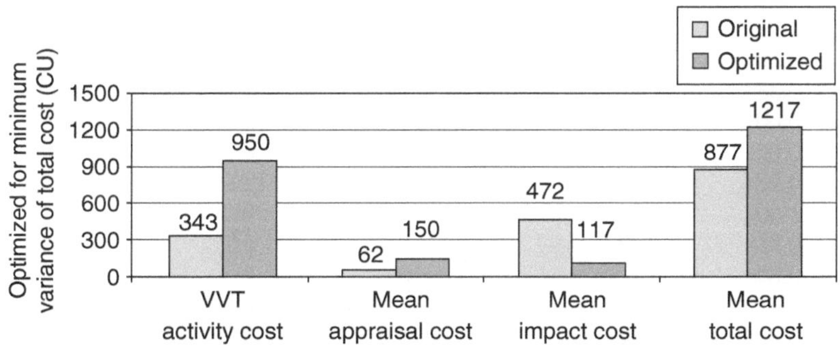

Figure 8.26 Comparison of minimum variance of expected total cost.

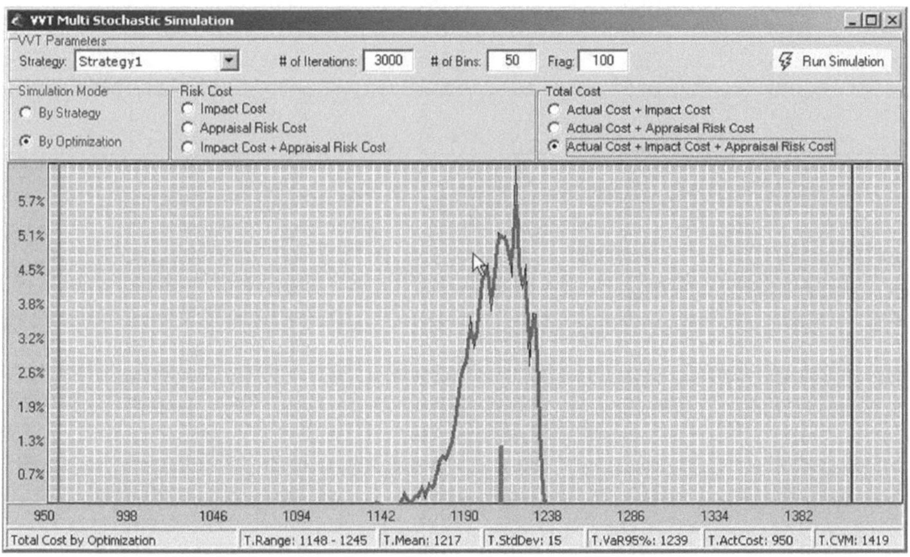

Figure 8.27 Stochastic simulation: minimum variance of expected total cost.

Optimizing for Minimum Loss Function We shall now optimize the VVT strategy in order to obtain a minimum loss function using the same raw VVT data. Our objective is to get optimal VVT performance levels ($X^*_{i,j}$) that minimize the total expected loss of the VVT process (see Table 8.14).

TABLE 8.14 Optimizing for Minimum Loss Function

VVT ID	Optimal Strategy, $X^*_{i,j}$	Minimum Expected Loss, $E[L(C_t(X^*_{i,j}))]$	VVT Cost	Expected Appraisal Cost	Expected Impact Cost	Expected Total Cost	Variance, $V[C_t(X^*_{i,j})]$
			$E[C_t(X^*_{i,j})]$				
V1.1	0.93	96.0	29.6	0.0	3.1	32.7	7.1
V1.2	0.44	322.6	11.5	0.0	29.7	41.2	180.1
V1.3	0.92	65.0	18.4	5.2	1.4	25.0	7.9
V1.4	0.99	50.3	18.8	2.5	0.2	21.5	1.9
V1.5	0.16	105.2	6.1	0.8	16.0	22.9	90.5
V1.6	0.38	91.9	9.9	1.0	8.2	19.1	57.8
V1.7	0.00	0.0	0.0	0.0	0.0	0.0	0.0
V1.8	0.00	0.0	0.0	0.0	0.0	0.0	0.0
V1.9	0.23	50.5	6.0	1.1	13.4	20.5	41.0
V1.10	1.00	0.1	13.0	1.0	0.0	14.0	0.0
V1.11	0.31	9.8	4.0	0.3	5.7	10.0	6.9
V2.1	0.84	2.5	5.1	0.6	1.3	7.0	0.6
V2.2	0.03	155.5	2.9	0.0	30.6	33.5	151.3
V2.3	0.01	20.9	0.3	0.0	19.8	20.1	20.9
V2.4	0.21	204.1	8.7	1.0	13.2	22.9	164.4
V2.5	1.00	3.8	26.0	1.5	0.0	27.5	2.3
V2.6	0.20	18.5	2.6	0.5	5.4	8.5	15.5
V2.7	0.00	0.0	0.0	0.0	0.0	0.0	0.0
V2.8	0.23	75.5	5.2	0.6	5.3	11.1	58.5
V2.9	0.00	118.5	0.0	0.0	34.0	34.0	118.5
V2.10	0.13	71.0	4.9	1.6	43.9	50.4	71.0
V2.11	1.00	75.1	13.0	12.4	0.0	25.4	8.0
V3.1	0.63	49.7	3.8	0.4	5.7	9.9	18.4
V3.2	0.68	112.5	21.9	3.4	7.8	33.1	41.2
V3.3	0.76	31.3	9.9	1.3	5.1	16.3	8.2
V3.4	0.00	4.7	0.0	0.0	9.5	9.5	4.7
V3.5	0.74	21.1	14.1	2.5	4.5	21.1	8.0
V3.6	0.78	18.9	10.1	1.2	4.3	15.6	6.0
V3.7	0.01	97.4	0.4	0.1	14.0	14.5	97.3
V3.8	0.73	51.1	9.5	4.2	7.1	20.8	17.9
V4.1	0.58	86.0	3.5	2.0	8.7	14.2	39.7
V4.2	0.12	26.6	2.6	0.6	14.6	17.8	25.1
V4.3	0.00	36.2	0.0	0.0	15.0	15.0	36.2
V4.4	0.01	22.1	0.2	0.0	6.7	6.9	22.0
V4.5	0.21	35.3	5.3	0.1	10.6	16.0	28.3
V4.6	0.00	0.0	0.0	0.0	0.0	0.0	0.0

TABLE 8.14 *Continued*

VVT ID	Optimal Strategy, $X^*_{i,j}$	Minimum Expected Loss, $E[L(C_t(X^*_{i,j}))]$	VVT Cost	Expected Appraisal Cost	Expected Impact Cost	Expected Total Cost	Variance, $V[C_t(X^*_{i,j})]$
			$E[C_t(X^*_{i,j})]$				
V5.1	0.50	11.8	5.0	0.7	3.4	9.1	6.5
V5.2	0.00	22.1	0.0	0.0	6.7	6.7	22.1
V5.3	0.00	0.0	0.0	0.0	0.0	0.0	0.0
V5.4	0.00	0.0	0.0	0.0	0.0	0.0	0.0
V5.5	0.00	0.0	0.0	0.0	0.0	0.0	0.0
V5.6	0.00	0.0	0.0	0.0	0.0	0.0	0.0
V5.7	0.00	0.0	0.0	0.0	0.0	0.0	0.0
Total		2163.6	272.3	46.6	354.9	673.8	1385.8

The optimal VVT strategy yields a minimal total expected loss of 2163.6. Additionally, this strategy produces an overall expected cost of 673.8 CU and a cost variance of 1385.8 CU or standard deviation of 37.2 CU. We can see that the overall expected cost is quite close to 574 CU, the result of the minimum expected cost strategy, whereas its variance is much lower than the 19,978 CU variance obtained by the latter strategy. We can now use this strategy and run the stochastic simulation which yields similar results (see Table 8.15 and Figures 8.28 and 8.29).

TABLE 8.15 Stochastic Simulation Results: Minimum Loss Function[81]

Cost Categories	Minimum Loss Function: Original (CU)	Optimized (CU)	Change (CU)	Change (%)
VVT activity cost	343	263	80	23
Mean appraisal cost	62	52	10	16
Mean impact cost	472	362	110	23
Mean total cost	877	677	200	23
CVM cost	1419	1419	0	0
SD	106	36	70	66
VaR95%	1126	736	390	35

[81]We can also see that the original (preoptimized) 95% Value-at-Risk (VaR95%) generated by the stochastic simulator (1126 CU) corresponds with the expected mean total VVT cost (877 CU) plus two standard deviations (106 CU), yielding $877 + 2 \times 106 = 1089$, which is 96.7% of our expectation. Similarly, for the optimized strategy, VaR95% generated by the stochastic simulator (736 CU) approximately corresponds with the expected mean total VVT cost (677 CU) plus two standard deviations (36 CU), yielding $677 + 2 \times 36 = 749$, which is 101.8% of our expectation.

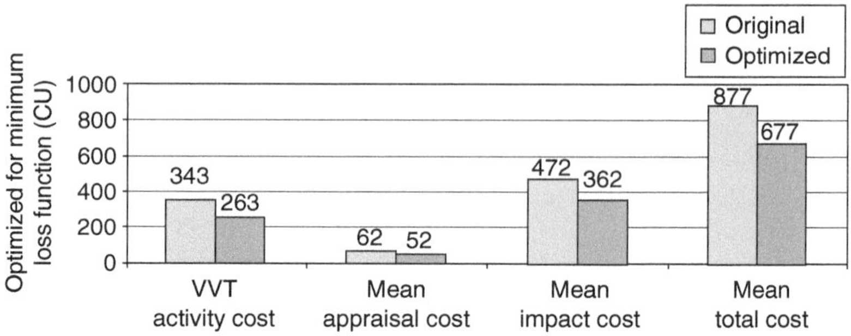

Figure 8.28 Comparison of results: minimum loss function.

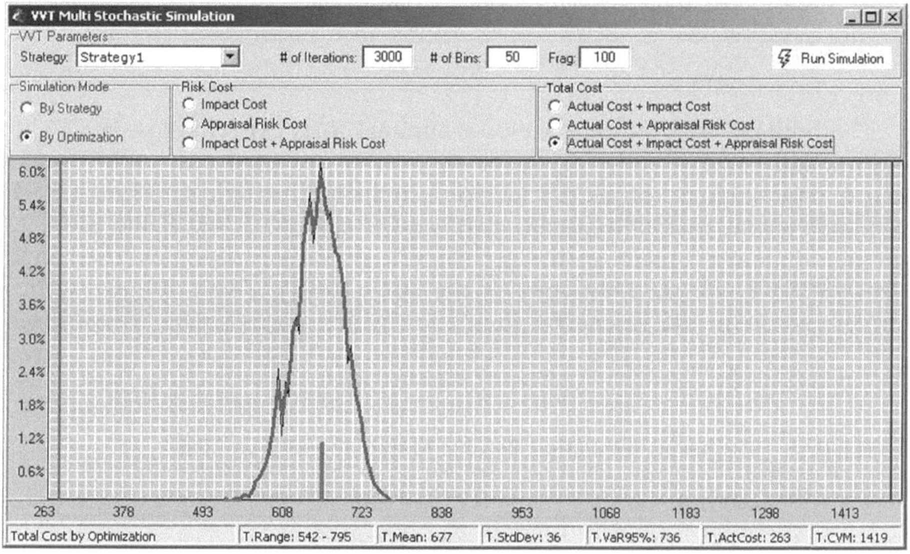

Figure 8.29 Stochastic simulation results: minimum loss function.

8.2.2 Cost Distribution by Phase

Table 8.16 and Figure 8.30 depict the quality cost distribution over the five development phases utilizing the VVT strategy based on a minimum loss function.

TABLE 8.16 Optimal Quality Cost Distribution by Phases (CU)

Cost Categories	Definition	Design	Implementation	Integration	Qualification
VVT activity cost	116	63	70	11	4
Mean appraisal cost	12	22	15	2	1
Mean impact cost	3	10	31	182	135
Mean total cost	132	95	114	195	142
CVM cost	254	455	224	176	310

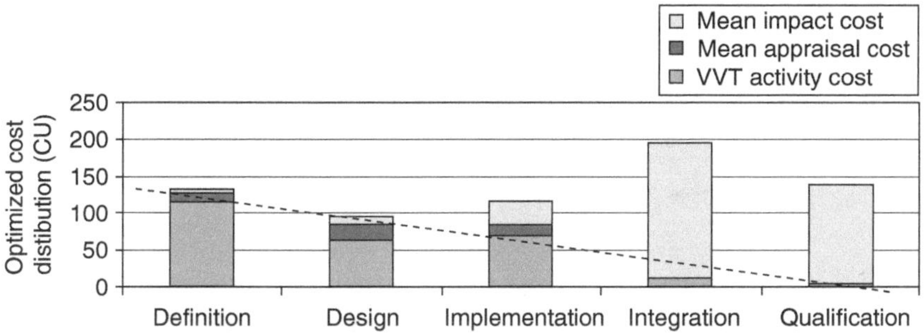

Figure 8.30 Optimal quality cost distribution by phases.

Note the near-linear decrease in the VVT activity phase cost. Indeed, the quality cost optimizer confirms our earlier supposition that significant increase in VVT performance during the early phases of the project will most likely yield substantially lower total development cost (i.e., 677 CU instead of 877 CU). The reader may compare this figure with Figure 8.7, which shows that the original VVT activity phase cost strategy tends to increase, probably due to skimping on VVT activities in the early phases of the development project.

8.2.3 Weight Optimization of Cost

We will continue to use the same VVT cost and risk data. However, from here on we shall (1) assume nonlinear impact risk cost behavior and (2) apply the free/fixed constraint for each VVT APL ($X_{i,j}$).

Weight Optimization Process We will now demonstrate how to improve the VVT strategy using the weight optimization method. This is a multiobjective optimization problem. We assume all VVT APLs ($X_{i,j}$) are independent of each other and therefore attempt to assign each one (provided it is free to change) a value ($0 \leq X_{i,j} \leq 1$) so that the various VVT cost components will be reduced according to their weights of importance (see Table 8.17 and Figure 8.31).

TABLE 8.17 Weighted Optimization Parameters

Cost Category	VVT Cost	Appraisal Cost	Impact Cost	SD	VaR95%
Weights	0.4	0.1	0.3	0.1	0.1

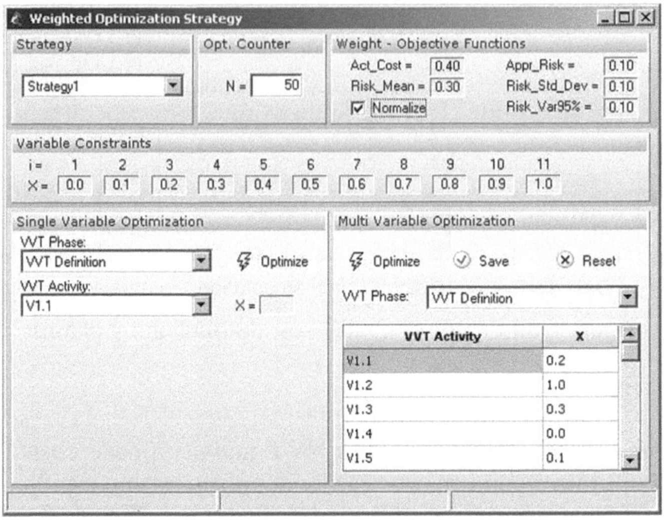

Figure 8.31 Weight optimization application.

The weight-optimized strategy is depicted in Table 8.18. (Note that certain VVT activities are fixed, that is, are not allowed to change due to external considerations.)

TABLE 8.18 Original Versus Weight Optimized VVT Strategies

System VVT Phase	Activity	Perform	Mode	Optimized Value
VVT Definition	V1.1	0.20	Free	0.20
	V1.2	0.10	Free	1.00
	V1.3	0.00	Free	0.30
	V1.4	**0.00**	**Fixed**	**0.00**
	V1.5	0.20	Free	0.10
	V1.6	0.05	Free	0.40
	V1.7	0.20	Free	0.00
	V1.8	0.10	Free	0.00
	V1.9	0.20	Free	0.00
	V1.10	**0.2**	**Fixed**	**0.2**
	V1.11	0.10	Free	0.00
VVT Design	V2.1	0.00	Free	0.10
	V2.2	0.05	Free	0.00
	V2.3	0.10	Free	0.00
	V2.4	0.20	Free	0.10
	V2.5	0.60	Free	0.10
	V2.6	0.10	Free	0.20
	V2.7	0.00	Free	0.00
	V2.8	0.10	Free	0.40
	V2.9	0.20	Free	0.00

TABLE 8.18 *Continued*

System VVT Phase	Activity	Perform	Mode	Optimized Value
	V2.10	**0.3**	**Fixed**	**0.3**
	V2.11	0.10	Free	0.20
VVT Implementation	V3.1	0.00	Free	0.80
	V3.2	0.30	Free	0.20
	V3.3	0.00	Free	0.10
	V3.4	0.05	Free	0.00
	V3.5	0.50	Free	1.00
	V3.6	0.40	Free	1.00
	V3.7	0.40	Free	0.00
	V3.8	0.00	Free	1.00
VVT Integration	V4.1	0.00	Free	1.00
	V4.2	0.50	Free	0.20
	V4.3	0.50	Free	0.00
	V4.4	**0.6**	**Fixed**	**0.6**
	V4.5	0.80	Free	0.00
	V4.6	0.10	Free	0.00
VVT Qualification	V5.1	0.00	Free	0.10
	V5.2	0.30	Free	0.00
	V5.3	**0.8**	**Fixed**	**0.8**
	V5.4	0.60	Free	0.00
	V5.5	**0.6**	**Fixed**	**0.6**
	V5.6	0.00	Free	0.00
	V5.7	0.10	Free	0.00

The VVT strategy performed during the pilot project as well as the optimized VVT strategy is illustrated in Figure 8.32. The X axis represents the set of VVT APL IDs and the Y axis represents the level of each VVT performance ($0 \leq X_{i,j} \leq 1$). The connecting lines are drawn for improved visibility only.

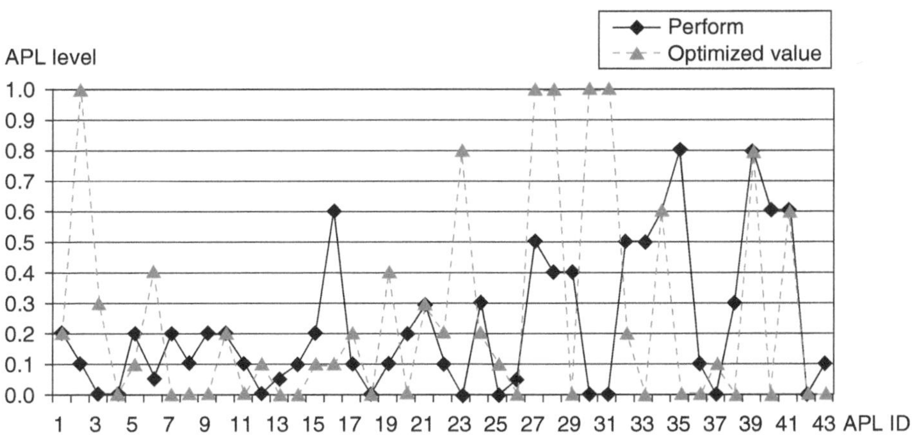

Figure 8.32 VVT strategies: performed versus weight optimized strategy.

Weight Optimization Results Figure 8.33 shows the overall VVT cost estimation modeled on this optimized VVT strategy. These results were obtained after performing 3000 Monte Carlo iterations within the VVT-Tool using the nonlinear VVT impact cost modeling approach.

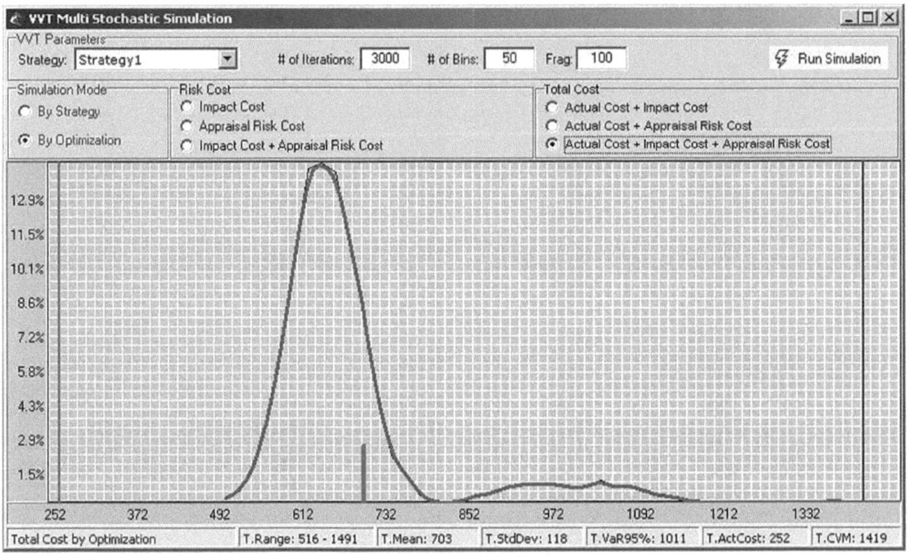

Figure 8.33 Overall weight optimized VVT cost estimation.

Table 8.19 and Figure 8.34 show the effects of the VVT strategy of weight optimization. As can be observed, all the VVT cost parameters have been improved. This result clearly indicates that the VVT strategy currently employed at IAI/Lahav is far from optimal and future projects can achieve an approximately 15% total VVT cost decrease.

TABLE 8.19 Original Versus Weight Optimized Cost Reduction

Cost Categories	Original (CU)	Optimized (CU)	Change (CU)	Change (%)
VVT activity cost	343	252	91	27
Mean appraisal cost	62	54	8	13
Mean impact cost	472	397	75	16
Mean total cost	877	703	174	20
CVM cost	1419	1419	0	0
SD	106	128	–22	–21
VaR95%	1126	1011	115	10

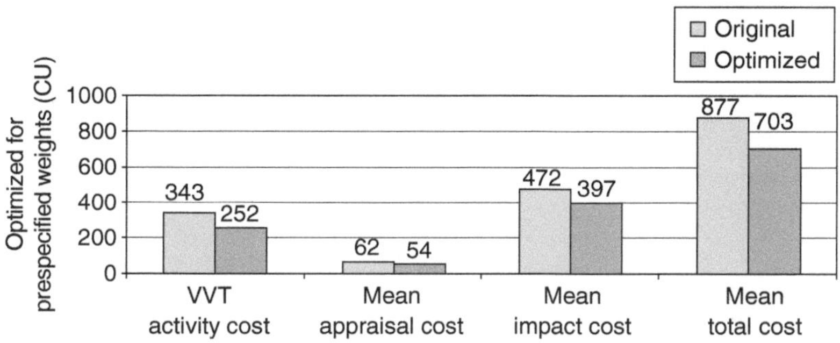

Figure 8.34 Original versus weight optimized cost reduction.

8.2.4 Goal Optimization of Cost

Goal Optimization Process We now demonstrate the improvement of the VVT strategy using goal optimization. This too is a multiobjective optimization problem. We have assumed the independence of each individual VVT APL ($X_{i,j}$) and we try to assign each APL (provided it is free to change) a value ($0 \le X_{i,j} \le 1$), so that the various VVT cost components will meet our goals. In this example, the VVT-Tool was executed mainly to reduce the VVT impact cost. This may be done by increasing the VVT investment and also, by necessity, increasing the overall VVT quality cost. Specifically, we are running the goal optimizer with the aim of identifying an appropriate VVT strategy in order to:

1. Reduce VVT impact cost to a target of 350 CU with a range of 330–380 CU.
2. Allow an increase in the VVT activity cost to a target of 600 CU with a range of 580–620 CU.
3. Set the VVT appraisal cost to a target of 100 CU with a range of 90–110 CU.

These requirements are summarized in Table 8.20. In order to estimate a realistic VVT impact cost, we have used a nonlinear modeling approach (see Figure 8.35).

TABLE 8.20 Goal Optimization Parameters

Objective	Priority	Goal	Minimum	Maximum
Impact risk	1	350	330	380
Actual VVT cost	2	600	580	620
Appraisal risk	3	100	90	110

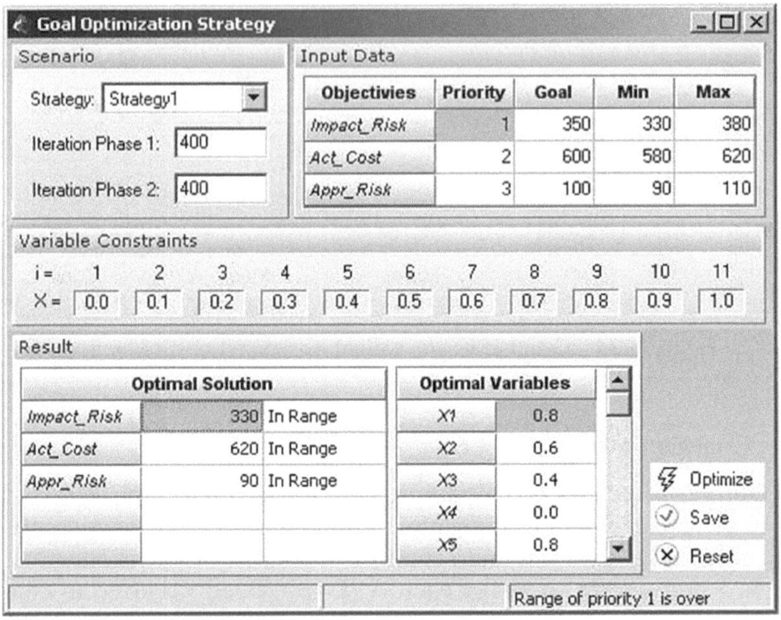

Figure 8.35 Goal optimization application.

The goal optimized strategy is shown in Table 8.21. (Note that certain VVT activities are fixed, i.e., are not allowed to be optimized due to external considerations.)

TABLE 8.21 Original Versus Goal Optimized VVT Strategies

System VVT Phase	Activity	Perform	Mode	Optimized Value
VVT Definition	V1.1	0.20	Free	0.80
	V1.2	0.10	Free	0.60
	V1.3	0.00	Free	0.40
	V1.4	**0.00**	**Fixed**	**0.00**
	V1.5	0.20	Free	0.80
	V1.6	0.05	Free	0.80
	V1.7	0.20	Free	0.80
	V1.8	0.10	Free	0.10
	V1.9	0.20	Free	0.20
	V1.10	**0.20**	**Fixed**	**0.20**
	V1.11	0.10	Free	0.00
VVT Design	V2.1	0.00	Free	0.70
	V2.2	0.05	Free	1.00
	V2.3	0.10	Free	0.00
	V2.4	0.20	Free	0.40
	V2.5	0.60	Free	1.00
	V2.6	0.10	Free	0.90
	V2.7	0.00	Free	0.90
	V2.8	0.10	Free	0.40

TABLE 8.21 ***Continued***

System VVT Phase	Activity	Perform	Mode	Optimized Value
	V2.9	0.20	Free	0.20
	V2.10	**0.30**	**Fixed**	**0.30**
	V2.11	0.10	Free	0.60
VVT Implementation	V3.1	0.00	Free	0.90
	V3.2	0.30	Free	0.30
	V3.3	0.00	Free	0.20
	V3.4	0.05	Free	0.10
	V3.5	0.50	Free	0.50
	V3.6	0.40	Free	0.30
	V3.7	0.40	Free	0.30
	V3.8	0.00	Free	0.40
VVT Integration	V4.1	0.00	Free	0.30
	V4.2	0.50	Free	0.20
	V4.3	0.50	Free	0.70
	V4.4	**0.60**	**Fixed**	**0.60**
	V4.5	0.80	Free	0.50
	V4.6	0.10	Free	0.40
VVT Qualification	V5.1	0.00	Free	0.40
	V5.2	0.30	Free	0.00
	V5.3	**0.80**	**Fixed**	**0.80**
	V5.4	0.60	Free	0.10
	V5.5	**0.60**	**Fixed**	**0.60**
	V5.6	0.00	Free	0.00
	V5.7	0.10	Free	0.50

The different VVT strategies can be seen graphically in Figure 8.36. The X axis represents the set of VVT APL IDs and the Y axis represents the VVT performance level ($0 \le X_{i,j} \le 1$).

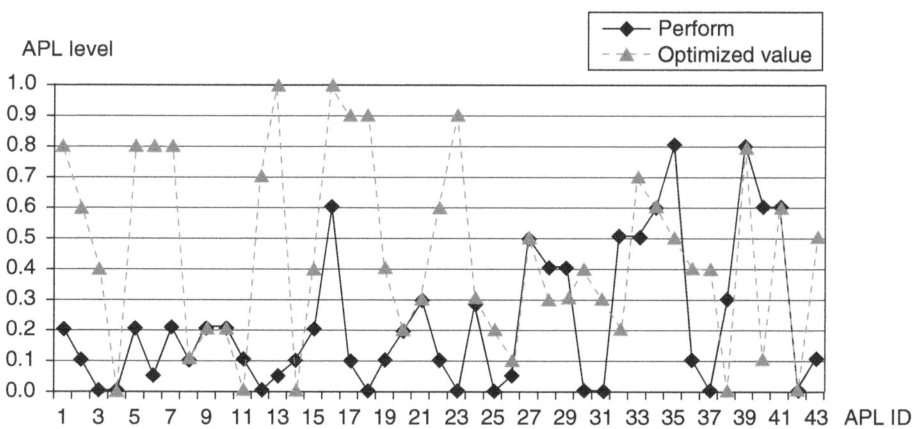

Figure 8.36 VVT strategies: performed versus goal optimized strategy.

Goal Optimization Results Figure 8.37 shows the overall VVT cost estimation model for the goal optimized VVT strategy. These results were obtained after performing 3000 Monte Carlo iterations within the VVT-Tool.

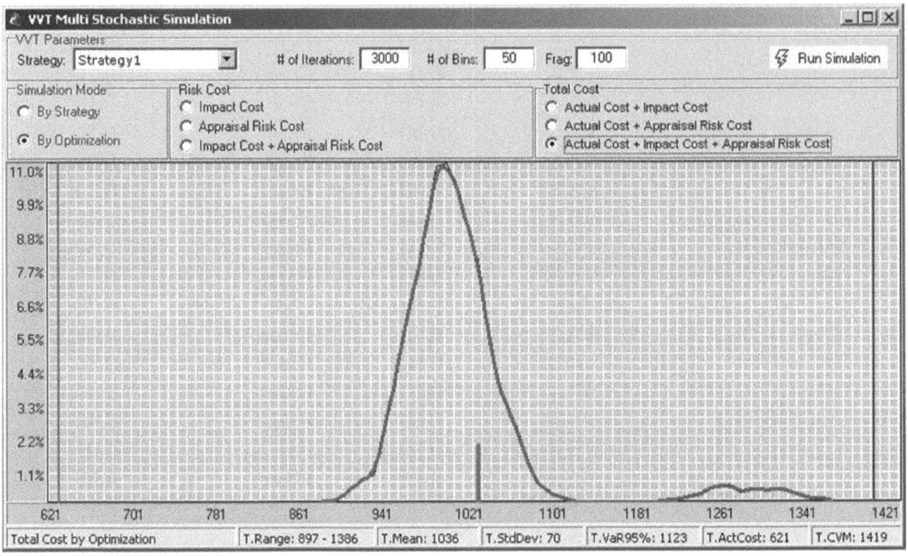

Figure 8.37 Overall goal optimized VVT cost estimation.

Table 8.22 and Figure 8.38 show the effects of the VVT strategy goal optimization. As expected, the optimized VVT strategy has improved the VVT impact cost by 31%; however, the VVT activity, appraisal and total costs increased. Here, a general business objective aiming at minimizing the potential impact by investing more in the VVT process is exemplified.

TABLE 8.22 Original Versus Goal Optimized Cost Reduction

Cost Categories	Original (CU)	Optimized (CU)	Change (CU)	Change (%)
VVT activity cost	343	621	–278	–81
Mean appraisal cost	62	89	–27	–44
Mean impact cost	472	326	146	31
Mean total cost	877	1036	–159	–18
CVM cost	1419	1419	0	0
SD	106	70	36	34
VaR95%	1126	1165	–39	-3

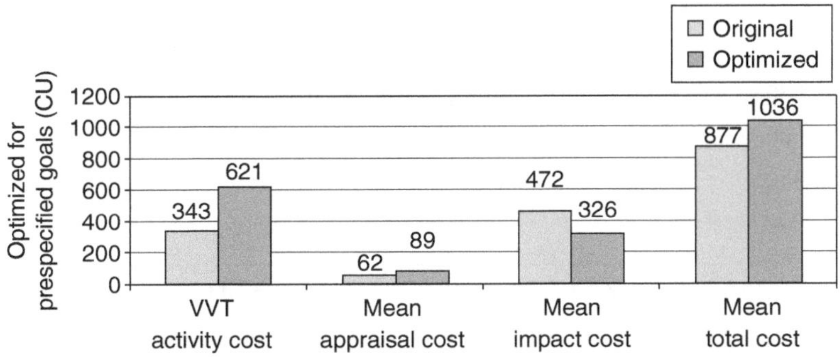

Figure 8.38 Original versus goal optimized cost reduction.

8.2.5 MPGA Optimization for Time

MPGA Optimization Process We now consider a different project-wide schedule optimization demonstration aimed at decreasing the total project time. We use a Multi Phase Genetic Algorithm (MPGA) where we do not need to assume independence of each individual VVT APL ($X_{i,j}$). The VVT-Tool has been applied to a population size of 43, an elitism of 1, crossover of 0.90, a mutation level of 0.05 and a project time goal of 380 days. As can be seen, the target time goal was achieved after 42 iterations (see Figure 8.39). As a point of interest, note the rate of project duration convergence under the MPGA optimization process (see Figure 8.40).

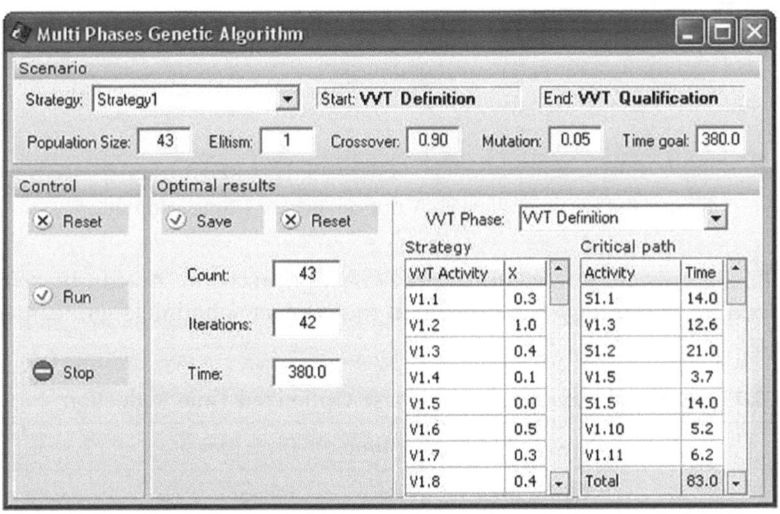

Figure 8.39 MPGA optimization application.

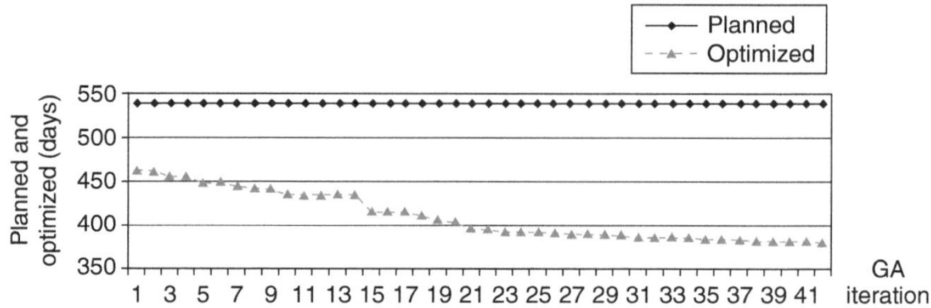

Figure 8.40 MPGA: project duration convergence.

MPGA Optimization Results Figure 8.41 shows the overall VVT time estimation based on the MPGA optimized VVT strategy.

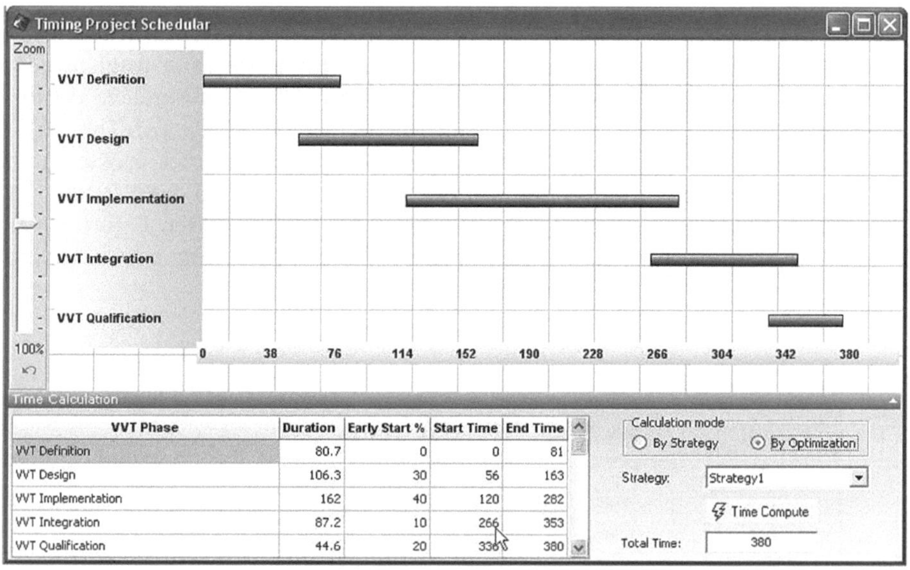

Figure 8.41 Overall MPGA optimized VVT time estimation.

Table 8.23 shows the effects of the MPGA optimization. As can be seen, the optimized VVT strategy has shortened the project schedule significantly.

TABLE 8.23 Original Versus GA Multiphase Optimized Time Reduction

Expected MPGA Optimized Time (days)			
Original	Optimized	Change	Change (%)
539	380	159	29

The project total schedule was reduced from 539 to 380 days, an improvement of 29%.

We ask ourselves how this shortening of the schedule affects the cost of the project. The surprising answer is: The mean total cost was in fact reduced by 9%, from 877 to 801 CU. The reader should note, however, that the variance of this VVT strategy is relatively large (Standard deviation 112 CU). Also the "tail of catastrophic risks" is much too long, extending beyond the CVM line. Both phenomena indicate a rather risky VVT strategy (see Figure 8.42).

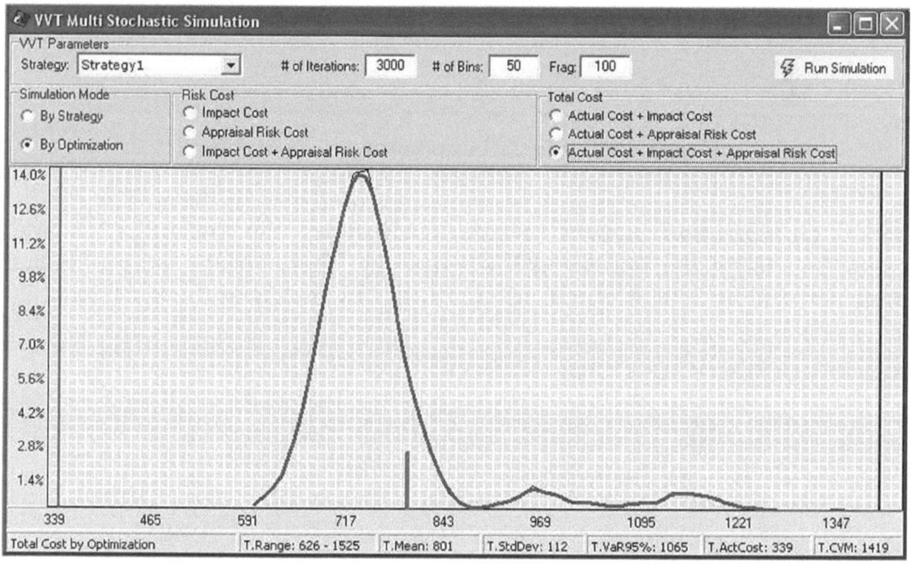

Figure 8.42 Overall MPGA optimized VVT cost distribution.

8.2.6 SSGA Optimization of Cost and Time

SSGA Optimization Process We now consider a different project-wide optimization demonstration aimed at decreasing project total VVT costs as well as project duration. We use a Steady-State Genetic Algorithm (SSGA) and therefore do not need to assume the independence of each individual VVT APL ($X_{i,j}$). The VVT-Tool has been applied with similar GA parameters as before but with a total VVT cost goal of 700 CU and a total project time goal of 420 days (see Figure 8.43 after 133 iterations).

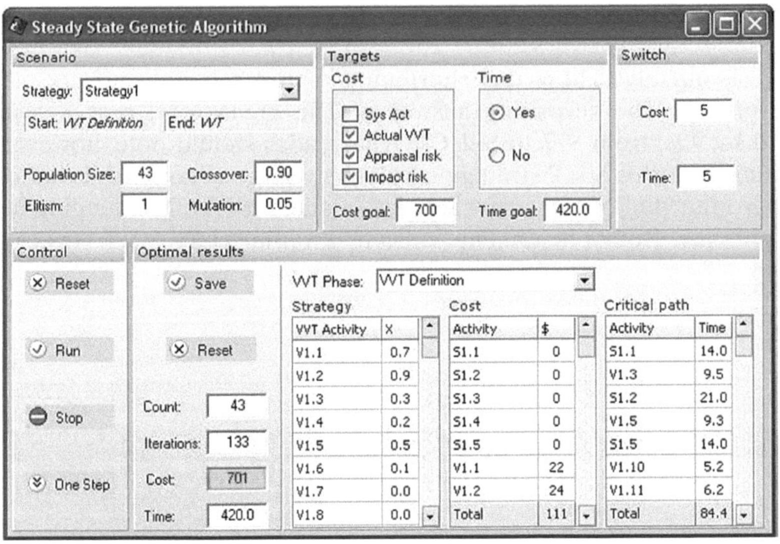

VVT Activity	X
V1.1	0.7
V1.2	0.9
V1.3	0.3
V1.4	0.2
V1.5	0.5
V1.6	0.1
V1.7	0.0
V1.8	0.0

Activity	$
S1.1	0
S1.2	0
S1.3	0
S1.4	0
S1.5	0
V1.1	22
V1.2	24
Total	111

Activity	Time
S1.1	14.0
V1.3	9.5
S1.2	21.0
V1.5	9.3
S1.5	14.0
V1.10	5.2
V1.11	6.2
Total	84.4

Figure 8.43 The SSGA optimization application.

SSGA Optimization Results Figure 8.44 shows the overall VVT cost estimation using the optimized VVT strategy. These results were obtained after performing 3000 Monte Carlo iterations by means of the VVT-Tool.

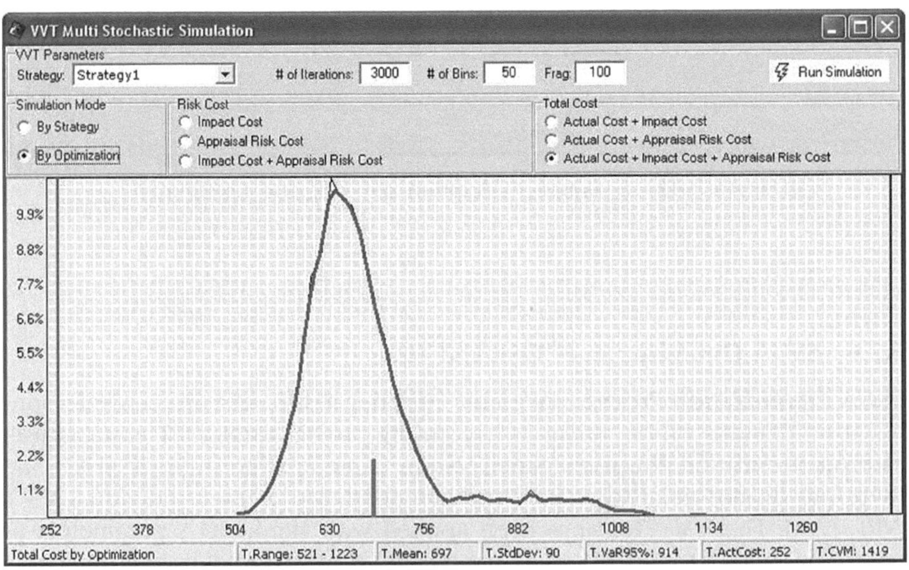

Figure 8.44 Overall SSGA optimized VVT cost estimation.

Figure 8.45 shows the overall VVT time estimation model using the optimized VVT strategy. As we have seen in the previous three figures, the goal of the SSGA was fully achieved after 133 iterations.

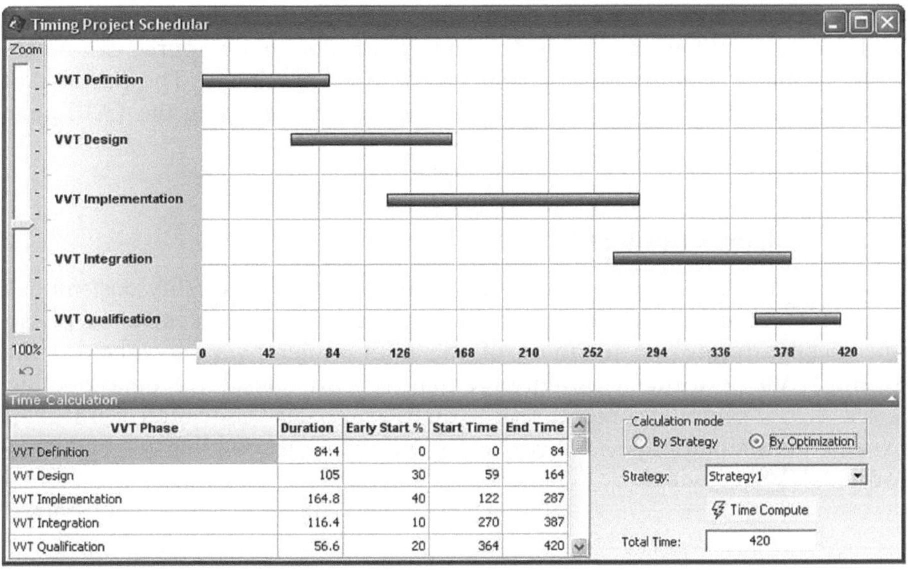

Figure 8.45 Overall SSGA optimized VVT time estimation.

Table 8.24 shows the overall effects of the SSGA optimization. As can be seen, the optimized VVT strategy has improved the total expected VVT cost by 20% and the overall expected project time by 22%.

TABLE 8.24 Original Versus Optimized Cost Reduction

Paremeter	Original	Optimized	Change	Change (%)
Project Cost	877 CU	700 CU	177 CU	20
Project Time	539 days	420 days	119 days	22

8.3 IDENTIFYING AND AVOIDING SIGNIFICANT RISKS

Modeling the cost, time and risk of the VVT process can also yield qualitative and quantitative information about potential risks associated with a given VVT strategy. Specifically, the test engineer can observe the type of risks associated with any particular VVT strategy as well as their estimated probability.

Such information is valuable for avoiding critical risks that may occur during the VVT horizon (in our case during the system development stage) as well as for obtaining a reasonable idea about future risk scenarios (in our case, during system production, use/maintenance and disposal).

Identifying expected costs of potential future impact risks is valuable in designing a VVT strategy. First, the VVT strategy itself can be altered in order to decrease the likelihood of such events. Second, either resource can be set aside for these eventualities or insurance can be purchased. The purpose of this section is to demonstrate this aspect of the model using the IAI/Lahav system and quality data.

8.3.1 Avoiding Critical Risks

Based on the risk scenarios defined as part of the VVT quality cost model, we can ascertain the nature of risks and their lifecycle occurrence phase. Specifically, during the planning of the pilot project, we can increase the level of VVT performance when potential risks are considered unacceptable. Table 8.25 exemplifies some risk scenario impacts associated with their expected monetary impacts in cost units and the corresponding, partially performed, VVT activities.

TABLE 8.25 Critical Risks During Development Timeframe

Risk ID	Risk Scenario	Expected Impact (CU)	VVT ID
	VVT Definition		
R1.1.1	Limited VVT activity was performed during early development stages due to lack of planning	10.8	V1.1
	VVT Design		
R1.3.1	Lack of requirement understanding caused a major redesign of the system	16.5	V1.3
R1.10.2	No external requirement review is conducted leading to incorrect/ incomplete design	6.5	V1.10
	VVT Implementation		
R3.1.1	No input from design stage was evaluated resulting in open issues not treated during implementation	10.0	V3.1
R2.2.1	No detailed VVT planning evaluation for subsystem resulted in inadequate testing procedures for subsystems	12.1	V2.2

TABLE 8.25 *Continued*

Risk ID	Risk Scenario	Expected Impact (CU)	VVT ID
	VVT Integration		
R1.1.2	VVT-related simulators, tools and equipment were not purchased/ developed, so integration was hampered	15.0	V1.1
R4.1.1	No input from implementation stage was evaluated, resulting in open issues not treated during integration	13.4	V4.1
R1.2.1	VVT-related support for subsystems was not developed, affecting the functional integration of subsystems	39.8	V1.2
R2.3.1	Internal review of subsystems not performed; design of subsystem interfaces does not meet requirements	16.0	V2.3
R3.3.2	VVT procedures for subsystems are not fully complete, causing responsibility arguments and delays at integration	16.6	V3.3
R2.9.2	Subsystem verifications are not reviewed, leading to untested subsystems delivered to integration, causing delays	13.0	V2.9
R2.10.1	Inadequate design review leads to incomplete design at the system level, causing loss of hardware and delays	17.4	V2.10
	VVT Qualification		
R2.2.2	Detailed acceptance procedures are not planned, resulting in insufficient budgets and logistics for proper qualification	15.0	V2.2
R3.8.1	Stakeholders are not consulted, resulting in major rework detected in system qualification	24.9	V3.8
R1.9.1	Lack of system review leads to inconsistencies with the customer, which is detected/corrected at qualification	10.8	V1.9
R2.10.2	Inadequate software (S/W) design review causes functionality gap between the customer and the developer, leading to S/W redesign substantial solution	11.6	V2.10
R2.11.2	Design not discussed with key stakeholders leading to major redesign of system at great cost	13.2	V2.11
Total (CU)		**262.6**	

8.3.2 Conjecture on Future Risk Scenarios

Similarly, we can speculate regarding the expected future impact and cost and can associate each potential impact with a specific, partially performed VVT activity during the development timeframe. Table 8.26 provides some examples of future risk scenarios with expected monetary impacts in cost units together with their associated, partially performed VVT activities. Note that the costs identified in the table represent impact costs expected during the future lifetime of the system.

TABLE 8.26 Conjecture Concerning Future Risk Scenarios

Risk ID	Risk Scenario	Expected Impact (CU)	VVT ID
	VVT Production		
R6.1.1	No input from the qualification stage was evaluated, resulting in open issues not treated during production, causing production problems	16.6	V6.1
R6.2.1	First article inspection was not conducted, causing product failures during manufacturing	16.6	V6.2
R3.4.1	No VVT procedure for production is generated, causing inadequate preparations causing production delays	12.1	V3.4
R2.7.1	No product review during production causes very expensive production process or redesign	16.6	V2.7
R1.8.1	Production requirements are not reviewed, resulting in expensive production processes and extra scrap	12.0	V1.8
	VVT Use/Maintenance		
R1.2.2	Environmental qualification was not performed, causing problems during use and maintenance	24.0	V1.2
R5.2.3	Minimal EMI/EMC testing was performed, causing failures during usage and necessitating numerous modifications after delivery	19.3	V5.2
R7.2.1	BIT operation was not tested, resulting in erroneous BIT calls requiring field modifications to all systems	16.6	V7.2
R1.3.2	Due to lack of stakeholder identification, the system was not accepted by customer after the end of production	20.1	V1.3

TABLE 8.26 ***Continued***

Risk ID	Risk Scenario	Expected Impact (CU)	VVT ID
R2.3.2	External review of subsystems not performed; design of subsystems does not meet user needs	33.2	V2.3
R4.3.3	Minimum VVT for subsystems performed, causing many defects, rework and commercial embarrassment at usage	33.2	V4.3
R1.4.2	No environmental plans were evaluated, contributing to unexpected costs during use phase	20.0	V1.4
R3.4.3	No VVT procedures for enabling products were generated, leading to bad documentation, delaying system delivery	15.0	V3.4
R7.4.1	User input was not obtained, leading to dissatisfied customers and major loss of market	33.0	V7.4
R2.5.2	Control degradation not checked, causing product modification (redesign, retest, modification) during usage	11.7	V2.5
R5.6.2	FCA for the subsystems was not conducted. As a result, unfulfilled requirements were identified during usage	16.6	V5.6
R2.7.2	No product review regarding maintenance causes lengthy/expensive maintenance, not meeting Mean Time To Repair (MTTR) requirements	16.6	V2.7
R3.7.1	Partial testing was carried out leading to late discovery of problems and dissatisfied customer	15.6	V3.7
R1.8.2	Maintenance requirements not reviewed resulting in high MTTR, low Mean Time Between Maintenance Actions (MTBMA) and High Mean Maintenance Time (MMT)	53.6	V1.8
R2.8.2	No product review regarding facility construction caused delay in system delivery	13.3	V2.8
R2.9.4	Integrated Logistics Support (ILS) test procedure not reviewed, causing inadequate ILS provisions affecting usage and product availability	19.5	V2.9
R1.11.2	Stakeholders did not provide input, resulting in missing requirements which caused usage problems	53.6	V1.11
R2.11.1	Design not discussed with key stakeholders leading to program cancellation due to incorrect design	20.4	V2.11

TABLE 8.26 *Continued*

Risk ID	Risk Scenario	Expected Impact (CU)	VVT ID
	VVT Disposal		
R6.2.2	Disposal requirements were neglected during FAI, wrong materials used for all systems resulting in heavy fines	41.5	V6.2
R8.2.1	Verification of proper system disposal was not conducted, causing environmental damage leading to litigation and fines	41.5	V8.2
R8.3.1	Verification of (production, deployment, disposal) products was not conducted, leading to litigation and fines	25.0	V8.3
R8.4.1	Stakeholders concerned with disposal were ignored, leading to litigation, fines and loss of reputation	41.5	V8.4
R5.6.3	No Physical Configuration Audit (PCA) for subsystems was conducted, causing unmet subsystem requirements to be detected at disposal	25.0	V5.6
Total (CU)		**683.7**	

8.4 IMPROVING SYSTEM QUALITY PROCESS

We suggest that by measuring quality costs incurred during system development one illuminates key process characteristics and this will lead to a better understanding of the process. Once the process is understood, it can be improved, which leads to reducing system quality costs and development time. Our main point is that the system quality cost (and thus time) in development projects is very high, probably in the range of 50–60% of the development expenditures. We argue that quality costs emanate, in a fundamental way, from poor understanding of needs as well as inferior design, implementation and management, in short waste. Unfortunately, many organizations do not measure their system quality costs. And system engineers by and large are unaware of the magnitude of waste within their organizations.

Quality gurus Juran and Deming claimed that there is "an optimal cost of quality." Another quality guru, Crosby believed that "quality is free." The more one invests in quality, the better one's corporate financial performance is. Our findings do not fit with either of these views. We look at each individual quality element in terms of cost (and time) and ask the question: "Is the cost of ensuring quality, in the long term, less than the cost of failure?" Only if the answer is yes do we advocate investing in quality. Quality strategy must be related to corporate business objectives. Sometimes investment in quality is paramount; at other times, market consideration dictates lower product quality.

It should be noted that very little research was carried out in the system VVT arena. As a result, limited system quality cost data are published. The SysTest project and the individual pilot projects conducted under it are a fresh promise for increases in studying this critical area. We consider the IAI/Lahav as a good, midsize development project in which to start measuring and analyzing quality costs. We used straightforward methods for gathering and aggregating the cost and time data, and we think other organizations can follow suit.

The IAI/Lahav project findings indicate that over 60% of development cost is spent on quality. In addition, about 10% of the total development effort was spent on the Definition phase and within that phase only 1–2% of the total development cost was invested in quality. Under the VVT strategy of the pilot project, the VVT model (embodied in the VVT-Tool) generated quality costs commensurable with real-life data. This gives us the confidence in the viability and accuracy of the model, at least for this project. Finally, the results of the optimization runs indicate a potential savings of 15–25% of total engineering expenditures in quality cost and project duration. Furthermore, the presented optimized VVT strategy significantly reduced the variance of the VVT strategy solution and its Value-at-Risk (VaR). Most importantly, the outliers, the catastrophic risks that may rarely occur, can be greatly reduced and perhaps eliminated. In addition, the VVT model shows quantitatively that additional investment in the upfront phases would reduce the overall quality costs. This then is not just an intuitive conjecture.

Some of the cost data presented here have also been reported in the software industry as well as under the SysTest project. However, this author is quick to admit that the above results reflect data emanating from a single system development project. Another limitation is the dependence on inexact historical database, as well as estimates made by domain experts. We think, however, that in systems engineering one data point is substantially more than no data at all. As system engineering is a young discipline, this data point is significant as a starting stage for better awareness of the system quality cost issue. Hopefully it will trigger extensive collection and publication of such data in the future.

Appendix A

SysTest Project

A.1 ABOUT SYSTEST

Those who are not engineers are often oblivious to the dramas occurring during the birth of each engineered system. All systems, as well as end products, carry with them engineers jockeying for position and seeking honor and appreciation by peers and managers. But the beauty of engineering stems from a group search for a single solution to a fuzzily specified problem, a desire for a predicted and encapsulated result through the application of a top-down problem-solving approach[82]. In short, engineering is a human endeavor.

Projects funded by the European Commission (EC) bring many engineers and managers from different nationalities, cultures, languages and industries to solve problems within a prespecified length of time and budget. Successful EC projects provide an environment for cross-cultural understandings, way beyond engineering, building shared dreams and lasting comradeships. SysTest (Full title: Developing Methodology for Advanced Systems Testing) was such a project. The SysTest team generated 20 formal deliverables and disseminated 24 presentations in 13 conferences plus 5 publications in refereed trade journals. Five SysTest team members obtained their Ph.D. in direct or indirect connection with this project. This appendix, however, deals with the technical aspects of SysTest. Readers interested in the engineering and political drama of engineered systems can find several books in this genre (see, e.g., Chaisson, 1998; Carroll, 1994). SysTest[83] was a five million Euro research and development project. It was partially funded by the EC under the Competitive and

[82]The above was derived from a Braha et al. (2006) observation.
[83]This Appendix is adapted with permission from Hoppe et al. (2007).

Sustainable Growth Fifth Framework Program (contract number G1RD-CT-2002-00683). The SysTest consortium was composed of eight companies, research institutes and organizations from six European and affiliated countries. The project was launched in March 2002 and ended in March 2005.

The overall objective of the SysTest project was to develop a Verification, Validation and Testing (VVT) methodology and economic model in order to realize improved product quality at reduced cost and increased availability. The strategy of SysTest is summarized as follows:

1. It addresses a broad variety of system categories emanating from diverse industrial segments, namely, aircraft avionics and jet engines, automobile engine castings and embedded software, liquid food packaging manufacturing and steel production.
2. It regards entire systems' lifetime from cradle (Definition phase) to grave (Disposal phase).
3. It identifies and describes numerous system level VVT activities and methods.
4. It provides heuristic tailoring rules for different industry domains, development cycles and project types.
5. It assesses the SysTest products by means of real-life pilot projects.

Based on the above approach, we consider the SysTest products as enablers sustaining the implementation of effective VVT processes by users from different engineering domains. A spectrum of industries in conjunction with academic and research centers enabled the project to develop generic VVT products and assess their effectiveness by means of real-life pilot projects. Figure A.1 depicts the SysTest project methodology and model concept which is applicable for the entire system lifecycle phases.

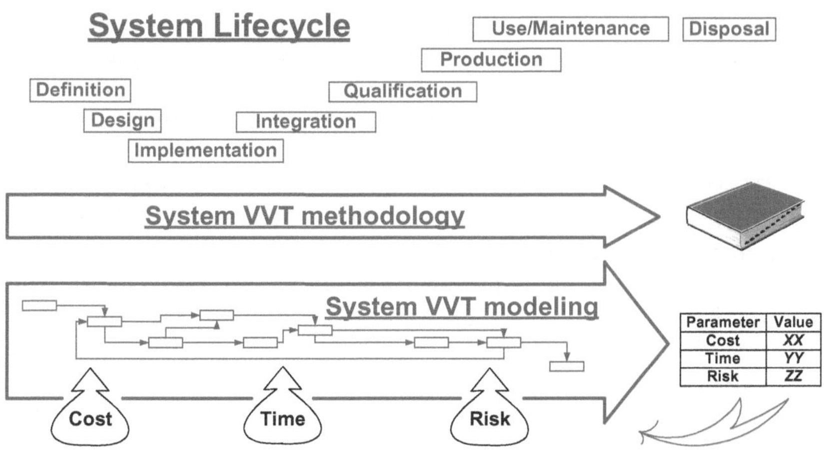

Figure A.1 SysTest methodology and model concept.

A.2 SYSTEST KEY PRODUCTS

The SysTest consortium developed two key products: (1) a VVT Methodology Guidelines (VVTMG), a document containing SysTest's collected body of knowledge related to system VVT and (2) a VVT Process Model (VVTPM), a quantitative model as well as a software embodiment to appraise system VVT strategies in terms of project time, cost, risk and product quality.

1. *VVT Methodology Guidelines.* The purpose of the VVTMG document (SysTest, 2005) was to provide a generic and customizable VVT methodology. The VVTMG describes a collection of 41 VVT activities and 31 methods applicable to the various system lifecycles. In addition, the VVTMG provides a collection of tailoring rules for different environments (industrial sectors, lifecycles and project types). The SysTest consortium agreed to make the VVTMG a public domain document. Therefore it was distributed widely around the globe. The VVTMG was a key source for writing Chapters 2–5 of this book and the author wishes to express deep appreciation and gratitude to the entire SysTest team.
2. *VVT Process Models.* Two quantitative VVTPM packages were developed during SysTest. These models embody an iterative planning technique which offers an effective method for smooth integration of the VVT planning with other system engineering activities. This technique supports maximal fulfillment of project goals and objectives. The specific objectives of this effort were to (1) predict the VVT cost, duration and resulting product quality as a function of a selected VVT strategy, (2) calculate the risk of not achieving cost, duration and quality targets based on the uncertainties inherent in future predictions and (3) define an optimal VVT strategy with respect to cost, duration and resulting product quality. Software packages were developed by SysTest for VVTPM that simulate the nondeterministic nature of the VVT optimization problem using a Monte Carlo approach. The VVTPM was developed in two variants, which are described below:
   - *Official SysTest VVTPM.* The official SysTest VVTPM was designed primarily as a risk management tool (Nguyen, 2008). It helps identify potential risk sources caused by insufficient verification and validation. It (1) calculates the likelihood of a product not satisfying the quality requirements, (2) calculates the likelihood of the VVT strategy exceeding cost and duration targets and (3) produces a list of the consequences of the above. Thus, the model establishes a decision structure for determining the best risk mitigation strategy. The SysTest VVTPM should be employed repeatedly during system development. Since VVT is an important means to reduce project

risk, VVT planning defines the optimal risk reduction strategy in the form of the VVT process and a measurement system to track project risks throughout its lifetime. This model and software tool is designed for estimating VVT cost and schedule and the resulting product quality (along with the accompanying risk) for a given VVT strategy. The strategy consists of identifying (1) the VVT goals and objectives, (2) the project VVT scope and (3) the key VVT parameters. The VVT strategy and planning procedure assists in the risk management efforts by feeding the results of risk analysis directly into the VVT planning and the system engineering measurement process (Hoppe et al., 2004)].

- *Unofficial VVTPM.* The unofficial SysTest VVTPM was designed primarily as a tool for identifying optimal VVT strategies in order to optimize quality cost (cost of performing the VVT activities plus cost emanating from system failure) and system availability throughout their lifetime. This model is meant to be used once at the inception of system development. It is based on modeling VVT risks and costs by means of a Canonical VVT Model (CVM). This is a hypothetical framework encapsulating the performance of a "complete and ideal" set of VVT activities designed to verify, validate, and test a system throughout its lifetime. The user may define a VVT Activity Performance Level (APL) for each activity such that the entire APL set delineates the VVT strategy. Capturing a VVT strategy yields a practical and realizable VVT process. The unofficial VVTPM was used in writing Chapters 6–8 of this book.

A.3 SYSTEST PILOT PROJECTS

Six real-life pilot projects were conducted by the SysTest industrial partners in order to assess the effectiveness of the SysTest products. For each pilot project, we provide in this section a short description of the project, its cost and its duration. For more information, see Hoppe et al. (2007).

1. *New Avionics Systems for Transport Helicopters at Israel Aerospace Industries (IAI).* The IAI/Lahav division is developing military and commercial aerospace systems. The pilot project at IAI is depicted in Figure A.2. The transport helicopters provided to IAI contained obsolete avionics and the task of IAI/Lahav was to replace it with a state-of-the-art display and control system as well as enhance the helicopters with improved mission capabilities (e.g., search-and-rescue mission):

Industry: Avionics

Pilot Project: Development of new avionics systems for transport helicopters

Project Cost: 20 Million Euros

Project Duration: 18 Months

Figure A.2 Pilot project at IAI.

2. *Engine Controller at Hispano-Suiza (HS), France.* HS manufactures aircraft and missile engines. The pilot project at HS, depicted in Figure A.3, is an engine controller for the M88 type 3 jet engine:

   Industry: Aircraft engines

   Pilot Project: Upgrading the control unit of the M88 type 3 jet engines.

   Project Cost: >2 Million Euros

   Project Duration: 33 Months

Figure A.3 Pilot project at HS.

3. *Cylinder Head Mold at Centro Ricerche Fiat (CRF), Italy.* CRF is a research institute of the Fiat Group with a mission to create and introduce new products, processes and methods, develop and use advanced engineering techniques for product development and production and provide state-of-the-art facilities, equipment and laboratories. The pilot project was undertaken by TEKSID Aluminum, a Fiat Group company (see Figure A.4). The main objective of TEKSID was the development and industrialization of adequate technology and tooling for the pouring of a cylinder head:

   Industry: Automobile manufacturing

   Pilot Project: Developing a mold for a new automobile engine cylinder head.

   Project Cost: 10 Million Euros

   Project Duration: 24 Months

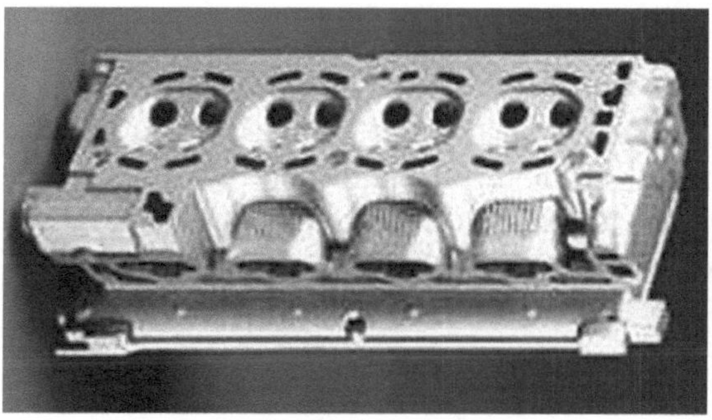

Figure A.4 Pilot project at CRF.

4. *Driver Assistance System at DaimlerChrysler (DC) Research and Technology Berlin, Germany.* DC is one of several research facilities of DaimlerChrysler (today Daimler) worldwide. The department involved in SysTest is mainly concerned with research in the area of software verification and validation. The pilot project, consisting of the development of a navigation-based driver assistance system, is depicted in Figure A.5:

   Industry: Automobile manufacturer

   Pilot Project: Development of a navigation-based driver assistance system for top-of-the-line-passenger cars

   Project Cost: 3 Million Euros

   Project Duration: 24 Months

Figure A.5 Pilot project at DC.

5. *Cap Applicator at Tetra Pak Carton Ambient (TPCA), Italy.* TPCA develops and manufactures liquid food packaging as well as packaging lines, filling machines and downstream equipment. The pilot project was the development of a flexible cap applicator for liquid food packaging (see Figure A.6):

Figure A.6 Pilot project at TPCA.

Industry: Liquid food-packaging manufacturing

Pilot Project: Development of a flexible cap applicator for liquid food packaging

Project Cost: 8 Million Euros

Project Duration: 12 Months

6. *Improved Tinplate Production Line at Arcelor Corporación Siderúrgica (ACS).* ACS is a large steel producer located in Aviles, Spain. The selected pilot project was an improvement in the efficiency of a tinplate production line (see Figure A.7):

   Industry: Steel production

   Pilot Project: Improvement of a tinplate production line

   Project Cost: N/A

   Project Duration: N/A

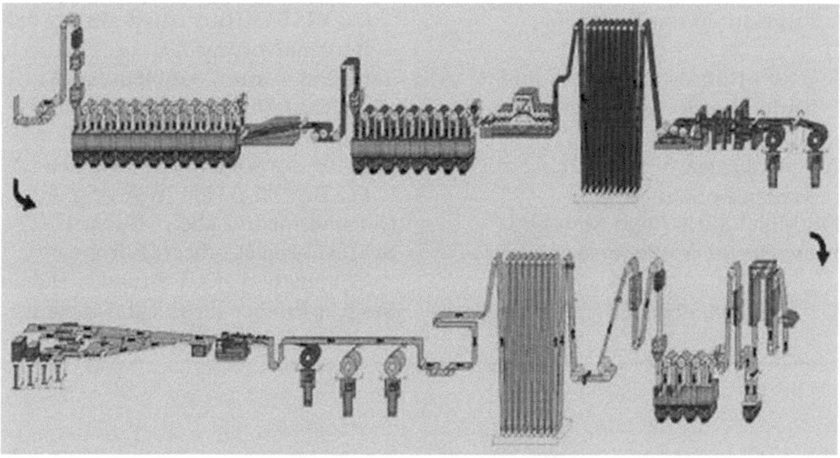

Figure A.7 Pilot project at ACE.

A.4 SYSTEST TEAM

SysTest partners represent five different industrial sectors (electronics/avionics, automobile, control systems for aerospace, steel producing, food packaging). Table A.1 depicts the names of the participants in the SysTest project, their affiliation and country. Figure A.8 portrays some members of the SysTest team.

TABLE A.1 SysTest Team Members and Affiliations

Name	Organization and country
Dr. Avner Engel, Dr. Michael Winokur, Izhak Bogomolni, Shalom Shachar, Arie Grinman, Ariav Savir	Israel Aerospace Industries (IAI), Ben Gurion International Airport, 70100, Israel
Anamul (Andy) Hoque	(Subcontractor: Alma Consulting Group, Lyon, France)
Hugues Granier, Alain Varizat, Julien Gandelot	Hispano-Suiza (HS), BP 42, 77552 Moissy Cramayel, Cedex France
Mario Gambera, Guido Scarafiotti, Zezza Vincenzo, Salvatore Asaro	Fiat Research Centre (CRF), Strada Torino, 50—10043 Orbassano (To), Italy
Professor Eduard Igenbergs, Dr. Andreas Vollerthun, Dr. Viktor Lévárdy, Dr. Markus Hoppe	Technische Universität München (TUM), Institute of Astronautics, Boltzmannstr. 15, 85748 Garching, Germany
Dr. Joachim Wegener, Dr. Harbhajan Singh, Andreas Kraemer, Eckard Lehmann, Frank Lammermann	DaimlerChrysler AG (DC), Research & Technology, Alt-Moabit 96 A, 10559 Berlin, Germany
Dr. Cecilia Haskins, Dr. Eric Honour, Professor Tyson Browning,	Norwegian Systems Engineering Council (NORSEC), Box 810 Sentrum, 5807 Bergen, Norway
Roberto Borsari, Andrea Sereni, Carlo Leardi	Tetra Pak Carton Ambient S.p.A. (TPCA), Via Delfini nr. 1, 41100 Modena, Italy
Juan Gonzalez, Nicolas De Abajo, Vanesa Lobato	Arcelor Corporación Siderúrgica (ACS), PO Box 90, 33400 Aviles, Spain
Dr. Mikel Sorli, Iñigo Mendikoa	(Subcontracor: Labein, Bilbao)
Hans-Hartmann Pedersen	SysTest project officer, European Commission (EC), Brussels, Belgium
Dr. Michel Seebacher	SysTest Project Technical Assistant (PTA), Wien, Austria

Figure A.8 Some members of the SysTest team.

A.5 EC EVALUATION OF SYSTEST PROJECT

August 3, 2005

Dear Avner:

Below are my comments on the SysTest evaluation which are agreed upon by Hans (Hans-Hartmann Pedersen):

1. *Satisfaction with Technical/Scientific Outcome of Project.* Despite minor adaptations, the technical/scientific outcome fulfilled 100% of the objectives stated in the description of work at the beginning of the project. Additionally some originally not foreseen tasks were performed during the pilot phase, proving the theory by more and better test results.
2. *Observations Regarding Project Management.* The project was very well managed by the coordinator IAI as well as by the supporting subcontractor ALMA. All efforts were set to ensure a successful performance of the project. All reports and deliverables were submitted on time.
3. *Potential Improvement on Part of SysTest Consortium/IAI/ Coordinator.* The technical and management performance of all partners was excellent from my point of view. Only the financial planning could have been performed better. At least for me it was in some way confusing that after the first project year and also at midterm the impression was given that the project might run out of budget, whereas after the second year the consortium was seeking for additional tasks to utilize the relatively high unspent budget.
4. *"SysTest Experience" Versus Experience Gained on Other EC Projects.* Compared to the other eight projects I am monitoring as PTA the SysTest project performed best. The consortium as well as the objectives regarding technical content, budget and timeframe for the overall project remained unchanged from the beginning, and these objectives were fully achieved. Also efforts for exploitation and communication of project results started comparably early with training units and a large number of publications.
5. *Conclusion.* I hope that all partners will be able to utilize the SysTest results to the best possible extent in order to achieve a maximum of economic and scientific advantage for their organizations.

Best regards,

Dr. Michael Seebacher Unternehmensberatung
Technical evaluator for the European Commission

REFERENCES

Braha, D., Minai, A. A., and Bar-Yam, Y. (Eds.), *Complex Engineered Systems: Science Meets Technology*, Springer, 2006.

Carroll, P., *Big Blues: The Unmaking of IBM*, Three Rivers Press, 1994.

Chaisson, J. E., *The Hubble Wars: Astrophysics Meets Astropolitics in the Two-Billion-Dollar Struggle over the Hubble Space Telescope*, Harvard University Press, 1998.

Hoppe, M., Engel, A., and Shachar, S., SysTest: Improving the Verification, Validation & Testing Process—Assessing Six Industrial Pilot Projects, *Syst. Eng. J.*, 10(4), 323–347, September 24, 2007, available online.

Hoppe, M., Levardy, V., Leardi, C., Mendikoa, I., and de Abajo, N., Application Experiences of the VVT Process Modeling Procedure at the Verification and Validation Planning, in *Proc. 14th Annual Int. Symp. INCOSE*, Toulouse, France, June 20–24, 2004.

Nguyen, H. H., Applicability of the SysTest VVT Process Model for the Die Manufacturing Process, Ph.D. Thesis in Computer Science, Department of Computer Science, Technical University of Munich, January 15, 2008.

SysTest Consortium, VVT Methodology Guidelines (VVTMG), Deliverable D7.1, SysTest-SYT901/021042, April 2005.

Appendix **B**

Proposed Guide: System Verification, Validation and Testing Master Plan

B.1 BACKGROUND

The Verification, Validation and Testing Master Plan (VVT-MP) is a proposed expansion to the *Test and Evaluation Master Plan* (TEMP), a U.S. Department of Defense (DoD) 5000.2-R directive[84]. It is the opinion of the author that the TEMP document does not constitute a satisfying basis for strategic VVT planning of engineered systems. This is because: (1) it deals only with the Qualification phase of systems development, (2) it is quite general thus difficult to put into practical use and (3) it contains great amount of jargon unique to the US-DOD. This VVT-MP guide provides users with guidance concerning the preparation of a master plan for planning strategic VVT processes in medium to large projects. Please note the following:

1. The scope of the document is to provide a template to help VVT engineers and managers responsible for VVT to plan the VVT process as well as determine the cost and other resources associated with all VVT activities in projects.
2. It is imperative to plan the VVT activity in an evolutionary way throughout the lifecycle of the system and to update the VVT-MP accordingly.
3. The project manager or any appropriate functionary should nominate a person who will be responsible for creating and maintaining the VVT-MP based on data gathered from all relevant stakeholders of the system.

[84]Mandatory Procedures for Major Defense Acquisition Programs (MDAPS) and Major Automated Information System (MAIS) Acquisition Programs, DOD, 2001.

Verification, Validation, and Testing of Engineered Systems, Avner Engel

B.2 CREATING THE VVT-MP

The VVT-MP generation process is presented in Figure B.1 and Table B.1.

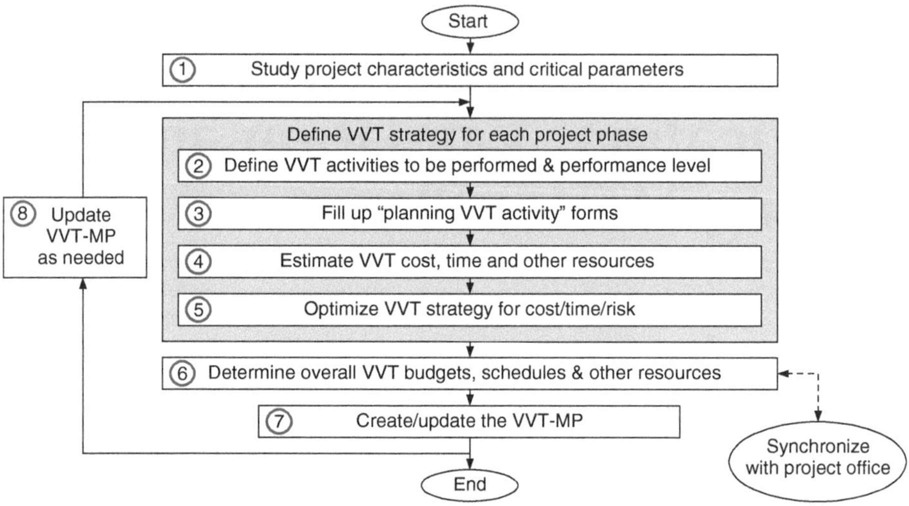

Figure B.1 VVT-MP generation flow chart.

TABLE B.1 Process of Creating VVT-MP

Stage	Activity
1	Study the project characteristics and the critical parameters that must be verified and validated.
2	Define the VVT strategy for each phase of the system lifecycle. Such definition determines the overall set of VVT activities and the level of performance (0.0 through 1.0) for each activity. This VVT performance level shall be captured in appropriate tables. Choosing the specific VVT strategy must be based on the characteristics of the project and must be translated into specific VVT activities. Such activities would be carried out either by VVT personnel organically attached to the VVT team or by other engineers and professionals who are doing a certain level of VVT activity as part of their regular assignments (e.g., software engineers often perform their own unit testing).
3	Filling up "planning VVT activity" forms for each VVT activity that will be performed at some level. These forms shall define all the relevant parameters associated with the performance of the specific VVT activity. However, during earlier stage of the VVT-MP, only the more rudimentary and available information about the testing should be filled in, such as the system and subsystem involved, the VVT activity name and objective and the responsible person.

TABLE B.1 *Continued*

Stage	Activity
4	Also, the above forms should include more complete information, including testing method and location of testing, expected schedule, cost and VVT/engineering hours as well as other required resources.
5	The VVT engineer should proceed by performing optimization of the VVT strategy in order to meet existing business objective, such as reducing cost, shortening schedule and improving reliability.
6	Next, the overall VVT budget and other resources must be calculated and an agreement must be reached with the project office and with other relevant engineering and non engineering players.
7	The VVT-MP should be created in an evolutionary manner.
8	The VVT-MP should be updated as needed.

B.3 CHAPTER 1: SYSTEM DESCRIPTION

Chapter 1 of the VVT-MP shall provide a brief description of the mission and the system as well as a set of critical technical parameters of the system.

B.3.1 Project Applicable Documents

This section shall include the list of project documents relevant to the VVT planning [such as Request for Proposal (RFP), operational requirements, system specifications].

B.3.2 Mission Description

This section shall describe the mission in terms of objectives and general capabilities. This section shall also include a description of the operational and logistical environment envisioned for the system.

B.3.3 System Description

This section shall describe the system design, including the following items:

1. Key features and subsystems, both hardware and software (such as architecture, interfaces and security levels), allowing the system to perform its required operational mission.
2. Interfaces with external systems (existing or planned) that are required for mission accomplishment. A diagram should show the system interfaces with the external environment.

3. Critical system characteristics or unique support concepts resulting in special test and analysis requirements (e.g., postdeployment software support, resistance to countermeasures, development of new threat simulators).

B.3.4 Critical Technical Parameters

This section shall describe the critical technical parameters associated with the system, including the following items:

1. A list of the main operational capabilities required for accomplishing the mission (e.g., mobility, area of operation, supportability).
2. A matrix depicting the critical technical parameters of the system (see Table B.2) that shall be evaluated during the phases of developmental testing:
   - Critical technical parameters must be measurable system characteristics that, when achieved, allow the attainment of desired operational performance capabilities.
   - Next to each technical parameter, a required threshold or value must be provided. This threshold or value should reflect the level of performance necessary to satisfy the desired capabilities.

TABLE B.2 Critical Technical Parameters (Example: UAV System)

Critical Technical Parameters	Parameter Threshold/Value
Mission range and direction	100 km from control center at 360°
Mission flight duration	Minimum 8 hours
Payload weight	Maximum 30 kg
Recognition range of man-size object	3000 meters ±5%

B.4 CHAPTER 2: INTEGRATED VVT PROGRAM SUMMARY

This chapter describes the integrated VVT program schedule and the VVT program management organization structure. These VVT planning steps must be synchronized with the project Gantt charts and the project budget as defined in the project Work Breakdown Structure (WBS).

B.4.1 Integrated VVT Program Schedule

Figure B.2 is a Gantt chart that illustrates the structure of an integrated VVT master schedule for a system's development. The chart has the following purposes:

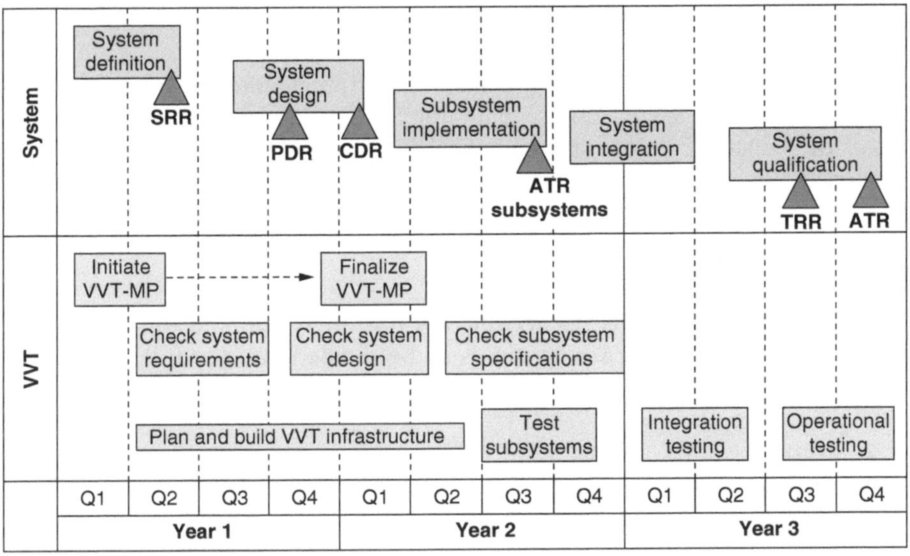

Figure B.2 Integrated VVT program schedule (example).

1. The schedule shall show how VVT activities synchronize with the project plan.
2. The main project engineering activities and the major milestones shall be identified.
3. The main VVT activities shall be identified, for example, development of VVT products, Validation and Verification (V&V) of requirements and design documents, subsystem and system tests and ground and flight test.

B.4.2 VVT Program Management

This section shall describe the VVT organization structure as implemented in the specific project and will include:

1. Subordination of different teams (system engineering, quality assurance, integration, etc.)
2. Responsibility of all participating organizations involved in the VVT process (e.g., VVT manager and staff, developers, testers, evaluators, users)
3. Identification of the test program structure shall include:
   - Subcontractor role/responsibility in testing major subsystems
   - Modeling and simulations required by different organizations (e.g., other organizational departments, subcontractors)
   - End-user involvement in test and evaluation phases
4. Identification of required VVT activities performed by "outside" organizations

B.5 CHAPTER 3: SYSTEM VVT

This chapter shall describe the specific VVT strategy adopted for the project and the resulting VVT activity planning, events and scenarios emanating from the selected VVT strategy. Finally, the VVT limitations shall be discussed.

B.5.1 VVT Strategy

This section shall describe the specific VVT strategy adopted for the project. In general, the set of VVT activities may be divided into the following three categories:

1. Preparing VVT products
2. Applying VVT to engineering products
3. Participating/conducting reviews

Appropriate VVT activities shall be performed during typical system lifecycle phases (see Figure B.3).

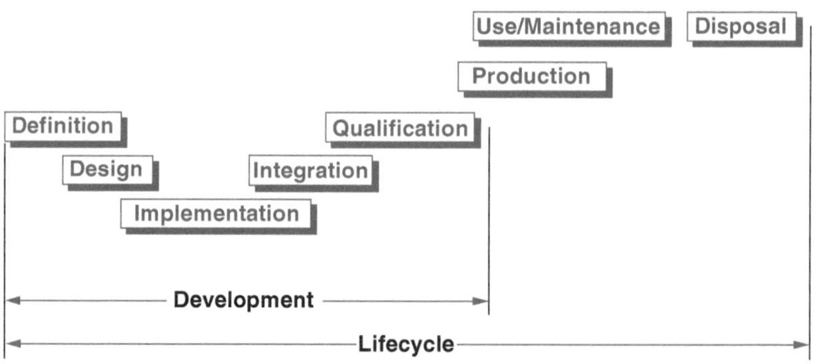

Figure B.3 Generic systems lifecycle phases.

Table B.3 could be considered as a CVM activity example for the VVT planners. It presents a generic (maximal) set of VVT activities, as described in this book (i.e., ID matching, Chapters 2 and 3). The VVT planner should identify a more specific set of VVT activities and then determine which VVT activities should be performed and at what performance level (0.0–1.0) they need to be performed.

TABLE B.3 CVM Activities Example for VVT Planners

ID	VVT Activity	Performance Level
2.2	**Definition phase**	
2.2.1	Generate Requirement Verification Matrix (RVM)	
2.2.2	Generate VVT Management Plan (VVT-MP)	
2.2.3	Assess Request For Proposal (RFP) document	
2.2.4	Assess System Requirement Specification (SysRS)	
2.2.5	Assess Project Risk Management Plan (PRMP)	
2.2.6	Assess System Safety Program Plan (SSPP)	
2.2.7	Participate in System Requirement Review (SysRR)	
2.2.8	Participate in System Engineering Management Plan (SEMP) review	
2.2.9	Conduct Engineering Peer Review (EPR) of VVT-MP document	
2.3	**Design phase**	
2.3.1	Optimize the VVT strategy	
2.3.2	Assess System/Subsystem Design Description (SSDD)	
2.3.3	Validate system design by means of virtual prototype	
2.3.4	Validate system design tools	
2.3.5	Assess system design for meeting future lifecycle needs	
2.3.6	Participate in System Design Review (SysDR)	
2.4	**Implementation phase**	
2.4.1	Prepar test cycle for subsystems and components	
2.4.2	Assess subsystem producer test documents	
2.4.3	Perform Acceptance Test Procedure (ATP)—subsystems/enabling products	
2.4.4	Assess system performance by way of simulation	
2.4.5	Verify design versus implementation consistency	
2.4.6	Participate in Acceptance Test Review (ATR)—subsystems/enabling products	
2.5	**Integration phase**	
2.5.1	Develop System Integration Laboratory (SIL)	
2.5.2	Generate System Integration Test Plan (SysITP)	
2.5.3	Generate System Integration Test Description (SysITD)	

TABLE B.3 *Continued*

ID	VVT Activity	Performance Level
2.5.4	Validate supplied subsystems in a stand-alone configuration	
2.5.5	Perform component, subsystem and enabling product integration tests	
2.5.6	Generate System Integration Test Report (SysITR)	
2.5.7	Assess effectiveness of the system Built-In Test (BIT)	
2.5.8	Conduct engineering peer review of SysITR	
2.6	**Qualification phase**	
2.6.1	Generate qualification/acceptance System Test Plan (SysTP)	
2.6.2	Create qualification/acceptance System Test Description (SysTD)	
2.6.3	Perform virtual system testing by means of simulation	
2.6.4	Perform qualification testing/ATP—system	
2.6.5	Generate qualification/acceptance System Test Report (SysTR)	
2.6.6	Assess system testability, maintainability and availability	
2.6.7	Perform environmental system testing	
2.6.8	Perform system Certification and Accreditation (C&A)	
2.6.9	Conduct Test Readiness Review (TRR)	
2.6.10	Conduct Engineering Peer Review of development enabling products	
2.6.11	Conduct engineering peer review of program and project safety	
3.2	**Production phase**	
3.2.1	Participate in Functional Configuration Audit (FCA)	
3.2.2	Participate in Physical Configuration Audit (PCA)	
3.2.3	Plan system production VVT process	
3.2.4	Generate a First Article Inspection (FAI) procedure	
3.2.5	Validate the production line test equipment	
3.2.6	Verify quality of incoming components and subsystems	

TABLE B.3 ***Continued***

ID	VVT Activity	Performance Level
3.2.7	Perform FAI	
3.2.8	Validate preproduction process	
3.2.9	Validate ongoing production process	
3.2.10	Perform manufacturing quality control	
3.2.11	Verify production operation strategy	
3.2.12	Verify marketing and production forecasting	
3.2.13	Verify aggregate production planning	
3.2.14	Verify inventory control operation	
3.2.15	Verify supply chain management	
3.2.16	Verify production control systems	
3.2.17	Verify production scheduling	
3.2.18	Participate in Production Readiness Review (PRR)	
3.3	**Use/Maintenance phase**	
3.3.1	Develop VVT plan for system maintenance	
3.3.2	Verify Integrated Logistics Support Plan (ILSP)	
3.3.3	Perform ongoing system maintenance testing	
3.3.4	Conduct engineering peer review on system maintenance process	
3.4	**Disposal phase**	
3.4.1	Develop VVT plan for system disposal	
3.4.2	Assess system disposal plan	
3.4.3	Assess system disposal strategies by means of simulation	
3.4.4	Assess ongoing system disposal process	
3.4.5	Conduct engineering peer review to assess system disposal processes	

B.5.2 Planning VVT Activities

This section shall tabulate the particular planning of each VVT activity in a set of "VVT planning activity" forms. The intent of these forms is to provide the means by which the VVT planner can identify the specifics of VVT activities as well as relevant information such as VVT method, support equipment, schedule and budget of each planned VVT activity. A VVT planning activity form should be filled for each VVT activity that will be performed at some specific level (i.e., no such form needs to be generated for VVT activities that will not be carried out). Table B.4 depicts a blank planning VVT activity form.

TABLE B.4 A Blank Planning VVT Activity Form

<table>
<tr><td colspan="2">Filling date:</td><td colspan="2">Revision:</td></tr>
<tr><td colspan="2">Generic VVT activity ID:</td><td colspan="2">Lifecycle phase:</td></tr>
<tr><td colspan="4">Generic VVT activity name:</td></tr>
<tr><td colspan="4">Specific VVT activity name:</td></tr>
<tr><td colspan="4">VVT performance level:
(Explain if partial)</td></tr>
<tr><td colspan="4">System/subsystem:</td></tr>
<tr><td colspan="4">Responsible person:________________________
Affiliation:________________________</td></tr>
<tr><td colspan="4"><u>VVT method:</u></td></tr>
<tr><td colspan="4"><u>Enabling/supporting equipment:</u></td></tr>
<tr><td colspan="4"><u>Relevant documents:</u></td></tr>
<tr><td colspan="4">VVT activity location:</td></tr>
<tr><td colspan="4">VVT activity schedule: From: _____________ To:____________</td></tr>
<tr><td colspan="4">Budget Estimation</td></tr>
<tr><td colspan="2">Engineering (hours)</td><td colspan="2">Cost (K$)</td></tr>
<tr><td>System Hours</td><td>VVT Hours</td><td>Purchasing Cost</td><td>Subcontract Cost</td></tr>
<tr><td></td><td></td><td></td><td></td></tr>
</table>

Table B.5 explains the meaning of each field in the planning VVT activity form.

TABLE B.5 Meaning of Fields in Planning VVT Activity Form

Field Number	Field Name	Meaning
1	Filling date	Latest date when VVT planning form was filled
2	Revision	Latest revision number of VVT planning form
3	Generic VVT activity ID	Generic identification (ID) of VVT activity (i.e., as defined in CVM activities list)
4	Lifecycle phase	Lifecycle phase in which VVT activity is to be carried out

TABLE B.5 ***Continued***

Field Number	Field Name	Meaning
5	Generic VVT activity name	Generic identification name of VVT activity (i.e., as defined in CVM activities list).
6	Specific VVT activity name	Specific VVT activity name associated with project at hand
7	VVT performance level	Level of VVT performance (0.0–1.0); explanation is desired if this level is less than 1.0
8	System/subsystem	Specific system or subsystem involved in VVT activity
9	Responsible person and affiliation	Name of person in charge of performing VVT activity together with his or her organization affiliation
10	VVT method	Method or methods to be used in performing VVT activity
11	Enabling/supporting equipment	List of facilities and enabling or supporting equipment required to accomplish VVT activity
12	Relevant documents	List of documents needed to accomplish VVT activity, including quantities and schedule of required availability
13	VVT activity location	Specific location where VVT activity shall be performed
14	VVT activity schedule	Planned schedule (starting and ending date) when VVT activity shall be performed
15–17	Engineering (hours) • System hours • VVT hours	Estimated engineering man-hours required to perform VVT activity: • Man-hour estimates for system engineers and other engineering disciplines • Man-hour estimates for organic VVT personnel
18–20	Cost (thousands of dollars) • Purchasing cost • Subcontract cost	Estimated overall dollar cost required to perform VVT activity: • Dollar cost of purchasing equipment or materials needed to carry out VVT activity • Dollar cost of subcontracting VVT efforts to external organizations and entities

Table B.6 illustrates an example of a filled planning VVT activity form related to aircraft radar subsystem tested during the implementation phase.

TABLE B.6 Example: Planning VVT Activity Form—Aircraft Radar Testing

Filling date: June 10, 2009		**Revision:** 002	
Generic VVT activity ID: 2.4.3		**Lifecycle phase:** Implementation	
Generic VVT activity name: Test subsystems, components and enabling products			
Specific VVT activity name: Laboratory and ground test			
VVT performance level: 0.8 This is a partial test. A complementary test will be performed during Qualification phase as part of the planned system test flights. **(Explain if partial)**			
System/subsystem: EL 827 Radar System			
Responsible person: Dan Levin **Affiliation:** Dept. 1234			
VVT method: Testing in the Radar Integration Laboratory will include detection chain, time and energy management and detection using injected targets.			
Enabling/supporting equipment: 1. Digital Integration Laboratory 2. Radar target simulator			
Relevant documents: 1. Radar system specification—7007A555-001 2. TBD			
VVT activity location: Radar integration Laboratory, Building 12, Room 10, IAI/ Lahav			
VVT activity schedule: From: January 10, 2010 **To:** March 22, 2010			
Budget Estimation			
Engineering (hours)		**Cost (K$)**	
System Hours	**VVT Hours**	**Purchasing Cost**	**Subcontract Cost**
300	500	20	60

B.5.3 VVT Limitations

This section shall describe the specific currently known VVT limitations that may significantly affect the VVT plan and the expected financial and schedule impact of these limitations. In particular this section will deal with the following issues:

- Resource availability (e.g., facilities, equipment, funding, schedule)
- Safety issues
- Operational environment limitations (e.g., system loading, climate considerations, simulation fidelity)

B.6 CHAPTER 4: VVT RESOURCE SUMMARY

This chapter shall summarize the resources and budget required in order to fulfill the planned VVT program. In particular, this section shall summarize the following resource categories.

B.6.1 Test Articles

This section shall identify the actual number and timing requirements for all test articles, including key support equipment required for testing according to the program plan. If key subsystems (components, assemblies, subassemblies or software modules) are to be tested individually before being tested in the final system configuration, then each subsystem and the quantity required shall be identified. In addition, when prototype, engineering models or production models shall be used for a given VVT activity, they shall be identified in this master plan.

B.6.2 Test Sites and Instrumentation

This section shall identify the specific test ranges or facilities to be used for each type of testing. The VVT planner shall compare the requirements for test ranges or facilities dictated by the scope and content of planned testing with existing capability and highlight any major shortfalls, such as inability to test under representative natural environmental conditions.

B.6.3 Test Support Requisition

This section shall identify test support equipment that must be acquired specifically to conduct the VVT master program.

B.6.4 Expendables for Testing

This section shall identify the type, number and availability requirements for all expendables (e.g., office supplies, fuel, equipment used in destructive tests) for each phase of the VVT master program.

B.6.5 Operational Force Test Support

For each lifecycle phase, this section shall identify the type and timing of operational force test support (e.g., aircraft flying hours, vehicle road test, customer personnel involvement).

B.6.6 Simulations, Models and Test Beds

For each lifecycle phase, this section shall identify the models and simulations to be used, including computer-driven simulation models and hardware/software-in-the-loop test beds. In addition, the VVT planner shall explain how these models and simulations shall be used as well as identify the resources required to accredit their usage.

B.6.7 Manpower/Personnel Needs and Training

This section shall identify the specific requirements for man-power and personnel as well as training needs that are required in order to implement the VVT process as defined in the master plan.

B.6.8 Budget Summary

This section shall depict a summary of the estimated budget required for performing all the identified VVT activities during the course of the VVT program (see Table B.7). Individual budget figures shall be extracted from the various VVT planning forms discussed above.

TABLE B.7 VVT Budget Summary

		Budget			
		Engineering (hours)		Cost (thousands of dollars)	
Activity ID	Activity Name	System	VVT	Purchasing	Subcontract
Total					

Appendix C

List of Acronyms

ABS	Anti-lock Braking System
ABS	Acrylonitrile, Butadiene and Styrene
ACR	American College of Radiology
ACS	Arcelor Corporación Siderúrgica
ADT	Air Data Terminal
AFD	Anticipatory Failure Determination
AI	Artificial Intelligence
ALARP	As Low As Reasonably Practicable
ANOVA	ANalysis Of VAriance
ANSI	American National Standards Institute
AOA	Activity-On-Arc
AON	Activity-On-Node
API	Application Programming Interface
APL	Activity Performance Level
ARM	Appraisal Risk Model
AS	Aerospace Standard
ASR	Alternative System Review
ATA	Air Transport Association
ATP	Acceptance Test Procedure
ATP	Acceptance Test Plan
ATR	Acceptance Test Review
AV	Air Vehicle
AVB	Air Vehicle Bus
AWACS	Airborne Warning And Control System

B2B	Business-to-Business
B2C	Business-to-Customers
BCET	Best-Case Execution Time
BFR	Brominated Flame Retardant
BIT	Built In Test
BITE	Built In Test Equipment
BRF	Basic Risk Factor
C&A	Certification and Accreditation
C&AIP	Certification and Accreditation Implementation Plan
C3I	Command, Control, Communications and Information
CAE	Computer Aided Engineering
CAIB	Columbia Accident Investigation Board
CAPA	Corrective And Preventive Action
CASRE	Computer-Aided Systems Reliability Estimation
CCB	Configuration Control Board
CD	Compact Disk
CDR	Critical Design Review
CE	Concept Exploration
CERT	Computer Emergency Response Team
CI	Configuration Item
CID	Controlled Impact Demonstration
CM	Configuration Management
CMM	Capability Maturity Model
CMMI	Capability Maturity Model Integration
CNC	Computer Numerically Controlled
COCOMO	COnstructive COst MOdel
COG	Center Of Gravity
COTS	Commercial Off-The-Shelf
CPM	Critical Path Method
CPU	Central Processing Unit
CRF	Centro Ricerche Fiat
CRT	Cathode Ray Tube
CSCI	Computer Software Configuration Item
CTE	Classification Tool Editor
CTM	Classification Tree Method
CU	Cost-Units
CVM	Canonical VVT Model
DAA	Designated Accrediting Authority
DC	DaimlerChrysler
DID	Data Item Description
DITSCAP	Defense Information Technology Security Certification and Accreditation Program

DMSO	Defense Modeling and Simulation Office
DNA	Deoxyribonucleic acid
DoD	Department of Defense
DOE	Design Of Experiments
DP	Disposal Plan
DPM	Defects Per Million
DSIL	Distributed System Integration Laboratory
DT&E	Development Test and Evaluation
EA	Evolutionary Algorithm
EAC	Election Assistance Commission
EC	European Commission
ECP	Engineering Change Proposals
EDD	Earliest Due Date
EDI	Electronic Data Interchange
EFSM	Extended Finite State Machines
EIA	Electronic Industries Alliance
ELV	End of Life Vehicles
EMC	Electro-Magnetic Compatibility
EMD	Engineering & Manufacturing Development
EMI	Electro-Magnetic Interference
EMSEC	Emissions Security
EPA	Environmental Protection Agency
EPP	Engineering Program Plan
EPR	Engineering Peer Review
ESA	European Space Agency
ESS	Environmental Stress Screening
EU	European Union
FAA	Federal Aviation Administration
FAAP	First Article Acceptance Plan
FAI	First Article Inspection
FAIP	First Article Inspection Plan
FAIR	First Article Inspection Report
F-ARM	Fuzzy-Appraisal Risk Model
FAV	First Article Verification
FCA	Functional Configuration Audit
FCFS	First-Come, First-Served
F-CVM	Fuzzy-Canonical VVT Model
FDA	Food and Drug Administration
FEIT	Flight Element Integration Testing
FEM	Finite Element Model
F-IRM	Fuzzy-Impact Risk Model
FMEA	Failure Modes and Effects Analysis

FSM	Finite-State Machine
FTS	Finish-To-Start
F-VSM	Fuzzy-VVT Strategy Model
GA	Genetic Algorithms
GCS	Ground Control System
GDM	Group Decision Making
GDT	Ground Data Terminal
GP	Goal Programming
GPS	Global Positioning System
GSIL	Generic Systems Integration Laboratory
GUI	Graphical User Interface
HALT	Highly Accelerated Life Testing
HASS	Highly Accelerated Stress Screening
HAZOP	Hazards and Operations Analysis
HCI	Hardware Configuration Item
HHM	Hierarchical Holographic Modeling
HIPAA	Health Insurance Portability and Accountability Act
HS	Hispano-Suiza
HSI	Human-System Interactions
HUD	Head-Up Display
HWCI	Hardware Configuration Items
I/O	Input Output
IAI	Israel Aerospace Industries
IAW	In Accordance With
IC	Integrated Circuits
IEC	International Electrotechnical Commission
IEEE	Institute of Electrical and Electronics Engineers
IIA	Independence of Irrelevant Alternatives
ILS	Integrated Logistics Support
ILSP	Integrated Logistics Support Plan
INCOSE	International Council on Systems Engineering
IP	Intellectual Property
IRM	Impact Risk Model
IRR	Integration Readiness Review
ISO	International Organization for Standardization
I-SRR	Internal Software Requirement Review
I-SysRR	Internal System Requirement Review
IV&V	Independent Verification & Validation
JIT	Just-In-Time
JSPS	Job Shop Production Schedule

LAI	Lean Aerospace Initiative
LAN	Local Area Network
LCD	Liquid Crystal Displays
LED	Light Emitting Diode
LRU	Lowest Replaceable Units
LSL	Lower Specification Limit
LTL	Linear Temporal Logic
M&S	Modeling and Simulation
MAD	Mean Absolute Deviation
MAIS	Major Automated Information System
MCDM	Multi-Criteria Decision Making
MDAPS	Major Defense Acquisition Programs
MDR	Medical Device Report
MDT	Mean Down Time
MEIT	Multi-Element Integration Testing
MIT	Massachusetts Institute of Technology
MMT	Mean Maintenance Time
MOEA	Multi-Objective Evolutionary Algorithms
MOOP	Multi-Objective Optimization Problem
MPGA	Multi Phase Genetic Algorithm
MPP	Maintainability Program Plan
MPS	Master Production Schedule
MRP	Material Requirements Planning
MSD	Mean Square Deviation
MSE	Mean Square Error
MT&SE	Maintenance, Test and Support Equipment
MTBF	Mean Time Between Failures
MTBMA	Mean Time Between Maintenance Actions
MTTR	Mean-Time-To-Repair
NASA	National Aeronautics and Space Administration
NASED	National Association of State Election Directors
NATO	North Atlantic Treaty Organization
NCAP	New Car Assessment Program
NHPP	Non-Homogeneous Poisson Process
NIAP	National Information Assurance Partnership
NIST	National Institute of Standards and Technology
NORSEC	NORwegian Systems Engineering Council
NVA	Non-Value-Adding
OT	Operational Test
OT&E	Operational Test and Evaluation

PAP	Production Acceptance Plan
PC	Personal Computer
PCA	Physical Configuration Audit
PD&RR	Program Definition & Risk Reduction
PDC	Propagation D Cubes
PDCF	Primitive D Cubes of Failure
PDR	Preliminary Design Review
PERT	Program Evaluation and Review Technique
PFD&OS	Production, Fielding/Deployment & Operational Support
PM	Program Manager
PMP	Program Management Plan
PNPS	Premature Next Phase Start
POC	Point Of Contact
PODEM	Path Oriented DEcision Making
PP	Production Plan
PPM	Process Planning Matrix
PROM	Programmable Read Only Memories
PRR	Production Readiness Review
PT	Penetration Test
PTP	Performance Test Plan
PTP	Production Test Procedure
PVC	Poly-Vinyl-Chloride
PYLD	Payload
QA	Quality Assurance
QC	Quality Control
QSIT	Quality System Inspection Techniques
RF	Radio Frequency
RFP	Request For Proposal
RFRM	Risk Filtering, Ranking, and Management
RMP	Risk Management Plan
RNVA	Required Non-Value Adding
ROI	Return On Investment
ROM	Read Only Memory
RPM	Revolutions Per Minute
RPN	Risk Priority Number
RPP	Reliability Program Plan
RVM	Requirements Verification Matrix
S/N	Signal-to-Noise
S/W	Software
SAE	Society of Automotive Engineers
SAR	Search-And-Rescue
SATC	Software Assurance Technology Center

SCADA	Supervisory Control And Data Acquisition
SEMP	System Engineering Management Plan
SFR	System Functional Review
SIL	System Integration Laboratory
SM	System Model
SMS	Short Message Service
SMV	Symbolic Model Verifiers
SOOP	Single Objective Optimization Problem
SOP	Standard Operating Procedure
SoS	Systems of Systems
SPC	Statistical Process Control
SPSS	Statistical Package for the Social Sciences
SPT	Shortest Processing Time
SQC	Statistical Quality Control
SRP	System Reporting Procedure
SRR	Software Requirement Review
SRS	Software Requirements Specification
SSDD	System Subsystem Design Description
SSGA	Steady State Genetic Algorithms
SSPP	System Safety Program Plan
SSR	Software Specification Review
SSS	System/Subsystem Specifications
ST	System Test
STP	Software Test Plan
SUT	System Under Test
SVR	System Verification Review
SWCI	Software Configuration Item
SWTP	Software Test Plan
SysDR	System Design Review
SysITD	System Integration Test Description
SysITP	System Integration Test Plan
SysITR	System Integration Test Report
SysRR	System Requirements Review
SysRS	System Requirements Specification
SysTD	System Test Description
SysTP	System Test Plan
SysTR	System Test Report
T&E	Test and Evaluation
TBC	Time-Based Competition
TCS	Trace Control System
TEMP	Test & Evaluation Management Plan
TEPP	Test and Evaluation Program Plan
TFN	Triangular Fuzzy Numbers
TGDC	Technical Guidelines Development Committee

TMMi	Test Maturity Model integration
TPCA	Tetra Pak Carton Ambient
TPM	Technical Performance Measures
TPP	Testability Program Plan
TRR	Test Readiness Review
TSS	Test Success Score
TTM	Time To Market
TUM	Technische Universität München
UAV	Unmanned Air Vehicle
ULSI	Ultra Large Scale Integration
US	United States
USAF	United States Air Force
USL	Upper Specification Limit
UUT	Unit Under Test
V&V	Verification and Validation
VA	Value Adding
VAD	Vehicle Autonomous Driver
VaR	Value-at-Risk
VSM	VVT Strategy Model
VSS	Voting System Standards
VT	Vulnerability Test
VV&A	Verification, Validation, and Accreditation
VV&T	Verification Validation & Testing
VVSG	Voluntary Voting System Guidelines
VVT	Verification Validation and Testing
VVTMG	VVT Methodology Guidelines
VVT-MP	VVT Master Plan
VVTPM	VVT Process Model
WBS	Work Breakdown Structure
WCET	Worst-Case Execution Time
WEEE	Waste Electrical and Electronic Equipment
WIP	Work In Process

Index

WILEY SERIES IN SYSTEMS ENGINEERING AND MANAGEMENT

Andrew P. Sage, Editor

ANDREW P. SAGE and JAMES D. PALMER
Software Systems Engineering

WILLIAM B. ROUSE
Design for Success: A Human-Centered Approach to Designing Successful Products and Systems

LEONARD ADELMAN
Evaluating Decision Support and Expert System Technology

ANDREW P. SAGE
Decision Support Systems Engineering

YEFIM FASSER and DONALD BRETTNER
Process Improvement in the Electronics Industry, Second Edition

WILLIAM B. ROUSE
Strategies for Innovation

ANDREW P. SAGE
Systems Engineering

HORST TEMPELMEIER and HEINRICH KUHN
Flexible Manufacturing Systems: Decision Support for Design and Operation

WILLIAM B. ROUSE
Catalysts for Change: Concepts and Principles for Enabling Innovation

LIPING FANG, KEITH W. HIPEL, and D. MARC KILGOUR
Interactive Decision Making: The Graph Model for Conflict Resolution

DAVID A. SCHUM
Evidential Foundations of Probabilistic Reasoning

JENS RASMUSSEN, ANNELISE MARK PEJTERSEN, and LEONARD P. GOODSTEIN
Cognitive Systems Engineering

ANDREW P. SAGE
Systems Management for Information Technology and Software Engineering

ALPHONSE CHAPANIS
Human Factors in Systems Engineering

YACOV Y. HAIMES
Risk Modeling, Assessment, and Management, Third Edition

DENNIS M. BUEDE
The Engineering Design of Systems: Models and Methods, Second Edition

ANDREW P. SAGE and JAMES E. ARMSTRONG, Jr.
Introduction to Systems Engineering

WILLIAM B. ROUSE
Essential Challenges of Strategic Management

YEFIM FASSER and DONALD BRETTNER
Management for Quality in High-Technology Enterprises

THOMAS B. SHERIDAN
Humans and Automation: System Design and Research Issues

ALEXANDER KOSSIAKOFF and WILLIAM N. SWEET
Systems Engineering Principles and Practice

HAROLD R. BOOHER
Handbook of Human Systems Integration

JEFFREY T. POLLOCK AND RALPH HODGSON
Adaptive Information: Improving Business Through Semantic Interoperability, Grid Computing, and Enterprise Integration

ALAN L. PORTER AND SCOTT W. CUNNINGHAM
Tech Mining: Exploiting New Technologies for Competitive Advantage

REX BROWN
Rational Choice and Judgment: Decision Analysis for the Decider

WILLIAM B. ROUSE AND KENNETH R. BOFF (editors)
Organizational Simulation

HOWARD EISNER
Managing Complex Systems: Thinking Outside the Box

STEVE BELL
Lean Enterprise Systems: Using IT for Continuous Improvement

J. JERRY KAUFMAN AND ROY WOODHEAD
Stimulating Innovation in Products and Services: With Function Analysis and Mapping

WILLIAM B. ROUSE
Enterprise Tranformation: Understanding and Enabling Fundamental Change

JOHN E. GIBSON, WILLIAM T. SCHERER, AND WILLAM F. GIBSON
How to Do Systems Analysis

WILLIAM F. CHRISTOPHER

Holistic Management: Managing What Matters for Company Success

WILLIAM B. ROUSE

People and Organizations: Explorations of Human-Centered Design

GREGORY S. PARNELL, PATRICK J. DRISCOLL, AND DALE L. HENDERSON

Decision Making in Systems Engineering and Management

MO JAMSHIDI

System of Systems Engineering: Innovations for the Twenty-First Century

ANDREW P. SAGE AND WILLIAM B. ROUSE

Handbook of Systems Engineering and Management, Second Edition

JOHN R. CLYMER

Simulation-Based Engineering of Complex Systems, Second Edition

KRAG BROTBY

Information Security Governance: A Practical Development and Implementation Approach

JULIAN TALBOT AND MILES JAKEMAN

Security Risk Management Body of Knowledge

SCOTT JACKSON

Architecting Resilient Systems: Accident Avoidance and Survival and Recovery from Disruptions

JAMES A. GEORGE AND JAMES A. RODGER

Smart Data: Enterprise Performance Optimization Strategy

YORAM KOREN

The Global Manufacturing Revolution: Product-Process-Business Integration and Reconfigurable Systems

AVNER ENGEL

Verification, Validation, and Testing of Engineered Systems